GLOBAL CLIMATE CHANGE
Convergence of Disciplines

GLOBAL
CLIMATE CHANGE
Convergence of Disciplines

ARNOLD J. BLOOM
University of California
Davis

 Sinauer Associates, Inc. • Publishers
Sunderland, Massachusetts U.S.A.

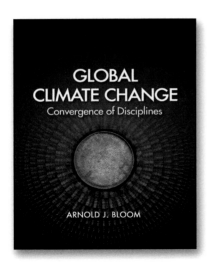

About the Front Cover

An interpretation of the photo montage, *Oil Barrels*, 2008, by Chris Jordan. The complete work depicts 28,000 42-gallon barrels, the amount of oil consumed in the United States every two minutes (equal to the flow of a medium-sized river). The original image is reproduced on the back cover. © Chris Jordan. www.chrisjordan.com

Global Climate Change is a Green Edition, printed on paper that includes 10% post-consumer waste and 10% pre-consumer waste. Other green initiatives include:

- Soy-based ink
- Covers printed on stock with 10% post-consumer waste
- Digitized work flow to reduce paper waste

All of which has earned this book Courier Corporation's Green Edition designation for reducing our environmental footprint. The savings we have achieved are:

Trees Saved: 23 • Air Emissions Eliminated: 1,983 pounds
Water Saved: 9,553 gallons • Solid Waste eliminated: 580 pounds

Planet Friendly Publishing
✔ Made in the United States
✔ Printed on Recycled Paper
GREEN EDITION   Text: 10%   Cover: 10%
Learn more: www.greenedition.org

Address inquiries and orders to:
Sinauer Associates, Inc., 23 Plumtree Road, Sunderland, MA 01375 U.S.A.
Fax: 413-549-1118
Email: publish@sinauer.com
Internet: www.sinauer.com

Library of Congress Cataloging-in-Publication Data

Bloom, Arnold J. (Arnold Jeffrey), 1950-
  Global climate change: convergence of disciplines / Arnold J. Bloom.
      p. cm.
  Includes bibliographical references and index.
  ISBN 978-0-87893-027-2 (pbk.)
  1. Climatic changes. 2. Global temperature changes. I. Title.
  QC903.B56 2010
  363.738'74—dc22
                              2009030957

5 4 3 2 1

CONTENTS

CHAPTER 11

GLOBAL CLIMATE CHANGE AND THE LAW   311

CHAPTER 12

GLOBAL CLIMATE CHANGE: THE HUMAN RESPONSE   337

PREFACE

Who, besides you, reads the preface of a textbook? Then again, who reads textbooks? My spouse suggested that I might just skip this endeavor and proceed directly to the CliffNotes or the *Global Climate Change for Dummies* edition. A college textbook, however, fits my purposes in several ways.

- I like using fancy words like "endeavor." A major portion of material on global climate change is oriented to readers with little science background and avoids such language. Most of the remainder of the material is oriented to experts on global climate change and avoids such language.

- A college textbook can assume that its readers have completed a high school education. A high school education includes the rudiments of natural sciences, social sciences, and languages. I expect that the readers know what an atom is, what a biological cell is, and what a public opinion survey is. The readers would benefit from some experience in interpreting numerical graphs, although I include a primer on graphs in an appendix.

- Readers of a textbook have ready access to an instructor for assistance on difficult topics. As such, I need not shy away from topics such as chemical isotopes, photosynthetic carbon fixation, energy efficiencies, and supply versus demand curves.

- I can support nearly every statement with data. Indeed, this textbook averages about two illustrations per page.

- We have a web site that provides even more documentation to support statements in the text.

- Modern textbooks are printed in full color. The wealth of information about global climate change warrants color illustrations. I apologize in advance to readers who are color-blind.

On the other hand, textbooks have several disadvantages, primarily those of cost and tone.

- The rising cost of college textbooks has become a major matter of concern. I have tried to contain the cost of this textbook by creating all the illustrations myself and by using photographs from the public domain whenever possible. The publisher offers the lower cost of a paperback edition and electronic distribution.

- Often textbooks are selected by someone who insists that the book will not offend anyone. As a result, textbooks tend to be as dull as possible. The use of visual and verbal humor in this textbook has already been the subject of extensive debate among the author, publisher, and reviewers. Some insist that the topic of global climate must be treated with utmost solemnity. The author takes full responsibility for all departures from such a treatment.

Finally, I wish to thank all the people who made this book possible. Sinauer Associates, from top to bottom, maintains the enterprise of textbooks as a noble profession. The University of California at Davis has supplied most of the resources. Students and colleagues have offered their encouragement, comments, and corrections. Stephen Gardiner (U. of Washington), Alan Kaufman (U. of Maryland), Richard Keen (U. of Colorado, Boulder), William Moomaw (Tufts U.), and Lowell Stott (U. of Southern California) reviewed the proposal for the book. Jacob Bloom (Humboldt State U.), Paul Bloom (Duke U.), Anne Britt (UC Davis), David Burdige (Old Dominion U.), Lynn Christenson (Vassar College), Daniel Douglass (Northeastern U.), Fred Fox (UC Davis), Tu Jarvis (UC Davis), Richard Keen (U. of Colorado, Boulder), Elise Pendall (U. of Wyoming), Kurt Schwabe (UC Riverside), Joel Thornton (U. of Washington), and E. J. Zita (Olympia, WA) reviewed one to several chapters of the text. The National Science Foundation has provided steady support for my research program. My family—Joy, Jacob, and Samuel—have provided the foundation of this effort.

In writing a book of this scope, I have finally realized the career goal that I indicated on my college admission forms: professional dilettante.

Arnold J. Bloom

INTRODUCTION TO GLOBAL CLIMATE CHANGE

"Everybody talks about the weather, but nobody does anything about it." This comment, attributed to Mark Twain in 1897, is no longer valid. Today, human activities influence climate, long-term trends in weather, on both local and global scales. Average temperatures are rising. Storms and forest fires seem to be increasing in severity (**Figure 1.1**). The vagaries of weather may obscure specific cause and effect relationships, but humans are definitely part of the problem.

Humans are also part of the solution. To diminish the potential damage from climate change and the extreme weather events that result, governments have implemented policies that range from limiting carbon emissions to reinforcing levees. As the public has become more aware about this issue, their behaviors in matters ranging from recycling of materials to the purchase of refrigerators, vehicles, and windows, increasingly reflects their concerns.

On contentious issues such as global climate change, a broad understanding of the subject generally contributes to the quality of debate. This book considers the factors responsible for climate change; the geophysical, biological, economic, legal, and cultural consequences of such change; and

FIGURE 1.1 **Extreme weather** Hurricane Katrina extends across the Gulf of Mexico as it approaches New Orleans on August 28, 2005. (Imagery from the GOES-12 weather satellite.)

various strategies to diminish some of the undesirable consequences. It compares methods that researchers in different disciplines employ to evaluate past and future conditions, and it highlights the complexity of developing effective policies when the information available may be sparse and vague.

Most textbooks concentrate on a single discipline (e.g., geophysics, biology, or economics) or subdiscipline (e.g., glaciology, plant physiology, or macroeconomics); they introduce the major concepts and then apply them to several examples involving a variety of issues. This book, by contrast, focuses on a single issue—global climate change—and relates concepts from a number of natural and social sciences to it. Nearly everyone will find certain topics from this wide spectrum to be challenging; nonetheless, stretching to maintain flexibility becomes critical as one matures.

Articles on environmental issues frequently evoke *f*ear, *u*ncertainty, and *d*oubt (FUD) that further exploitation of natural resources might cause irrevocable damage. Excessive exposure to FUD, however, desensitizes the public to such issues ("crying wolf") or, worse, elicits fatalistic despair. This book strives to present a more balanced perspective and occasionally unbridled optimism, but will deserve the subtitle *Doom and Gloom with Bloom* (**Figure 1.2**) if this attempt fails.

FIGURE 1.2 **"Doom and Gloom with Bloom"** The author, Arnold J. Bloom, is shown marching in Times Square, New York City in this manipulated image.

This chapter recalls the last 70 years of research on global climate change. The next three chapters constitute a geophysical section that examines the past, present, and future of Earth's climate: Chapter 2 presents historical reconstructions of temperature and a few other indicators of climate, Chapter 3 details factors that influence climate, and Chapter 4 describes global climate models and the changes they predict during the next century. Subsequent sections of the book introduce additional aspects about climate change: direct and indirect effects on organisms, mitigation strategies and the economics thereof, international cooperation and accords, and, finally, the interplay of culture and public opinion. The book strives to support every statement with data in graphical or tabular form. The number of graphs and tables that results from such an effort disrupts the continuity of the text and may overwhelm readers who are less quantitative in orientation. For this reason, some of this documentation is presented on the book's web site, www.sinauer.com/bloom, for readers who are more inquisitive and quantitatively inclined.

Climate

Long-term weather patterns, as mentioned in the previous section, characterize the climate. The weather page in your local newspaper includes information on daily maximum and minimum temperatures; humidity; precipitation; and wind speed and direction. Long-term averages of these measures define the climate in your area. For example, a Mediterranean climate is characterized by relatively hot, dry summers and cool, wet winters. In Davis, California, more than 80% of the rainfall occurs during the winter months (**Figure 1.3**), thus Davis is considered to have a Mediterranean climate.

"Climate is what we expect; weather is what we get," is another statement attributed to Mark Twain. In other words, weather conditions are highly variable from day to day or year to year. Over an 11-year period, total precipitation during the month of December in Davis ranged from 0 to 250 millimeters (see Figure 1.3). Consequently, predicting daily weather based on climatic trends over time is seldom worthwhile. Subtle changes in climate over a few decades are sometimes difficult to discern from normal fluctuations in weather.

The first person to note the recent warming trend in Earth's climate and associate it with fossil-fuel emissions was Guy Stewart Callendar (1898–1964). Callendar's father, Hugh Longbourne Callendar, a professor of physics at the Imperial College of Science, London, had developed the platinum resistance thermometer, an

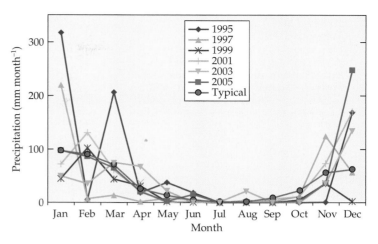

FIGURE 1.3   Precipitation (millimeters) in Davis, California
Plotted are data for each month of every other year during the last decade and typical precipitation values (long-term averages).

instrument that permitted precise, continuous recording of temperatures. Guy Stewart Callendar worked as a steam engineer for the British Electrical and Allied Industries Research Association, but he had inherited his father's interest in temperature measurement and, as a hobby, scrutinized weather records from around the world.

Callendar examined historical trends in global average temperatures by grouping temperature data from the most reliable weather stations in given regions of the world and weighting the importance of each group according to the geographic area represented by its stations (Callendar 1938). He calculated 10-year moving averages (the average of the values 5 years before and 5 years after a given date) to smooth out year-to-year fluctuations (**Figure 1.4**). Callendar's analysis suggested that world temperatures had increased by more than 0.2°C between 1890 and 1935. Based on crude measurements of carbon dioxide (CO_2) concentrations in the atmosphere and a simplistic model, Callendar proposed that rising CO_2 levels were responsible for over half of this warming.

The ideas of Callendar, an amateur encroaching on the domain of climatologists, licensed professionals who focused on climate, were not well received (Weart 2003). Most climatologists of the day believed that temperature data, because they were so variable, could be statistically manipulated to support nearly any conclusion. For example, Helmut E. Landsberg (1906–1985), perhaps the most renowned climatologist of the twentieth century (Baer 1992), did not acknowledge any significant historical changes in global average temperatures and declared, "There is no scientific

FIGURE 1.4 A graph from Callendar's 1938 publication
This graph shows temperature patterns (°C) for various climatic zones and of Earth. Plotted here are 10-year running averages (the average of 5 years before the date and 5 years after) with respect to the average temperatures from 1901–1930. (From Callendar 1938).

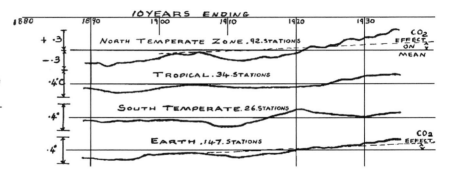

reason to believe that our climate will change radically in the next few decades, hence we can safely accept the past performance as an adequate guide for the future" (Landsberg 1946).

The larger scientific establishment also doubted whether atmospheric CO_2 concentrations had changed significantly (Weart 2003). Readings of CO_2 concentrations would fluctuate with the winds because local **sources** that released CO_2, such as nearby factories, and **sinks** that absorbed CO_2, such as nearby forests, influenced every sample. The consensus of the scientific community was that nearly all the CO_2 released from fossil-fuel burning would dissolve in the immense volume of Earth's oceans, and thus atmospheric changes would be negligible.

With the dawn of the nuclear age at the end of World War II, atmospheric and oceanic scientists became preoccupied with other products of human ingenuity, namely radioactive wastes. In 1954, fallout from an American nuclear bomb test injured the crew of a Japanese fishing vessel. Later that year came the release of *Gojira*, the first in a long series of horror movies to feature Godzilla, a fictional monster created by an American nuclear bomb test. Anxiety was escalating. Would radioactive carbon dioxide ($^{14}CO_2$), which was generated in the atmosphere during nuclear explosions, dissolve in the oceans and widely contaminate sea life and seafood?

Roger Revelle (1909–1991) and Hans Suess (1909–1993) of the Scripps Institution of Oceanography in San Diego, California, analyzed the exchange of $^{14}CO_2$ between the atmosphere and the oceans. They published a seminal work in 1957 showing that only a thin, upper layer of seawater rapidly exchanged materials with the atmosphere (Revelle and Suess 1957). These results had broad implications. On the positive side, contamination of sea life from nuclear testing would be highly localized; but on the negative side, the oceans would remove only a small portion of the $^{14}CO_2$ being released into the atmosphere; most of the radioactive gas would remain airborne. By analogy, these results indicated that only a

small portion of the CO_2 released from fossil-fuel burning would dissolve in the oceans: the bulk would remain in the atmosphere, and atmospheric CO_2 concentrations would increase substantially.

By the mid-1950s, technological advances had increased the precision of CO_2 measurements tenfold. C. D. (Dave) Keeling (1928–2005), also of Scripps, obtained funds sufficient to equip two weather stations with instruments that monitored atmospheric CO_2 with unprecedented accuracy. To minimize the influence of local disturbances, he chose sites that were remote from industrial and biological sources of CO_2 and were subject to strong prevailing winds (**Figure 1.5A**). One site was at the South Pole and the other was on the Island of Hawaii, at the Mauna Loa Observatory atop the northern flank of the Mauna Loa volcano, at an elevation of 3,397 meters (**Figures 1.5B and C**).

Monitoring at the South Pole began in September 1957 and at Mauna Loa 6 months later. Concentrations of CO_2 at the South Pole rose slowly, but steadily. The CO_2 measurements at Mauna Loa, however, oscillated from month to month (**Figure 1.6**), raising doubts about whether this instrument was performing properly (Keeling 1978). Fortunately, with more observations, Keeling realized that the oscillations at Mauna Loa reflected the annual cycle on nearby continents of relatively rapid plant **photosynthesis** in summer, which removes CO_2 from the atmosphere, and relatively rapid biological **respiration** in winter, which releases CO_2 to the atmosphere. Funding for the South Pole station ran out after about 2 years, during which time data showed a rise in CO_2 concentrations from 311 ppm to 314 ppm (parts per million; 311 ppm = 0.0311%). The Mauna Loa station has provided a nearly continuous record of rising atmospheric CO_2 concentrations from about 315 ppm in 1957 to about 386 ppm in 2009, an increase of about 20%. This record has become known as the Keeling curve.

As evidence accumulated, the scientific establishment became more receptive to the idea of global warming and its relationship to atmospheric CO_2 concentrations (**Figure 1.7**). Even Landsberg, who by 1958 had

(A)

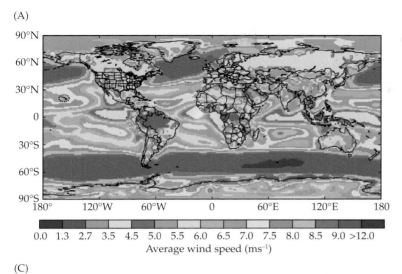

Average wind speed (ms⁻¹)
0.0 1.3 2.7 3.5 4.5 5.0 5.5 6.0 6.5 7.0 7.5 8.0 8.5 9.0 >12.0

(B)

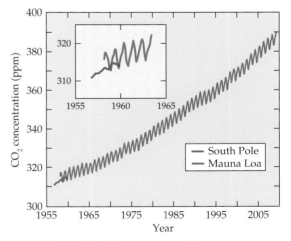

(C)

FIGURE 1.5 **Sites to monitor atmospheric CO₂ concentration** (A) Average wind speeds (meters per second) at ground level around the globe. The black dotted oval on the left demarcates the Hawaiian Islands. (B) Satellite photo of Hawaii showing the location of the Mauna Loa Observatory. (C) The observatory in 1982 shown against the backdrop of the neighboring peak Mauna Kea.

become the director of the Office of Climatology in the U.S. Weather Bureau, shifted his stance on the subject:

> For nearly a half century, a general warming trend has been noted…For the moderate latitudes, 30° to 50°N in the area around the Atlantic, the natural rise can be estimated at about 2°F (1.1°C) per century …For the latest temperature change, there is an important contender as cause: atmospheric carbon dioxide. There are some interpretations of historical and current observations pointing toward a gradual increase of this atmospheric constituent…Carbon dioxide is an absorber of outgoing long-wave radiation, and hence has an influence…often referred to as the 'greenhouse effect' (Landsberg 1958).

Current State of Affairs

Disagreements still remain about the extent to which the recent warming in global temperatures deviates from normal climatic cycles. Direct measurements of

FIGURE 1.6 **Monthly average CO₂ concentration** (parts per million: 1 ppm means 1 microliter of CO₂ is dissolved in each liter of total air; 1 ppm = 0.001%) in the atmosphere at the South Pole and near the summit of Mauna Loa in Hawaii. The inset in the upper left shows data from the first few years on an expanded scale.

FIGURE 1.7   Rising temperatures A graph from Landsberg's 1958 publication shows temperature patterns over nearly 50 years (°F and °C) for the summer (June–August) and winter (December–February) at Winthrup College, South Carolina. Dashed lines show the general temperature trend for the region. (From Landsberg 1958).

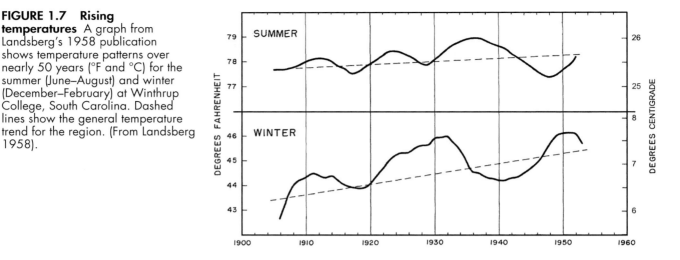

temperature have been available from weather stations around the world only since 1861. To reconstruct temperature patterns before 1861 requires the use of **proxy measures**, measurements such as the width of tree rings that are strongly dependent on temperature and can be dated with accuracy. Chapter 2 considers different types of proxy measures.

In 1999, Michael Mann of Pennsylvania State University and his coworkers reconstructed the average annual temperatures in the Northern Hemisphere over the last 1000 years from a variety of direct measurements and proxy measures (Mann et al. 1999). The resulting graph (**Figure 1.8**) became affectionately known as the "hockey stick" because the "shaft" (representing temperatures from A.D. 1000 to 1900) was relatively straight, whereas the "blade" (representing temperatures during the current century) was abruptly bent upward about 0.6°C (1.1°F). Mann and his colleagues proposed, as had Callendar and Landsberg several decades earlier, that emissions of CO_2 and other greenhouse gases from the burning of fossil fuels were responsible for the dramatic warming trend (Mann et al. 1998).

By 2002, the political climate of the United States had changed, and fossil-fuel companies assumed a larger role in governmental policies on energy. ExxonMobil, the largest supplier of fossil fuels, distributed over $2 million per year from 2000 through 2003 to organizations promoting the message that the scientific evidence linking global warming and fossil-fuel burning was unsound (McKibben et al. 2005; The Royal Society 2006). The "hockey stick" data became even more contentious, and the U.S. Congress requested that the National Academy of Sciences, a body of prestigious scientists, verify Mann's research.

Eight years after Mann and his coworkers published their 10-page article, the committee appointed by the National Academy released a 196-page report (National Research Council 2006). This report upheld the major premise of the hockey stick: Global temperatures have warmed more than 0.6°C during the last century, and such changes are without precedent during the preceding 4 centuries and probably much longer. In particular, the year 2005 was the hottest on record, followed in descending order by 2007, 1998, 2002, 2003, 2006, 2004, 2001, and 2008 (Goddard Institute for Space Studies 2009). All indications are that this warming trend will continue and perhaps even accelerate.

About 130 stations around the world now monitor atmospheric CO_2 concentrations and have affirmed the trends first observed in Keeling's data from the South Pole and Mauna Loa. Atmospheric concentrations of CO_2 have increased worldwide (**Figure 1.9**). Concen-

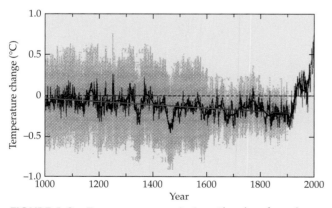

FIGURE 1.8   Temperature variation This data from the 1999 publication of Mann et al. is dubbed the "hockey stick." It shows deviations of average annual temperatures (°C) in the Northern Hemisphere from the long-term average from 1902–1980 (temperature change). It is a composite estimate that the authors reconstructed from a variety of sources. Gray shading indicates the extent of variation in the estimate, and the red line in the middle of the data represents the long-term trends. (After Mann et al. 1999).

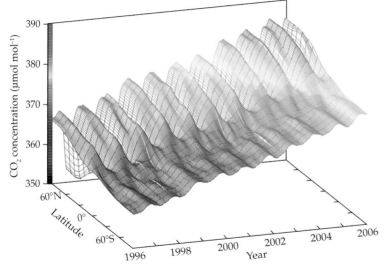

FIGURE 1.9 **Global distribution of atmospheric** CO_2 This is a three-dimensional representation of the latitudinal distribution of atmospheric CO_2 based on data from the Global Monitoring Division (GMD) cooperative air-sampling network. Notice that the overall global trend is that CO_2 concentrations are rising. CO_2 concentration varies more during a year at higher latitudes because seasonal variation is greater there. This concentration also varies more in the northern hemisphere because it has more land area than the southern hemisphere. Moreover, the seasonal peak in concentration shifts with date from the northern hemisphere to southern hemisphere because photosynthesis, which absorbs CO_2, is faster during the summertime of each hemisphere. (From Dr. Pieter Tans and Thomas Conway, NOAA ESRL GMD Global Carbon Cycle, Boulder, CO.)

trations are lower in the summer, when plants incorporate CO_2 from the atmosphere into their organic carbon compounds via photosynthesis, and higher in the winter, when biological respiration exceeds photosynthesis and releases CO_2 from organic carbon compounds. Seasonal variation is greater in the Northern Hemisphere than the Southern Hemisphere because the Northern Hemisphere has substantially more landmass (see Fig-

ure 3.13) and thus more terrestrial organisms that conduct rapid photosynthesis and respiration.

Global temperatures and atmospheric CO_2 concentrations have tended to change in conjunction with one another (i.e., show a positive **correlation**) both in the current century and during the last 650,000 years (see Chapter 2). Admittedly, correlation does not imply causality (**Box 1.1**), but most in the scientific commu-

BOX 1.1 Alternative explanations for global warming

The following examples about global warming demonstrate that correlation does not imply causality. In a parody of scientific method, Bobby Henderson—self-described as an unemployed, amateur pirate with a physics degree—found a negative correlation between the number of pirates and global average temperature (**Figure A**) and advocates that people become pirates to stop global warming (Henderson 2006). In another

spoof, Connie M. Meskimen, a bankruptcy lawyer from Arkansas, suggested that daylight savings time exacerbates global warming by setting sunrise at an earlier hour (**Figure B**). Only a few scientists, however, have turned to piracy or turned back their clocks prematurely.

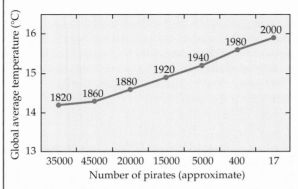

Figure A Global average temperatures as a function of the number of pirates. This parodies standard presentations of scientific data. Notice that the labels on the *x* axis are not consistent. (After Henderson 2006).

LETTERS

Daylight exacerbates warning

You may have noticed that March of this year was particularly hot. As a matter of fact, I understand that it was the hottest March since the beginning of the last century. All of the trees were fully leafed out and legions of bugs and snakes were crawling around during a time in Arkansas when, on a normal year, we might see a snowflake or two.

This should come as no surprise to any reasonable person. As you know, Daylight Saving Time started almost a month early this year. You would think that members of Congress would have considered the warming effect that an extra hour of daylight would have on our climate. Or did they?

Perhaps this is another plot by a liberal Congress to make us believe that global warming is a real threat. Perhaps next time there should be serious studies performed before Congress passes laws with such far-reaching effects.

CONNIE M. MESKIMEN

Figure B A tongue-in-cheek letter to an Arkansas newspaper.

nity agree that global temperatures are rising and that human-generated emissions of CO_2 and other greenhouse gases are contributing to this rise. Alternative explanations for the current temperature trends conflict with a growing body of evidence. Even organizations with strong vested interests in fossil fuels have modified their message.

For instance, in 2005, ExxonMobil's "Corporate Citizenship Report" acknowledged that "the accumulation of greenhouse gases in the Earth's atmosphere poses risks that may prove significant for society and ecosystems. We believe that these risks justify actions now, but the selection of actions must consider the uncertainties that remain" (ExxonMobil 2005). The report presents ExxonMobil's view of the uncertainties regarding the causes of global warming but then touts the $200 million that ExxonMobil had just bequeathed to the Global Climate and Energy Project at Stanford University in California as "the largest-ever privately funded research effort in low-greenhouse-gas energy."

Other fossil-fuel companies have taken similar approaches. In 2006, BP (formerly British Petroleum) and Chevron announced plans to allocate $500 million and $400 million, respectively, for research on **biofuels**. All these companies now advertise their efforts in developing energy resources while minimizing environmental degradation.

Telltale Signs

Global warming has altered a broad range of geophysical and biological phenomena. These phenomena are the focus of several chapters in this book. Recent changes in ice cover, however, are so visually striking as to warrant a place in this first chapter.

Mount Kilimanjaro reaches 5,895 meters above sea level in equatorial Tanzania. Not only is it the highest peak in Africa, it is also the only place on the continent covered with snow year-round; hence its name, which translates as "Shining Mountain." Satellite photographs show the mountain in 1993 and 2000 (**Figure 1.10A**). A compilation of maps outlining the ice fields near the summit documents the changes over the last century (**Figure 1.10B**). If the current rate of decline continues, the snows of Kilimanjaro will disappear during the next few decades (Thompson et al. 2002).

The far end of Earth provides another visual example of changes in ice cover over time. Since 1950, the Arctic has experienced an increase in average temperatures of about 2°C, more than twice that observed at lower latitudes (ACIA 2005). In response, the polar ice cap is receding at a rate of around 10% per decade (**Figure**

FIGURE 1.10 **Dramatic changes in ice cover** (A) Mt. Kilimanjaro on February 17, 1993 (top) and February 21, 2000 (bottom). Satellite images of the summit indicate the minimum amount of snow cover. (B) Outlines of the ice fields near the summit of Mt. Kilimanjaro in 1912, 1953, 1976, 1989, and 2000. The inset illustrates the near-linear decrease in ice area over time. (B after Thompson et al. 2002.) ▶

1.11). Sometime in the not-too-distant future, the Arctic Ocean will have an ice-free season and realize the long-sought Northwest Passage, a sea route between the Atlantic Ocean and the Pacific Ocean through the Canadian archipelago (see Chapter 4). This might prove to be a financial windfall for Pat Broe, a Denver entrepreneur who bought the port of Churchill on Hudson Bay at auction for $10 Canadian in 1997: An ice-free Northwest Passage could bring up to $100 million of shipping business to Churchill each year.

In the Course of Human Events

The effects of climate have determined the outcome of many human endeavors. An appropriate ending to this introduction is the famous figure drawn by Charles Joseph Minard (**Figure 1.12**), a testament to climate and the fate of empires (Tufte 2001). On June 24, 1812, Napoleon invaded Russia, crossing the Niemen River from Poland into Russia with 422,000 men. Six months later, after experiencing temperatures as low as –38°C, Napoleon's *Grande Armée* departed Russia with a mere 10,000 men.

Climate again played a pivotal role in the disastrous German invasion of the Soviet Union in 1941. German forces were trapped outside Moscow during the Russian winter with inadequate shelter, clothing, fuel, and food. All in all, more than 4 million German and 8 million Soviet troops lost their lives on the Eastern Front.

As these examples illustrate, insufficient consideration of climate can have dire consequences. In 2007, military and spy agencies in the United States warned that anticipated changes in the world's climate pose a serious threat to the security of nations because of substantial shifts in the availability of natural resources and the forced migration of millions of people (Mazzetti 2007). In their fourth assessment, the Intergovernmental Panel on Climate Change (IPCC), an organization established by the World Meteorological Organization and United Nations, agreed that further global warming is already unavoidable due to past human activities and a major international effort is required to mitigate the impacts (IPCC 2007a, 2000b,c).

(A)

(B)

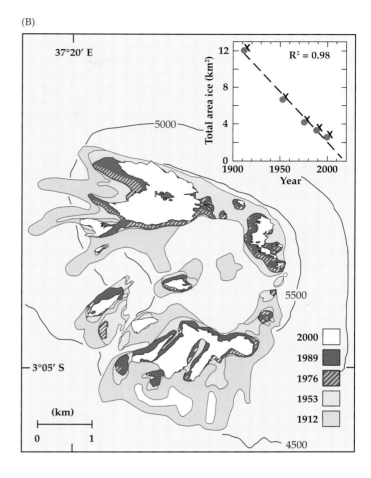

This book outlines the causes and possible consequences of global climate change and then presents various strategies for addressing these possible consequences. Climate change is a complex problem, probably one without a simple solution. This book cannot provide definitive answers to many of the questions raised while considering the issue of global climate change, but it can provide a broad context from which readers may draw their own insights and conclusions.

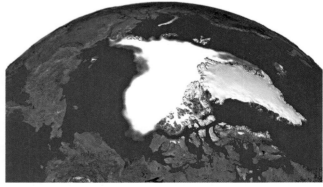

FIGURE 1.11 Minimum amount of Arctic sea ice in 1979 (left) and 2007 (right). This is based on data collected by NASA satellites at the end of the Arctic summer.

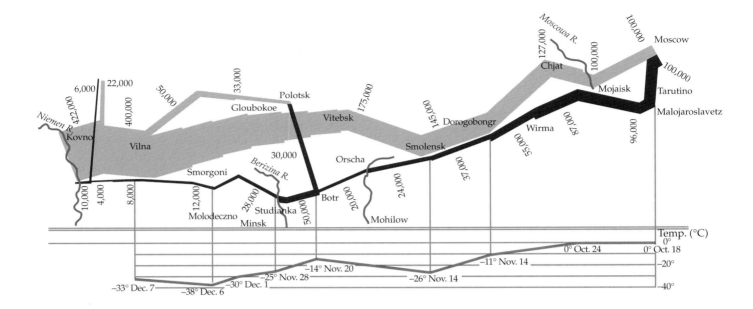

FIGURE 1.12   A map of Napoleon's invasion of Russia in 1812 The tan band depicts the French army's crossing the Niemen River into Russia and advancing east toward Moscow; the black band depicts the army's westward retreat from Moscow. The thickness of the bands reflects the size of the French army at various locations. The branch points indicate where some troops separated during the advance and later rejoined the retreat. Temperatures (°C) in red are linked to the path of retreat.

Summary

Climate refers to the general weather patterns in a particular location. Long-term changes in such patterns are often difficult to differentiate from the large fluctuations in weather that normally occur. Nonetheless, long-term measurements of temperatures from around the world indicate that Earth is warmer today than it has been for several centuries. This warming has been associated with human activities, especially the release of carbon dioxide from the burning of fossil fuels. Although a small portion of this carbon dioxide dissolves into the oceans, the vast majority remains airborne and increases atmospheric carbon dioxide concentrations.

Global warming has produced noticeable changes in a broad range of phenomena including the disappearing snows of Kilimanjaro and a decrease in the extent of the polar ice caps. Such changes are likely to influence the fate of human civilizations.

Review Questions

1. Over the last century, global average temperatures have
   (a) cooled about 0.1°C.
   (b) not changed significantly.
   (c) warmed about 0.1°C.
   (d) warmed about 0.6°C.
   (e) warmed about 6.0°C.

2. Atmospheric CO_2 concentrations were about _____ ppm in 1957 and about _____ ppm in 2009.
   (a) 200
   (b) 270
   (c) 315
   (d) 386
   (e) 550

3. Which of the following researchers made major contributions to our understanding about climate trends? (Select one or several.)
   (a) Guy Stewart Callendar
   (b) C. D. Keeling
   (c) Roger Revelle
   (d) Pat Broe
   (e) Charles Joseph Minard

4. Most of the carbon dioxide released from the burning of fossil fuels
   (a) dissolves in the oceans.
   (b) remains in the atmosphere.
   (c) is consumed by animals via respiration.
   (d) becomes radioactive.
   (e) freezes and becomes dry ice.

5. Evidence of current global temperature trends include (select one or several)
   (a) the disappearance of year-round snow in Africa.
   (b) Napoleon's successful invasion of Russia.
   (c) the Northwest Passage through the Canadian archipelago.
   (d) the proliferation of movies featuring Godzilla.
   (e) the extension of daylight savings time.

Suggested Readings

National Research Council. 2006. *Surface Temperature Reconstructions for the Last 2,000 Years*. The National Academies Press, Washington, D.C., *http://www.nap.edu/catalog/11676.html*.

The National Academy of Sciences has a mandate from the U.S. Congress to advise the federal government on scientific and technical matters. The Academy appointed a committee of twelve distinguished scholars to prepare this report on global temperature changes over the past two millennia.

Weart, S. R. 2003. *The Discovery of Global Warming*. Harvard University Press, Cambridge, Mass.

A fascinating account about the history of research on climate change. Dr. Weart regularly updates the version available for free at *http://www.aip.org/history/climate/*.

2

HISTORY OF
EARTH'S CLIMATE

Collisions of Earth with extraterrestrial bodies, volcanic eruptions, and other cataclysmic geophysical events that may dramatically shift our planet's climate capture the attention of climatologists and most other people. Living organisms, however, have influenced Earth's climate to an equal extent, and throughout Earth's history, climate has derived from a balance between geophysical and biological factors. As we will see throughout the first section of this book, climate depends increasingly on the dynamic interplay between natural climatic cycles and human activities. Our ability to predict how climate may change during the coming decades requires an understanding of past climates.

Unfortunately, human recorded history provides only limited assistance. It extends back only 5 millennia and focuses more on struggles for political power than on forces of nature. As mentioned in Chapter 1, reliable weather records are available for less than the past 2 centuries. Therefore, reconstruction of Earth's climate over geological time must rely on

proxy measures, measurements of physical, chemical, or biological processes that are functions of weather. Consider an oak tree that has been growing in Sherwood Forest for 1000 years (**Figure 2.1**). Temperature affects the growth of wood in this tree, and so an annual temperature cycle produces a distinct layer of wood known as a tree ring. As we will explore later in the chapter, the size of a tree ring reflects the temperature regime, and the ring's relative position in the trunk indicates the date when it was formed. Taken together this tree provides information about the temperature in a given year in Nottinghamshire. Components of a proxy measure often differ in their accuracy. For tree rings, temperature estimates have a low certainty because factors other than temperature, including the availability of nutrients and water, influence growth. Dates are more certain because counting tree rings is a relatively straightforward procedure. Location is most certain because trees rarely move. Many reconstructions of climate rely on multiple types of proxy measures to improve reliability.

This chapter begins with a lesson in some simple chemistry that influences Earth's climate. It then presents a brief account of Earth's ancient climate history, with little detail about methods. It covers the major climatic events during the past billion (10^9) years in more detail and introduces the relevant proxy measures. Finally, it proceeds slowly over the most recent million years and discusses several more precise measurements of climate.

CO_2 Reactions and Climate

Central to our ability to interpret changes in climate are several biochemical and geochemical reactions involving the molecule carbon dioxide (CO_2). The following describes each reaction and then gives the corresponding **chemical equation**.

- Photosynthetic organisms such as plants use solar energy to convert a low-energy carbon in CO_2 into a high-energy carbon in a carbohydrate (CH_2O) such as the sugar glucose. These reactions of **photosynthesis** usually split water (H_2O) and release gaseous oxygen (O_2) to the atmosphere.

$$CO_2 + H_2O + light \rightarrow CH_2O + O_2 \qquad (2.1)$$

- Organisms such as animals and plants, when they conduct cellular **respiration** to generate the chemical energy that they require for maintenance and growth, convert a high-energy carbon in CH_2O into a low-energy carbon in CO_2. **Aerobic respiration** occurs in the presence of O_2, whereas **anaerobic respiration** (or fermentation) occurs in the absence of O_2. In aerobic respiration, CH_2O reacts with O_2.

$$CH_2O + O_2 \rightarrow CO_2 + H_2O + chemical\ energy \qquad (2.2)$$

- One pathway of anaerobic respiration involves the breakdown of CH_2O to methane (CH_4).

$$6\ CH_2O \rightarrow 3\ CH_4 + 3\ CO_2 + chemical\ energy \qquad (2.3)$$

- Methane (CH_4) reacts with atmospheric O_2, according to

$$CH_4 + 2O_2 \rightarrow CO_2 + 2H_2O + \text{chemical energy} \quad (2.4)$$

This reaction occurs at a rate whereby half of the CH_4 that is initially present breaks down to CO_2 in an average of 12 years.

If respiration were to balance photosynthesis, the net result (the sum of Equations 2.1 and 2.2) would be the conversion of solar energy (light) into chemical energy. Normally, however, a portion of the carbohydrate produced during photosynthesis is transformed into organic carbon compounds such as cellulose and lignin that decompose slowly (see Chapter 7). Excess photosynthesis over respiration leads to accumulation of organic carbon compounds in soils and depletion of CO_2 from the atmosphere, a process known as **carbon sequestration** (see Chapter 5).

- In the weathering of silicate rock ($CaSiO_3$), CO_2 first dissolves in H_2O to form carbonic acid (H_2CO_3), which then reacts with $CaSiO_3$ to produce silicon dioxide (SiO_2; silica, quartz, or sand) and calcium carbonate ($CaCO_3$; aragonite, calcite, chalk, limestone, marble, or travertine).

$$CO_2 + H_2O \rightarrow H_2CO_3 \quad (2.5)$$

$$H_2CO_3 + CaSiO_3 \rightarrow CaCO_3 + SiO_2 + H_2O \quad (2.6)$$

In the Beginning…

The Big Bang, the single point in space and time from which all matter and energy in the universe supposedly emanated, is thought to have occurred sometime around 13.7 billion years ago (13.7×10^9 years ago). Our solar system, including planet Earth, began to coalesce about 4.5 billion years ago (**Figure 2.2**). When the surface of primitive Earth cooled to a temperature below the boiling point of water (100°C), sometime around 4.1 billion years ago (Hopkins et al. 2008), the atmosphere consisted primarily of gases released from volcanoes. These gases included high concentrations of carbon dioxide (CO_2), carbon monoxide (CO), water vapor (H_2O), nitrogen (N_2), hydrogen chloride (HCl), and, perhaps, small amounts of methane (CH_4). Within a relatively short time, perhaps as early as 3.8 billion years ago, photosynthetic organisms called **cyanobacteria** (**Figure 2.3**) appeared (Buick 2008).

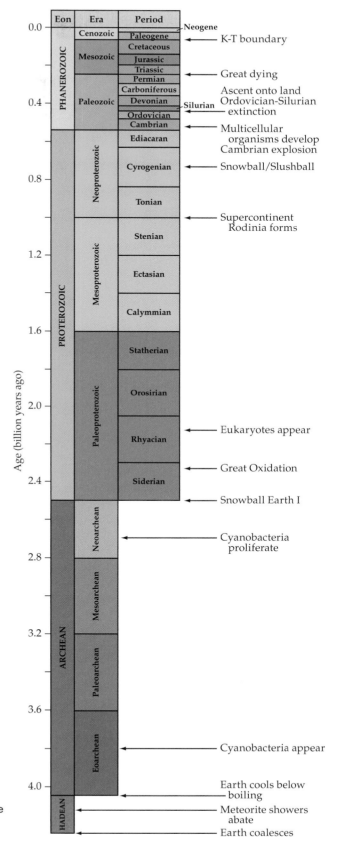

FIGURE 2.2   Geological timeline Major events in the climate history of Earth. Timeline is in billion years ago.

FIGURE 2.4 Banded iron in a boulder A 2.1 billion-year old black-band ironstone (2 meters high × 3 meters wide × 1 meter thick) found in North America. The red stripes are oxidized iron, and the black stripes are rock that does not contain iron.

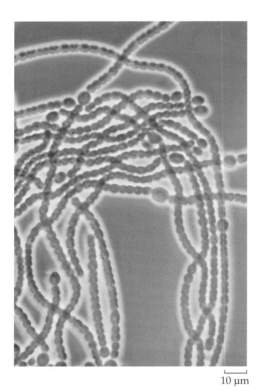

10 µm

FIGURE 2.3 Cyanobacteria Shown is the free-living, nitrogen-fixing cyanobacterium *Nostoc* sp.

Cyanobacteria proliferated widely during the next few billion years, especially more recently than 2.7 billion years ago. Their photosynthesis produced a substantial shift in Earth's atmosphere, depleting the CO_2 concentration to below 10% and releasing sufficient O_2 to rust (oxidize) iron near Earth's surface. Evidence for this includes the banded iron formations, rocks of this age that contain layers of oxidized iron (**Figure 2.4**). The photosynthetic release of O_2 by cyanobacteria also exhausted minerals such as elemental sulfur near Earth's surface and methane (CH_4) in the atmosphere (see Equation 2.4).

At this time, energy emitted from our young sun amounted to only about 83% of the energy it emits today, and so, at about 2.5 billion years ago, a relatively small loss of CO_2 and CH_4 sent global temperatures plummeting. As we will see in Chapter 3, CO_2 and CH_4 are two of the major **greenhouse gases** that warm the Earth by absorbing a significant portion of the **infrared radiation** emitted by Earth's surface. Without this warming effect, portions of early Earth became covered with ice and reflected more solar energy. All climatologists believe that Earth became colder; some believe

that the surface of the entire planet froze, an episode they have named Snowball Earth (Kopp et al. 2005; Kasting and Ono 2006).

Recovery from this frozen state was slow. Volcanoes continued to release CO_2 at a steady rate, but weathering of silicate rock and photosynthesis, processes that deplete CO_2, virtually stopped at the low temperatures; thus, CO_2 gradually accumulated in the atmosphere. After about 30 million years, atmospheric CO_2 reached a level at which Earth retained sufficient solar energy to thaw ice at the equator. Once sunlight struck open water and bare ground instead of ice, Earth absorbed more solar energy. Additional water vapor, another greenhouse gas, entered the atmosphere and further accelerated warming. Snowball Earth melted.

Indeed, after this melting, Earth experienced a heat wave that persisted from 2.4 billion years ago to 2.0 billion years ago. High temperatures plus high CO_2 levels stimulated a bloom of cyanobacteria that released massive amounts of O_2. Sometime between 2.45 and 2.22 billion years ago, atmospheric O_2 concentrations jumped from less than 0.02% to around 3% in what is called the Great Oxidation Event (Bekker et al. 2004; Canfield 2005; Holland 2006).

The Great Oxidation Event had major biological repercussions. Oxygen, as the name implies, is a strong **oxidizing agent**, one that removes electrons from many other molecules and thus interferes with several vital biochemical reactions. Many simple life forms that had up to that time been prevalent on Earth could not tolerate exposure to O_2. Today, descendants of such organisms are limited to the few remaining anaerobic (oxygen-deficient) environments on Earth, such as bogs.

Eukaryotes, organisms composed of cells that have nuclei and other specialized compartments (organelles), appeared after the height of the Great Oxidation Event (Rasmussen et al. 2008). This was no coincidence. Some of the specialized compartments in eukaryotic cells, such as mitochondria, peroxisomes, and chloroplasts, isolate reactions involving O_2 from more sensitive parts of the cell. Aerobic respiration, O_2-based breakdown of organic compounds, generates far more energy than its anaerobic alternatives. The energy bonanza of aerobic respiration fostered the development of more complex organisms.

Additionally, O_2 in the presence of sunlight produces ozone (O_3).

$$O_2 + \text{sunlight} \rightarrow 2\,O \qquad (2.7)$$

$$O + O_2 \rightarrow O_3 \qquad (2.8)$$

Higher O_2 levels thus eventually gave rise to an ozone layer in the upper atmosphere that screened out **ultraviolet radiation** harmful to life. This enabled eukaryotes to complete relatively longer life cycles without extensive ultraviolet damage.

The next period, 1.8 to 0.8 billion years ago, has been called the "Boring Billion" (Holland 2006) for its lack of major catastrophes. Notwithstanding, Earth's landmasses were in motion and coalesced into one supercontinent named **Rodinia** (*Russian*; homeland) near the equator at about 1.0 billion years ago, uplifting vast amounts of silicate rock as well as burying vast amounts of organic matter. Rodinia disassociated into eight smaller continents by about 0.8 billion (800 million) years ago (**Figure 2.5**). Algae (photosynthetic eukaryotes) proliferated in the nutrient-rich, equatorial continental shelves.

Together, these processes—weathering of silicate rock, burial of organic matter, and algal photosynthesis—depleted atmospheric CO_2 concentrations to about 0.03% (Schrag et al. 2002b). For a while, Earth's climate remained uniform despite low CO_2 concentrations because anaerobic respiration of the buried organic matter (see Equation 2.3) released lots of methane (CH_4), and CH_4 is 23 times more effective than CO_2 per molecule as a greenhouse gas (see Table 3.1). The situation was unstable given the transient nature of CH_4 (a half-life of about 12 years; see Equation 2.4), and Earth was again at the brink of a major cold spell.

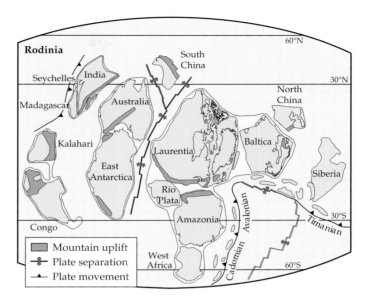

FIGURE 2.5 Supercontinent Rodinia Breakup and dispersal of the Rodinia supercontinent about 750 million years ago. The brown areas indicate mountain belts that arose between 1300 million and 1000 million years ago (After Torsvik 2003).

Isotopic Measures

Several lines of evidence indicate that a significant climate shift occurred at about 730 million years ago. To appreciate this evidence requires a basic understanding of chemical **isotopes**, distinct forms of a chemical **element**. Isotopes are involved in proxy measures for age, atmospheric CO_2 concentration, photosynthetic activity, oceanic pH, and temperature. The following is a brief introduction to isotopes.

Definitions

ELEMENTS Each chemical element has a unique number of protons (a sub-atomic particle) in the nucleus of an atom. For example, a carbon atom (C) has 6 protons.

MOLECULES A molecule is a stable group of two or more atoms held together by strong chemical bonds.

IONS Molecules become ions when they become electrically charged through the gain or loss of electrons (another sub-atomic particle). Ions tend to be soluble in water. For example, carbonic acid (H_2CO_3) tends to disassociate in water producing two ions, a proton (H^+) and a bicarbonate ion (HCO_3^-).

$$H_2CO_3 \rightarrow H^+ + HCO_3^- \qquad (2.9)$$

FIGURE 2.6 Isotopes of carbon ^{12}C and ^{13}C are stable with 6 and 7 neutrons, respectively, in the nucleus. ^{14}C with 8 neutrons in the nucleus is radioactive.

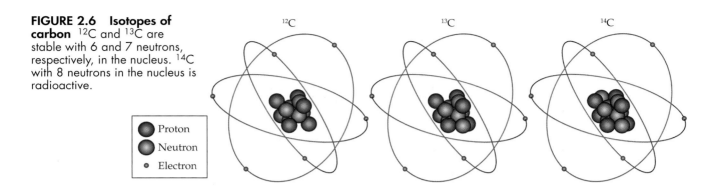

ISOTOPES Isotopes are forms of a chemical element with different numbers of **neutrons** (another sub-atomic particle) in the nucleus. Usually, but not always, a nucleus will have equal numbers of **protons** and **neutrons**. The number that precedes an elemental abbreviation signifies the sum of its protons and neutrons. For example, the common form of carbon is ^{12}C, which has 6 protons and 6 neutrons. Carbon has 14 other isotopes, ranging from ^{8}C to ^{22}C; these have from 2 to 16 neutrons, respectively (**Figure 2.6**).

STABLE ISOTOPES The most common isotopes are stable unless subjected to high-energy bombardment. The carbon isotopes ^{12}C and ^{13}C are stable and comprise 98.9% and 1.1%, respectively, of the carbon found in nature.

RADIOACTIVE ISOTOPES Many isotopes are unstable, meaning they breakdown or decay at a predictable rate, emit sub-atomic particles or **electromagnetic radiation**, and afterwards become a different element. The original isotope is called the parent, and the element produced is called the daughter. The unstable carbon radioisotope ^{14}C is produced when cosmic rays strike the atmosphere or during a nuclear bomb explosion. It accounts for 0.0000000001% (1×10^{-10}%) of all carbon on Earth. The **half-life** of ^{14}C is 5730 years; that is, half of the ^{14}C atoms present at any one time will decay within 5730 years. When ^{14}C decays, it emits sub-atomic particles (an electron and anti-neutrino) and transforms into a stable isotope of nitrogen, ^{14}N. The other radioisotopes of carbon are much more transient; their half-lives range from 20 minutes for ^{11}C to 2×10^{-21} seconds for ^{8}C.

RADIOACTIVE DATING Decay of long-lived radioisotopes is the primary method for dating materials (**Table 2.1**). The following mathematical expression relates radioactive decay to geologic time (**Figure 2.7**):

$$t = \frac{T_{\frac{1}{2}}}{\ln(2)} \ln\left(1 + \frac{D}{P}\right) \qquad (2.10)$$

where

t = age of the material

D = current amount of daughter product

P = current amount of parent isotope

ln = natural logarithm

$T_{\frac{1}{2}}$ = half-life

Mass spectrometry and other analytical methods determine the amounts of parent and daughter isotopes in a specimen. The principal difficulties with this approach are measuring tiny quantities of isotopes with precision

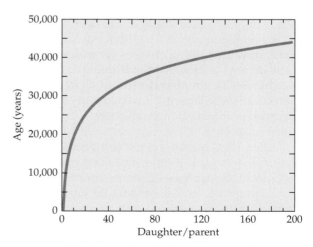

FIGURE 2.7 Radiocarbon dating The age of a rock as determined by the relative amounts of nitrogen (daughter) to ^{14}C (parent). Because nitrogen is so ubiquitous (78% of the atmosphere is N_2, nitrogen gas), radiocarbon dating measures the ratio of ^{14}C to ^{12}C in the rock and compares this value to what this ratio would be if the rock was newly formed.

TABLE 2.1 Radioisotopes used in dating ancient rocks, the stable products they turn into, and rates of decay in terms of half-life

Parent isotope	Stable daughter product	Half-life (billions of years)
^{235}Uranium	^{207}Lead	0.704
^{40}Potassium	^{40}Argon	1.25
^{238}Uranium	^{206}Lead	4.5
^{232}Thorium	^{208}Lead	14.0
^{87}Rubidium	^{87}Strontium	48.8
^{147}Samarium	^{143}Neodymium	106.0

Source: http://pubs.usgs.gov/gip/geotime/radiometric.html

and establishing that no parent or daughter isotopes mistakenly entered or exited the specimen.

ISOTOPE DISCRIMINATION Some biochemical reactions have a higher affinity for one isotope than another. In particular, the photosynthetic reaction that incorporates CO_2 into an organic carbon (C) compound (see Equation 2.1) discriminates against $^{13}CO_2$, the heavy form, in favor of $^{12}CO_2$, the light form. The degree of discrimination depends on the availability of CO_2.

When CO_2 is plentiful, photosynthetic organisms discriminate strongly against $^{13}CO_2$ and, thereby, organic C becomes strongly enriched in ^{12}C. When CO_2 is limiting, photosynthetic organisms use whatever is available, and organic C becomes less enriched in ^{12}C. This is analogous to a hungry person choosing something to eat in a cafeteria: When the cafeteria is well stocked, the person eats more of the item he/she likes best (i.e., $^{12}CO_2$), but when the cafeteria is running low on everything, the person may be forced to eat his/her second choice (i.e., $^{13}CO_2$). Consequently, the ratio of ^{13}C to ^{12}C ($\delta^{13}C$; pronounced delta-13-C) in the organic C found in the sediments derived from these organisms varies inversely with atmospheric CO_2 concentrations (Royer et al. 2001); that is, lower $\delta^{13}C$ values for sediments indicate they were formed during periods of higher atmospheric CO_2 concentrations.

As algae photosynthesize and preferentially assimilate $^{12}CO_2$, the CO_2 remaining in the surrounding water becomes depleted of ^{12}C; that is, the water becomes enriched in ^{13}C. Thus, the ratio of ^{13}C to ^{12}C ($\delta^{13}C$) in carbonate rock that precipitates from this water (see Equation 2.6) increases with photosynthetic activity.

BORON ISOTOPES Boron dissolves in water to form $B(OH)_3$ (boric acid) and $B(OH)_4^-$ (borate ion).

$$B + 3 H_2O \rightarrow B(OH)_3 + 3 H^+ \quad (2.11)$$

$$B(OH)_3 + H_2O \rightarrow B(OH)_4^- + H^+ \quad (2.12)$$

The balance between these reactions depends on the activity of hydrogen ions (H^+) in the water, that is, the **pH** of the water (**Figure 2.8A**). Exchange of the stable isotopes ^{10}B and ^{11}B between $B(OH)_3$ and $B(OH)_4^-$ is described in the following equation.

$$^{10}B(OH)_3 + {}^{11}B(OH)_4^- \rightarrow {}^{11}B(OH)_3 + {}^{10}B(OH)_4^- \quad (2.13)$$

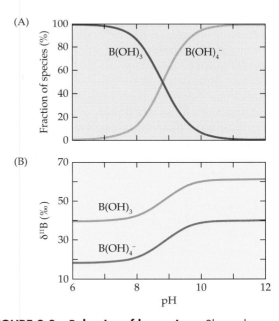

FIGURE 2.8 Behavior of boron ions Shown here are the pH dependence of (A) the balance between boric acid and borate, and (B) the $\delta^{11}B$ of each species. Marine organisms incorporate primarily $B(OH)_4^-$; therefore, the $\delta^{11}B$ of sediments formed by such organisms increases as oceanic pH increases. (After Hemming and Hanson 1992.)

Together, these reactions result in a pH dependence of the ratio of ^{11}B to ^{10}B ($\delta^{11}B$) in which $\delta^{11}B$ increases as the pH of the medium increases (**Figure 2.8B**). Marine organisms incorporate mostly $B(OH)_4^-$; therefore, the $\delta^{11}B$ of sediments formed by marine organisms tends to increase as oceanic pH increases (Pagani et al. 2005). Oceanic pH decreases with higher atmospheric CO_2 concentrations because higher atmospheric CO_2 concentrations lead to more CO_2 being dissolved in the oceans and greater carbonic acid formation (see Equation 2.5). As a consequence, $\delta^{11}B$ in sediments may serve as a rough indicator of atmospheric CO_2 concentration (Royer et al. 2001).

PHASE CHANGES Most chemicals can occur in one of three phases—a gas, liquid, or solid—depending on temperature and pressure. When a chemical is in the gas phase, its molecules move more randomly and at higher speeds than when the molecules of same chemical are in the liquid phase. Similarly, when the chemical is in the liquid phase, its molecules move more randomly and at higher speeds than when the molecules of the same chemical are in the solid phase. At a given temperature, molecules have the same average amount of kinetic energy ($\frac{1}{2} mv^2$, where m is mass and v is velocity), and so heavy molecules will move slower than lighter molecules (for heavy isotope a and lighter isotope b, if $\frac{1}{2} m_a v_a^2 = \frac{1}{2} m_b v_b^2$ and $m_a > m_b$, then $v_a < v_b$). Therefore, molecules of a heavy isotope (those having more neutrons in their nucleus) are less likely than lighter ones to move fast enough to evaporate (change phase from a liquid to a gas) or melt (change phase from a solid to a liquid) and more likely to condense (change phase from a gas to a liquid) or freeze (change phase from a liquid to a solid). Higher temperatures stimulate the motion of all molecules, and a greater percentage of the molecules with a heavy isotope gain sufficient velocity to change phase.

OXYGEN ISOTOPES Oxygen has three stable isotopes: ^{16}O (the most abundant, at 99.76%), ^{17}O (0.04%), and ^{18}O (0.20%). Water that has the light oxygen isotope, $H_2{}^{16}O$, is more likely to evaporate than water that has the heavy isotope, $H_2{}^{18}O$. Water vapor, thus, has a lower ratio of ^{18}O to ^{16}O ($\delta^{18}O$) than the liquid water left behind. Conversely, heavy water ($H_2{}^{18}O$) is more likely to condense than light water ($H_2{}^{16}O$).

Taking the average $\delta^{18}O$ value of ocean water as a reference point and setting it to zero, water vapor in a cloud over the equator has a $\delta^{18}O$ value of about $-14‰$ ($1‰$ equals one part per mil or 0.1%) (**Figure 2.9**), meaning it is more depleted in the heavy isotope

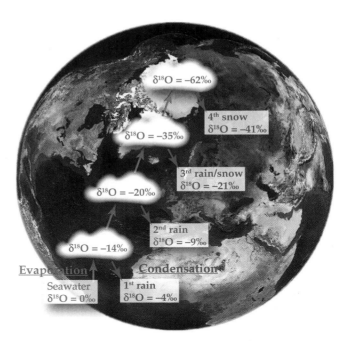

FIGURE 2.9   Discrimination against the heavy oxygen isotope ^{18}O during evaporation and precipitation Seawater is defined as having an average ratio of heavy and light oxygen isotopes ($\delta^{18}O$ value) of zero. Water that evaporates from the oceans near the equator is depleted in ^{18}O. Subsequently, several cycles of cooling and precipitation as water vapor drifts toward the North Pole depletes it of ^{18}O even more.

than seawater by a factor of 1.4%. As the cloud drifts to higher latitudes, some of the water vapor precipitates. This rain water or snow is enriched in the heavy isotope (higher $\delta^{18}O$), and the water vapor that remains in the cloud is further depleted (lower $\delta^{18}O$). Additional precipitation events further deplete the cloud of its heavy isotope. The snow that falls and accumulates near the poles may have a $\delta^{18}O$ value of about $-40‰$.

As temperatures rise, all water molecules become agitated, a larger proportion of $H_2{}^{18}O$ evaporates, and the liquid becomes less depleted of $H_2{}^{16}O$. Therefore, at higher temperatures, the $\delta^{18}O$ value of water vapor increases (deviates less from zero) as does the $\delta^{18}O$ value of precipitation that forms when this vapor condenses or freezes (**Figure 2.10A**). Moreover, the $\delta^{18}O$ value of the water left behind in the oceans decreases with temperature as does the $\delta^{18}O$ value of shells of marine organisms living in this water (**Figure 2.10B**).

Climate influences $\delta^{18}O$ values in another manner. During a glacial period when temperatures are colder than normal, the ice masses expand because the snow that falls upon them fails to melt in the summer. This snow is isotopically lighter than average sea water, and

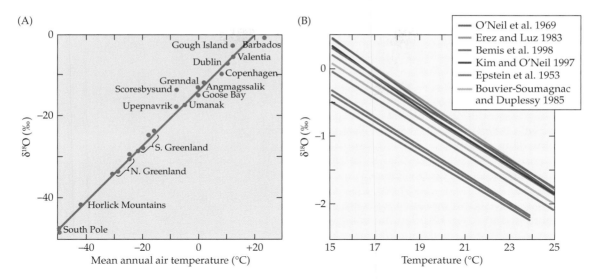

FIGURE 2.10 **Ratio of heavy and light oxygen isotopes (δ^{18}O)** (A) Variation of δ^{18}O in precipitation as a function of mean annual air temperature for various sites around the world. (B) Variation of δ^{18}O in carbonates of marine shells as a function of temperature. Different colored lines represent the trends observed in different studies. (A after White 2005.)

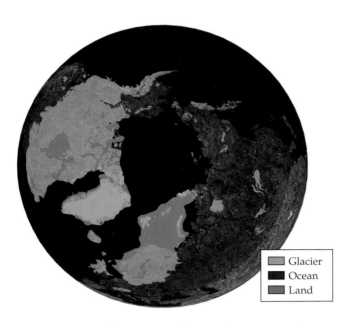

FIGURE 2.11 **Glaciation in the Northern Hemisphere during the Wisconsin Ice Age**, about 20,000 years ago. Ice sheets are designated by turquoise shading. Development of ice sheets 3-km to 4-km thick caused sea level to lower by about 125 m. Also the Alps, the Himalayas, and much of North America, the United Kingdom, Scandinavia, and Russia were covered by glaciers. Winter sea ice extended much farther south than it does today.

so as the snow compresses and becomes incorporated into the ice masses, the ice masses incorporate water that is isotopically lighter than average ocean water. The oceans, although they are vast, decrease in volume and rapidly become isotopically heavier because the temperature effect on evaporation is removing the lighter isotope from this shrinking volume. For example, at the height of the most recent ice age (called the Wisconsin Glaciation, 20,000 years ago), sea level fell 125 meters, and the ice pack contained approximately 50% more water than it does today (**Figure 2.11**). If the δ^{18}O value of glacial ice averaged −30‰, as it does today, the δ^{18}O of the oceans would have increased by about 1.0‰ (Schrag et al. 2002a). Shells of marine organisms, which were formed in the ocean water at this time, have higher δ^{18}O values than do shells formed in modern-day ocean water.

HYDROGEN ISOTOPES Hydrogen has two stable isotopes: the common [1]H (accounting for 99.985%) and deuterium [2]H (0.015%), often abbreviated as "D." As with oxygen isotopes, water composed of the light hydrogen isotope ([1]H$_2$O) is more likely to evaporate than is water composed of deuterium (heavy water denoted by [2]H$_2$O or D$_2$O). At higher temperatures, the δD value of water vapor increases as does the δD value of the precipitation from this vapor. Discrimination between hydrogen and deuterium is much more pronounced than between the oxygen isotopes [16]O and [18]O because the difference in mass is twofold between hydrogen and deuterium, $(2 - 1)/1 = 2$, whereas the difference in mass for the oxygen isotopes is only $(18 - 16)/16 = 1/8$. In specific, the δD value of ice cores is a more sensitive temperature proxy than the $\delta^{18}O$ value.

Now, equipped with a rudimentary knowledge of isotopes, a major tool in geophysics, let us return to a historical account of Earth's climate.

The Past Billion Years

About 730 million years ago, Earth underwent severe glaciations that eliminated much of life. Sedimentary and paleomagnetic evidence for this is manifold (Evans 2000; Schrag et al. 2002b). The high $\delta^{18}O$ values of sedimentary rocks dated to this time (rocks formed out of the isotopically heavier oceans) are consistent with a scenario that includes low temperatures and more of Earth's water being sequestered in glaciers. Rock layers deposited immediately after this time show evidence of a sudden and severe decline in $\delta^{13}C$, indicating a cessation of photosynthetic CO_2 consumption (see Equation 2.1). These same layers have nonoxidized iron deposits, a likely result if atmospheric O_2 levels had plunged from a lack of photosynthetic O_2 production. The next-deposited rock layers contain sediments deposited by glaciers (glacial till) covered with a carbonate cap. Such layers form when silicate rock weathers in an atmosphere of high CO_2 concentration (see Equations 2.5 and 2.6).

What occurred? Was it another Snowball Earth that covered the entire planet in ice or was it a Slushball Earth where ice covered most of the planet, but the equatorial oceans remained ice-free? Snowball and Slushball events would have profoundly different consequences. A Snowball with complete ice coverage would have caused extinction of many ancient lineages of life, and only the hardiest bacteria would survive. A Slushball with some open water would permit more complex life forms to survive.

One piece of evidence that distinguishes between Snowball and Slushball Earth comes from extraterrestrial sources (Bodiselitsch et al. 2005). Iridium (Ir) is a metal 10,000 times more prevalent in cosmic dust and meteors than in Earth's crust. A cosmic rain distributes Ir over Earth's surface at steady rate. In open ocean, Ir sinks to the bottom continuously and is distributed evenly through the sedimentary rock that is formed. If oceans freeze, Ir accumulates on and within the icepack. If an icepack suddenly melts, all of this accumulated Ir sinks at once to the bottom and forms a distinct layer of sediment enriched in Ir.

FIGURE 2.12 Drill cores These drill cores from South America are part of a paleomagnetism study to determine the movement of continents. The age of these cores of sedimentary rock increases with depth. The magnetic field of a core from a particular depth indicates the geographical location of the rock when it was formed.

At some equatorial sites, cores removed while drilling through sedimentary rock to depths of several hundred meters (**Figure 2.12**) support the occurrence of a Snowball Earth II, not a Slushball. These drill cores have a compact layer enriched in iridium (Ir) sandwiched between layers of the glacial till that marks the beginning of this period and a carbonate cap that marks the end. From the amount of Ir in the layer and an estimation of the cosmic Ir deposition rate, Snowball Earth II appears to have lasted about 12 million years.

 WEB TOPIC 2.1 Extraterrestrial evidence of Snowball Earth Iridium (Ir), one of the rarest chemical elements in Earth's crust, is relatively prevalent in some rock layers formed during the Cryogenian.

Drill cores at other equatorial sites, however, provide conflicting evidence (Allen and Etienne 2008). These cores indicate that dynamic glaciers and ice streams continued to deliver large amounts of sediment to open oceans throughout this period. Therefore, this evidence indicates that at least some of the tropical oceans remained essentially ice-free, a Slushball Earth.

Despite the diversity of opinion about Snowball versus Slushball Earth, there is general agreement that this was a very cold period of severe glaciation, which fully deserves its name, the Cryogenian (*Greek*; ice + birth). Several conditions initiated this period (Schrag et al.

2002b). The breakup of the supercontinent Rodinia exposed organic matter to anaerobic respiration, creating an atmosphere in which the major greenhouse gas was methane (CH_4) (see Equation 2.3). Atmospheric CH_4, in contrast with CO_2, reacts with O_2 (see Equation 2.4) and is not buffered by a massive ocean reservoir; therefore, atmospheric CH_4 levels decreased rapidly once continental movements diminished and less organic matter was exposed. As CH_4 declined, Earth cooled. Cooler temperatures inhibited biological production of CH_4, accelerating the cooling. Ice appeared on the landmasses, increasing the reflectance of solar energy (a phenomenon called **albedo**) and decreasing the solar energy retained at Earth's surface. Temperatures sank until even some tropical areas were covered with ice (Pollard and Kasting 2005).

Recovery from these cold conditions, as during the previous Snowball episode, depended on the buildup of atmospheric CO_2 levels through volcanic emissions of CO_2 and inhibition by low-temperatures of processes that deplete atmospheric CO_2, namely, silicate rock weathering (see Equations 2.5 and 2.6) and photosynthesis (see Equation 2.1). Atmospheric CO_2 concentrations reached as high as 1.2% before the ice began to crack, exposing bare ground and open water, decreasing albedo, increasing retention of solar energy, and creating a cascade of events that warmed the planet (Pollard and Kasting 2005).

Another Snowball Earth may have occurred between 605 million and 585 million years ago. Sediments from this period show a marked decline in $\delta^{13}C$ that suggests diminished photosynthesis. These layers are followed by glacial deposits and capped with carbonates that indicate an abrupt increase in atmospheric CO_2 concentrations. Yet the transitions between these rock layers are not as pronounced as those of the previous glaciation episode in the Cryogenian, indicating that the glaciation about 600 million years ago was not as severe.

The question of how many Snowball Earths transpired throughout history (at 2.5 billion, 730 million, or 600 million years ago) may be of less interest than the possibility that such an event might occur in the near future. Our sun has matured since 600 million years ago, increasing the amount of solar energy reaching Earth by 6%. This factor alone, however, would not prevent another planetary freeze.

A more important factor is the change in distribution of Earth's landmasses (Schrag et al. 2002b). During the geological times about 730 million and 600 million years ago, when Snowball Earths were a possibility, all the continents were situated near the equator (see

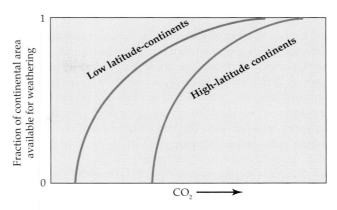

FIGURE 2.13   Influence of latitude on weathering of rock Glaciation occurs at higher atmospheric CO_2 concentrations on continents at high latitudes than on continents at low latitudes. Glaciers cover silicate rock and inhibit weathering that consumes CO_2. (see Equation 2.6). Therefore, for those continents at high latitudes, more CO_2 is more likely to accumulate in the atmosphere and reverse any cooling trend. (After Schrag et al. 2002b.)

Figure 2.5). Glaciation on these continents commenced only after atmospheric CO_2 concentrations reached such low levels and global average temperatures had declined so much that once the cooling process began, it continued to intensify (**Figure 2.13**). The probability of another Snowball Earth occurring in modern times is very low because most of Earth's landmasses are now situated at high latitudes, and glaciation commences at higher atmospheric CO_2 concentrations and at higher global average temperatures than those that existed during the previous Snowball events. This glaciation at high latitudes inhibits CO_2 sequestering processes, such as weathering of silicate rock and photosynthesis, and causes atmospheric CO_2 concentrations to increase and reverse any cooling trend.

The Cambrian Explosion

By 570 million years ago, Earth's climate became warm, and the glaciers retreated. The continents were primarily low-lying deserts and alluvial plains. Rising sea level encroached upon these areas and covered about 85% of Earth's surface (today, water covers 71%). Without continental landmasses at the poles, oceans circulated freely, and ice formation was negligible.

Photosynthetic organisms that had escaped extinction during the severe glaciations of the Cryogenian proliferated, and atmospheric O_2 concentrations climbed to about 15% (Berner 2003). Multicellular organisms appeared. Again, this was not a coincidence.

At higher O_2 concentrations, parts of organisms could be situated deeper within the organism and still receive sufficient O_2 to conduct aerobic respiration (Budd and Jensen 2000). Also, the thicker ozone (O_3) layer in the upper atmosphere afforded greater UV protection, so organisms could complete longer life cycles (a development analogous to the rise of the ozone layer in the upper atmosphere during the Great Oxidation Event mentioned earlier in the chapter).

A rapid expansion of biological diversity known as the Cambrian Explosion becomes evident in the fossil record at about 540 million years ago. More than 900 animal species appear, including representatives of all modern animal **phyla** (the taxonomic classification ranking below the kingdom animalia) (Marshall 2006). Earlier life forms may have been equally diverse, but they were soft-bodied and seldom left their mark in the fossil record. In contrast, Cambrian animals had shells or other hard body parts that left a stronger impression, and thereby these animals are better preserved as fossils in the rock.

A supercontinent, **Gondwana**, coalesced during the Cambrian period. This continent included the landmasses that comprise all the continents of today's Southern Hemisphere (Antarctica, South America, Africa, Madagascar, Australia–New Guinea, and New Zealand), plus two continents of today's Northern Hemisphere (Arabia and India). At about 510 million years ago, Gondwana began to drift southward toward the South Pole (**Figure 2.14**). This scenario about continental movements, as well as those scenarios introduced earlier (see Figure 2.2), are based on studies of **paleomagnetism**, the study of residual magnetism in ancient rocks.

Paleomagnetism

Iron in rock has a magnetic orientation and intensity that reflect Earth's magnetic field during the time and place at which the rock was formed. Earth's magnetic field depends on latitude and, to a lesser extent, longitude.

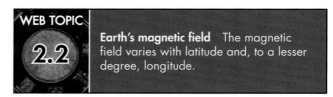

WEB TOPIC 2.2 **Earth's magnetic field** The magnetic field varies with latitude and, to a lesser degree, longitude.

Consequently, precise measurements of a rock's magnetic field via a sensitive **magnetometer** (**Figure 2.15**) in conjunction with estimates of age from radioisotopic dating (see Equation 2.10) can indicate where a landmass was at a particular time and how the positions of continents have shifted over time. These analyses are not straightforward. Earth's magnetic field changes over geological time scales. Every navigator knows

FIGURE 2.14 The supercontinent Gondwana Merger of major landmasses (yellow and purple) to form Gondwana at about 550 million years ago.

FIGURE 2.15 A magnetometer This highly sensitive scientific instrument measures the strength and direction of the magnetic field in rocks. Notice the large magnets on the instrument.

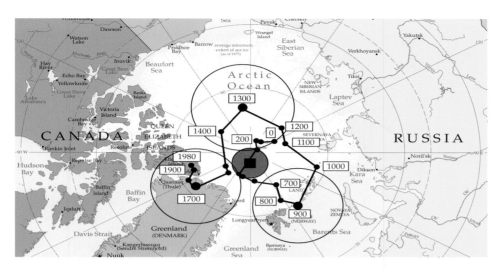

FIGURE 2.16 Positions of the North Magnetic Pole over the past 2000 years Means for each century, with numbers indicating the century. Circles at A.D. 900, A.D. 1300, and A.D. 1700 are 95% confidence limits. The black square is the mean magnetic pole position over the past 2000 years; the red circle indicates the 95% confidence limit (After Butler 1992).

that a compass points to the North Magnetic Pole, and its current location (82°07' north, 114° 04' West) deviates about 8° away from the geographic North Pole, the spot around which Earth rotates. The North Magnetic Pole has meandered through the ages, but paleomagnetic researchers postulate that its long-term average is near the geographic North Pole (**Figure 2.16**).

A further complication is that Earth's entire magnetic field completely reverses between zero and five times every million years (**Figure 2.17**). Fortunately, the rate

of geomagnetic field reversals is rapid in comparison to the movement of continents, and so an abrupt change in the magnetic fields of adjacent layers in a drill core indicates a field reversal at the time of formation rather than continental movement. Because geomagnetic field reversals are a worldwide phenomenon, they serve as distinct markers to synchronize layers in cores from around the world.

The Ordovician–Silurian Extinction Event

Devastating mass extinctions that affected all major life forms occurred during the late Ordovician, at about 440 million years ago (Sheehan 2001). Approximately 85% of the species previously found in the fossil record disappeared.

The primary cause appears to be a brief glacial period. Ice began to accumulate on the polar areas of Gondwana as the supercontinent slid southward. This initiated falling sea level, a higher global albedo, cooler temperatures, and glaciation. As before, inhibition of CO_2-consuming processes by cold temperatures led first to a gradual rise in atmospheric CO_2 concentrations from about 0.4% to 0.6%, and then, 0.5 to 2 million years later, to an acute global warming. Yet the severity of this climate event does not seem commensurate with the scale of extinctions.

Thus, geologists have postulated several additional contributing factors to the extinctions. One is that the sea-level decline was large, on the order of 50 meters to 100 meters. This would have drained the shallow seas that extended into the interior of landmasses, resulting in a harsh climate at low and mid-latitudes. Low sea level also could have activated deep ocean currents

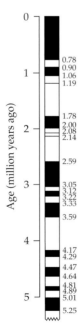

FIGURE 2.17 Earth's magnetic field reversal during the late Cenozoic Era. Dark areas denote periods of normal polarity, light areas denote reverse polarity. (After Mankinen and Wentworth 2003.)

(A)

(B)

FIGURE 2.18 **Early terrestrial life** (A) Liverworts, descendants of the first land plants. (B) A forest of tree ferns, giant horsetails, and scale trees during the Carboniferous period.

that aerated the seabed and brought up toxic materials from the depths.

A Landmark Event

Every major catastrophe seems to provide opportunities for the development of new biological life forms. So it was with the Ordovician–Silurian Extinction. Up to that point, life had been bound to the seas, an environment that strongly limits both the penetration of sunlight and the diffusion of carbon dioxide (CO_2). In the mild climate that followed the extinction, plants emerged from the aquatic habit at about 430 million years ago and colonized land.

The ascent to land was a formidable undertaking. In their ancestral aquatic habitat, all cells were bathed by, or in close proximity to, water and the nutrients dissolved in it. The restlessness of the surrounding fluid promoted mixing as well as diminished the boundary layer between an aquatic organism and its medium. Thus, absorption of water and nutrients by an aquatic organism did not quickly deplete their immediate surroundings.

Encroaching on land, plants left behind this relatively benign habitat. On land, a desiccating atmosphere "sucked" water from organisms. Water and mineral nutrients were available primarily on the surface of or embedded in particles of soil. Soil was a difficult medium, one that was often hard to penetrate and that suffered from limited mixing and local depletion of water and nutrients.

Plants in a terrestrial environment did, however, gain greater access to the critical resources of sunlight and CO_2. In bodies of water, the amount of sunlight available for photosynthesis halves every 0.2 meters to 5 meters in depth, depending on the roughness and opacity of the surface and the density of photosynthetic organisms. Diffusion of CO_2 is roughly 9000 times slower through still water than through air. On land, freed from these limitations, primary productivity literally skyrocketed. Vast jungles soon covered the continents.

These first jungles, however, were only millimeters tall. Primitive plants such as liverworts (**Figure 2.18A**) must rely on relatively slow cell-to-cell mechanisms to move water and nutrients from the soil to aerial parts of the plants and to move carbohydrates from aerial parts to the light-deprived belowground parts. Slow movement of materials thus restricted the height of such plants to a few millimeters.

With the advent of a vascular system at about 380 million years ago, plants were able to extend their reach.

Vascular plants have specialized tissues that serve as pipes through which water, nutrients, and carbohydrates rapidly flow for distances of meters and beyond. Competition for light became fierce, and plants grew in stature. Towering forests of vascular plants (**Figure 2.18B**) dominated the Carboniferous period (360 million to 290 million years ago), so named because the organic matter from these forests eventually became the coal deposits that powered our Industrial Age.

The appearance of more complex plant life enhanced the amount of photosynthesis worldwide. Atmospheric concentrations of O_2 rose from 15% to 30%, while those of CO_2 declined from 0.5% to less than 0.05%. Lower concentrations of this important greenhouse gas were associated with falling global temperatures. By the end of the Carboniferous, from about 310 million to 290 million years ago, a new supercontinent **Pangaea** (*Greek*; all earth) was developing from the merger of Gondwana and Euramerica, and glaciers covered all but tropical areas (**Figures 2.19 and 2.20A**).

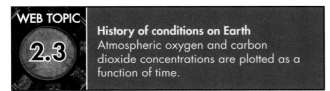

WEB TOPIC
2.3
History of conditions on Earth
Atmospheric oxygen and carbon dioxide concentrations are plotted as a function of time.

During the Permian period (290 million to 250 million years ago), a greater percentage of Pangaea straddled the equator (**Figure 2.20B**). This ice-free landmass decreased the albedo (reflectance) of Earth and warmed the planet. Eventually, Pangaea extended from pole to pole. The interior of the continent, without the moderating influence of the ocean, developed wide temperature fluctuations, highly seasonal rainfall patterns, and expansive deserts. The dry conditions favored

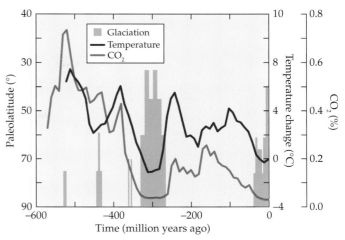

FIGURE 2.19 History of conditions on Earth
Latitudinal extent of glaciers, temperature changes, and CO_2 concentration as a function of time. The estimates for temperature changes are based on the $\delta^{18}O$ of shallow-marine carbonates from the shells of marine organisms. The CO_2 estimates are based on a model of global carbon cycling. (After Royer et al. 2004.)

gymnosperms—plants (e.g., conifers) with seeds that enclose the embryos in a desiccation-resistant cover—over plants such as ferns that disperse spores that are vulnerable to drought.

The Great Dying at 251 million years ago, the most severe extinction event in the history of the planet, marks the end of an era (Paleozoic). Over 96% of marine and 70% of terrestrial species disappeared. Possible causes include an impact event in which a large meteor struck Earth (Basu et al. 2003), major volcanic eruptions (Mundil et al. 2004), or methane release (Krull and Retallack 2000). Because no single cause for this devastation is evident, researchers suspect the convergence of several factors.

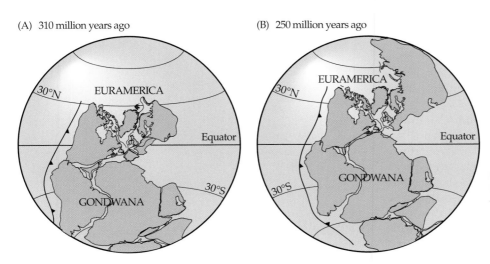

(A) 310 million years ago

(B) 250 million years ago

FIGURE 2.20 Supercontinent Pangaea at (A) 310 million years ago (late Carboniferous) and (B) 250 million years ago (late Permian). (After Torsvik and Cocks 2004.)

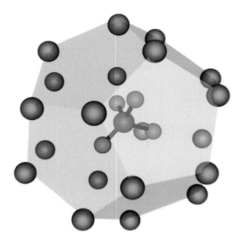

FIGURE 2.21 Methane hydrate The red spheres represent the oxygen atoms in the water molecules that surround the methane molecule depicted by the gray carbon and green hydrogen spheres.

In one scenario, massive volcanic eruptions released vast amounts of CO_2 to the atmosphere and initiated global warming (Wignall 2005). This warming diminished the temperature gradient between the equator and the poles, which in turn disrupted circulation of the ocean surrounding the supercontinent Pangaea. The stagnant water heated sufficiently to melt methane **hydrate**, ice that contains large amounts of CH_4 within its crystal structure (**Figure 2.21**), in deep seabeds and produced a massive release of CH_4 to the atmosphere. Global temperatures soared. Oxygen availability in the warm, stagnant ocean surrounding Pangaea dropped, and sea life suffocated.

The Age of Dinosaurs

Reptiles occupied the void left by the Great Dying, grew in stature, and dominated animal life from 251 million to 66 million years ago, an era known as the Mesozoic, or the Age of Dinosaurs. Angiosperms, true flowering plants, also first appeared and became prevalent. All of this biological activity undoubtedly benefited from a relatively stable, warm climate.

The Triassic period (251 million to 200 million years ago), like the periods that preceded it, was dry and had large seasonal variations in temperature and precipitation. This is deduced from the widespread occurrence of red beds (reddish-colored sedimentary rocks that were deposited in hot climates under oxidizing conditions) and evaporites (sediments formed when mineral-rich water evaporated rapidly under arid conditions) in rock layers formed during this period. Low sea level probably intensified seasonal variations.

During the Jurassic period (200 million to 146 million years ago), Earth's crust expanded on the ocean floor, displacing water and causing sea level to rise. Pangaea began to rift into smaller continents, bringing more land area in contact with the ocean (**Figure 2.22**). With closer proximity to water, these continents had more stable temperatures and higher humidity.

Climate during the Cretaceous period (146 million to 65 million years ago) was uniform across the planet. The poles were ice free, and the deep oceans were at least 10°C warmer than today. Temperature gradients between the ocean's surface and deeper water were small, diminishing currents and thereby the circulation of oxygen. In this anaerobic environment, decomposition of organic matter was slow, and large volumes of organic matter accumulated on the ocean floor, forming deposits known as "black shale." Such deposits often contain fossils of exquisite detail; in some cases, fossils of insects show wing patterns and venation, antennae, and even hair.

 WEB TOPIC 2.4 **High-quality fossils in black shale rock** The fossil of a wing of *Megatypus schucherti*, a dragonfly precursor, in Elmo Kansas, was formed during the Cretaceous period.

The "K-T boundary," at about 65 million years ago, marks the end of the Age of Dinosaurs. In a relatively short time frame, all non-avian dinosaurs went extinct. The K-T boundary, like many other geological events, is still shrouded in mystery.

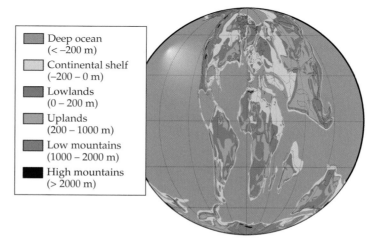

	Deep ocean (< –200 m)
	Continental shelf (–200 – 0 m)
	Lowlands (0 – 200 m)
	Uplands (200 – 1000 m)
	Low mountains (1000 – 2000 m)
	High mountains (> 2000 m)

FIGURE 2.22 Landmasses on Earth during the late Mesozoic era. (After MacLeod et al. 2005.)

FIGURE 2.23   K-T boundary Badlands near Drumheller, Alberta, Canada where erosion has exposed the K-T boundary, indicated by the abrupt transition between the light brown and dark brown rock layers (see arrow).

First, the name itself *K-T boundary* defies simple logic.

- The letter "K" stands for the *Cretaceous* period to avoid confusion with the *Carboniferous* period, which had already laid claim to the letter "C."

- The letter "T" stands for the *Paleogene* period, which was previously called the *Tertiary* period.

- "Boundary" is for the event separating the *Cretaceous* and *Paleogene* periods.

Then there is the matter of what killed the dinosaurs. A closer examination of the fossil record reveals that a broad range of organisms, not just dinosaurs, went extinct within a time span of 500,000 years. For example, mammals also suffered, with marsupials and multituberculates (rodent-like mammals) experiencing heavy losses. More than 50% of all plant species vanished.

Rock layers at the K-T boundary (**Figure 2.23**) incriminate extraterrestrial forces. These layers contain 10-fold to 100-fold more iridium than adjacent ones; iridium (as mentioned earlier in the chapter) is generally more prevalent in asteroids and meteorites than on Earth's surface. Also, rock layers near the K-T boundary that encircle the Yucatan Peninsula of Mexico contain glass spherules 0.3 millimeters to 4.0 millimeters in diameter; such spherules are formed by the melting and quenching (rapid cooling by immersion in water) of terrestrial rocks during a high-velocity impact. Finally, a circular depression 180 kilometers in diameter, centered in the waters off the Yucatan Peninsula, is discernible from space. These and other pieces of evidence support the hypothesis that an asteroid struck the Yucatan and produced an impact crater.

WEB TOPIC

2.5

Impact crater in the Yucatan Peninsula A trough measuring about 3 to 5 meters deep by about 5 kilometers wide marks the area where the asteroid struck approximately 65 million years ago.

Was one asteroid sufficient? Clearly, an impact of the size that occurred in the Yucatan could blast enough dust and vapor into the atmosphere to block sunlight from reaching the surface of Earth for months, if not years (**Figure 2.24**). The planet would cool rapidly.

FIGURE 2.24   A depiction of the asteroid striking the Yucatan Peninsula at around 65 million years ago.

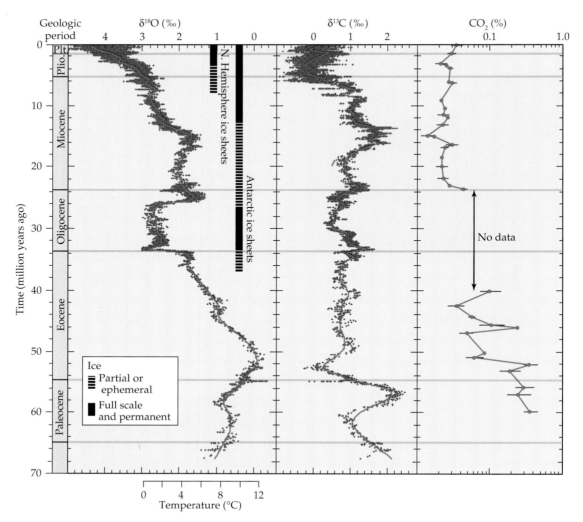

FIGURE 2.25 Conditions during the Cenozoic Era Oxygen and carbon isotope ratios ($\delta^{18}O$ and $\delta^{13}C$, respectively) (Zachos et al. 2001) and estimates of atmospheric CO_2 concentrations (from boron isotope ratios, $\delta^{11}B$) (Pearson and Palmer 2000) through the Cenozoic Era (65 million years ago to the present), based on deep-sea cores. The red lines through the $\delta^{18}O$ and $\delta^{13}C$ data are the averages from many cores. The $\delta^{18}O$ data during the ice-free periods earlier than 35 million years ago roughly correspond to the temperature scale at the bottom. The black vertical bars indicate the times when each hemisphere had significant amounts of sea ice. Note that the CO_2 data are plotted on a logarithmic scale and provide error bars for the analyses of $\delta^{11}B$.

Photosynthesis would be severely compromised, plant growth would stop, and animals would starve. Such an impact could also trigger volcanic eruptions and firestorms from incendiary fragments. As organisms died and decayed and as volcanoes and firestorms spewed gas, atmospheric CO_2 concentrations would rapidly increase and thereby induce a temperature increase of several degrees once the dust settled.

Several issues remain controversial (Keller et al. 2007, 2008; Schulte et al. 2008). Some researchers date the impact on the Yucatan about 300,000 years before the K-T boundary. They also believe that the mass extinctions during this period transpired over many millennia, not just a few years. In their view, rock layers from around the world, which contain iridium at high concentrations and impact glass spherules and which date to the K-T boundary, indicate that a series of major asteroids struck Earth at around this time rather than a single impact off the Yucatan Peninsula. Other researchers view all these data as consistent with a single asteroid impact.

Recent Geological History

Drill cores from deep-sea beds provide a detailed history of oxygen, carbon, and boron isotopes that extends

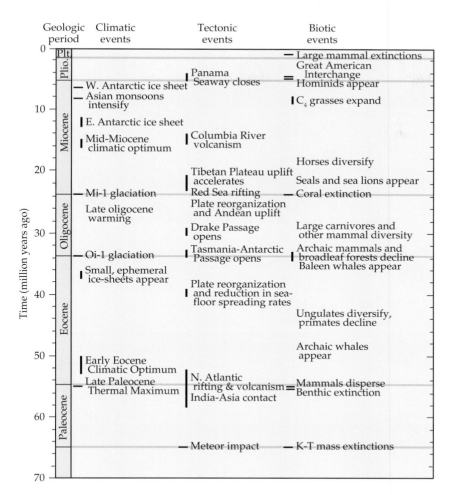

Geologic period	Climatic events	Tectonic events	Biotic events

FIGURE 2.26 Major events of the Cenozoic era (Zachos et al. 2001). The horizontal bars indicate the timing of short-term events. "Climatic Optimum" were periods of benign warm climate. "Great American Interchange" denotes the exchange of fauna and flora that occurred when the Ismus of Panama rose up from the seas and bridged North and South America. Vertical black bars indicate ranges of time. (After Zachos et al. 2001.)

from 65 million years ago to the present (Cenozoic era). Recall that $\delta^{18}O$ (ratio of oxygen isotopes) is a function of both water temperature and volume of water incorporated into glacial ice, $\delta^{13}C$ is a function of photosynthetic activity, and $\delta^{11}B$ is a function of CO_2 concentration. Aligning the isotope data (**Figure 2.25**) with major events (**Figure 2.26**) provides a perspective on Earth's climate history since the K-T boundary.

The most extreme climate event of the modern era was the Late Paleocene Thermal Maximum, which occurred at around 55 million years ago (Zachos et al. 2001). Apparently, deep-sea temperatures were warming gradually when methane hydrates (ice containing CH_4) in seafloor sediments melted and belched massive amounts of CH_4 into the atmosphere (Katz et al. 1999). Addition of this greenhouse gas to the already high levels of CO_2 (approximately 0.2%) triggered a global warming of 5°C to 7°C over about 10,000 years (Panchuk et al. 2008). Recovery was slow, taking over 100,000 years from the onset of the event. Support for this interpretation derives from the gradual decline in

$\delta^{18}O$ and $\delta^{13}C$ values during the Paleocene epoch (early part of the Paleogene period), interrupted by a precipitous drop in both measures (see Figure 2.26); the drop in $\delta^{18}O$ values reflects a temperature spike, whereas that in $\delta^{13}C$ values reflects that the CH_4 released from the seafloor was enriched in the lighter isotope, ^{12}C.

The two other major climatic deviations that occurred during the modern era were global cooling events, indicated by positive shifts in $\delta^{18}O$ values (Zachos et al. 2001). The first of these (the Oi-1 Glaciation) involved the sudden growth of ice sheets on Antarctica at about 34.0 million years ago (the Eocene–Oligocene boundary, separating the middle and late epochs of the Paleogene period). This event inaugurated 400,000 years of glaciation and caused worldwide shifts in the distribution of marine biogenic sediments and an overall increase in ocean photosynthetic productivity (thereby increasing $\delta^{13}C$). The second climatic deviation (the Mi-1 Glaciation) was a brief but intense glacial maximum that occurred at about 23 million years ago (the Oligocene–Miocene boundary, separat-

(A)

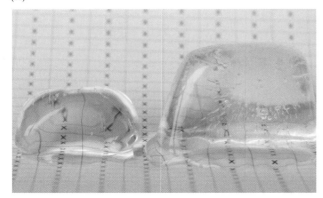

(B)

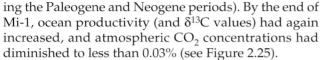

FIGURE 2.27 Air bubbles in ice (A) Fancy, clear ice cubes (left) and normal, gas-filled ice cubes (right). (B) Ice core taken on the Clark Glacier in Antarctica.

ing the Paleogene and Neogene periods). By the end of Mi-1, ocean productivity (and $\delta^{13}C$ values) had again increased, and atmospheric CO_2 concentrations had diminished to less than 0.03% (see Figure 2.25).

The Late Paleocene Thermal Maximum, Oi-1, and Mi-1 events precipitated the loss of certain organisms and accelerated speciation of others (see Figure 2.26). Of particular note are the extinction of benthic protozoans (single-celled animals who lived in sea or lake bottoms) at the Thermal Maximum, the appearance of baleen whales (see Chapter 6) and decline of broad-leaf forests at the Oi-1 Glaciation, and the extinction of Caribbean corals at the Mi-1 Glaciation. Although the isotope records and other geological evidence indicate climatic variation throughout the modern era, including numerous glacial and interglacial periods, no other events approach in magnitude the Thermal Maximum, Oi-1 Glaciation, or Mi-1 Glaciation.

The Past Million Years

Ice cores from Greenland and Antarctica offer even more detail about past climate, extending as far back as one million years ago. For a better perspective about these measurements, examine the ice cubes in your glass the next time you are in a bar or restaurant. If you are in an elegant establishment, the ice cubes will look clear because they are made in a special machine that excludes dissolved gases from the water during freezing. By contrast, the ice cubes in most establishments (all the ones that I frequent) are riddled with gas bubbles (**Figure 2.27A**). So it is with ice from glaciers (**Figure 2.27B**).

As snow accumulates on top of a glacier, the weight of the upper layers compacts porous snow of the lower layers into ice, trapping the air as bubbles within the ice. These bubbles are time capsules, samples of the

atmosphere at the time the ice was formed. Analysis of these air bubbles provides a direct assessment of how the composition of Earth's atmosphere has changed over the history of the glacier (Brook 2005).

Drilling and analyzing ice cores requires great dedication and attention to detail. The most promising drilling sites are located in the coldest and most remote places on Earth. Drilling of a deep ice core involves many hundreds of work cycles:

1. Lower the drill bit assembly.

2. Cut another 3 m in depth.

3. Raise the assembly to the surface.

4. Empty the drill core barrel.

5. Prepare the drill bit assembly for another round.

To prevent the core hole from collapsing under the surrounding pressures, it is filled with a drilling fluid of special oils. Ice cores removed from the core barrel are cleaned of the drilling fluid, sealed in plastic bags, and kept frozen. In a clean room maintained at temperatures below freezing, disks less than a millimeter thick are sliced off cores. Every disk is divided into many segments and sent to various research groups, each specializing in a particular analysis.

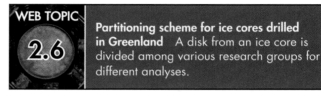

WEB TOPIC 2.6 Partitioning scheme for ice cores drilled in Greenland A disk from an ice core is divided among various research groups for different analyses.

Dating a disk taken from an ice core entails keeping track of the number of annual cycles from the very top layer of ice. Researchers determine annual cycles from

(A)

(B)

South Pole

Dome C

FIGURE 2.28   Dome C in Antarctica (A) Concordia base at Dome C.
(B) Location of Dome C on the continent of Antarctica.

seasonal variations in the isotopic composition of the ice, the concentrations of contaminants in the ice, and the visual appearance of the ice.

- Because of their temperature dependence, the isotope ratios δD and $\delta^{18}O$ of the ice are higher in summer.

- Cosmic rays produce the radioactive isotopes ^{10}Be and ^{36}Cl (with half-lives of 1.51 million years and 0.308 million years, respectively) as charged particles in the upper atmosphere. Their leakage to Earth's surface and contamination of the ice decreases in summer, when more sunlight increases the extent to which the upper atmosphere becomes charged and retains these isotopes.

- Hydrogen peroxide (H_2O_2) is created when ultraviolet (UV) radiation from the sun strikes the atmosphere. UV radiation and the resultant H_2O_2 concentrations in the ice are higher in summer.

- In some ice cores, summer layers are visibly different from winter layers. Winter ice layers contain only snowflakes, whereas summer ice layers contain snow and fern-like, hoar-frost crystals. These layers can be counted, without magnification, with the human eye.

Usually, several of these parameters are measured for independent verification of annual cycles. Still, counting the years in ice cores presents a challenge, particularly as the ice becomes compressed with depth or distorted from sideward flows. For example, the deepest sections of ice cores in Antarctica are so com-pressed that, on average, each year is represented by a disk less than 2 millimeters thick.

To summarize ice-core methodology: Assessments of gas concentrations in an ice core are most certain. Those of temperature and date are less certain.

At Dome C (**Figure 2.28A**), located on the high Antarctic plateau 1650 kilometers southeast of the South Pole (**Figure 2.28B**), a European consortium has drilled an ice core extending to a depth of 3270 meters. The deepest sections are about 890,000 years old. Analyses of sections as old as 650,000 years are complete for the greenhouse gases methane (CH_4), carbon dioxide (CO_2), and nitrous oxide (N_2O) and for δD, a proxy measure for temperature.

The historical temperature record based on measurements of δD at Dome C (**Figure 2.29**) indicate that Earth's climate during the past million years has been punctuated with about 8 cycles of relatively long, colder periods (glacials) interrupted by relatively short, warmer periods (interglacials). Each of nine interglacial periods correlates with higher concentrations of the greenhouse gases CH_4, CO_2, and N_2O. Current levels of these gases, however, greatly exceed those at any time during the past 650,000 years, an observation discussed more fully in Chapter 4.

The Past Millennium

Climate reconstructions for the past 1000 years employ several proxy measures in addition to ice cores. These include tree rings, corals, glacier lengths, and bore holes. The following describes each of these approaches and shows reconstructions derived from them.

FIGURE 2.29 A record of methane (CH₄), carbon dioxide (CO₂), and nitrous oxide (N₂O) concentrations and a temperature proxy (δD, the deuterium/hydrogen ratio) over the past 650,000 years, based on data from the ice core at Dome C in Antarctica. Data for the past 200 years are derived from other ice cores and direct measurements. Gaps in the N₂O record correspond to periods of elevated dust levels that interfere with N₂O evaluation. Gray-shaded sections denote warm, interglacial periods. The δD value increases with increasing temperatures. (After Brook 2005; Spahni et al. 2005.)

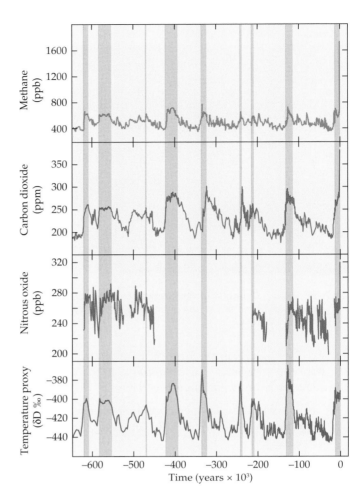

Tree rings

Dendrochronology (*Greek*; tree + time) is the study of tree rings to evaluate past environmental conditions. It is based on the phenomenon that each year a tree adds a layer of new wood between its old wood and bark. When growing conditions are favorable, a tree produces large cells. If conditions worsen, growth slows and cell size decreases. Under unfavorable conditions, growth and cell division ceases. The contrast between the small cells formed late in the season one year and the large cells formed early the following season appear like a ring when observed in a cross section (**Figure 2.30**).

Trees that grow in a single region develop a similar pattern of ring widths during a given time

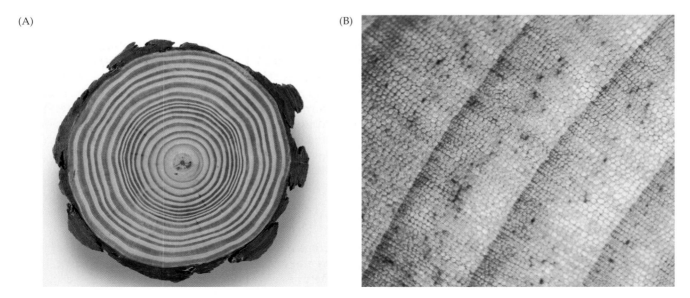

FIGURE 2.30 (A) Tree rings of basswood. (B) Close-up of tree rings in bald cypress.

(A)

(B)

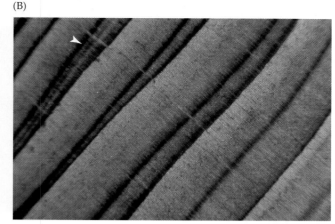

FIGURE 2.31 Problems with tree rings (A) Locally absent rings in Scots pine (*Pinus sylvestris*). The ring at the middle left side (indicated by the white arrowhead) disappears on the right side. (B) False rings in Mexican cypress (*Cupressus lusitanica*). For example, the darker band near the upper left-hand corner is not a true annual ring.

period. Comparing these patterns, matching them ring for ring, and averaging over multiple trees establishes a chronology. This minimizes errors that can occur when a stressful condition during the growing season, such as drought, prevents the formation of a growth ring in some trees, resulting in locally absent rings (**Figure 2.31A**) or causes the temporary cessation of growth and subsequent formation of another, false ring (**Figure 2.31B**). Samples collected from trees that are on relatively flat terrain or are close to rivers or lakes are less prone to drought that may lead to poor tree-ring formation.

Sometimes wood from ancient structures, when aligned with chronologies from living trees, extends a chronology even further back. The date of the structure provides an approximate date of when the trees were felled. Years of extraordinary growth rings offer additional clues for alignment. One chronology for river oaks from southern Germany extends more than 10,000 years, and another for bristlecone pine in the White Mountains, California extends more than 8500 years. Of course, as one goes further back in time, appropriately aged samples become fewer in number and farther between, and so chronologies become less certain.

To sample a tree, the researcher screws a tool called an increment borer (**Figure 2.32**) into the trunk and extracts a core about 5 millimeters in diameter. This procedure does not permanently injure the tree. To count rings, the researcher sections and polishes the extracted core and examines it under a microscope.

The width of growth rings, especially for trees found near their altitudinal or latitudinal limits, increases

In 1964, a graduate student studying glaciation on Wheeler Peak in Nevada came upon what he suspected was a very old bristlecone pine tree. With his increment borer broken and winter nigh upon him, the student sought and received permission from the U.S. Forest Service to cut down the tree. Examining the trunk of the felled tree, he counted 4844 rings. He had just killed the oldest living organism on Earth (Hall 1998).

with air temperature (National Research Council 2006); for instance, at higher elevations near treeline, tree growth responds mostly to temperature. Furthermore, temperatures during the middle of a growing season have the strongest influence on ring width. Calibration for a given tree species requires sites that have records of both temperatures and ring widths for individual trees of various ages. A compilation of tree ring data from the Northern Hemisphere indicates that air temperatures today are as warm as any time during the past 1200 years (**Figure 2.33**).

Corals

Corals are marine animals that form exoskeletons of calcium carbonate ($CaCO_3$). Colonies of corals produce reefs in clear, shallow waters. These animals generate denser layers in their exoskeletons during months with severe weather and less dense layers during months with more benign weather. As a result, corals develop

(A)

(B)

FIGURE 2.32 Increment tree borer (A) Borer is drilled into the trunk of a tree. (B) The borer is unscrewed from the trunk, and the plunger extracts the tree core from the barrel of the plunger.

discernible annual bands (**Figure 2.34**) that can be counted to establish the age of a sample.

The ratio of heavy to light oxygen isotopes ($\delta^{18}O$) in shells of marine organisms, as discussed earlier, decreases with the temperature of the surrounding seawater. In shallow waters where corals grow, the $\delta^{18}O$ values of the seawater and, thereby, of the corals, also changes with rainfall, evaporation, and river input. Therefore, the $\delta^{18}O$ record is often supplemented with other proxy measures of temperature, such as the strontium-to-calcium ratio (Sr/Ca).

Strontium (Sr) can substitute for calcium (Ca) in certain biological processes because both elements dissolve in water to form ions with the charge +2, and

these ions are of similar size. The Sr/Ca ratio in coral skeletons decreases with temperature for reasons that may involve discrimination between the two elements during their transport from seawater into the coral (Corrège 2006).

Cores from coral reefs at the Great Barrier Reef, Australia dating as far back as 1565 show that temperatures near the corals during the last half of the twentieth century are as warm as they have been in over 400 years (**Figure 2.35**). Contrasting the Sr/Ca and $\delta^{18}O$ data suggests that changes in the $\delta^{18}O$ values of coral exoskeletons derive from both rising temperatures and increasing input of fresh water from human activities onshore, particularly after 1850.

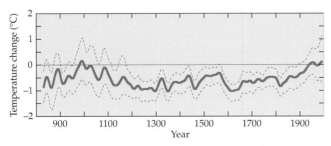

FIGURE 2.33 Temperature changes in the Northern Hemisphere over the past millennium, based on tree-ring data. Tree-ring chronologies from 14 sites in the Northern Hemisphere were selected and calibrated against average temperatures from 1900–1977. Data were smoothed with a mathematical function. Dotted lines indicate the range of error in the estimates. (After Esper et al. 2002).

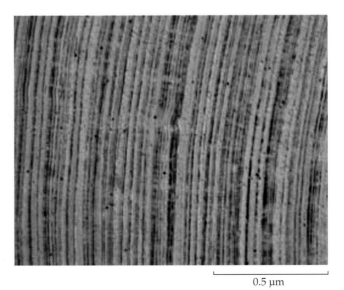

0.5 μm

FIGURE 2.34 Thin section of a black coral showing the annual banding patterns.

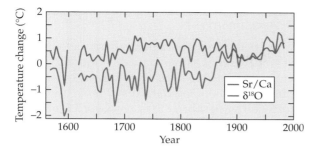

FIGURE 2.35   Temperature changes around the Great Barrier Reef, Australia Temperatures were estimated from the Sr/Ca of corals and from the $\delta^{18}O$ of coral carbonates. The $\delta^{18}O$ also reflects the dilution of seawater with fresh water. Shown are five-year averages with a sample size of one to two cores for dates before 1730 and four to eight cores after 1830. (After Hendy et al. 2002.)

FIGURE 2.36   Easton Glacier on Mt. Baker, Washington in 2003 The red line marks the extent of the glacier in 1985.

Glacier length

Quantitative assessments of the size of glaciers for the past 400 years are available. These include direct measurements, paintings, etchings, and, more recently, photographs and satellite images (**Figure 2.36**; see Figure 1.10A). Assuming a simple relationship between glacier length and average air temperature (ignoring precipitation, which varies less over time and space than temperature), one can reconstruct the temperatures near various glaciers over the past 400 years (Oerlemans 2005). These reconstructions indicate that temperatures were relatively uniform through the seventeenth, eighteenth, and nineteenth centuries but began to rise worldwide during the twentieth century (**Figure 2.37**). The reconstruction for North America shows a marked cooling after 1940, but this anomaly derives mainly from the dearth of recent measurements for North America glaciers.

Boreholes

Temperatures deep in the ground respond to changes near the surface. For example, a sustained heat wave at the surface will cause warming to propagate slowly downward, taking roughly 100 years for the perturbation to reach a depth of 150 meters. Therefore, the vertical distribution of temperatures in boreholes from drilling operations contains information about past air temperatures. This approach has the advantage that it measures temperature directly, and is not a proxy for temperature; the disadvantage is that various soils and rocks differ in how they transfer heat and distort the temperature signal from the surface; there-

fore, the precision of the measurement with respect to time and temperature declines rapidly with depth. An examination of present-day temperatures in 695 boreholes from the Northern Hemisphere indicate that temperatures have risen about 1°C since 1500, with a 0.5°C rise in the past century alone (**Figure 2.38**).

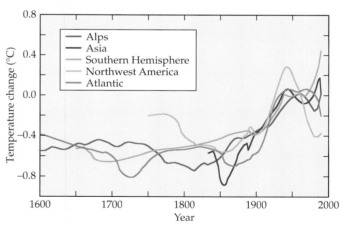

FIGURE 2.37   Reconstruction of temperature changes based on lengths of glaciers in various parts of the world. The "Atlantic" region includes Greenland, Iceland, and Scandinavia. Sample size ranged from less than 20 glaciers before 1800 to more than 150 glaciers after 1885. The apparent cooling in North America after 1940 derives mainly from the dearth of recent measurements for North America glaciers. (After Oerlemans 2005.)

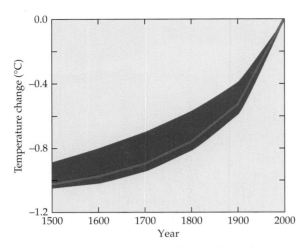

FIGURE 2.38 Reconstruction of global temperature changes based on 695 boreholes in the Northern Hemisphere. Blue-shaded area indicates the range of values for various estimates. (After Pollack and Smerdon 2004.)

Ice cores

This chapter previously presented temperature reconstructions as far back as 650,000 years ago based on δD and $\delta^{18}O$ values in sections of deep ice cores from the high Antarctica plateau (see Figure 2.29). For shorter-term reconstructions, on the order of the past 2000 years, shallower ice fields are suitable and provide a more global perspective on climate. For example, $\delta^{18}O$ values of ice cores from low-latitude, high-altitude sites in the Andes and Himalayas suggest that temperatures in the tropics have risen abruptly during the past 150 years (**Figure 2.39**).

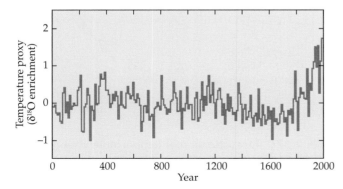

FIGURE 2.39 Trends in $\delta^{18}O$ from ice cores taken at low-altitude, high-elevation sites in the Andes and Himalayas. The values were averaged over ten years. (After Thompson et al. 2006.)

Direct measurements

Reliable temperature measurements from a global network of land-based weather stations (**Figure 2.40A**), plus readings from ships and fixed buoys, have been available since 1850. One data set of monthly-averaged surface temperatures has reached 3.7 million observations (Brohan et al. 2006). Analysis of this data set shows that average global temperatures have significantly increased during the past 25 years (**Figure 2.40B**), with temperatures over most landmasses rising at about 0.25°C per decade and over landmasses at polar latitudes warming more than 0.5°C per decade (**Figure 2.40C**).

Sea Level

Global sea levela rise and fall depending on the volume of the ocean basins versus that of the water in them. Changes in the volume of ocean basins occur over millions of years and are not directly responsive to climate, whereas changes in water volume may occur relatively rapidly (less than 100,000 years) and depend on global temperatures (Miller et al. 2005). The following factors have an effect on the volume of ocean water.

- Temperature directly changes the volume of water in the oceans. When ice melts to a liquid at 0°C, water decreases in volume by 9% (ice = 19.651 milliliters per mole, whereas water = 18.019 milliliters per mole; a mole = 6.0221 × 10²³ atoms or molecules of a chemical). For this reason, icebergs float with about 9% of their volume above the surface. The volume of liquid water contracts to a minimum at 4°C (18.016 milliliters per mole) and then expands with temperature (**Figure 2.41**).

- Fresh water in lakes, rivers, groundwater, and continental ice sheets sequester water that would otherwise be found in the oceans. In particular, the ice sheets of Antarctica and Greenland contain, respectively, about 2% and 0.2% of the water on Earth, and their accretion or attrition alters sea level.

- Sea ice floating on the ocean, such as the ice pack around the North Pole, has a mass equal to the water it displaces. When a mass of ice melts, the amount of liquid it forms is equal in volume to the water it displaces. Therefore, melting of this sea ice does not significantly contribute to the volume of the oceans. Next time you have an ice cube floating in a drink, check whether the level of fluid changes as the ice cube melts.

(A)

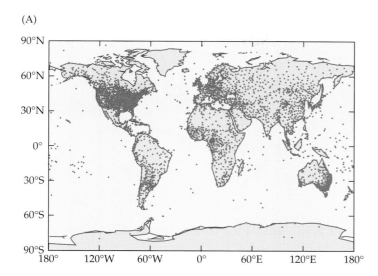

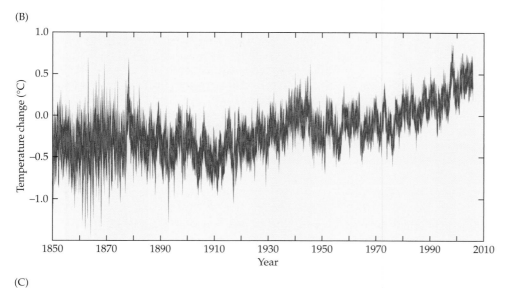

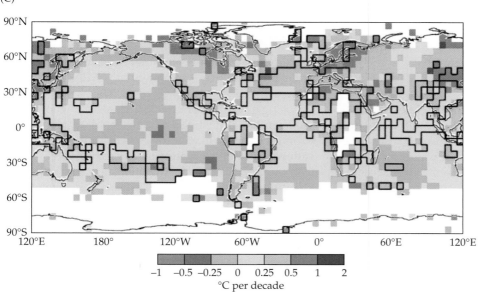

FIGURE 2.40 Worldwide direct temperature measurements (A) Red dots indicate the locations of the 4349 land-based weather stations used for estimating global temperatures. (B) Trends in monthly averages of global surface temperatures, with respect to the average from 1961–1990. The red line is the best-estimate value; the green line (barely visible) denotes the 95% uncertainty range caused by station, sampling, and measurement errors; and the blue line adds the 95% error range caused by limited coverage because some geographical regions have fewer weather stations. (C) Spatial patterns of changes in surface temperature (°C per decade) for 1977–2001. Thick black lines around a grid box signifies that its temperature trend was significant at the 95% level. The white areas are those for which information is inadequate. (A, B, after Brohan et al. 2006; C after Jones and Moberg 2003.)

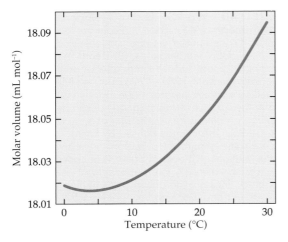

FIGURE 2.41   The molar volume of water, in milliliters per mole, as a function of temperature.

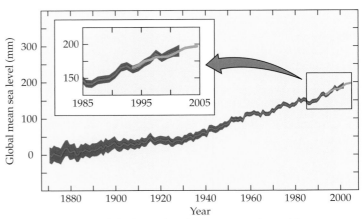

FIGURE 2.42   Global sea level over time The red line presents the yearly averages derived from tide gauges; the blue area indicates the standard deviation. The green line beginning in 1993 presents satellite measurements. The inset shows an expanded view of the past two decades. (After Church and White 2006.)

Warmer global temperatures (see Figure 2.40B) and the melting of glaciers (see Figure 2.37) over the past century have increased the global volume of water in the oceans. These factors potentially could raise sea level roughly 0.7 millimeters per year (mm y^{-1}) (Miller and Douglas 2004; Woodworth 2006). Two independent research approaches, however, indicate that sea level is rising at more than twice this rate.

One approach involves tide gauges. These instruments are usually located on piers and monitor the height of the sea relative to a nearby geological marker. Periodic surveys determine whether the pier has shifted or whether the elevation of land in the area has changed from natural forces, such as tectonic movements, or anthropogenic (resulting from human activities) ones, such as groundwater pumping and mining. Extensive and reliable data from tide gauges are available worldwide since 1870 and show that sea level has risen an average of 1.7 mm y^{-1} over the past century and 3.0 mm y^{-1} over the past decade (**Figure 2.42**).

The other approach involves the satellites TOPEX/Poseidon (in operation from 1993 through 2005) and Jason-1 (in operation from 2001 through the present) that were specially designed for monitoring sea level. These satellites follow a path that transects the globe uniformly from 66°S to 66°N every ten days. They bounce microwaves off the oceans and monitor the time it takes between sending and receiving the microwaves (i.e., their time of flight) to determine their distance from the surface (Cazenave and Nerem 2004). Global Positioning System (GPS) receivers on the satellites, along with laser range finders, establish their orbital height. Sea level is taken as the difference between the orbital height of a satellite and its distance from the sea surface, and such a determination has an error of less than 1 centimeter. The satellite data show that average sea level has risen 2.9 mm y^{-1} over the past decade (see Figure 2.42).

Why do theory and practice differ? In this case, the theory predicts that changes in the volume of ocean water should produce a sea level rise of only 0.7 mm y^{-1}, whereas both tide gauges and satellites indicate an average rate of greater than 1.7 mm y^{-1}. Most likely, the Antarctic and Greenland ice sheets are melting faster than previously estimated (Cazenave and Nerem 2004; Church and White 2006; Miller and Douglas 2006; Woodworth 2006).

Stormy Weather

The amount of ocean water that can evaporate into the atmosphere increases exponentially with temperature (**Figure 2.43**).

Rising global temperatures thus result in more water vapor in the atmosphere (**Figure 2.44**), which in turn promotes the formation of clouds and precipitation. Indeed, annual precipitation over land increased an average of 2% during the twentieth century (Dore 2005; Huntington 2006). Regional variation, however, has been high; for example, the United States experienced a 7% increase in precipitation (Groisman et al. 2004), more than three times the global average, whereas sub-Saharan Africa experienced more than a 30% decline. Moreover, this variation is increasing. Some regions have been subject to more severe flooding and others to

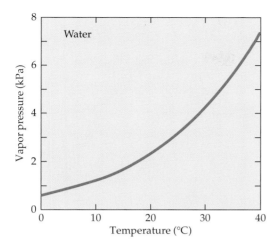

FIGURE 2.43 Vapor pressure of water (kPa) as a function of temperature when atmospheric pressure equals 101.325 kPa, the mean value in Paris. This is the maximum amount of water vapor that air will absorb at a given temperature.

more severe drought, during the past two decades than during the early part of the twentieth century (Dai et al. 1998). Many of the regions experiencing more precipitation are already wet, whereas those facing declining precipitation are already semiarid (**Figure 2.45**).

As Earth becomes warmer, the temperature gradient between the higher atmosphere and the surface of seas in the tropics becomes more extreme. This gradient is highly correlated with the intensity of tropical storms (Hoyos et al. 2006) that are called hurricanes, typhoons, or cyclones depending on location. While the number of tropical storms occurring worldwide has not changed significantly during the past three decades, the percentage of these storms that fall into categories 4 and 5—that is, with winds of speeds greater than 210 kilometers per hour (131 miles per hour)—has doubled (**Figure 2.46**). These extreme weather events, which are much more damaging than smaller storms, heighten public concern about the degree to which human activities are influencing climate, a topic we will address in Chapters 4 and 12.

Self Control

NASA (the National Aeronautics and Space Administration), when they were designing spacecraft for a landing on Mars, asked James Lovelock, a noted British atmospheric chemist, to evaluate the evidence for life in the solar system. From such considerations, he developed the **Gaia hypothesis** that life on Earth has served as an active control system that stabilizes the physical environment and chemical composition of the planet (Lovelock 1979, 1988). Some proponents of this hypothesis have applied it to theological issues (Milne 1990). This hearkens back to a belief common in many human cultures: Constancy in the natural world is

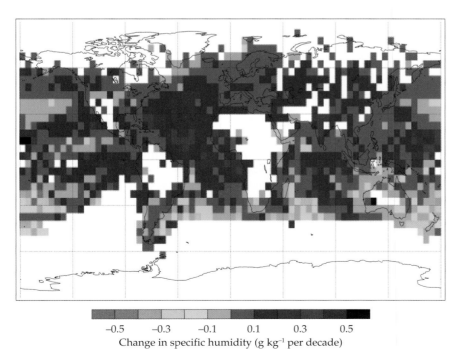

FIGURE 2.44 Trends in specific humidity from 1973–1999 in grams of water vapor per kilogram of air per decade. (After Willett et al. 2007.)

FIGURE 2.45 Change (% per decade in the amount of precipitation over land from 1979–2005. Areas in gray lack sufficient data to estimate changes. A "**+**" in an area indicates that the trend is significant at the 95% level. (After Trenberth et al. 2007.)

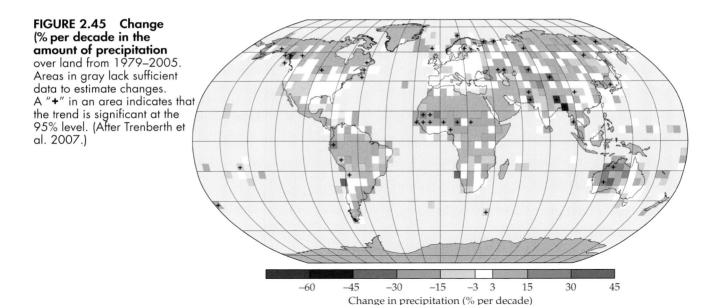

Change in precipitation (% per decade)

guaranteed, if not by divine providence, then by the "balance of nature" (Weart 2003).

Does the history of Earth's climate support this view? Has Mother Nature ever "lost control"? The temperature reconstructions presented in this chapter do not provide a definitive answer. Our planet has fluctuated a dozen times from thick ice sheets covering the majority of its surface to little or no permanent ice anywhere. Temperature regulation of Earth, if it exists, seems to allow deviations on the order of 10°C. Yet, on a cosmic scale, such deviations in Earth's climate have never been as extreme as those of its neighboring planets, where surface temperatures on Venus and Mars appear to have fluctuated in recent history more than 100°C and 60°C, respectively (Bullock and Grinspoon 2001; Head et al. 2008). Moreover, Earth has had liquid water somewhere on its surface throughout its recent history.

The rise and fall of temperatures on Earth, as well as on Mars and Venus, are strongly correlated with changes in greenhouse gas concentrations, and these gases serve as regulating mechanisms. Four billion years ago, when the sun was less than 80% as bright as it is today, the temperature of Earth should have been about 0°C, but an atmosphere filled with greenhouse gases generally kept the planet from freezing. Over time, as the sun became brighter, the temperature of Earth began to warm, but this sped up the hydrolysis reaction through which atmospheric CO_2 dissolves into the oceans (see Equation 2.5). The resulting bicarbonate ions in the oceans created calcite and other carbonate minerals (see Equation 2.6) that formed the sedimentary rock that covers continents. The net result is that the atmospheric greenhouse gas concentrations such as CO_2 have been gradually decreasing (see Web Topic 2.3) as the sun has been getting brighter.

This process has also operated in reverse. When the planet became very cold, weathering rates decreased,

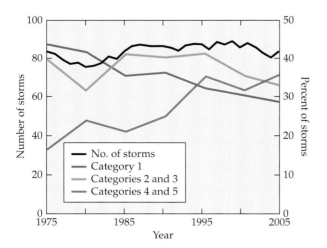

FIGURE 2.46 Number of tropical storms (five-year running averages) occurring worldwide over the past three decades (black). The Saffir–Simpson Hurricane Scale is a 1–5 rating based on a hurricane's intensity. The graph shows the percentage of these storms (five-year averages) that fall into Saffir–Simpson category 1, categories 2 and 3, and categories 4 and 5. (After Webster et al. 2005.)

less CO_2 was removed from the atmosphere, and the accumulation of additional CO_2 in the atmosphere promoted global warming. This is thought to have been responsible for the defrosting of Snowball Earth I and Snowball Earth II.

At other times, greenhouse gases have exacerbated temperature perturbations, such as when CH_4 released from seafloor sediments increased global warming during the Great Dying and the Late Paleocene Thermal Maximum. Chapter 3 examines the causal relationships among climate, greenhouse gases, and other factors.

Here, in this chapter, we have seen how changes in climate and biota are dynamically linked. Organisms serve as major sources or sinks of greenhouse gases and thereby have influenced climate change. In the reverse direction, climatic shifts including Snowball Earth, the Great Oxidation Event, the Ordovician-Silurian Extinction, the Great Dying, the K-T Boundary, and the Late Paleocene Thermal Maximum extinguished many existing life forms but created opportunities for the proliferation of different ones.

Such considerations inspire questions with broad ramifications. Have human activities altered the stabilizing effect of other living organisms on Earth's climate? To what extent is the current distribution of species on Earth merely happenstance? But for one large asteroid, might a dinosaur of another kind be writing this book?

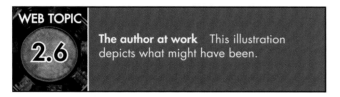

WEB TOPIC

2.6

The author at work This illustration depicts what might have been.

Summary

To determine whether our current climate deviates from normal trends requires some knowledge of the history of Earth's climate. Direct measurements of climate are available for only a century or two, and so climatologists must rely on proxy measures that are associated with climate. For example, the relative amount of heavy and light isotopes of oxygen or hydrogen in a rock or ice indicates the temperature at the time when that substance formed. These proxy measures indicate that Earth's average temperature has fluctuated many times as much as $10°C$ in concert with changes in atmospheric concentrations of greenhouse gases such as carbon dioxide and methane. Such temperature fluctuations, according to the fossil record, have buffeted life on Earth.

Based on many types of evidence—tree rings, corals, glacial extents, boreholes, ice cores, and direct measurements—temperatures today are as warm and probably warmer than any the Earth has experienced in the past several centuries. Sea level is rising significantly. Major storms are becoming more intense.

Review Questions

1. A proxy measure *cannot* provide an estimate of (select one)
   (*a*) the date.
   (*b*) precipitation.
   (*c*) location.
   (*d*) the composition of the atmosphere.
   (*e*) Earth's climate during the Big Bang.

2. Select all of the following changes that are thought to have occurred during Earth's last 4 billion years.
   (*a*) At times, the planet was completely or nearly covered with ice.
   (*b*) At times, the planet was completely free of ice.
   (*c*) At times, climate shifts were responsible for the extinction of many species.
   (*d*) At times, there were far fewer continents than there are today.
   (*e*) At times, continents were located in different places with respect to one another.

3. Which one of the following does *not* serve as a proxy measure of temperature?
   (*a*) Relative amounts of oxygen isotopes in water
   (*b*) Relative amounts of oxygen isotopes in marine shells
   (*c*) Magnetic orientation of iron in rocks
   (*d*) Tree rings
   (*e*) Length of glaciers

4. Snowball Earth might have been Slushball Earth if (select one or several)
   (*a*) tropical oceans had open water.
   (*b*) an asteroid distributed iridium only around the impact site.
   (*c*) most of the continents were located at higher latitudes.
   (*d*) the sun had the same strength as when the Earth first formed at 4.5 billion years ago.
   (*e*) ice cores from mid-latitudes were analyzed.

5. Major species extinction events in Earth's history usually (select one)
   (*a*) resulted from several factors.
   (*b*) were associated with the freezing of methane hydrates.
   (*c*) depended on the existence of a single super-continent.

(*d*)   caused a major shift in climate.
(*e*)   are undetectable in the fossil record.

6. The K-T Boundary refers to (select one or several)
(*a*)   when a large asteroid struck Earth.
(*b*)   when atmosphere concentrations of oxygen increased several fold.
(*c*)   the merging of landmasses to form super-continents.
(*d*)   when many species of dinosaurs went extinct.
(*e*)   the event separating the *Cretaceous* and *Paleogene* periods.

7. Ice cores from Antarctica do *not*
(*a*)   provide a climate record that extends back in history for more than 100,000 years.
(*b*)   indicate that Earth, during the history of the ice sheet, has experienced a number of cycles of relatively long, colder periods interrupted by relatively short, warmer periods.
(*c*)   suggest that warmer periods during the history of the ice sheet are associated with higher concentrations of greenhouse gases.
(*d*)   demonstrate that cosmic rays and hydrogen peroxide are correlated with greenhouse gas levels.
(*e*)   show that current levels of greenhouse gases exceed those found at any previous time during the history of the ice sheet.

8. Annual bands in corals derive from (select one)
(*a*)   drought.
(*b*)   differential growth during severe versus benign weather.
(*c*)   methane exposure.
(*d*)   dust accumulation.
(*e*)   predation by octopi.

9. Sea level may rise because (select one or several)
(a)   icebergs floating on the oceans melt.
(*b*)   ice sheets on the continents Greenland and Antarctica melt.
(*c*)   the volume of water expands with temperature above 4°C.
(*d*)   the volume of ocean basins expands through continental movements.
(*e*)   the values from tide gauges and satellite measurements disagree.

10. Select all of the following climate changes that have been observed during the last 150 years.
(*a*)   Global average temperatures has increased 0.6°C (1.1°F).
(*b*)   Global average temperatures has increased 6.0°C (10.8°F).
(*c*)   Average sea level has risen by 0.2 meters (7.9 inches).
(*d*)   Average sea level has risen by 2.0 meters (78.7 inches).
(*e*)   Antarctica has become entirely free of ice in the summer.

Suggested Readings

Jones, P. D. and M. E. Mann. 2004. Climate over past millennia. *Reviews of Geophysics* 42: RG2002, doi:10.1029/2003RG000143.

This article provides a thorough review of the proxy measures used to characterize climate and what they indicate about the climate over the last 2000 years.

Ruddiman, W. F. 2008. *Earth's Climate: Past and Future*, 2nd Ed. W.H. Freeman, New York.

An introductory textbook on climatology.

Williams, P. 2009. *Don't Be an Isodope, Learn about an Isotope.* UC Santa Barbara, *http://www.geog.ucsb.edu/~williams/Isotopes.htm* (accessed May 27, 2009).

A one-page explanation about isotopes and their uses in geology and life sciences.

3

CAUSES
OF CLIMATE

Various causes, or **forcing factors**, are responsible for the climatic fluctuations that Earth has experienced throughout its history. To predict the future of Earth's climate requires a thorough understanding of these factors and how they are likely to change. Chapter 2 introduced a few of them, particularly greenhouse gases. This chapter details the mechanisms behind greenhouse gases and other major forcing factors.

Forcing factors fall into two categories: external and internal. External forcing factors involve agents outside of Earth and its atmosphere. In contrast, internal forcing factors originate right here on Earth. Variations in solar energy, for example, are external, whereas greenhouse gases in Earth's atmosphere are internal.

FIGURE 3.1 Photograph of our galaxy, the Milky Way But how was such a picture taken? Actually, it is a photo of NGC 7331, a galaxy that is a twin in size and shape to our galaxy. The arrow indicates the approximate position of our sun, if this galaxy were the Milky Way.

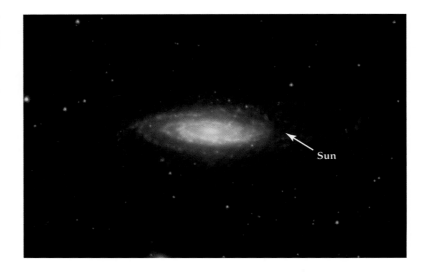

External Forcing Factors

Three external forcing factors—galactic variations, orbital variations, and sunspots—influence our planet's climate.

Galactic variations

Our sun lies in the Milky Way, a swirling spiral galaxy of 200 billion to 400 billion stars (**Figure 3.1**). Every 150 million to 350 million years, our solar system rotates around the Milky Way. The quantity and quality of energy reaching Earth from nearby star systems and from the gases and dust that pervade interstellar space undoubtedly varies during this rotation. The cyclical fluctuations in this energy are so long and so uncertain that they obscure the influence of galactic rotation on Earth's climate. Nonetheless, several major climatic events, such as tropical temperature changes (**Figure 3.2**), the K-T boundary, the Great Dying, and the Ordovician-Silurian Extinction (see Chapter 2) are separated by about 150 million to 190 million years and might be coincidental with the period of galactic rotation (Scherer et al. 2006).

Orbital variations

The major driver of Earth's climate is the amount of solar energy it receives. Here we introduce three laws—the cosine law, Kepler's second law, and the inverse square law—that relate the solar energy reaching Earth to the planet's orientation with the sun.

The first is the **cosine law**. Sunbathers instinctively obey this law in orienting themselves perpendicular to

the sun to maximize the interception of solar rays (**Figure 3.3A**) or more parallel to the sun to decrease interception (**Figure 3.3B**). The more slanted or oblique the angle between the sunbather and the sun, the greater the area over which a given amount of solar energy is distributed and, thus, the smaller the energy per unit area. Mathematically,

$$I = I_0 \cos \theta$$

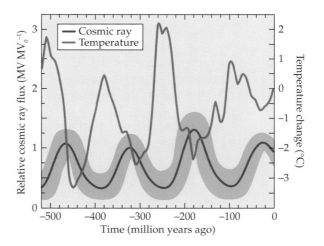

FIGURE 3.2 Relative cosmic ray flux and tropical temperature changes over time Cosmic ray flux relative to a low-level is based on iron meteorite record, and that of temperature changes is based on oxygen isotope ratios $\delta^{18}O$. Blue band indicates the error range of the cosmic ray flux. (After Shaviv and Veizer 2003.)

(A)   (B)

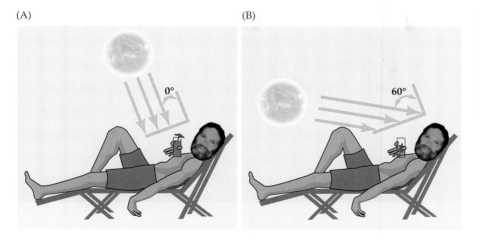

FIGURE 3.3 Sunbathing
(A) Sunbather is aligned perpendicular to solar rays ($\theta = 0°$), maximizing incident energy. (B) Sunbather is aligned more obliquely to solar rays (e.g., $\theta = 60°$), diminishing incident energy.

where

I_0 = incoming solar energy
θ = angle of incidence

For the sunbathing example in Figure 3.3,

$I = I_0 \cos 0° = I_0$, whereas $I = I_0 \cos 60° = \frac{1}{2} I_0$

The cosine law explains many of the differences in climate between the equator and the poles. At the March 20 and September 22 equinoxes, the sun at midday is directly over the equator and focused upon a relatively small area, whereas near the poles it is at an oblique angle and spread over a relatively large area (**Figure 3.4**). Moreover, at the oblique angle near the poles, solar energy passes a longer distance through the atmosphere and is more likely to be reflected or absorbed by gases or particles in the atmosphere before striking Earth's surface. The reflectance of many materials increases at an oblique angle (e.g., sunsets reflecting off water). For these reasons, the poles receive less solar energy per unit area than the equator and are colder.

The cosine law also explains differences among the seasons. Earth rotates daily around an axis that is oriented at an angle of 23.4° away from its plane of orbit around the sun (**Figure 3.5**). At the equinoxes, the sun is directly over the equator, but at the solstices, June 21 and December 21, the sun is directly above the Tropic of Cancer (latitude 23.4°N) or the Tropic of Capricorn (latitude 23.4°S), respectively. Therefore, on June 21, the Tropic of Cancer receives full sunlight ($I_{\text{Cancer}} = I_0 \cos 0° = I_0$) and experiences summer; the Tropic of Capricorn receives 68.5% of full sunlight

($I_{\text{Capricorn}} = I_0 \cos 46.8° = 0.685\, I_0$) and experiences winter. On December 21, the reverse applies.

The second law introduced here is **Kepler's second law**, which states that gravitational bodies rotate through equal areas in equal time. Consequently, a line from Earth to the sun sweeps an equal amount of area in the **orbital plane** (the plane on which Earth is orbiting) every day, regardless of where the planet is in its orbit (**Figure 3.6**). Consequently, Earth moves faster when it is near the sun than when it is far from the sun because at the smaller distance it needs to trace a greater arc to

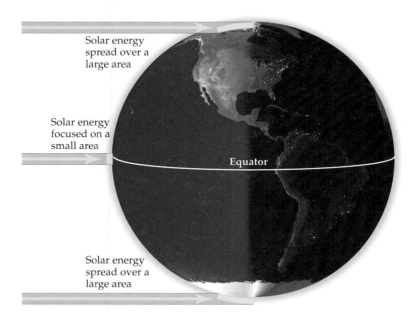

Solar energy spread over a large area

Solar energy focused on a small area

Equator

Solar energy spread over a large area

FIGURE 3.4 Cosine law Earth at the equinoxes when the sun aligns directly over the equator. Yellow arrows designate beams of equal solar energy. Solar energy per unit area at the equator is full strength ($\theta = 0°$), whereas that at the poles is diminished ($\theta > 80°$).

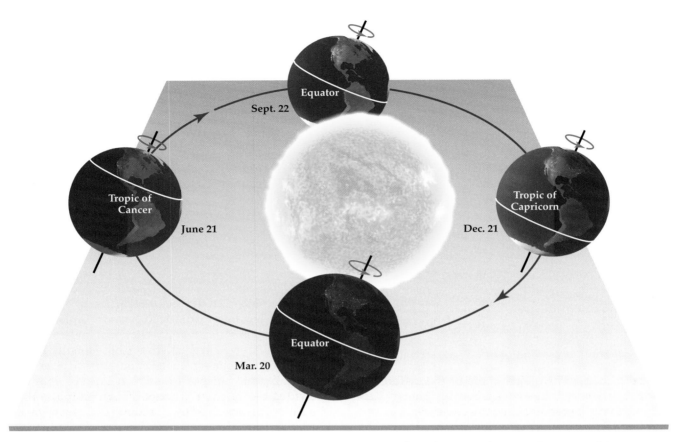

FIGURE 3.5 Orbits of Earth around its own axis and around the sun meet an angle of 23.4° Consequently, the sun is directly overhead at the Tropic of Cancer (latitude 23.4° north) at the June solstice, at the Tropic of Capricorn (latitude 23.4° south) at the December solstice, and at the equator (latitude 0°) at the March and September equinoxes.

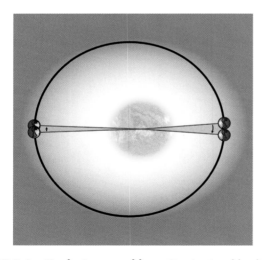

FIGURE 3.6 Kepler's second law Gravitational bodies in their orbits transcribe equal areas in equal amounts of time. Earth requires the same amount of time to rotate through the gray areas and, thus, moves faster when it is closer to the sun.

cover the same area. This means that Earth spends less time orbiting close to the sun than far from it.

Third, and last, is the **inverse square law**. It treats the sun as a point source of light, one that distributes light energy evenly in all directions. Therefore, at a given distance from the sun (r), the sun's energy is distributed over the surface area of a sphere ($4\pi r^2$). At twice the distance, its energy is distributed over 4 times the area, and so on (**Figure 3.7**). The inverse square law states that the amount of solar energy striking an area decreases with the square of the distance from the sun or, in mathematical terms,

$$I = \frac{I_0}{4\pi r^2}$$

where

I_0 = initial solar energy
r = distance from the sun

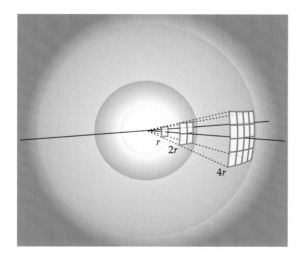

FIGURE 3.7   Inverse square law The sun distributes energy evenly in all directions. The letter "*r*" designates a certain distance. At twice this distance (2*r*), this energy is distributed over four times the area; at four times the distance (4*r*), it is distributed over 16 times the area.

The distance between the sun and Earth increases from 147 million km to 152 million km between January 3 and July 4. Putting these values into this equation,

$$I_{Jan} = \frac{I_0}{4\pi(147)^2} \text{ , } I_{July} = \frac{I_0}{4\pi(152)^2}$$

$$\frac{I_{July}}{I_{Jan}} = \frac{(147)^2}{(152)^2} = 0.935$$

This means that, at its farthest distance from the sun (**aphelion**), during the summer in the Northern Hemisphere, Earth receives only 93.5% of the solar energy that it receives at its closest distance (**perihelion**), during winter in the Northern Hemisphere. Nonetheless, the more perpendicular angle of the Northern Hemisphere to the sun during July more than compensates for the greater distance. For example, at the Tropic of Cancer, in the Northern Hemisphere,

$$\frac{I_{June}}{I_{Dec}} = \frac{I_0 \cos(0°)}{I_0 \cos(46.8°)} = 146\%$$

and together 0.935 × 1.46 = 128%.

Three characteristics of Earth's orbit around the sun—obliquity, eccentricity, and precession—change periodically and influence the amount of energy that Earth receives from the sun.

OBLIQUITY   As discussed above, Earth's daily rotation around its own axis currently has an angle of 23.4° with respect to its orbital plane around the sun. This angle is called the axial tilt or **obliquity**. The planet, however, wobbles like a spinning top (**Figure 3.8A**), and the obliquity oscillates between 22.1° and 24.5° (**Figure 3.8B**) every 41,000 years (**Figure 3.9**). This means that the seasonal differences in the amount of solar energy reaching Earth varies with a period of 41,000 years. At the extremes, the current Tropic of Cancer (latitude 23.4°N) would receive

$$\frac{I_{June}}{I_{Dec}} = \frac{I_0 \cos(1.3°)}{I_0 \cos(45.5°)} = 142.6\%$$

or

$$\frac{I_{June}}{I_{Dec}} = \frac{I_0 \cos(1.1°)}{I_0 \cos(47.9°)} = 149.1\%$$

about a 4.5% change in solar energy from minimum to maximum tilt.

ECCENTRICITY   The elliptical shape of Earth's orbit around the sun is characterized by its **eccentricity**, a measure of the deviation of an orbit from a perfect circle (**Figure 3.10**). Eccentricity in mathematical terms is

$$e = \frac{a - p}{a + p}$$

(A)   (B)

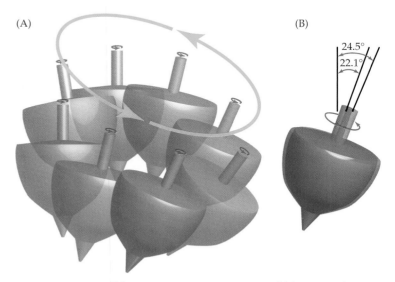

FIGURE 3.8   Obliquity (A) A spinning top wobbles around its axis of rotation. (B) The spinning top shows the range of inclination angles for Earth. Every 41,000 years, the axial tilt of Earth's daily rotation oscillates between an angle of 22.1° and an angle of 24.5° from its orbital plane around the sun. Currently, the angle measures 23.4°.

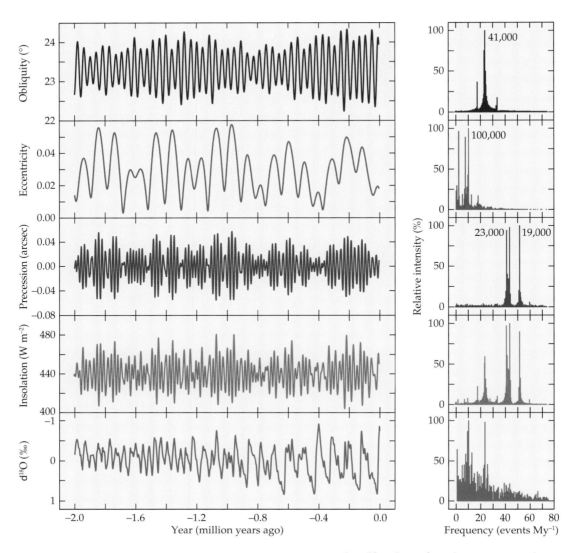

FIGURE 3.9 Influence of orbital variations on solar insolation and historical temperature changes
(A) Variations in Earth's solar orbit (obliquity, eccentricity, and precession), total solar energy impinging at latitude 65° north during July (insolation), and changes in benthic (seabed-level) $\delta^{18}O$ (inversely related to temperature) over the last 2 million years; a plot of 10,000-year running averages calculated from the orbital data of Laskar et al. (2004) and theoretical insolation data of Berger and Loutre (1991). The $\delta^{18}O$ data of Huybers (2006) is the average of 12 sea-bottom core records. Eccentricity (unitless) equals 0 for a perfect circle. (B) Spectral analyses of these data showing the frequencies (number of events per million years) of the oscillations (like the spectral display on a sound system that shows how much bass or treble is in the music). The numbers by the peaks are in years. These data indicate that past temperature changes were more sensitive to variation in obliquity and eccentricity.

where e is the eccentricity, a is the aphelion (farthest distance from the sun), and p is the perihelion (closest distance to the sun). Currently, Earth's orbit has an eccentricity of $(152 - 147) / (152 + 147) = 0.0167$, but it oscillates between 0.005 (nearly circular) and 0.0617 (more elliptical) with an average periodicity of about 100,000 years (see Figure 3.9).

As the eccentricity of Earth's orbit increases, the amount of solar energy reaching Earth fluctuates more from summer to winter because of greater seasonal differences both in the distance between Earth and the sun (calculated via the inverse square law) and in the time Earth spends at the aphelion and perihelion (calculated via Kepler's second law). When the eccentricity is minimal (i.e., the orbit is near circular), seasonal variation derives solely from differences in the angle between the axis of Earth's diurnal rotation and its orbital plane around the sun (calculated via the cosine law).

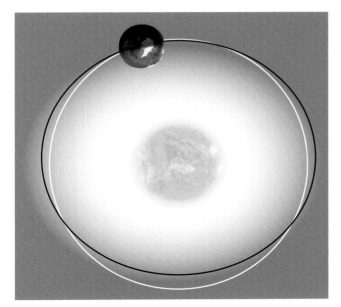

FIGURE 3.10 Eccentricity (e) of Earth's solar orbit, as seen from above the plane of the orbit The white line is a near-circular orbit (e = 0.005), and the black line is an elliptical one (e = 0.062); Earth's orbit ranges between these values.

PRECESSION At present, the summer and winter solstices and aphelion/perihelion (farthest/closest) positions of Earth relative to the sun nearly coincide; the solstices are June 21 and December 21, while the aphelion and perihelion are, respectively, July 4 and January 3. Earth, however, behaves like a wobbling top, and its **precession**, the alignment of its axis of diurnal rotation with its distance from the sun, oscillates with an average period

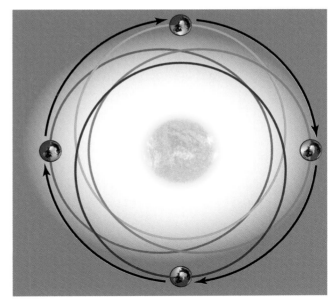

FIGURE 3.11 Precession of Earth's solar orbit The planet is shown at the aphelion (farthest distance) of various orbits. The red ellipse represents the current situation, in which the aphelion occurs during the Northern Hemisphere's summer. The green ellipse represents the situation 10,500 years ago, when the aphelion occurred during the Southern Hemisphere's summer. The orange and blue ellipses represent intermediate stages. Arrows indicate the rotation of the orbit ellipse over 21,000 years.

of about 21,000 years (**Figure 3.11**). That is, 10,500 years ago, Earth was closest to the sun on July 4, when the Northern Hemisphere was tilted toward the sun, and farthest from the sun on January 3, when the Northern Hemisphere was tilted away (**Figure 3.12**).

(A)

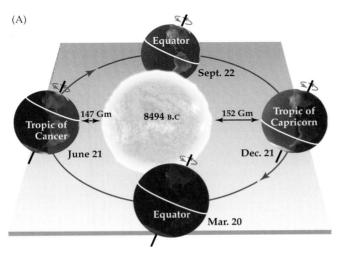

(B)

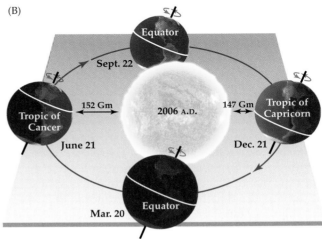

FIGURE 3.12 Precession of Earth's solar orbit (A) This alignment occurred 10,500 years ago, when Earth was closest to the sun, during June–July (147 Gm = 147 million km). (B) This alignment is the present situation, where Earth is closest to the sun during December–January.

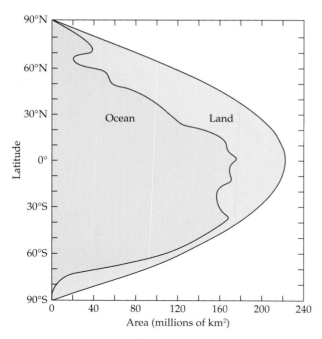

FIGURE 3.13 Areas of ocean and land at different latitudes The Southern Hemisphere has relatively little landmass at the mid-latitudes.

How well does this theory fit the observed climatic variations? The $\delta^{18}O$ signature of sea-bottom sediment cores, a temperature proxy (see Chapter 2), fluctuates at regular intervals. This periodicity reflects changes in obliquity and eccentricity; the influence of precession on $\delta^{18}O$ is less evident (see Figure 3.9). More specifically, the Northern Hemisphere experienced a double whammy about 10,500 years ago: As previously mentioned, it was tilted away from and was farther from the sun in the months of December and January and was tilted toward and closer to the sun in the months of June and July (see Figure 3.12). Colder winters and warmer summers in the Northern Hemisphere produced less snowfall in winter and more snowmelt in summer and thereby led to an interglacial period (**Figure 3.14**) (Rind 2002).

Currently, the obliquity of Earth's orbit is intermediate, its eccentricity is small, and its precessional alignment is such that Earth is farthest from the sun during June and July. These conditions lead to moderate and stable solar insolation in the Northern Hemisphere and explain the long intergalacial period that we are now experiencing. If orbital variations were the sole forcing factors, Earth's climate should remain about the same for the next 40,000 years or so (see Figure 3.14).

Milankovitch theory By combining the influences of the three orbital factors of obliquity, eccentricity, and precession, one can reconstruct a history of **solar insolation**, the amount of solar energy reaching Earth per unit area (see Figure 3.9). Solar insolation oscillates with the periods of its components: 41,000 years (obliquity), 100,000 years (eccentricity), and 21,000 years (precession). At the beginning of the twentieth century, estimates for the timing of these periods and of Earth's glacial cycles became sufficiently accurate to suggest a causal relationship between Earth's solar radiation balance and climate.

Solar insolation in the Northern Hemisphere has a greater effect than that in the Southern Hemisphere because the Northern Hemisphere currently has about 65% of Earth's landmass (**Figure 3.13**), and land absorbs much more solar energy than ocean (see the following section on albedo). Therefore, when the Northern Hemisphere receives more solar energy, so does the planet as a whole. The Serbian astrophysicist Milutin Milankovitch (1879–1958) proposed that when the solar insolation in the Northern Hemisphere was relatively high during December and January, and relatively low during June and July, more snow accumulated during the winter and less melted during the summer, causing glaciers to advance.

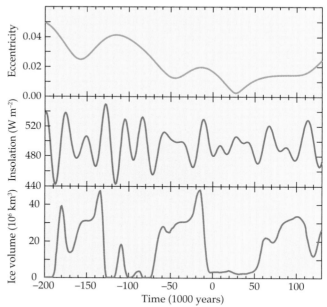

FIGURE 3.14 Past and predicted variation in natural forcing and climate Shown here are Earth's eccentricity, total solar energy impinging at latitude 65° north during June (insolation), and simulated Northern Hemisphere ice volume over time (thousands of years from the present). These calculations assume no human influence on the concentrations of greenhouse gases in the atmosphere. (After Berger and Loutre 2002.)

(A)

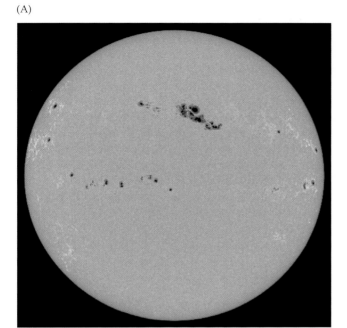

(B)

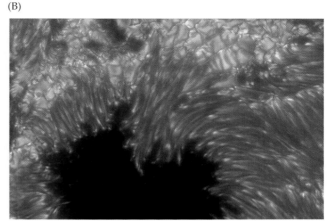

FIGURE 3.15 Sunspots (A) Largest sunspots of the current solar cycle. On March 29, 2001, sunspots spanned an area more than 13 times that of Earth. (B) Dark central region of a planet-sized sunspot. The bright filaments that extend into the sunspot have dark cores that are thousands of kilometers long but only about 100 km wide.

Sunspots

Periodic changes in the alignment between the sun's rotational axis and the gravitational center of the solar system produces intense fluctuations in vertical magnetic fields of the sun. These divert heat flow from deeper layers in the sun and generate patches of fluc-tuating temperatures on the surface that manifest as **sunspots** (**Figure 3.15**). Although Chinese astronomers recorded the presence of sunspots as early as 28 B.C., systematic counts of sunspots began with the invention of the optical telescope. These counts with telescopes show that the number of sunspots varies with about an 11-year period (**Figure 3.16**) as well as with some less-predictable longer cycles (**Figure 3.17**).

FIGURE 3.16 Number of sunspots A 1-year running average and 11-year running average compiled from various sources between 1612–1749 (Hoyt and Schatten 1998) and from the Zurich Observatory since 1749. (SIDC, RWC Belgium, World Data Center for the Sunspot Index, Royal Observatory of Belgium).

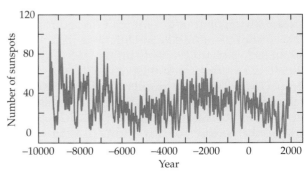

FIGURE 3.17 Sunspot numbers (10-year averages) reconstructed from $\Delta^{14}C$ values of tree rings Sunspot activity reflects the strength of solar magnetic fluxes that divert cosmic rays from striking Earth. Cosmic rays, when they collide with nitrogen atoms, generate ^{14}C that reacts with oxygen to produce radioactive $^{14}CO_2$. Trees may incorporate this $^{14}CO_2$. Thus, sunspot numbers are inversely proportional to ^{14}C levels in trees. (After Solanki et al. 2004.)

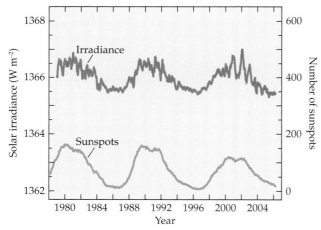

FIGURE 3.18 **Solar irradiance (energy per unit area)** **and sunspots** Total solar irradiance since 1979, as measured by radiometers on several spacecraft; shown are 81-day running averages. Number of sunspots at the Zurich Observatory; yearly running averages. (After Foukal et al. 2006; SIDC, RWC Belgium, World Data Center for the Sunspot Index, Royal Observatory of Belgium.)

One might presume that solar energy reaching Earth would decline during times of high sunspot activity, but the opposite is true. Heightened solar magnetic field activity that produces sunspots increases the overall brightness of the sun and more than compensates for the dark areas within the sunspots. Measurements from spacecraft since 1978 affirm that total solar energy oscillates 0.05% to 0.07% in synchrony with sunspot number (**Figure 3.18**).

Variations in solar energy of this magnitude should account for only about a 0.03°C change in global temperatures. By contrast, solar energy reaching Earth varies 6.5% between the perihelion and aphelion, 28% between seasons in the tropics, and 4.5% during the 41,000-year cycle of obliquity. Differences from other external factors should thus dwarf the 0.05% to 0.07% oscillations resulting from sunspots.

Nonetheless, sunspot number and mean surface temperatures of the terrestrial Northern Hemisphere are positively correlated (Usoskin et al. 2005). For example, the period from 1645 to 1715—the middle of the Little Ice Age when Europe, North America, and perhaps much of the world suffered bitterly cold winters—was largely devoid of sunspots (see Figure 3.16). One possible explanation is that during periods of low solar magnetic fluxes (i.e., few sunspots), up to 18% more cosmic rays penetrate the solar system and strike Earth (Usoskin et al. 2005). Cosmic rays, as they pass through Earth's atmosphere, may leave a trail of ions (charged atoms) that could serve as condensation centers on which cloud droplets form. Therefore, cosmic rays might stimulate the formation of clouds that reflect solar energy and cool the planet. Recent studies, however, have found little correlation between cosmic rays and cloud formation (Kristjánsson et al. 2008; Sloan and Wolfendale 2008).

Do sunspots significantly affect Earth's climate, and might they be responsible for some of the recent changes in climate (Kanipe 2006)? The current consensus is that the influence of sunspots, which is an external forcing factor, is smaller than the influence of internal forcing factors such as Earth's atmospheric composition (Foukal et al. 2006). The remainder of this chapter is devoted to a discussion of internal forcing factors.

Internal Forcing Factors

Processes intrinsic to Earth and its atmosphere also alter climate. They include **orogeny** (the process of mountain formation), **epeirogeny** (the formation and distribution of landmasses), volcanism, **albedo** (reflectance of solar energy), and atmospheric composition. Often these factors operate in conjunction with one another. For example, volcanic activity may affect both Earth's albedo and its atmospheric composition. The following sections introduce each type of internal forcing factor and evaluate its current influence.

Orogeny

Orogeny (*Greek*, mountain building) is the process in which **tectonic** movements of Earth's crust or volcanic activities form mountains (**Figure 3.19**). Mountains, especially those with a north–south orientation (e.g.,

FIGURE 3.19 **Origami orogeny** A "mountain fold" in origami where the paper is creased upward, creating a tent-like form.

the Sierra Nevada, Rocky, Appalachian, Andes, and Ural mountains), disrupt global atmospheric circulation patterns that generally move east–west because of Earth's rotation. Uplifting of mountains also newly exposes rock that undergoes chemical weathering and absorbs CO_2 (see Equation 2.6). Moreover, mountains tend to accumulate ice and snow that increases Earth's albedo. For these reasons, times of relatively rapid mountain building—say, from 40 million years ago, when the Himalayas and Sierra Nevada first arose, until today, as these ranges continue to uplift—are usually cooler periods.

FIGURE 3.20 June 12, 1991 eruption of Mount Pinatubo, in the Philippines

Epeirogeny

Epeirogeny (*Greek*, continent formation) is the formation of continents and ocean basins through deformations of Earth's crust from the movement of tectonic plates. As discussed previously (see Chapter 2), global distribution of landmasses determines the amplitude of glacial-interglacial cycles and, at the extreme, may foster a Snowball Earth (see Figure 2.13). Mid-ocean ridges, where most of the new plate material is produced, release large amounts of energy and greenhouse gases. Sea level rises and falls as new plate materials modify the shape of ocean basins.

Volcanism

In comparison to orogeny and epeirogeny, which proceed over many millions of years, volcanoes can have an explosive effect on climate. For example, Mount Pinatubo in the Philippines erupted on June 12–15, 1991 (**Figure 3.20**), after over four centuries of inactivity, and spewed vast amounts of sulfur dioxide and fine particles into the upper atmosphere. These materials quickly spread over most of the world (**Figure 3.21**) and formed a haze of aerosols in the upper atmosphere that reflected enough sunlight to lower global temperatures by as much as 0.5°C for nearly 2 years (see Figure 2.40B).

Volcanic eruptions as large as Pinatubo occur only every century or so, but smaller, yet significant, eruptions such as Mount St. Helens and El Chichón occur every decade (**Figure 3.22**). All of these eruptions release carbon dioxide (CO_2), a greenhouse gas that contributes to global warming, as well as sulfur dioxide and fine particles that reflect sunlight and contrib-

ute to global cooling. The amount of CO_2 released from volcanoes, however, averages about 10,000-fold less than that from the modern-day burning of fossil fuels. Therefore, on balance, volcanic eruptions cool, rather than warm, the planet.

Albedo

Albedo (*Latin*, white) is the percentage of solar energy reflected by Earth (see Chapter 2). The albedo of various materials ranges from about 85% for pure, fresh snow to 5% for asphalt parking lots or deep, still water (**Figure 3.23**). The global average is about 29%.

Satellite measurements provide a worldwide perspective on albedo (**Figure 3.24**). Clearly, deserts and snow-covered regions have a high albedo, whereas forests and agricultural fields have a low albedo. The overall trend seems to be that global albedo is decreasing (**Figure 3.25**), most likely due to diminishing snow and ice cover (see Figure 2.37). This change in albedo will increase the amount of solar energy absorbed by the planet.

Atmospheric composition

Electromagnetic radiation is the dominant form of energy that is exchanged among the sun, Earth, and void of space. The interaction between electromagnetic radiation and gases in Earth's atmosphere is the basis of the so-called greenhouse effect. Therefore, an exposi-

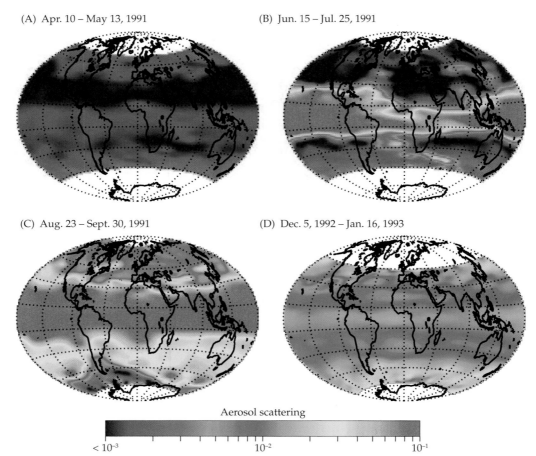

(A) Apr. 10 – May 13, 1991

(B) Jun. 15 – Jul. 25, 1991

(C) Aug. 23 – Sept. 30, 1991

(D) Dec. 5, 1992 – Jan. 16, 1993

Aerosol scattering

$< 10^{-3}$ 10^{-2} 10^{-1}

FIGURE 3.21 Atmospheric aerosols Measurements of aerosol scattering by the SAGE (Stratospheric Aerosol and Gas Experiment) II satellite. (A) The atmosphere was relatively aerosol-free before the eruption of Mount Pinatubo in the Philippines. (B) Aerosols in the tropics increased by almost a factor of 100 immediately following the eruption. (C) Aerosols had spread into Earth's mid-latitudes three months later. (D) Volcanic aerosols decreased slowly over the next several years.

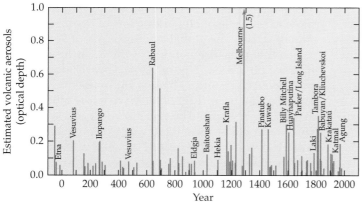

FIGURE 3.22 Estimated aerosols from volcanic eruptions Amount of aerosols (in terms of the negative logarithm of the relative amount of light scattering) estimated from sulfur deposited in the GISP2 ice core from Greenland. These data indicate the timing and size of major volcanic eruptions, although the locations of the eruptions are only sometimes known from historical accounts. (After Zielinski 2000; with additional volcano names from Kurbatov et al. 2006.)

tion on energy and electromagnetic radiation is necessary before embarking on an explanation of the greenhouse effect.

ENERGY AND ELECTROMAGNETIC RADIATION

Words seem to fail certain concepts from physics and chemistry. These concepts, however, fit into a consistent theoretical framework and provide the core equations of mathematical models that both accurately reconstruct Earth's climate history (see Chapter 2) and successfully predict tomorrow's weather (see Chapter 4). Energy is one of these concepts.

Energy, according to a classical definition, is the ability to do work, where work is a force applied over a distance, and a force is that which has the ability to change the motion of a particle. An alternative definition of energy, which is perhaps more pertinent to climate change, is that it is the capacity to change environmen-

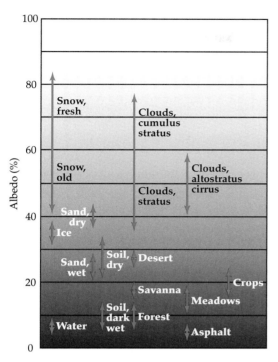

FIGURE 3.23   Albedo, reflection of solar energy, of various surfaces on Earth

tal conditions; for example, the capacity to change the temperature of a substance. The **conservation of energy law**, or the first law of thermodynamics, states that this capacity, this energy, is never created nor destroyed; it is just transformed from one form into another. Nearly all of the energy impinging on Earth is electromagnetic radiation. The conservation of energy law dictates that the amount of electromagnetic radiation entering Earth and its atmosphere must equal the amount leaving.

Electromagnetic radiation has a split personality. It exhibits properties of a particle (e.g., reflection and **photoelectric effect**, the emission of electrons from matter after it absorbs electromagnetic radiation) as well as those of a wave (e.g., **diffraction**, the bending of electromagnetic radiation around small objects, and **refraction**, the bending of electromagnetic radiation as it passes through different media). Thus, we declare electromagnetic radiation to be composed of particle waves.

As a particle, electromagnetic radiation always travels at the same speed in a vacuum: the speed of light ($c = 299,792,458$ meters per second). As a wave, electromagnetic radiation has a characteristic wavelength, the distance (λ, in meters) between successive peaks. It also has a characteristic frequency, the number of successive

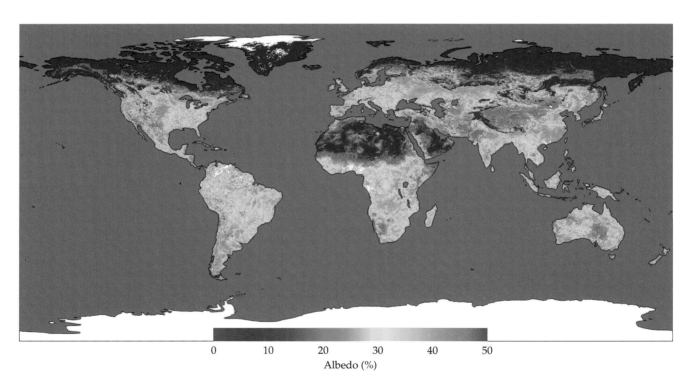

FIGURE 3.24   Albedo of Earth's landmasses, measured by the Terra satellite Shown are the average values from April 7, 2002 through April 22, 2002. White areas signify that data were not available.

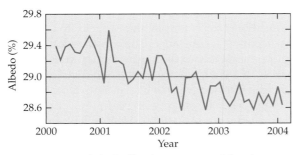

FIGURE 3.25 **Global albedo, measured by the Terra satellite** Shown are monthly values over a 4-year period after removing seasonal variation. The horizontal line indicates the average value over this period. (After Wielicki et al. 2005.)

peaks per unit time (ν, in cycles per second or Hz) (**Figure 3.26**). Either wavelength or frequency will suffice to define electromagnetic radiation because the product of wavelength and frequency must equal the speed of light ($c = \lambda \times \nu$): Given a wavelength, one can calculate its frequency ($\nu = 299{,}792{,}458 / \lambda$) and vice-versa.

Electromagnetic radiation extends from **radio waves** with long wavelengths (slow frequencies), on the order of kilometers (10^3 m), to **gamma rays** with short wavelengths (high frequencies), on the order of picometers (10^{-12} m). Visible light has a wavelength between 4×10^{-7} m and 7×10^{-7} m, and thus falls in a narrow band near the middle of the spectrum (**Figure 3.27**).

The energy of an electromagnetic particle wave is proportional to its frequency:

$$E = h \times \nu \qquad (3.1)$$

where
    E = energy in joules (J),

where one joule is the work done to produce the power of one watt continuously for one second
    h = Planck's constant
        = $6.6260693 \times 10^{-34}$ joule-seconds

    ν = cycles per second (Hz)

A radio particle wave due to its relatively slow frequency contains far less energy than a particle wave of visible light (a photon), which in turn contains far less energy than a gamma ray.

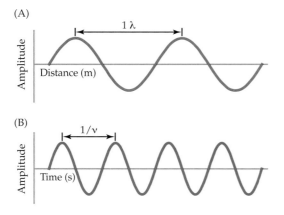

FIGURE 3.26 **Waves** (A) Wavelength (λ) is the distance covered by a full wave cycle. (B) Frequency (ν) is the inverse of the time period required per cycle.

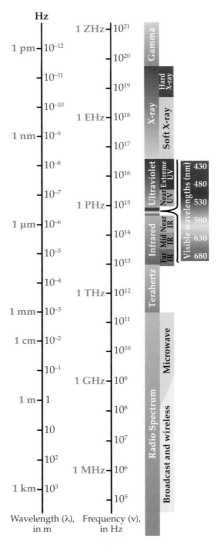

FIGURE 3.27 **Spectrum of electromagnetic radiation** Wavelengths and frequencies are shown on a logarithmic scale. Hertz (Hz) is an abbreviation for cycles per second, a measure of frequency.

Temperature is a measure of the extent to which molecules in a substance move around. Vibrating molecules in a substance emit electromagnetic radiation depending on their temperature (to the fourth power) according to the equation

$$\frac{P}{A} = \varepsilon\sigma T^4$$

(3.2)

where

P = power = energy per unit time in watts ($J\ s^{-1}$)

A = area in m^2

ε = **emissity,** the efficiency with which a substance emits electromagnetic radiation, ranging from 0 (inefficient) to 1 (efficient)

σ = Stefan-Boltzmann constant
= 5.6704×10^{-8} W m^{-2} K^{-4}

T = temperature in degrees kelvin (K)

Molecular movement ceases in an object at absolute zero (0 K = –273.15°C), and it no longer emits electromagnetic radiation.

For a less extreme example, take human beings. In the parlance of physics, we all behave like nearly perfect **blackbodies**. This means that people are almost ideal emitters of electromagnetic radiation, having an emissity of 0.98. An average person has a surface area of about 2 m^2, and naked skin has a temperature of about 32°C (305 K). Therefore, the net electromagnetic radiation from a naked person in a 20°C (293 K) room is

$$\begin{aligned}
P_{\text{net}} &= A\varepsilon\sigma\ [(T_{\text{skin}})^4 - (T_{\text{room}})^4]\\
&= (2\ \text{m}^2)(0.98)(5.67 \times 10^{-8}\ \text{W m}^{-2}\ \text{K}^{-4})\\
&\quad \times [(305\ \text{K})^4 - (293\ \text{K})^4]\\
&= 143\ \text{W}
\end{aligned}$$

In other words, each of us is only slightly brighter than a 100-watt lightbulb.

Rest assured that you differ from a lightbulb, at least in respect to your spectral qualities. The wavelength at which a substance emits the most electromagnetic radiation is inversely proportional to its temperature.

$$\lambda_{\text{max}} = \frac{2.8977685 \times 10^{-3}\,\text{m K}}{T\ (\text{K})}$$

(3.3)

A standard 100-watt incandescent bulb has a color temperature of about 2870 K and thereby emits most of its electromagnetic radiation around a wavelength (λ_{max}) equal to 1.01×10^{-6} m (1.0 μm) in the near **infrared**, but also emits enough radiation in shorter wavelengths from 4×10^{-7} m to 7×10^{-7} m (that is, 0.4 μm to 0.7 μm) that are visible to the human eye to make it useful as a lighting source. A person with a skin temperature of

FIGURE 3.28 Infrared image of a person running through a woods at night.

305 K has a $\lambda_{\text{max}} = 9.5 \times 10^{-6}$ m, a wavelength in the mid-infrared and emits negligible amounts of radiation in visible wavelengths. For this reason, only cameras sensitive to mid-infrared radiation can capture the image of a person in the dark (**Figure 3.28**).

The sun has an effective color temperature of 5780 K and emits a maximum amount of energy at 5.01×10^{-7} m (0.501 μm), a wavelength that we perceive as the color green (**Figure 3.29**). Over half of the electromagnetic radiation from the sun is in wavelengths from 0.4 μm

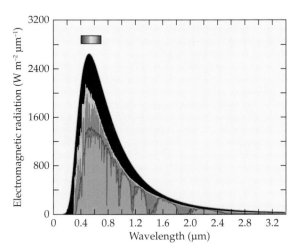

FIGURE 3.29 Solar electromagnetic radiation Relative intensity at various wavelengths for a perfect blackbody at 5780 K (black), for sunlight when monitored just outside Earth's atmosphere (orange-blue), and for sunlight at Earth's surface (red-green). The rainbow inset shows the wavelengths of visible light. Differences in the spectra derive from absorption by gases in the atmosphere.

FIGURE 3.30   A Texas blind salamander (*Eurycea rathbuni*), an endangered species found only in the Edwards Aquifer underneath the city of San Marcos, Texas.

to 0.7 µm that are visible to the human eye. This is not happenstance. Not only do particle waves with these wavelengths have energies suitable to empower well-behaved biochemical reactions, but having an organ (i.e., eye) that is most sensitive to the prevalent solar wavelengths serves surface dwellers like us very well. In contrast, cave dwellers have no need for a human type of eye because their natural habitat is devoid of solar wavelengths (**Figure 3.30**).

Electromagnetic radiation is subject to one or more of three fates when it strikes an object: It may be transmitted/refracted, reflected, or absorbed (**Figure 3.31**). Transmitted/refracted particle waves may change direction depending on the material through which they are passing and their angle of entry into the material, but they do not change frequency (energy). Reflected parti-

cle waves change direction, but not frequency. Absorbed particle waves change direction and frequency, although the total energy must remain the same (according to the conservation of energy law). All three processes can occur concurrently. For example, when sunlight strikes a piece of glass, some might be reflected (depending on the **angle of incidence**), some will be absorbed (the glass will become warmer), and most will be transmitted/refracted (pass through the glass).

Another perspective of these processes comes from quantum chemistry. An object transmits or refracts (lets through) certain wavelengths of electromagnetic radiation because its molecules fail to interact with particle

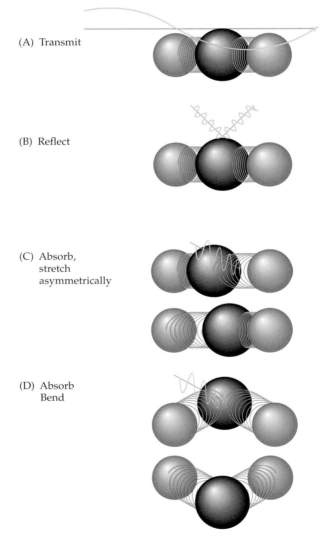

(A) Transmit

(B) Reflect

(C) Absorb, stretch asymmetrically

(D) Absorb Bend

FIGURE 3.32 CO_2 molecules interacting with particle waves Red spheres depict oxygen atoms, black spheres depict carbon atoms, green "springs" depict the chemical bonds between them, and the yellow/gray lines depict particle waves.

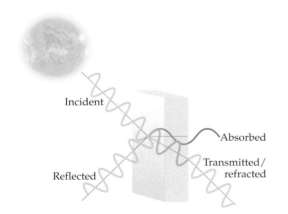

Incident

Absorbed

Transmitted/ refracted

Reflected

FIGURE 3.31   Fate of sunlight when it strikes an object Light from the sun (incident light) may be transmitted/refracted, reflected, or absorbed. When an object absorbs sunlight, the electromagnetic radiation that it then emits has longer wavelengths than the incident radiation.

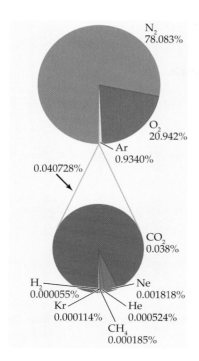

FIGURE 3.33 Composition of normal dry air
The lower pie chart is an expansion of the yellow sliver representing 0.040728% on the upper chart. The charts do not show water vapor (1% – 4%), nitrous oxide (0.00003%), xenon (0.0000009%), or ozone (up to 0.0000007%).

waves of those wavelengths (**Figure 3.32A**). Reflection is an elastic interaction whereby particle waves bounce off the molecules in an object (**Figure 3.32B**). An object absorbs electromagnetic radiation when particle waves of particular wavelengths resonate with molecules in an object and thereby change the electrical, vibrational (**Figure 3.32C,D**), rotational, or translational energies of molecules.

Most of the gases in Earth's atmosphere are inert (**Figure 3.33**). The bulk (78% of dry air) are nitrogen molecules (N_2), and substantial portions are the **noble gases** argon (0.93%), neon (0.0018%), helium (0.00052%), krypton (0.00011%), and xenon (0.0000009%). These gases remain prevalent in the atmosphere because they neither engage in chemical reactions with other elements, nor do they interact with solar electromagnetic radiation; that is, they are totally transparent. Of the remaining gases,

- Oxygen molecules (O_2), which constitutes 21% of dry air, absorb in the visible and near infrared (**Figure 3.34**),

- Water vapor (H_2O), which normally constitutes anywhere from 1% to 4% of the atmosphere, absorbs at a number of wavelengths in the visible, near infrared, and mid-infrared,

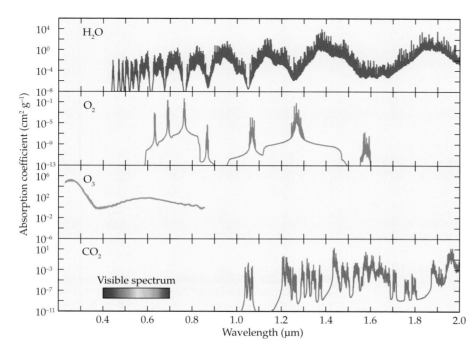

FIGURE 3.34 Absorption coefficients for atmospheric gases as a function of wavelength Shown are the coefficients for H_2O, O_2, O_3, and CO_2 in the near ultraviolet, visible, and near infrared regions. Notice that the absorption coefficients are plotted on a logarithmic scale and are proportional to the amount of the gas (g^{-1}). Other atmospheric gases (methane and nitrous oxide in particular) do not significantly absorb electromagnetic radiation in these wavelengths. (After Ehrenreich et al. 2006.)

FIGURE 3.35 Absorbance of mid-range infrared wavelengths by greenhouse gases in Earth's atmosphere Shown are the spectra for CO_2, CH_4, N_2O, O_3, and H_2O. (Data from the NIST Standard Reference Collection.)

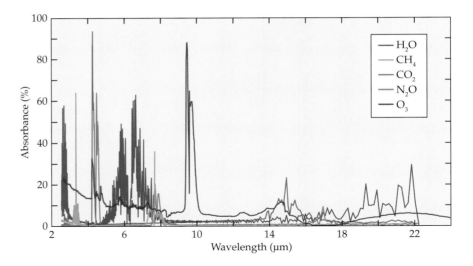

- Ozone (O_3), which constitutes up to 0.0000007% of dry air, absorbs broadly throughout the ultraviolet, visible, near infrared, and mid infrared wavelengths, and

- Carbon dioxide (CO_2; 0.038%), methane (CH_4; 0.00018%), and nitrous oxide (N_2O; 0.000032%) transmit visible wavelengths and absorb at various wavelengths in the near- and mid-infrared (**Figure 3.35**).

Absorption by all of these gases as solar radiation passes through the atmosphere accounts for the notches in the solar radiation that ultimately reaches Earth's surface (see red-green line in Figure 3.29).

Earth acts approximately like a blackbody with an effective temperature of about 287 K (14°C) and thereby emits a maximum amount of energy at 10.1 μm (**Figure 3.36**), a wavelength that falls in mid-infrared. Several gases in the atmosphere absorb some of the wavelengths that Earth emits. In particular, CO_2, CH_4, and N_2O are considered greenhouse gases because they transmit the strongest wavelengths from the sun (0.3 μm to 1 μm) and are thereby transparent to the human eye, but they absorb major wavelengths emitted from Earth (2 μm to 8 μm).

When gas molecules absorb infrared radiation, their vibrational energy increases, and so does their temperature. Warmer gas molecules emit electromagnetic radiation at shorter, more energetic wavelengths than do cooler gas molecules (see Equations 3.2 and 3.3). In accordance with the conservation of energy law (which states that energy of the incident electromagnetic radiation equals the energy of the emitted electromagnetic radiation plus the kinetic energy in the motion of the gas molecules), the wavelengths emitted by the warmer gas molecules are longer and less energetic than those originally absorbed. These gas molecules, however, move more rapidly after absorbing electromagnetic radiation and collide with

other gas molecules that, in turn, move more rapidly and emit additional electromagnetic radiation.

Putting this all together, a volume of gas molecules emits electromagnetic radiation proportional to its temperature. When illuminated with electromagnetic radiation of appropriate wavelengths, some of the molecules in the volume will absorb this energy and increase their movements. Through collisions of these molecules with their neighbors (molecular "moshing"), movement of the volume as a whole increases. The gas volume, therefore, will increase in temperature and emit electromagnetic radiation at wavelengths that are shorter than the gas volume emitted before illumination, but longer than the incident radiation. Energy is conserved because the energy in the shorter-

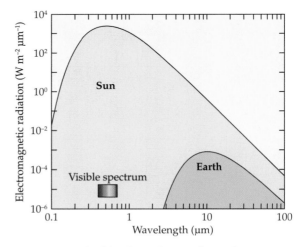

FIGURE 3.36 Blackbody radiation from the sun (5780 K) and Earth (287 K) Note that both axes are presented on logarithmic scales; electromagnetic energy from the sun is over 3 million times greater than that from Earth.

FIGURE 3.37 A thought experiment in which the sun illuminates Earth in a bottle.

wavelength electromagnetic radiation absorbed by a few molecules equals the total energy in the longer-wavelength radiation emitted by many molecules.

GREENHOUSE EFFECT Imagine placing Earth in a glass bottle where the only possible energy transfers are through electromagnetic radiation (**Figure 3.37**). Solar radiation either passes through the bottle with only slight changes in direction (is transmitted/refracted); bounces off the bottle, changing direction but not energy (is reflected); or is absorbed by the content of the bottle. The radiation absorbed stimulates the motion of molecules in the bottle, increasing the temperature of the bottle and its contents.

At a higher temperature, the bottle and its contents emit both more and shorter wavelength electromagnetic radiation than at a cooler temperature. This thermal radiation from the bottle and its contents has a broad spectrum. Depending on wavelength, it either passes out of the bottle, bounces back off the inside of the bottle and remains inside, or is absorbed by the bottle and its contents. Radiation absorbed by the bottle and its contents will warm the bottle and its contents, and so they will emit still more electromagnetic radiation.

Eventually, all of the solar energy that enters the bottle escapes. The conservation of energy law dictates that the energy coming in must equal the energy going out. Most of the energy enters as visible light, of relatively short wavelengths (0.4 μm to 0.7 μm). Most of the energy escapes as infrared wavelengths that are not absorbed by the bottle or its contents; in specific, most of it escapes through the "windows" between 8 μm and 9 μm, between 10 μm and 14 μm, and between than 23 μm and 500 μm (see Figure 3.35).

If solar energy increases or the contents of the bottle absorb more infrared radiation, say, from higher CO_2 concentrations, the bottle and its contents will warm and emit more thermal energy at wavelengths that they themselves do not absorb. If the bottle reflects more solar radiation (i.e., its albedo increases), less solar energy enters the bottle, and the bottle and its contents will cool until the thermal energy that they emit comes into equilibrium with this decrease. If liquid in the bottle evaporates and the vapor floats upward, energy will transfer from the bottom to the top of the bottle. If warm air rises in the bottle, energy will transfer from the bottom to the top.

An annual **energy budget**, which divides the planet into atmospheric components (i.e., atmosphere and clouds) and surface, takes into account these major energy transfers (Kiehl and Trenberth 1997; Trenberth et al. 2008). On average, Earth receives 341 W m^{-2} of electromagnetic radiation from the sun annually (**Figure 3.38**). Of this radiation, about 30% (102 W m^{-2}) is reflected, 79 W m^{-2} by atmospheric components and 23 W m^{-2} by Earth's surface. Atmospheric components and Earth's surface absorb 78 W m^{-2} and 161 W m^{-2}, respectively, of the 70% (239 W m^{-2}) that is not reflected.

Earth's surface receives a total of 494 W m^{-2}: 161 W m^{-2} from the sun plus 333 W m^{-2} of longer wave radiation from the atmosphere and clouds. Water on the surface absorbs some of this energy and evaporates (changes state from liquid to gas); the resultant water vapor may rise in the atmosphere (a process called **evapo-transpiration**). Another portion of the energy received by the surface heats the air near the surface and generates winds or thermal air currents (**thermals**) that rise in the atmosphere. Consequently, of the 494 W m^{-2} received, Earth's surface transfers 356 W m^{-2} of longer wave radiation to the atmosphere and clouds, 40 W m^{-2} of longer wave radiation to outer space, 80 W m^{-2} via evapo-transpiration to the atmosphere, and 18 W m^{-2} via thermals to the atmosphere (Trenberth et al. 2008).

Earth's atmosphere and clouds receive a total energy input of 532 W m^{-2}. This includes 78 W m^{-2} of solar energy, 356 W m^{-2} of longer wave radiation from Earth's surface, 80 W m^{-2} via evapo-transpiration, and 18 W m^{-2} via thermals (see Figure 3.38). The energy output to outer space from the atmosphere is 169 W m^{-2} and from clouds is 30 W m^{-2}; this, together with the 333 W m^{-2} of infrared radiation emitted to Earth's surface, equals a total energy output of 532 W m^{-2} from the atmosphere and clouds.

Notice that the energy output equals the energy input for each compartment, in accordance with the conservation of energy law. Also notice that energy exchanges between Earth's surface and atmospheric components can exceed those between the sun and Earth. This may

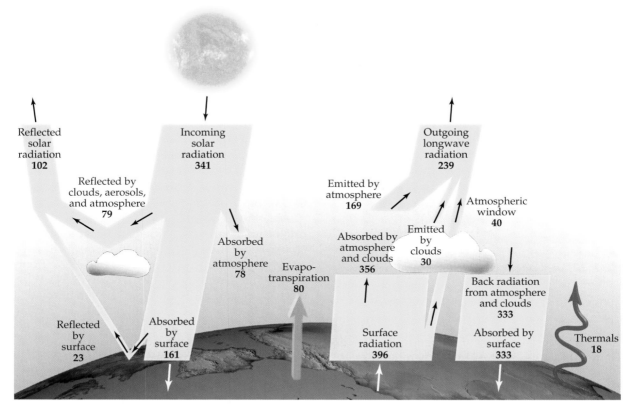

FIGURE 3.38 Earth's annual energy budget Values are in W m^{-2}. Arrows indicate the direction of energy flow. (After Kiehl and Trenberth 1997; Trenberth et al. 2008.)

be analogous to the relatively small number of transactions that occur between the federal treasury and a local bank, but the relatively large number of transactions that occur between the local bank and its borrowers and depositors; in all these transactions, the accounting books need to balance.

Without greenhouse gases such as CO_2, CH_4, or N_2O, Earth (its atmosphere and surface) would still absorb 239 W m^{-2} of shortwave solar radiation and would still emit 239 W m^{-2} of longwave infrared radiation that would pass through a transparent medium without further ado. Under such circumstances, Earth's surface would receive less back-radiation from the atmosphere and clouds, and its temperature would cool to an average of about –18°C (255 K), rather than the 14°C (287 K) that we currently enjoy. Such a greenhouse-gasless world would approximate a perfect blackbody having a maximum energy emission at a wavelength of 11.4 μm, in contrast to the 10.1 μm actually observed (**Figure 3.39**).

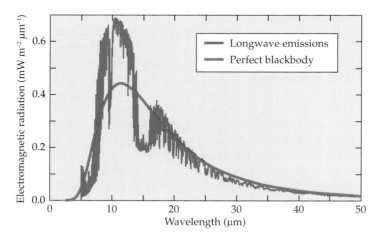

FIGURE 3.39 Earth's typical longwave emissions at the top of the atmosphere versus emissions of a perfect blackbody at 255 K. Values are in mW m^{-2} μm^{-1}. Note that the total amounts of radiation are similar for the two sources although the distribution among wavelengths differs. (After Dykema and Anderson 2006; longwave emissions data courtesy of John A. Dykema.)

TABLE 3.1 Properties of the major anthropogenic greenhouse gases

Gas	Common name	Concentration (ppb)[a]	Lifetime (years)[b]	Warming potential[c]
CO_2	Carbon dioxide	380,000	120	1
CH_4	Methane	1,850	12.0	25
N_2O	Nitrous oxide	324	114	298
CO	Carbon monoxide	130	0.25	2
CCl_3F	CFC-11[d]	0.26	45	4,750
CF_2Cl_2	CFC-12	0.54	100	10,900
$CHClF_2$	HCFC-22[d]	0.16	12	1,810
CCl_4	Carbon tetrachloride	0.10	26	1,400
CCl_2FCClF_2	CFC-113	0.08	85	6,130
CH_3Cl	Methyl chloride	0.06	1.3	16
CH_3CCl_3	Methyl chloroform	0.05	5	146
CH_2Cl_2	Dichloromethane	0.04	0.46	10
CH_3CClF_2	HCFC-142b	0.02	17.9	2,310
$CHCl_3$	Chloroform	0.01	0.51	30
SF_6	Sulfur hexafluoride	0.006	3,200	22,800

Source: IPCC 2001; Forster et al. 2007.
[a]Parts per billion.
[b]Mean residence time in Earth's atmosphere.
[c]Total radiative effect per molecule over 100 years relative to CO_2.
[d](CFC) chlorofluorocarbon and (HCFC) hydrochlorofluorocarbon, respectively, gases being phased out in compliance with the Montreal Protocol.

Earth with its greenhouse gases emits more radiation at shorter wavelengths than a blackbody at 255 K, but the planet emits less radiation at certain wavelengths, for example, those around 4 μm and 15 μm, at which carbon dioxide (CO_2) absorbs energy.

Greenhouse gases differ not only in the wavelengths of electromagnetic radiation that they absorb and the degree to which they absorb a particular wavelength (see Figure 3.35), but also in their concentrations and chemistry in Earth's atmosphere (Table 3.1). All greenhouse gases participate in chemical reactions that limit their residence time in the atmosphere. For example, CO_2 dissolves in water and forms carbonic acid (see Equation 2.5). We can compare the "warming potential" of greenhouse gases relative to CO_2 based on their mean residence time and absorption characteristics. The total effect of a greenhouse gas on Earth's energy budget thus depends on the concentration and warming potential of the gas.

In summary, greenhouses gases do not influence the total amount of electromagnetic energy that Earth emits to outer space. Rather they alter the temperature of the surface and the wavelengths that escape. Higher concentrations of greenhouse gases warm the surface and shift outgoing radiation to wavelengths at which the greenhouse gases are transparent. For example, higher concentration of CO_2 in the atmosphere would increase the depth of the "notch" at 15 μm (see Figure 3.39), but this would be compensated by a warmer planet that would emit more radiation at shorter wavelengths.

The term "greenhouse effect" is somewhat a misnomer. Real greenhouses are warmer than the surrounding air predominantly because they limit the exchange of energy through convection (exchange through the movement of air masses). For example, indium gallium arsenide glass is transparent to both infrared and visible radiation and is used in lenses for infrared cameras costing tens of thousands of dollars. Were we to construct a greenhouse from such an expensive material, it would still limit convection and be nearly as warm as one constructed from common borosilicate glass that transmits visible radiation but absorbs infrared. Nonetheless, this book will continue to use the term "greenhouse effect" in reference to the differential absorption of visible and infrared radiation that is partly responsible for global warming.

Clouds

Clouds exhibit schizophrenic tendencies with respect to Earth's energy budget (see Figure 3.38). They reflect incoming solar radiation and thereby promote global cooling. Simultaneously, they absorb longwave radiation from Earth's surface and radiate some of it back

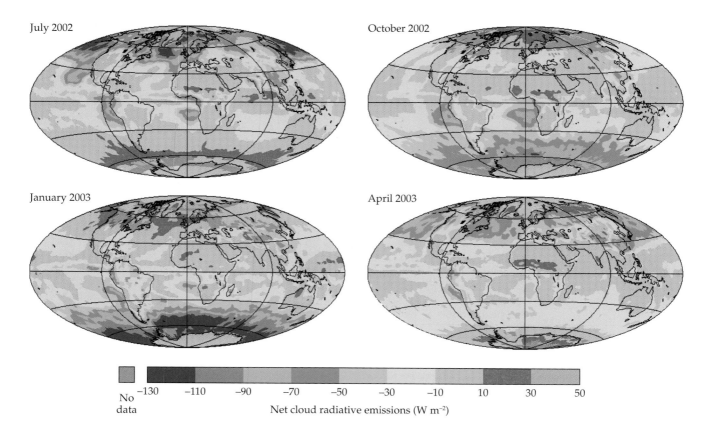

July 2002

October 2002

January 2003

April 2003

No data / −130 −110 −90 −70 −50 −30 −10 10 30 50

Net cloud radiative emissions (W m⁻²)

FIGURE 3.40 Net radiative emissions from clouds to Earth's surface Depicted are the monthly means for July, October, January, and April.

to the surface, thereby promoting global warming. On balance, however, clouds reflect more solar energy than they emit back to the surface. Therefore, the **net forcings** or the balance of electromagnetic emissions that transfer energy from clouds to the planet's surface—although they vary with location and season—are negative on average (**Figure 3.40**): Clouds generally cool Earth more than they warm it (Ramanathan et al. 1989).

As global temperatures rise, evaporation of water from the oceans increases exponentially (see Figure 2.43). This is partly responsible for the heavier cloud cover that has resulted over oceans (**Figure 3.41**) and has decreased net forcing from clouds (**Figure 3.42**), counteracting to some degree global warming, although warming still has occurred. General trends with clouds are apparent, yet mathematically describing cloud formation remains a major challenge in developing accurate computer models to predict global climate change. We will discuss this topic in Chapter 4.

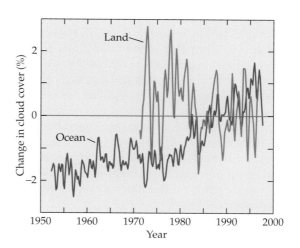

FIGURE 3.41 Trends in total cloud cover over the oceans (latitude 60° south to latitude 60° north) and over land in the mid-latitudes of the Northern Hemisphere (latitude 30° north through latitude 60° north). Plotted are 72-day averages. (After Norris 2005.)

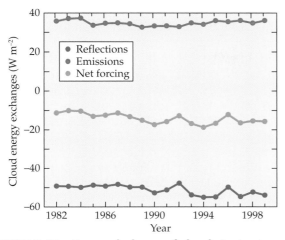

FIGURE 3.42 Energy balance of clouds in Arctic regions Depicted are the annual averages for cloud reflection of shortwave solar radiation, cloud emissions of longwave radiation back to the surface, and net energy exchange from clouds or cloud forcing, based on satellite measurements. (After Wang and Key 2003, 2005.)

Ocean currents

Ocean waters circulate around the globe in established patterns or currents (**Figure 3.43**) that derive from the following factors:

- Differences in solar energy received by the equator and the poles

- Topography of the ocean floor and coastal landmasses
- Changes in seawater density
- Rotation of Earth around its axis
- Atmospheric winds

Seawater becomes denser as it cools or as its salinity increases from the removal of fresh water during evaporation or ice formation. Conversely, the density of seawater diminishes as it warms or as its salinity decreases with the influx of fresh water from rivers or melting of ice. The prevailing atmospheric winds develop from the interactions among Earth's rotation around its axis, the heating of air near Earth's surface near the equator, the cooling of this air as it rises in the atmosphere or drifts to higher latitudes, and the sinking of this air as it cools (**Figure 3.44**).

One major pattern of ocean currents (**Figure 3.45**), the **Thermohaline circulation** or so-called Global Conveyor Belt, involves the northward flow of warm surface waters from the Caribbean along the Atlantic coast of the United States. This is known as the Gulf Stream. The flow continues across the Atlantic Ocean toward Great Britain, a current called the North Atlantic drift. These warm currents contribute to the higher average temperatures of the East Coast of the United States and Europe, which are about 5°C warmer than other landmasses at the same latitude. In the North Sea, these waters become denser as their temperature

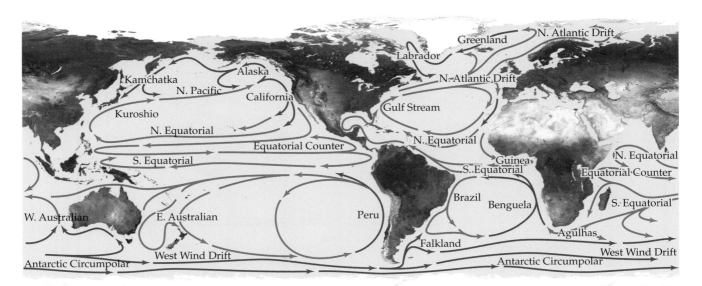

FIGURE 3.43 Currents near the surface of Earth's oceans Red lines depict the major warm currents; blue lines depict the major cool currents.

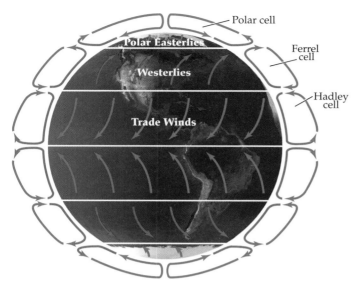

FIGURE 3.44 Major circulation patterns in the atmosphere Convective cells (Hadley cells near the Equator, Ferrel cells at mid-latitudes, and Polar cells) drive the prevailing warm (red) and cool (blue) winds.

drops and as they become more saline from evaporation of surface water and ice formation. The denser waters sink to between 1200 m and 5000 m in depth and flow south as the Greenland and Labrador Cur-

rents. Eventually they reach Antarctica and veer eastward. This deep, cold flow, the Antarctic Circumpolar Current, splits both around the Horn of Africa and around New Zealand and heads northward. Passing across the equator, these waters warm, rise to the surface, and join the general westward flow of the Equatorial Currents across the Pacific, the Indian, and, finally, the Atlantic oceans.

Polar regions have experienced rapid warming over the last 25 years, averaging an increase of between 1°C and 2°C per decade (see Figure 2.40C). This results from changes in the albedo when snow or ice melts, reveals bare ground, and increases the solar radiation absorbed. This increase is small in absolute magnitude but large in relative terms at the poles, where the total energy input is smaller. This warming trend at the poles has not only diminished the amount of fresh water captured in the Arctic Sea Ice (see Figure 1.15) but also has added to the volume of fresh water from rivers flowing into the Arctic Ocean. Both processes decrease the salinity of the waters in the North Atlantic and, thereby, their density.

Might these changes interfere with the Global Conveyor Belt? Would this wreak havoc with the Gulf Stream and North Atlantic drift that temper the climates of the East Coast and Europe? Might Manhattan

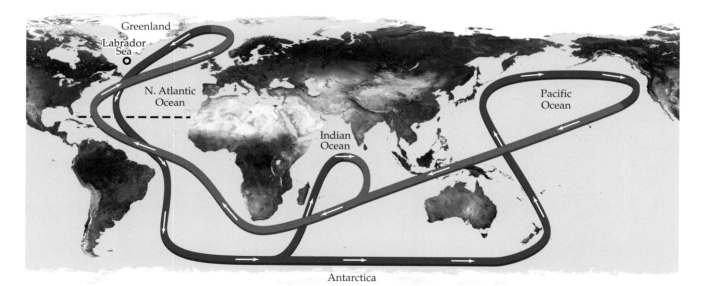

FIGURE 3.45 Global Conveyor Belt of ocean currents through which warm surface currents of the North Atlantic (red line) cool around Greenland and become more salty as water evaporates and icebergs form. Both processes increase the density of the water, which sinks in the Atlantic, forming a cold bottom current (blue line). This current circulates to Antarctica, until it warms in the Indian or Pacific ocean and again rises to the surface. The black circle in the central Labrador Sea denotes where satellite measurements of sea level since 1994 indicate decreased southward current flows. The dashed line between the North American and African continents denotes a transatlantic section where measurements over the last 50 years show slowing southward currents in the mid-ocean.

FIGURE 3.46   Scene from the Twentieth-Century Fox film *The Day After Tomorrow*, in which Manhattan freezes when the Global Conveyor Belt stops.

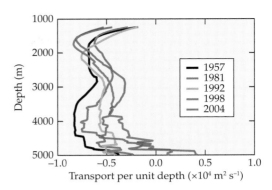

FIGURE 3.48   Flow of deep Atlantic currents Flow of cold water across latitude 25° north in the Atlantic Ocean between 1200 m and 5000 m in depth. Transects were conducted in five different years. Negative transport values indicate southward water movement, whereas positive values indicate northward movement. (After Bryden et al. 2005.)

experience the near-instantaneous freezing depicted in the Hollywood fantasy *The Day After Tomorrow* (**Figure 3.46**)?

Several lines of evidence indicate that the Global Conveyor Belt in the Northern Atlantic has slowed in recent years. Satellites have monitored sea surface heights all over the world with high precision for over a decade (see Figure 2.42). Because water flows downhill, differences in sea surface heights from place to place indicate the direction and magnitude of currents. The Labrador Current by this measure has been declining (**Figure 3.47**) (Han and Tang 2001; Häkkinen and Rhines 2004; Böning et al. 2006). Moreover, oceanographers have conducted transects across the Atlantic Ocean at latitude 25° north five times since 1957, measuring water temperatures, current flow, and current direction at various depths. Deep (below 3000 m), cold, southward currents were slower in the more recent samplings (**Figure 3.48**).

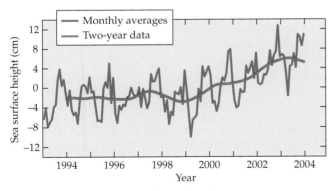

FIGURE 3.47   Flows of the Labrador Current, as indicated by surface height of the central Labrador Sea High sea level correlates with slow southward flows, because faster flows dissipate sea-level differences. Monthly averages and two-year data are merged from satellite-based altimeters. (After Böning et al. 2006.)

These recent changes in North Atlantic currents may merely reflect the large fluctuations in climate that naturally occur in this region, however. In particular, the North Atlantic Oscillation refers to the long-term fluctuations (ranging in period from decades to centuries) in the relative strengths and positions of a low atmospheric pressure system sitting near Iceland versus a high-pressure system sitting near the Azores (**Figure 3.49**). Large differences in the pressure between these two systems (a condition known as a "high index"), especially during the winter months, bring warm, moist westerly winds into Europe and, consequently, are associated with cool summers and mild winters in the region. By contrast, a small pressure difference (a "low index") suppresses westerly winds and induces more severe seasonal temperature swings in Europe.

The North Atlantic Oscillation since 1994 appears roughly correlated with changes in the surface height of the Labrador Sea (see Figure 3.49). The surface height is an indicator of the strength and direction of the Labrador Current. Detailed analyses of these data, however, have lead to divergent conclusions: Some investigators believe the contribution of the North Atlantic Oscillation to changes in the Global Conveyor Belt to be large (Latif et al. 2006), some consider it medium (Böning et al. 2006), and others believe it to be small (Häkkinen and Rhines 2004). Additional measurements over a longer time should provide more definitive answers about the relationships among currents, temperatures, and pressures in the North Atlantic and other regions.

In conclusion, most climatologists consider that the quick-freeze scenario for New York City depicted in natural disaster films (see Figure 3.46) is highly improbable

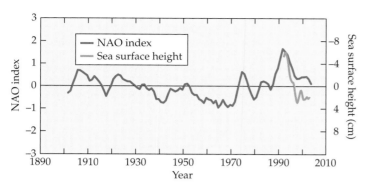

FIGURE 3.49 Index of the North Atlantic Oscillation (NAO) and surface height of the Labrador Sea over time This NAO index is based on an analysis of seasonal changes (December through March) in air pressure at sea level over a portion of the Atlantic Ocean. Depicted are smoothed 5-year running averages of the NAO index and satellite measurements of sea surface height. (Data courtesy of James W. Hurrell, National Center for Atmospheric Research.)

(Schiermeier 2006). Indeed, despite the weakening of the deep North Atlantic currents, temperatures in the North Atlantic and neighboring coastal regions are warming, rather than cooling (**Figure 3.50**). It would, thus, seem premature to give my regards to Broadway.

Storms

On November 30, 2006, the Atlantic hurricane season of 2006 ended without a single hurricane striking the United States. Contrast this with the 2005 season, when an unprecedented 15 hurricanes struck the United States, nearly a third of which attained category 5 status, in which winds exceed 280 km h^{-1} (174 mph). Worldwide, major storms—called hurricanes, typhoons, or cyclones, depending on location—numbered less than 60 in 2006, down from an annual average of about

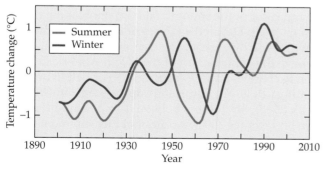

FIGURE 3.50 Temperature changes in southern Scandinavia in summer and winter The baseline (0) is the average for the entire period. Plotted are data smoothed by a mathematical function. (After Moberg et al. 2006.)

80 (see Figure 2.46). Clearly, the number and severity of storms vary widely from year to year (**Figure 3.51**), and an understanding of the multiple factors involved in their development is necessary for accurate prediction.

The primary source of energy that empowers storms is the condensation of water. When water condenses (that is, when it changes phase from a gas to a liquid), it releases large amounts of energy (539 cal per gram) to its surroundings. The 80 W m^{-2} (19 calories per second per m^2) released annually to the atmosphere through this process exceeds the energy that the atmosphere absorbs from the sun (see Figure 3.39). Conversely, when water evaporates, it removes energy (539 cal per gram) from its surroundings. (In comparison, metabolizing sugar to water and CO_2 releases 4000 cal per gram: To dissipate the heat generated from burning 1 gram of sugar, an athlete may sweat a little over 7 grams of water, if the sweat comes off as vapor.

Most major storms originate in the tropics (**Figure 3.52**) when water evaporates from the warm sea surface, rises in the atmosphere, and condenses in the relatively cool higher altitudes. A major storm siphons vast amounts of water vapor up along the wall of its eye (**Figure 3.53**) that releases roughly 130 × 10^{12} (trillion) calories per second when it condenses. The majority of this energy generates updrafts that lift storm clouds even higher, to cooler parts of the atmosphere, accelerating condensation. Another portion of this energy drives the ferocious winds that rotate in response to the spin of the planet (the Coriolis effect) like water swirling down a drain. The winds kick up water spray, and this, along with the low atmospheric pressures near the eye, enhances surface evaporation. This positive feedback—higher updrafts accelerating condensation and faster winds and lower pressures enhancing surface evaporation—spurs the growth of storms until they pass over land and are cut off from warm surface waters.

Six conditions foster the development of major tropical storms (Gray 1979):

1. Sea temperatures must be above 26.5°C (80°F) to a depth of 50 m. Warm temperatures promote evaporation of water from sea surfaces.

2. Air temperatures must cool rapidly with altitude. The amount of water vapor that an air mass can hold decreases exponentially with temperature (see Figure 2.43), and so declining temperatures promote condensation as an air mass rises.

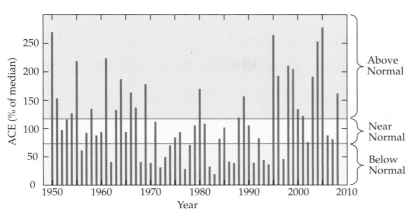

FIGURE 3.51 **North Atlantic hurricane season** The National Oceanographic and Atmospheric Administration's Accumulated Cyclone Energy (ACE) Index equals the sum of squares of the 6-h maximum sustained wind speed, in knots, for all storm periods. Blue bars are yearly values as a percentage of the 1951–2000 median value. (After Levinson and Lawrence 2008.)

3. Relative humidity must be high. An air mass must become saturated with water vapor before water will condense, and moist air is already close to this threshold.

4. A location must be more than 500 km (300 miles) away from the equator (Figure 3.52). At such a distance from the equator, the angle of Earth's rotation (**Coriolis force**) deflects winds from directly blowing into the eye of the storm and dissipating the low pressures there.

5. Vertical wind shears between the sea surface and the middle atmosphere must be less than 10 meters per second (22 mph). Faster vertical air movements disrupt the temperature gradient from the surface to higher altitudes and dries out air at mid altitudes, both of which prevent storms from expanding.

6. A storm must be already brewing. Without a preexisting storm, winds do not begin to spin, and air pressures do not begin to drop.

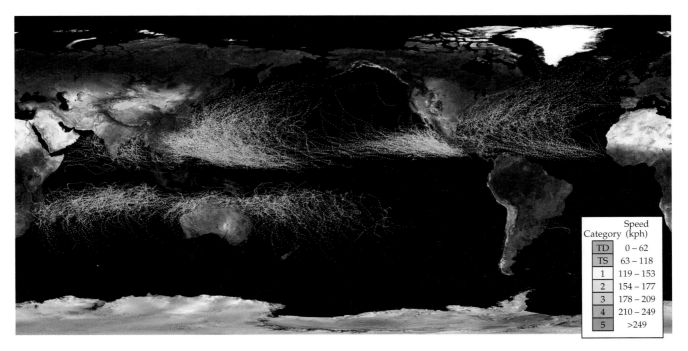

Category	Speed (kph)
TD	0 – 62
TS	63 – 118
1	119 – 153
2	154 – 177
3	178 – 209
4	210 – 249
5	>249

FIGURE 3.52 **Paths of major storms from 1985–2005** Storms originate over tropical waters, move westward, and then veer to the higher latitudes, where they lose strength. Points show the locations of the storms at 6-hour intervals, and their colors follow the key at the bottom right, where "TD" stands for tropical depression, "TS" for tropical storm, and "1" through "5" for categories of the Saffir-Simpson Hurricane scale.

FIGURE 3.53 Cross section of a major storm Red arrows indicate the upward movement of warm air; blue arrows indicate the movement of cool air.

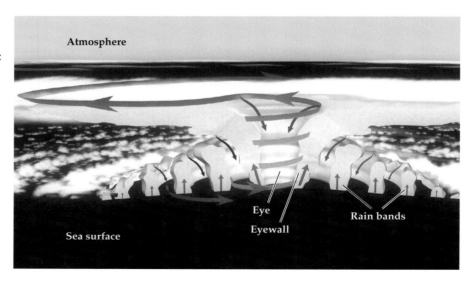

Climate models try to account for these factors, but the effects of wind shears and the need for a preexisting storm present formidable challenges to their accuracy in predicting the number of major tropical storms in a year. We address climate modeling further in Chapter 4.

Global climate change, through its warming of sea surface temperatures (see Figure 2.40C), should increase the severity of storms. Indeed, several studies have detected such a trend (**Figure 3.54**). Neverthe-

less, the record-breaking 2005 North Atlantic hurricane season followed by the 2006 season, or lack thereof, only serves to cloud the relationship between global temperatures and storms, and controversy still swirls around this issue (Witze 2006).

Summary

Forcing factors are responsible for changes in Earth's climate. External forcing factors are those that originate outside of Earth and its atmosphere and include galactic variations, orbital variations, and sunspots. Galactic variations derive from changes in energy received as our solar system rotates through the galaxy. The extent of such variations is highly uncertain. In contrast, orbital variations that derive from changes in the orientation between Earth and the sun are highly predictable via the cosine law, Kepler's second law, and the inverse square law. The cosine law dictates that solar energy per unit area of Earth depends on the angle between the sun's rays and the area being illuminated. This law helps explain climate differences between the equator and the poles and seasonal variations in weather. Kepler's second law explains why Earth spends less time orbiting close to the sun than far from it. Finally, the inverse square law defines the amount of solar energy received by Earth as a function of its distance from the sun. Earth's orbit around the sun oscillates in its obliquity (angle between the axis of Earth's daily rotation and orbit around the sun), eccentricity (elliptical shape), and precession (alignment of distance from sun and axial tilt). Past climate changes such as Ice Ages are highly

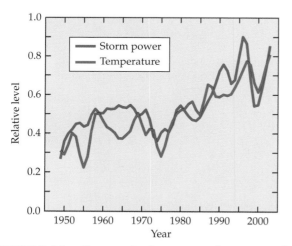

FIGURE 3.54 Changes in the power of storms and sea surface temperatures over 54 years Storm power was assessed as the integral over time of the maximum wind velocity cubed for all storms in the Atlantic and Western Pacific during a given year. Plotted are data smoothed by a mathematical function. (After Emanuel 2005.)

associated with these orbital oscillations. Sunspots, manifestations of fluctuations in the sun's magnetic field, are associated with periods of warmer weather.

Internal forcing factors, processes intrinsic to Earth and its atmosphere, also alter climate. Volcanoes may spew large amounts of particles high into the atmosphere that reflect sunlight and cool the planet's surface. Human activities are having a large influence on the internal factors of albedo and atmospheric composition. Albedo is the extent to which Earth reflects incoming solar energy. Melting of ice and snow is decreasing Earth's albedo and thus accentuating the amount of solar energy absorbed by the planet and global warming.

Nearly all of the energy that Earth exchanges with outer space is electromagnetic radiation. The conservation of energy law dictates that the amount of electromagnetic radiation entering Earth's sphere of influence must equal the amount leaving the sphere. Sunlight consists of electromagnetic radiation centered on the visible wavelengths (0.4 μm to 0.7 μm). Some of this radiation is reflected, some transmitted/refracted, and some absorbed by molecules in Earth's atmosphere and surface. When molecules absorb electromagnetic radiation, their movements increase and thereby their temperatures rise.

All molecules emit electromagnetic radiation in proportion to their temperature: as their temperatures rise, they emit shorter, higher-energy wavelengths. Earth, which has an average temperature of 287 K, emits electromagnetic radiation centered on infrared wavelengths of 10 μm. Greenhouse gases in Earth's atmosphere such as carbon dioxide, methane, and nitrous oxide transmit visible wavelengths from sunlight, but absorb the infrared wavelengths from Earth's surface. Higher atmospheric concentrations of greenhouse gases absorb more of the emissions from Earth's surface, warm the planet, and emit shorter, higher-energy wavelengths to outer space. This greenhouse effect — Earth retains more of the longer wavelengths and heats up, but emits more of the shorter wavelengths — still obeys the conservation of energy law.

Clouds both reflect sunlight to Earth during the day and absorb infrared radiation from the surface to warm it at night, but on balance clouds cool Earth more than they warm it. Ocean currents influence the distribution of temperatures across the face of the planet, and melting sea ice may be altering the strength and direction of these currents. Major storms such as hurricanes are empowered by evaporation from warm tropical seas followed by condensation of this vapor in the upper atmosphere. Warmer sea surface temperatures from global warming seem to enhance the severity of these storms.

Review Questions

1. Which one of the following is an example of an external forcing factor?
   (a) Reflections from clouds
   (b) Reflections from snow
   (c) Reflections from particles that volcanoes emit
   (d) Changes in Earth's albedo
   (e) Changes in Earth's axial tilt

2. The cosine law explains (select all that apply)
   (a) the relative amount of electromagnetic radiation per meter at Earth's perihelion (nearest distance to the Sun) versus aphelion (furthest distance from the Sun).
   (b) the rate of Earth's rotation at the perihelion and aphelion.
   (c) the orientation of sunbathers to the sun.
   (d) the relative amount of electromagnetic radiation per meter at the poles versus the equator.
   (e) part of the seasonal variation in temperature at a geographical location.

3. The Tropic of Cancer
   (a) is where the sun is directly overhead at noon during the fall and autumn equinoxes.
   (b) changes latitude between 22.1° and 24.5°N with a period of 41,000 years.
   (c) is located next to the Tropic of Capricorn.
   (d) is where melanomas are most likely to develop.
   (e) is the point of origin for the Global Conveyor Belt.

4. As the eccentricity of Earth's orbit increases,
   (a) the amount of solar energy reaching Earth fluctuates more from summer to winter.
   (b) seasonal variation derives solely from differences in the angle between the axis of Earth's diurnal rotation and its orbital plane around the sun.
   (c) the arctic ice cap will completely melt.
   (d) the Earth's orbit will align the planet closest to the sun on July 4 of each year.
   (e) the Northern Hemisphere will retreat into darkness.

5. Solar insolation in the Northern Hemisphere has a greater effect on global climate than that in the Southern Hemisphere because (select all that apply):
   (a) the Northern Hemisphere currently has about 65% of Earth's landmass.
   (b) the Arctic has more sea ice than the Antarctic.
   (c) land absorbs much more solar energy than ocean.
   (d) when the Northern Hemisphere receives more solar energy, so does the planet as a whole.
   (e) the Northern Hemisphere has more volcanoes than the Southern Hemisphere.

6. Albedo of Earth decreases as global warming
   (a) melts more polar ice.
   (b) enhances evaporation from warmer oceans.
   (c) increases average cloud cover.
   (d) expands mid-latitude deserts.
   (e) clears tropical rainforests.

7. Energy (select all that apply)
   (a) is neither created nor destroyed.
   (b) may be transformed from one form to another.
   (c) is the ability to do work.
   (d) is the capacity to change environmental conditions.
   (e) includes electromagnetic radiation.

8. Sunlight
   (a) consists of chemical energy that strikes the surface of Earth.
   (b) has a color temperature of 1000 K (blackbody radiation at 1000 K).
   (c) contains wavelengths only between 400 and 700 nm.
   (d) is not absorbed by atmospheric gases.
   (e) is the primary source of energy that supports life on Earth.

9. Earth's atmosphere contains
   (a) 78.1% oxygen, 20.9% nitrogen, 0.9% carbon dioxide, 0.038% argon, 0.002% neon, 0.0005% helium, and 0.0002% methane.
   (b) 78.1% argon, 20.9% oxygen, 0.9% nitrogen, 0.038% neon, 0.002% carbon dioxide, 0.0005% helium, and 0.0002% methane.
   (c) 78.1% nitrogen, 20.9% argon, 0.9% oxygen, 0.038% methane, 0.002% neon, 0.0005% helium, and 0.0002% carbon dioxide.
   (d) 78.1% nitrogen, 20.9% oxygen, 0.9% argon, 0.038% carbon dioxide, 0.002% neon, 0.0005% helium, and 0.0002% methane.
   (e) 78.1% nitrogen, 20.9% oxygen, 0.9% carbon dioxide, 0.038% methane, 0.002% neon, 0.0005% helium, and 0.0002% argon.

10. Earth's surface
   (a) releases substantial amounts of chemical energy to the void of space.
   (b) has a color temperature of 1000 K (blackbody radiation at 1000 K).
   (c) emits wavelengths only between 400 and 700 nm.
   (d) emits wavelengths that are not absorbed by atmospheric gases.
   (e) reflects a little over 10% of incident sunlight.

11. Cloud cover (select all that are *true*)
   (a) decreases on average with global warming.
   (b) reflects sunlight during the day.
   (c) absorbs electromagnetic radiation emitted from Earth's surface.
   (d) tends to warm Earth more than it cools it.
   (e) is straightforward to simulate in computer models.

12. Select all the factors that influence ocean currents.
   (a) Differences in solar energy received by the equator and the poles
   (b) Acidity of seawater
   (c) Topography of the ocean floor and coastal land-masses
   (d) Changes in seawater density
   (e) Rotation of Earth around its axis

13. The Thermohaline Circulation (select all that are *true*)
   (a) explains some of the climate differences between the east and west coasts of continents.
   (b) is slowing down where ocean salinity decreases with the melting of the arctic ice.
   (c) is speeding up at depths in the tropics.
   (d) is likely to become part of a Broadway musical.
   (e) influences the North Atlantic Oscillation.

14. Major storms do *not*
   (a) depend on the energy released when water vapor condenses.
   (b) spin because of Earth's rotation around its axis.
   (c) increase dramatically in numbers with global warming.
   (d) increase dramatically in intensity with global warming.
   (e) start to form over tropical waters.

Suggested Readings

Earth Observatory. 2009. *Climate and Earth's Energy Budget*. NASA, *http://earthobservatory.nasa.gov/Features/Energy-Balance/page1.php* (accessed July 4, 2009).

This series of web pages covers the basics of Earth's energy budget and its influence on climate.

Emanuel, K. 2006. Hurricanes: Tempests in a greenhouse. *Physics Today*, August: 74–75.

This short article discusses some of the controversy about trends in hurricanes.

Lee, J. A. 2009. *Milankovitch Cycles*. National Council for Science and the Environment, Encyclopedia of Earth, *http://www.eoearth.org/article/Milankovitch_cycles* (accessed May 28, 2009).

This is an example of the many informative entries available on the Encyclopedia of Earth web site.

Schlosser, P., S. Pfirman, M. Ting, and J. Smerdon. 2008. *The Climate System*. Columbia and Barnard Universities, *http://eesc.columbia.edu/courses/ees/climate/index.html* (accessed May 28, 2009).

These course materials provide an excellent introduction to climate.

4

WORLD OF TOMORROW: COMPUTER SIMULATION MODELS

The Old Farmer's Almanac has predicted the weather of the continental United States since 1792. "We derive our weather forecasts according to a secret formula that was devised by the founder of this Almanac, Robert B. Thomas.... Over the years, we have refined and enhanced that formula with state-of-the-art technology and modern scientific calculations.... Although neither we nor any other forecasters have as yet gained sufficient insight into the mysteries of the universe to predict the weather with *total* accuracy, our results are almost always very close to our traditional claim of 80 percent." (*The Old Farmer's Almanac* 2006).

Weather forecasting has a day-to-day and town-by-town focus. For example, TV weather reporters each evening must address the question, "Should you carry an umbrella to work tomorrow?" In contrast, climate forecasting takes a broader perspective. Climatologists may spend their entire career pondering, "Will Manhattan and the rest of the North Atlantic coast freeze solid during the coming decades?" Indeed, climatic forcing factors discussed in Chapter 3 range from the cosmic (spinning around the galaxy) to the mundane (sunbathing). Have

(A)

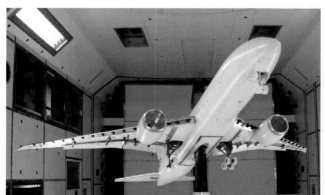

(B)

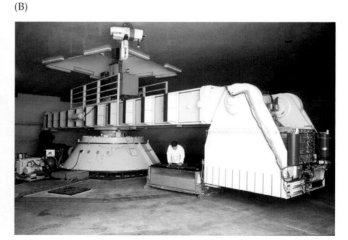

FIGURE 4.1 Physical models
(A) Model of a Boeing 787 airliner in the QinetiQ wind tunnel, Farnborough, U.K. (B) Giant centrifuge at the University of California, Davis with a 9.1m arm that can spin 4500 kg at 80 times normal gravity. (C) A view of the San Francisco Bay Model administered by the U.S. Army Corp of Engineers in Sausalito, California.

(C)

climate forecasters gained sufficient insight into the mysteries of the universe to predict Earth's climate with an accuracy that approaches *The Old Farmer's Almanac*'s claim of 80%?

One impediment to accurate climate forecasting is that some important factors are so slow, so large, so unapproachable, or so costly as to prohibit direct experimental manipulations. Study of such phenomena requires the development of models, facsimiles that are shorter in duration, smaller in size, more accessible, or less expensive than the originals. These facsimiles may be physical entities or mathematical abstractions.

Physical models are common. They include playthings such as dolls and toy trucks that mimic the functions of their full-size counterparts. Other physical models require additional adult supervision. For example, the Boeing Company conducts tests on scaled-down replicas

of their new 787 airliner (notice the holes in the leading edge of the wings that facilitate measurements of wind speeds) in large wind tunnels to simulate the airplane's performance under extreme conditions (**Figure 4.1A**). Professor Bruce L. Kutter (of the University of California, Davis), spins 4500 kg of dirt in a giant centrifuge at speeds of about 320 kph to help predict which buildings might survive the next major California earthquake (**Figure 4.1B**). In Sausalito, California, the U.S. Army Corp of Engineers built a model of the San Francisco Bay and Delta region to a 1/1000 scale in length and width and 1/100 scale in water depth to determine how diverting water from these bodies might influence water quality (**Figure 4.1C**).

When emulating the details of a complex system in a physical model becomes difficult, mathematical models may supplement or replace physical models. The wind

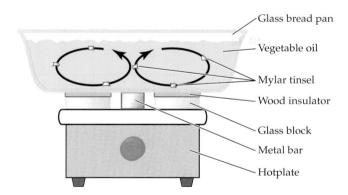

FIGURE 4.2   **Simple physical model of convective circulation**   A glass bread pan is filled with vegetable oil and placed on a hot plate, with a metal bar under the middle and an insulating block under each side. Bits of Mylar tinsel circulating in the oil help the observer visualize the convective cells that develop. (After Ebert et al. 2005.)

FIGURE 4.3   **A rotating tank used for studies of atmospheric motion**   at the U.S. Weather Bureau in Washington, D.C. in 1926 with Dr. Carl Gustav Rossby, who first explained the large-scale motions of the atmosphere in terms of fluid mechanics.

tunnel tests on the Boeing 787 and the giant centrifuge simulations of earthquakes are tightly integrated with mathematical models calculated on digital computers. Research on water quality in the San Francisco Bay and Delta region has relied entirely on computer simulations since 2001, and the 12,000-square-meter physical model in Sausalito has become primarily a tourist attraction.

General Circulation Models

Climate models have undergone a similar transition from physical to mathematical. Early studies of atmospheric and oceanic circulation patterns examined simple physical models. Sometimes, a container filled with fluid was placed on a hotplate that represented the warming of the tropics; cooler areas of the container represented the poles (**Figure 4.2**). In more elaborate physical models, a heated tank was mounted on a turntable (**Figure 4.3**) to mimic Earth's rotation (Coriolis force). These efforts became known as **general circulation models** (**GCMs**).

Physical GCMs, however, were limited in their abilities to account for the intricacies of atmospheric convection or ocean currents (see Figures 3.43 and 3.44), and climatologists soon turned to mathematical GCMs. Contemporary mathematical models depict Earth's climate in its entirety, and the acronym GCM also now stands for **global climate model**.

Digital computers, which were developed near the end of World War II, are well suited to the repetitive nature of the calculations in GCMs. The first large-scale digital computer, named ENIAC for *e*lectronic *n*umerical *i*ntegrator *a*nd *c*omputer (**Figure 4.4**), was used to

calculate trajectories of artillery shells and to design the first hydrogen bomb before it was applied to climate models. In 1950, ENIAC achieved real-time climate predictions whereby 1 minute of calculations estimated air movements for the next minute.

As digital computers became more sophisticated, so did the GCMs (**Table 4.1**). GCMs, for convenience, divide the world into boxes having a certain breadth and width (grid size) and a certain height determined by the number of vertical levels. They also treat the passage of time in a stepwise fashion whereby they cal-

FIGURE 4.4   **The ENIAC computer,** which began operation in 1947, was 2.4 m × 0.9 m × 30 m in size, and contained 17,468 vacuum tubes.

TABLE 4.1 History of GCMs and computers in the Met Office of the United Kingdom (Hadley Center)

Year	Computer	Calculations per second	Main memory	Grid (km)	Vertical levels
1959	Ferranti Mercury	3.0×10^3	1.0×10^3	Local only	2
1965	Eng. Electric KDF9	5.0×10^4	1.2×10^4	500	3
1972	IBM 360/195	4.0×10^6	2.5×10^5	300	10
1982	CDC Cyber 205	2.0×10^8	1.0×10^6	150	15
1991	Cray Y-MP C90/16	1.0×10^9	2.6×10^8	90	19
1997	Cray T3E 900/1200	1.5×10^{12}	3.6×10^{10}	60	38
2004	NEC SX-8/SX-6	2.0×10^{13}			

Source: http://www.metoffice.gov.uk/research/nwp/numerical/computers/history.html

culate conditions only at discrete moments separated by a certain time interval (time step). Generally, the accuracy of such models improves with a smaller grid size, more vertical levels, or a shorter time step, but these require additional computation time.

In 1969, one of the first mathematical models to couple atmospheric and oceanic processes engaged the most powerful computer of its time for 46 straight days (Weart 2003). Six years later, in 1975, the next generation of computer churned for 50 straight days on the next generation of this model, to simulate 300 years of global climate at a low resolution (500 km grid size, 3 vertical levels, 10-minute time step) (**Figure 4.5**). In 1994, another model running on a cluster of 352 computers, required 10 minutes to calculate the climate for 1 day at a higher **resolution** (200 km grid size, 9 vertical levels, 7.5-minute time step) (Mechoso et al.1994); at this rate, reconstructing 300 years of climate would

Solving the mathematical equations in GCMs seems to tax whatever computational resources are available (Weart 2003). Some of the first mathematical GCMs were developed by Lewis Fry Richardson (1881–1953). While serving as an ambulance driver during World War I, Richardson would work on differential equations in his head as a diversion from the horrors of the Western Front. In one such effort, he developed a primitive atmosphere model and, using just paper and pencil, took 6 weeks to calculate air movements across Western Europe over an 8-hour period.

require 760 days of computer time. In 2004, a cluster of 2560 computers in the Japanese Earth Simulator Center

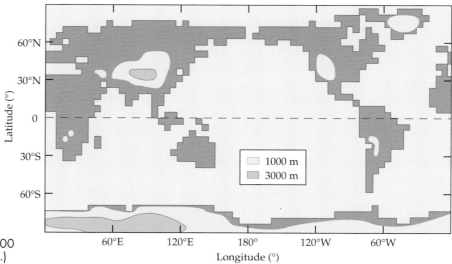

FIGURE 4.5 Grid size and vertical levels in a 1975 GCM Horizontal resolution was 500 km × 500 km, and elevations were divided by height: sea level (green), 1000 m (yellow), and 3000 m (orange). (After Manabe et al. 1975.)

FIGURE 4.6 **Figure 4.6 Main floor of the Earth Simulator Center in Yokohama, Japan** that has 160 processor nodes, each containing 8 vector-type arithmetic processors.

(**Figure 4.6**) required as little as 3 hours to simulate 1 day of climate at a very high resolution (3.5 km grid size, 54 vertical levels, 25-second time step) (Satoh et al. 2004); at this rate, a 300-year reconstruction, were it ever attempted, would require over 37 years of computer time. Clearly, GCMs have expanded to take full advantage of advances in computer science, and their insatiable appetite has been a major driving force spurring the development of even faster and larger machines.

Contemporary GCMs continue to balance the finite nature of computer resources against the need for simulating climate over long periods with a grid size that contains adequate spatial detail and a time step that provides acceptably smooth transitions from moment to moment. One approach is to run different parts of a model simultaneously on separate computer systems linked by high-speed networks. Another is to feed the results from a coarse-scale model that covers a large area such as the continent of North America to a fine-scale model that covers with a smaller grid and shorter time step a small region such as California (Collins et al. 2006). Such combined models expend only about 6 seconds of computer time per simulated day (Drake et al. 2005) and can thus calculate long-term climate reconstructions for a particular region in a more reasonable amount of time.

General Circulation Models: Core Concepts

Potentates from King Philip II of Macedon (382 B.C.–336 B.C.) to the present day have espoused a philosophy of divide and conquer as a solution to the world's prob-

lems. In computer science, a divide-and-conquer algorithm is one that splits a problem into an ever greater number of related subproblems until the solution to each becomes relatively simple. GCMs follow this philosophy in several respects.

First, GCMs divide the planet into a minimum of two compartments, atmosphere and ocean (**Figure 4.7**). Many GCMs also treat sea ice and land separately. Each compartment has its own model that characterizes processes intrinsic to the compartment and transfers of material and energy with other compartments. These models are in themselves large computer programs (having more than 100,000 lines of code) that undergo independent development. For example, atmosphere models may derive from any of ten distinct lineages (**Figure 4.8**), ocean models from at least five other lineages (Baum 2006), and sea-ice models from yet again six others (Washington and Parkinson 2005). Integrating all the pieces into a coherent semblance of Earth requires another large computer program called a coupler that coordinates both the timing and the format of data exchange among the atmosphere, ocean, sea ice, and land models (Craig et al. 2005).

Second, each compartment model is further divided into submodels (see Figure 4.7). In particular, an atmosphere model usually has discrete submodels for the dynamical core of fluid motions described in the following paragraphs; precipitation and cloud formation; atmospheric chemistry; aerosols such as smoke, smog, and haze; and characterization of physical processes such as radiation and convection. An ocean model usually has submodels for dynamical core of interior ocean fluid motions, wind stress on surface mixing layers, and, sometimes, fresh water inputs, such as river

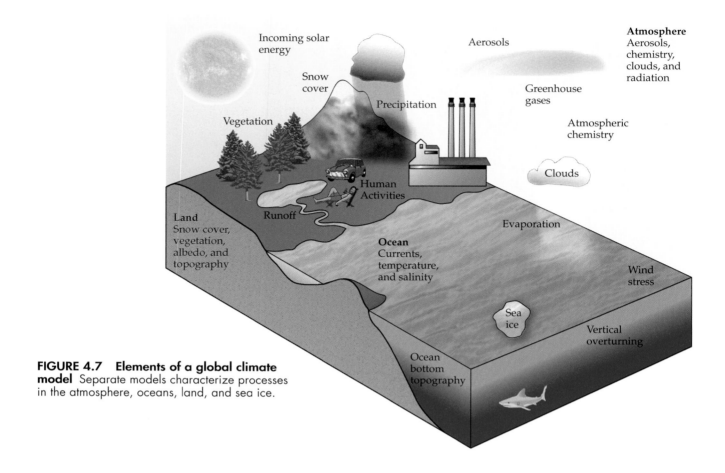

FIGURE 4.7 **Elements of a global climate model** Separate models characterize processes in the atmosphere, oceans, land, and sea ice.

runoff. A land model may have submodels for carbon cycling, vegetation, and albedo. Each submodel may again be divided into sub-submodels, and so on until the processes that are being simulated sufficiently diminish in complexity to become mathematically straightforward.

Third, as mentioned previously, GCMs divide the world into a patchwork of small boxes operating over a relatively short period of time. There are several types of grid box systems (**Figure 4.9**) and time steps, and sometimes a single GCM will use a different one for each submodel. No matter what grid box system or time step a GCM uses, the classical laws of physics apply to every grid box during every time step. These laws are the conservation of momentum (a mass changes its velocity only in proportion to the magnitude of outside forces), conservation of mass (the mass of a closed system will remain constant, regardless of the processes acting inside the system), and conservation of energy (the total amount of energy in a closed system remains constant).

Warning!

The following section contains "M" or "MA" material, namely, mathematical equations. If you find such material objectionable, please (1) focus on the descriptive text, (2) just view the equations as a form of hieroglyphics, (3) let your eye follow the sweeping forms of the symbols, and (4) notice the variables and constants that influence models.

Together, the following equations constitute the fundamental, or primitive, equations of fluid mechanics that are at the core of every GCM.

Conservation of momentum (Newton's second law) asserts that a mass changes its velocity only in proportion to the magnitude of outside forces:

$$F = m \cdot a \qquad (4.1)$$

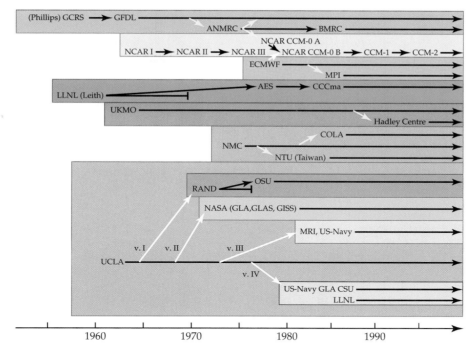

FIGURE 4.8   Genealogy of atmosphere GCMs A black arrow indicates a direct line of development, a white arrow indicates a major influence, and a bar indicates that development stopped. AES: Atmospheric Environment Service (Canada), ANMRC: Australian Numerical Meteorology Research Centre, BMRC: Bureau of Meteorology Research Centre (Australia), CCCma: Canadian Centre for Climate Modelling and Analysis, COLA: Center for Ocean-Land Atmosphere Studies (USA), ECMWF: European Center for Medium Range Weather Forecasts, GFDL: Geophysical Fluid Dynamics Laboratory (USA), GISS: Goddard Institute for Space Studies (USA), GLA: Goddard Laboratory for Atmospheres (USA), LLNL: Lawrence Livermore National Laboratories (USA), MPI: Max Planck Institut (Germany), MRI: Meteorological Research Institute (Japan), NCAR: National Center for Atmospheric Research (USA), NMC: National Meteorological Center (USA), NTU: National Taiwan University (Taiwan), OSU: Oregon State University (USA), UKMO: United Kingdom Meteorological Office (UK), and UCLA: University of California Los Angeles (USA). (After Edwards 2000; Canadian Centre for Climate Modelling and Analysis 2006.)

where

F = force

m = mass

a = acceleration = change in velocity over time

The major forces driving movements of the atmosphere, oceans, and sea ice are pressure differences on separate sides of a volume of air, water, or ice; Earth's rotation (Corolis force); gravity; and friction.

Several assumptions simplify the mathematical representation of air, water, and ice movements. A common assumption is that above a friction layer, the Coriolis force is responsible for the horizontal differences in air pressure across a volume of air (**geostrophic balance**). In such a case, the differential equations governing the conservation of momentum reduce to:

$$v = \frac{1}{a\cos\phi}\frac{1}{f\rho}\frac{\partial p}{\partial \lambda} \qquad (4.2)$$

$$u = -\frac{1}{a}\frac{1}{f\rho}\frac{\partial p}{\partial \phi} \qquad (4.3)$$

where

v = velocity in north/south direction tangent to Earth

a = Earth's mean radius

ϕ = latitude

f = Coriolis parameter

ρ = density

∂ = partial derivative

p = pressure = force per unit area

λ = pressure = force per unit area

u = velocity in east/west direction tangent to Earth

(A)

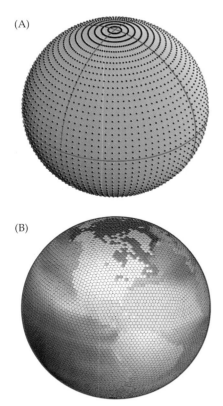

(B)

FIGURE 4.9 **Examples of grid box systems** (A) A latitude–longitude grid with the pole on top, red lines at the equator and at two constant longitudes, and dots representing grid cell centers. (B) A geodesic grid with the continents in white and a color-coded plot of observed sea-surface temperatures. This grid has 10,242 cells, each of which is roughly 240 km across. Twelve of the cells are pentagons; the rest are hexagons.

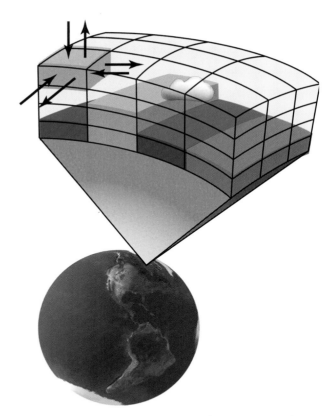

FIGURE 4.10 **GCMs characterize fluxes into and out of a grid box during a time step** Each box represents atmosphere, ocean, or land.

A further assumption is that the force of gravity is responsible for the vertical differences in air pressure across this volume of air (**hydrostatic balance**):

$$g = -\frac{1}{\rho}\frac{\partial p}{\partial z} \tag{4.4}$$

where

    g = gravitational acceleration

    ρ = density

    ∂ = partial derivative

    p = pressure

    z = altitude

Of course most people would not consider these, or any, differential equations simple, but the equations are even more daunting without these assumptions.

Conservation of mass or energy dictates that during each time step, the mass or energy entering a grid box minus the mass or energy leaving the box equals changes to the mass or energy within the box (**Figure 4.10**). The continuity equation is a mathematical expression of the conservation of mass:

$$\frac{\partial \rho}{\partial t} = -\frac{1}{a\cos\phi}\left[\frac{\partial}{\partial\lambda}(\rho u) + \frac{\partial}{\partial\phi}(\rho v\cos\phi)\right] - \frac{\partial}{\partial z}(\rho w) \tag{4.5}$$

where

    ∂ = partial derivative

    ρ = density

    t = time

    a = Earth's mean radius

    ϕ = latitude

λ = longitude

u = velocity in east/west direction tangent to Earth

v = velocity in north/south direction tangent to Earth

w = velocity in radial direction from Earth's center

z = altitude

The first law of thermodynamics is a mathematical expression of the conservation of energy:

$$C_p \frac{dT}{dt} - \frac{1}{\rho} \frac{dp}{dt} = Q \qquad (4.6)$$

where

C_p = specific heat at constant pressure

d = total derivative

T = temperature

t = time

ρ = density

p = pressure

Q = net heat gain

Finally, there are the **equations of state** (equations that relate density, pressure, temperature, and water balance). In the atmosphere, the ideal gas law provides the equation of state for dry air:

$$p = \rho RT \qquad (4.7)$$

where

p = pressure

ρ = density

R = gas constant

T = temperature

In the ocean, the equation of state takes the form:

$$\rho_w = f(T, S, p) \qquad (4.8)$$

where

ρ_w = seawater density

f = function of the enclosed parameters

T = temperature

S = salinity

p = pressure

This function for the equation of state for oceans is determined empirically.

Both atmosphere and ocean models incorporate versions of the preceding equations, but the spatial scales are larger in the atmosphere because air is thinner than water, and the temporal scales are longer in oceans because of greater thermal inertia. Moreover, seawater is nearly incompressible, and this simplifies the equations for continuity and the first law of thermodynamics in ocean models (see Equations 4.5 and 4.6).

In theory, a system containing six equations (Equation 4.2 through Equation 4.6, plus Equation 4.7 or Equation 4.8) and six dependent variables (u, v, w, ρ, p, and T) should be solvable once the constants (a, f, g, C_p, λ, ϕ, and R), the independent variable (Q), and the initial conditions (u_0, v_0, w_0, ρ_0, p_0, and T_0) are known. In practice, solving these equations proves challenging. Conditions often deviate from geostrophic and hydrostatic balances. Air is not dry. Water is not pure. Net heat gain (Q) depends on the multifaceted energy budget of Earth's surface and its atmosphere (see Figure 3.38). Appropriate initial conditions may not be obvious. Despite these difficulties, climate modelers approximate solutions to these differential equations via several approaches.

One approach, that of **finite differences**, examines changes in a variable such as temperature over small increments of space or time (e.g., $\Delta T/\Delta x$ or $\Delta T/\Delta t$, where T is temperature in degrees, x is distance in meters, t is time in seconds, and Δ indicates a small change in one of these variables) instead of infinitesimal ones (e.g., dT/dx or dT/dt, where d indicates an infinitesimal change in one of these variables). These changes correspond to the slopes of the curve in a plot of temperature as a function of position or time. If temperature is a well-behaved function of position or time (i.e., does not shift abruptly in value over space or time and is therefore infinitely differentiable), then a finite difference expression provides a reasonable approximation of a **derivative** (a measure of how a function instantaneously changes as its input changes), particularly as the interval for the approximation diminishes (**Figure 4.11**). Calculating slopes over a grid-box interval or a time-step interval can be far less demanding than calculating instantaneous slopes.

Another approach for solving differential equations is via **spectral transforms**. Any variable that changes periodically over space or time can be approximated by a Fourier series, which is the sum of a sequence of sine and cosine functions (**Figure 4.12**). This is the same process as the spectral analyzer on a music system that divides the sound into its various low (bass), medium (mid), and high (treble) frequency components. Equa-

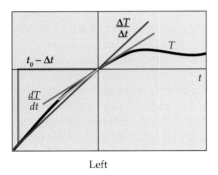

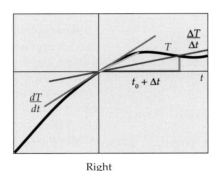

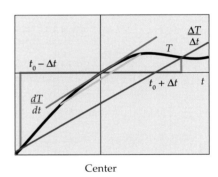

Left   Right   Center

FIGURE 4.11   Finite difference approximation of a derivative Plotted is the temperature (T) as a function of time (t) (black curve). The vertical axis in the middle of each plot represents a starting time (t_0). The derivative with respect to time (dT/dt) at t_0 equals the slope of the curve at that point (red line). A finite difference approximation ($\Delta T/\Delta t$) equals the slope of the line (blue line) for an interval (Δt) just before (left), just after (right), or on both sides (center) of t_0. Halving Δt provides a more accurate estimate (orange line) of the derivative.

tions based on sine and cosine functions are relatively easy to differentiate. When the dependent variables (u, v, w, ρ, p, and T) are well behaved (infinitely differentiable), spectral methods converge on the solution to a set of differential equations faster than finite difference methods. Moreover, spectral methods use a spherical coordinate system that may be less distorted than the x-y grid systems of finite difference methods, especially near the poles.

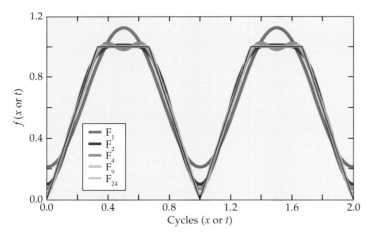

FIGURE 4.12   Fourier series approximation of a "flat tooth" function *f* (*x* or *t*) that is periodic over space (*x*) or time (*t*) Shown are the function (black) and approximations by a single cosine function (red, F_1), the sum of two cosine functions (blue, F_2), the sum of four cosine functions (green, F_4), the sum of nine cosine functions (orange, F_9), and the sum of 24 cosine functions (cyan, F_{24}). The more cosine functions in the series, the better the approximation. As a result of "spectral ringing," even the approximation with nine cosine functions (orange) has difficulty following the function when values change abruptly at the corners.

On the other hand, spectral methods may fail when the dependent variables become poorly behaved, as is the case with liquid water in clouds and surface topography. Under these circumstances, the number of sine and cosine functions needed to approximate the situation dramatically increases, and **spectral ringing** develops at the points where the variables change abruptly (see Figure 4.12). With spectral ringing, variables that should always be positive, such as mass, may become negative (Ringler et al. 2000). Moreover, distributing calculations over a cluster of computers (parallel processing) is often more difficult for spectral methods than for finite difference ones.

Many GCMs, such as the Hadley Climate Model, use finite difference methods. Other GCMs, such as that of the Canadian Centre for Climate Modelling and Analysis (CCCma), use spectral methods. Still others such as the Community Climate Model (CCM) from NCAR (National Center for Atmospheric Research, Boulder, Colorado) take advantage of both approaches, using finite differences for vertical dimensions and temporal dimensions and spectral transforms for horizontal dimensions.

Forcing Factors and Emission Scenarios

Climate forecasting requires insight into the predictability of forcing factors that influence Earth's climate (see Chapter 3). External forcing factors run the gamut in terms of predictability. The orbital variations of obliquity, eccentricity, and precession follow patterns set by the law of gravity that should continue indefinitely (see Figure 3.9). Sunspot numbers oscillate with an 11-year period (see Figure 3.16). In contrast, external factors such as comets crashing into Earth seem to be exceedingly rare, random occurrences (see Chapter 2) and therefore are often ignored in forecasts or treated as a very slight possibility.

Internal forcing factors also run the gamut. Albedo, the extent to which Earth's surface reflects incoming solar electromagnetic radiation, declines as global warming melts snow or ice packs and raises the sea level. Albedo increases with deforestation and desertification and with greater cloud cover. GCMs incorporate the interactions between albedo and these changes. On the other hand, future encounters with fire and brimstone—orogeny, epeirogeny, and volcanism—while always in the realm of possibilities, are highly uncertain or infrequent, and so most GCMs either do not include such factors or treat them as a small probability.

The future of greenhouse gas concentrations in the atmosphere proves difficult to anticipate because of its strong dependence on human activities and the unpredictability of such activities. The Intergovernmental Panel on Climate Change (IPCC), an organization established by the United Nations Environment Programme and the World Meteorological Organization in 1988, developed 40 scenarios about future concentrations of greenhouse gases. These scenarios fall into four different narrative storylines about possible demographic, social, economic, technological, and environmental circumstances that will drive greenhouse gas emissions.

One storyline (A1) assumes slowing human population growth but rapid economic development based on either fossil fuels (A1FI), alternative fuels (A1T), or a mix of both (A1B). Another storyline (A2) emphasizes regional autonomy (instead of international cooperation) and sustained human population growth, fragmented economic development, and slow technological adoption. A third storyline (B1) assumes a worldwide shift to a service and information economy with international agreements that promote general adoption of clean and resource-efficient technologies. A fourth storyline (B2) focuses on local solutions to economic, social, and environmental sustainability, with intermediate economic growth and diverse technological change. All of these scenarios depend upon an understanding of the global cycles for greenhouse gases with particular sources and sinks for each one.

The IPCC considers that any of these emission scenarios is equally likely. The scenarios are not predictions, but assumptions that GCMs require as input.

They serve as "if–then" statements: If human activities follow scenario W (which includes factors such as amounts and types of economic activity, adoption of evolving technologies, and rates of human population growth), then a GCM predicts that the amount of greenhouse gas emissions will be X (in terms of amount of carbon equivalents per year), atmospheric concentrations of the greenhouse gases will be Y (in terms of carbon dioxide equivalents), and the climate will be Z (in terms of temperature, precipitation, storms, etc.).

Anthropogenic Greenhouse Gases

Human activities are largely responsible for the rise in atmospheric concentrations of the following greenhouse gases. These gases are listed in descending order, based on the magnitude of their effect.

Carbon dioxide

Carbon dioxide (CO_2) is foremost among anthropogenic greenhouse gases in concentration and effect (see Table 3.1). It is responsible for about one-third of the global warming that derives from human activities (Hansen and Sato 2001). Concentrations of CO_2 in the atmosphere depend on the global carbon cycle that accounts for the fluxes of carbon among various storage pools (**Figure 4.13**). The vast majority of carbon on Earth is part of limestone ($CaCO_3$) and other sedimentary rocks. Weathering of rock (see Equations 2.5 and 2.6) or the high temperatures used during cement production release CO_2 from limestone (see Chapter 9), and this CO_2 enters the atmosphere or dissolves in bodies of water.

Photosynthetic organisms absorb CO_2 from the atmosphere or bodies of water and then use solar radiation to convert this low-energy carbon into the high-energy carbon in organic compounds (see Equation 2.1). This process reverses when organisms respire (see Equations 2.2 and 2.3) or when they die and decompose. The CO_2 thus released once more enters the atmosphere or dissolves in bodies of water. Fossil fuels derive from organic carbon compounds formed during decomposition of prehistoric organisms. The burning of fossil fuels rapidly oxidizes these compounds and releases CO_2.

Carbon exchanges among the atmosphere, soils, vegetation, and changes in land use (e.g., clearing of rainforest for agriculture) are balanced; that is, they result in no net change in the amount of CO_2 present. Exchanges among the atmosphere, surface ocean, marine biota, and dissolved organic carbon are simi-

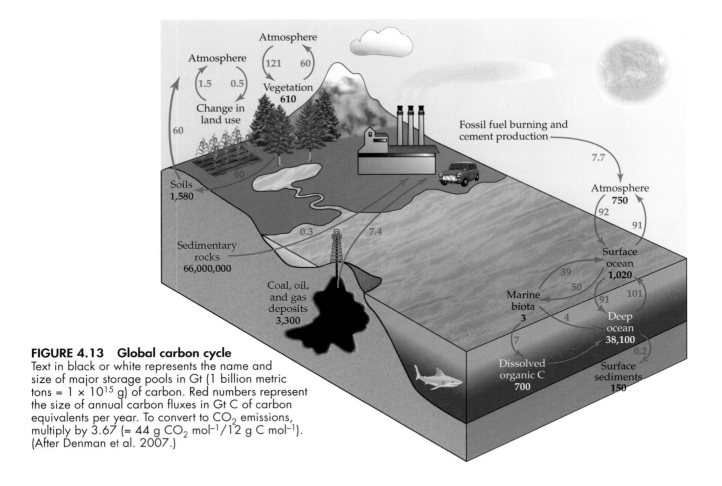

FIGURE 4.13 Global carbon cycle
Text in black or white represents the name and size of major storage pools in Gt (1 billion metric tons = 1×10^{15} g) of carbon. Red numbers represent the size of annual carbon fluxes in Gt C of carbon equivalents per year. To convert to CO_2 emissions, multiply by 3.67 (= 44 g CO_2 mol^{-1}/12 g C mol^{-1}). (After Denman et al. 2007.)

larly balanced (see Figure 4.13). Carbon extractions from sedimentary rocks and from coal, oil, and gas deposits and the subsequent carbon emissions from fossil-fuel burning and cement production are out of balance; that is, they result in a net release of CO_2. Tallies of purchases of petroleum, natural gas, and coal indicate that global CO_2 emissions from the burning of fossil fuels increased by 58% from 1980 to 2006.

All areas of the world participated in this increase with the exception of Eurasia, where emissions declined in 1991 after the breakup of the Soviet Union but are now rising again (**Figure 4.14**). China surpassed the United States in total carbon emissions during 2006, although the United States still leads in emissions per person (Energy Information Administration 2008e). Carbon emissions from China tripled between 1980 and 2006, and China's contributions to the world's total emissions jumped from 8% to 21%. Emissions from the United States increased 23% over the same period, yet its contributions to the world's total dropped from 26% to 20%.

Predictions of carbon emissions must take into account the contributions of various economic sectors and their potential expansion or contraction.

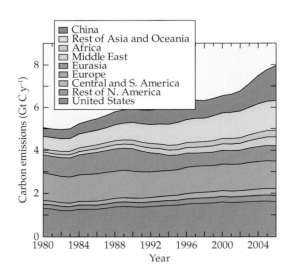

FIGURE 4.14 Annual carbon emissions from the burning of fossil fuels (Gt C y^{-1}) Each colored band represents the contribution of the designated area of the world. The United States is separated from the rest of North America and China from the rest of Asia and Oceania because of their large contributions. The sum of all the color bands indicates total global emissions. Between 1980 and 2006, total emissions rose from 5.0 to 7.9 Gt C y^{-1}. (After Energy Information Administration 2008d.)

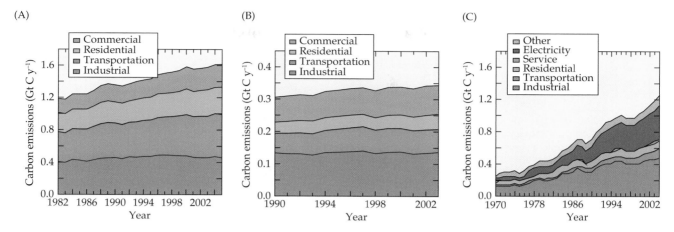

FIGURE 4.15 **Greenhouse gas emissions from different countries.** (A) United States annual carbon emissions (Gt C y⁻¹). Each colored band represents the contribution of the designated economic sector. (B) Japan's annual carbon emissions (Gt C y⁻¹) according to economic sector. (C) China's annual carbon emissions (Gt C y⁻¹) according to economic sector. Note that in both (A) and (B) emissions from "Electricity" are divided into the appropriate sectors. (A after Energy Information Administration 2006d; B after Ministry of the Environment 2007; C after van Vuuren et al. 2003.)

The transportation and residential sectors contribute a relatively large fraction of the carbon emissions from the United States (**Figure 4.15A**). In contrast, the industrial sector constitutes the largest fraction of the carbon emissions from Japan (**Figure 4.15B**) and China (**Figure 4.15C**). Emissions from the industrial sector have held steady in the United States and Japan because these countries have adopted more service-oriented economies, whereas emissions from the industrial sector (e.g., manufacturing) in China have continued to climb.

Various scenarios about future demographic, social, economic, technological, and environmental conditions (**Figure 4.16A**) result in a range of predicted global carbon emissions (**Figure 4.16B**). Scenarios that assume rapid economic growth and reliance on fossil fuels (A1FI) or regional autonomy (A2) predict a near tripling of the rate of carbon emissions during the twenty-first century. In contrast, scenarios that assume the adoption of alternative energy sources (A1T or B1) predict a rise in emissions until the middle of the century and then a decline. For the intermediate scenarios (A1B and B2), carbon emissions are predicted to approximately double and then level off.

The carbon released by human activities ends up in several places. Most of this carbon remains as CO_2 in the atmosphere (**Figure 4.17**), and atmospheric CO_2 concentrations may increase by 50% to 300% to somewhere between 550 ppm and 970 ppm during the twenty-first century (**Figure 4.18**). A portion of this carbon dissolves in the oceans, 111 ± 25 Gt (1 Gt = 1 × 10^{15} g) from 1800 to 1994 (Keeling 2005), and more is

likely to do so. Vegetation assimilates some of this carbon through photosynthesis, and organic carbon compounds have accumulated on land and ocean floors (see Figure 4.13). Global warming, however, will accelerate respiration and thereby will release more CO_2 from this organic carbon. Some GCMs, such as the Hadley Climate Model, predict that by the latter half of the twenty-first century, respiration will exceed photosynthesis, and so landmasses will become sources rather than sinks for carbon.

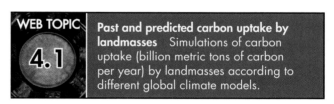

WEB TOPIC 4.1 **Past and predicted carbon uptake by landmasses** Simulations of carbon uptake (billion metric tons of carbon per year) by landmasses according to different global climate models.

Methane

Methane (CH_4) is second only to CO_2 in its contribution to the greenhouse effect and accounts for about 15% of anthropogenic warming (Hansen and Sato 2001). Moreover, sudden release of CH_4 from the melting of seabed methane hydrate is implicated in such cataclysmic events as the Great Dying and the Late Paleocene Thermal Maximum (see Chapter 2). Once in the atmosphere, CH_4 oxidizes to CO_2 (see Equation 2.4) in about a dozen years (see Table 3.1).

Contrary to popular belief, the majority of humans do not release CH_4 when they pass flatus gas (Suarez et al. 1997)—this provides yet another reason why attempts to ignite these emissions are seldom worth-

FIGURE 4.16 Emissions scenarios (A) Schematic of 40 emissions scenarios based on the four different storylines (A1, A2, B1, B2) and six different scenario. "HS" denotes harmonized scenarios, which share assumptions about global population, economic growth, and final energy use. "OS" denotes other scenarios, which explore alternative assumptions. The A1 storyline assumes rapid economic growth based on fossil fuels (A1FI), alternative energy sources (A1T), or a mix of the two (A1B). The A2 storyline assumes regional autonomy, B1 assumes large-scale adoption of clean technologies, and B2 assumes local solutions. (B) Predictions of total annual global carbon emissions (Gt C y^{-1}) for the four storylines and six scenario groups. Each colored band depicts the range of values for the harmonized and non-harmonized scenarios within each group. The red lines indicate the predictions for an illustrative scenario within each group. (After IPCC Working Group III 2000.)

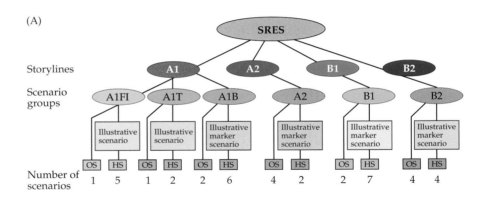

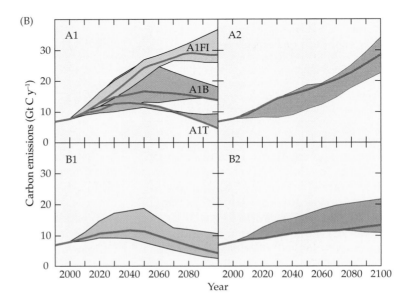

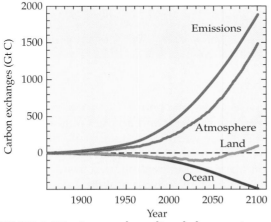

FIGURE 4.17 Past and predicted changes in carbon sinks Distribution (Gt C) of carbon emissions among the atmosphere, land, and ocean predicted by a GCM (Hadley-CM3LC) that assumes an emissions scenario similar to B2 and includes feedback between climate and the biosphere. (After Cox et al. 2000.)

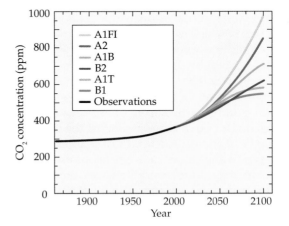

FIGURE 4.18 Past and predicted changes in global average atmospheric CO$_2$ concentration Future global atmospheric concentrations (ppm) under different emissions scenarios. (After IPCC 2001.)

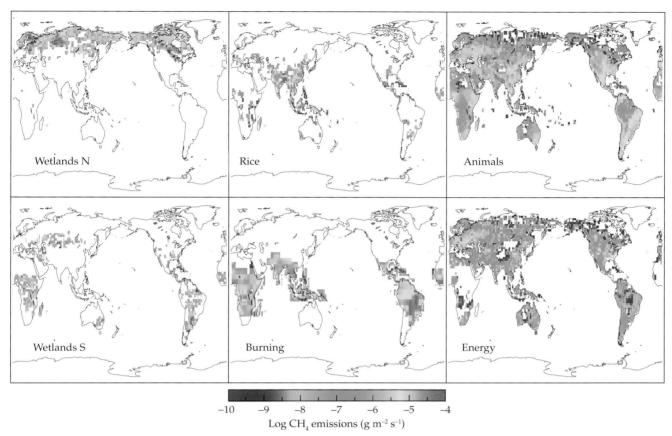

$$\text{Log CH}_4 \text{ emissions (g m}^{-2} \text{ s}^{-1})$$

FIGURE 4.19 Geographical distribution of average annual methane emissions (g CH_4 m^{-2} s^{-1}) from northern wetlands, mid-latitude and southern wetlands, rice fields, biomass burning, animals and their waste, or natural gas and coal energy production. Notice that the scale is logarithmic.

while. Other human activities, however, do emit large amounts of CH_4 (**Figure 4.19**). Over 75% of the world's rice is grown in flooded paddies. Soils during flooding soon become anaerobic, and soil microbes generate CH_4, or "marsh gas," through anaerobic respiration (see Equation 2.3). Clearing of agricultural land through burning also produces CH_4 because of incomplete combustion. Animals, particularly ruminants such as cows, sheep, and goats, have bacteria in their digestive tracts that generate CH_4. A primary component of natural gas is CH_4, and losses of CH_4 occur during natural gas extraction, processing, storage, transmission, and distribution. Coal deposits also contain trapped CH_4 that is released during normal mining operations in both underground and surface mines.

All told, the human activities of agriculture, biomass burning, and energy production release somewhere around 320 Gt CH_4 per year (240 Gt C y^{-1}) (Chen and Prinn 2006). These sources, together with CH_4 produced by natural wetlands and non-domestic animals such as wild ruminants and termites, emit a total of between 410 and 660 Gt CH_4 per year (Wuebbles and Hayhoe 2002). Various scenarios about future demographic, social, economic, technological, and environmental conditions (see Figure 4.16A) predict that global CH_4 emissions will be between 500 and 1150 Gt CH_4 y^{-1} by the end of the century (**Figure 4.20A**) and that atmospheric concentrations will reach between 1.4 and 3.6 ppm (**Figure 4.20B**).

Ozone

Ozone (O_3) may either warm or cool the surface of Earth, depending on whether it accumulates in the upper atmosphere (the **stratosphere**, between 10 km and 50 km above sea level) or the lower atmosphere (the **troposphere**, from Earth's surface to about 10 km above sea level) (**Figure 4.21**). The following reactions generate ozone in the stratosphere:

$$O_2 + \text{ultraviolet radiation} \rightarrow 2\,O \qquad (4.9)$$

$$O + O_2 \rightarrow 2\,O_3 \qquad (4.10)$$

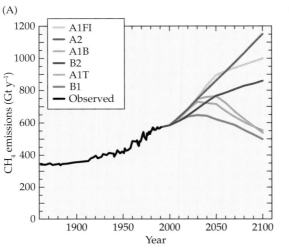

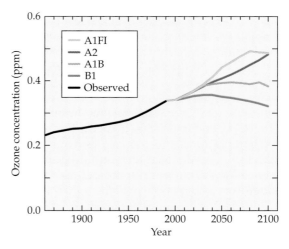

FIGURE 4.20 Atmospheric methane (A) Past and projected total methane emissions (Gt CH_4 y^{-1}) under the different scenario groups for carbon emissions. (B) Past and projected atmospheric methane concentrations (ppm) under the different scenario groups for carbon emissions. (A after IPCC 2001, and Wuebbles and Hayhoe 2002; B after IPCC 2001, Wuebbles and Hayhoe 2002.)

This stratospheric ozone blocks ultraviolet radiation from penetrating the atmosphere (see Figure 3.35) and thereby cools Earth's surface. As we discuss in the next section, certain chemicals that humans have created are depleting this ozone layer.

In contrast, most of the ozone in the troposphere derives from photochemical smog, a process that involves a complex set of reactions among nitrogen oxides (NO_X), carbon monoxide (CO), and volatile organic compounds (VOCs), such as xylene. Vehicles, coal-burning power plants, and industrial manu-facturing sites emit NO_X and CO. Volatile organic compounds derive from both synthetic sources (e.g., gasoline, paints, solvents, and pesticides) and natural sources (e.g., pine and citrus trees). The summary reaction for photochemical smog is:

$$CO + 2\,O_2 + sunlight \rightarrow CO_2 + O_3 \qquad (4.11)$$

Ozone produced in the troposphere acts as a greenhouse gas (see Figure 3.35) that contributes about 13% to global warming (Hansen et al. 2005; IPCC 2007c). Human activities are increasing tropospheric ozone (**Figure 4.22**).

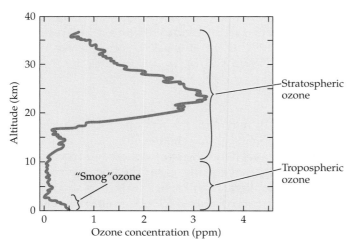

FIGURE 4.21 Typical concentrations (parts per million, ppm) of ozone as a function of altitude in kilometers. (After Schoeberl 1999.)

FIGURE 4.22 Past and future concentrations (parts per million, ppm) of ozone in the troposphere under different emission scenarios. (After Horowitz 2006.)

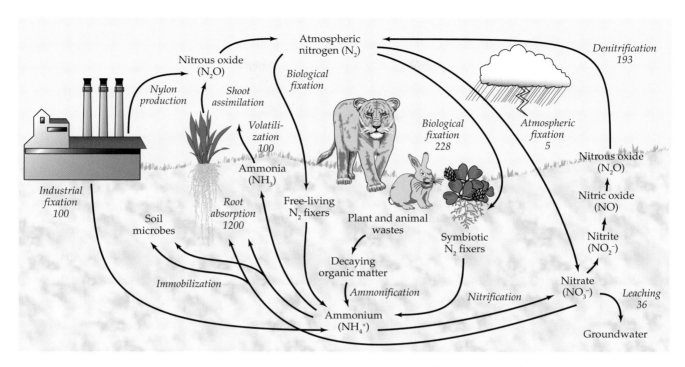

FIGURE 4.23 Nitrous oxide emissions and the terrestrial nitrogen cycle During microbial denitrification and plant nitrogen assimilation, nitrite (NO_2^-) is converted into nitrous oxide (N_2O) and, subsequently, nitrogen gas (N_2). Fluxes are in Tg y^{-1} (10^{12} g per year). (After Galloway et al. 2004; Epstein and Bloom 2005.)

Nitrous oxide

A fourth major greenhouse gas is nitrous oxide (N_2O), which is responsible for about 6% of anthropogenic warming (Hofmann et al. 2006). N_2O is perhaps best known as "laughing gas" because inhalation not only acts as an anesthetic, but also induces a state of euphoria. Drag racers inject N_2O instead of gaseous oxygen into their engines because combustion of N_2O releases both oxygen and extra energy. Normal concentrations of N_2O in the atmosphere, however, are only one-millionth of those required to induce laughing or boost combustion.

Atmospheric N_2O is part of the global nitrogen cycle (**Figure 4.23**). Most N_2O derives from **denitrification**, in which microbes conduct anaerobic respiration and generate energy using nitrogen compounds such as nitrite (NO_2^-) or nitric oxide (NO), instead of oxygen, to accept electrons during the breakdown of organic compounds. Agricultural irrigation and fertilization greatly enhance these emissions (Matson et al. 1998). In addition, plants release tiny amounts of N_2O when they convert NO_2^- into ammonium (NH_4^+) during the synthesis of organic nitrogen compounds, and this may comprise 5% of N_2O emissions (Smart and Bloom 2001). Manufacture of nylon and nitric acid also

releases N_2O. Once in the atmosphere, N_2O slowly converts (over approximately 112 years) to nitrogen gas, N_2 (see Table 3.1).

Most of the world's N_2O emissions come from land areas under intensive agricultural cultivation (**Figure 4.24**). Improved fertilizer and irrigation practices—as is assumed in scenario groups A1B, B2, A1T, and B1—are expected to lower N_2O emissions during the twenty-first century. Without such measures (scenarios A1FI and A2), N_2O emissions and atmospheric concentrations are likely to rise by 63% and 41%, respectively, during the century (**Figure 4.25**).

Chlorofluorocarbons and hydrochlorofluorocarbons

The last greenhouse gases that we will consider are the chemicals called chlorofluorocarbons (CFCs) and hydrochlorofluorocarbons (HCFCs). These compounds have carbon atoms bound to chlorine, fluorine, and—in the case of HCFCs—hydrogen. They are highly effective as refrigerants, propellants for aerosol sprays, cleaning solvents, and bubbles to expand foams. Unfortunately, they also are greenhouse gases with warming potentials more than a thousand times greater than CO_2 (see

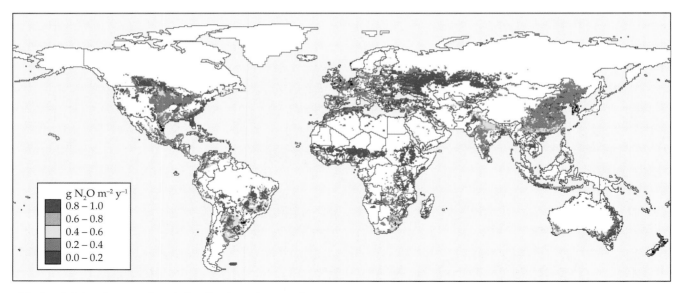

FIGURE 4.24 Geographical distribution of annual N₂O emission rates for agricultural lands Values are weighted averages over the crops and grasslands within an area.

Table 3.1). If this were not enough, these compounds also deplete Earth's stratospheric ozone layer. Consequently, the Montreal Protocol, an international treaty to abate loss of the ozone stratospheric layer (see Chapter 11), severely limited the use of these chemicals in 1989.

Emissions of CFCs dropped precipitously after the Montreal Protocol came into force (**Figure 4.26A**). The treaty encouraged certain users to substitute HCFCs for CFCs because HCFCs have shorter life spans in the atmosphere (see Table 3.1); as a result, HCFC emissions continued to rise after 1989, but now appear to have

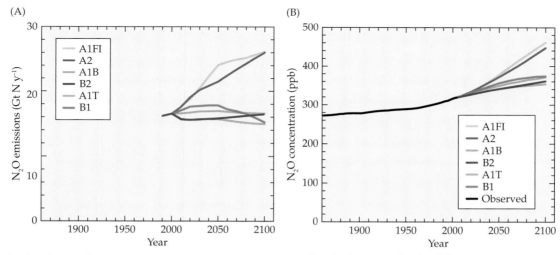

FIGURE 4.25 Atmospheric nitrous oxide
(A) Projected nitrous oxide emissions (Gt nitrogen y⁻¹) under the different scenario groups for greenhouse gas emissions.
(B) Atmospheric nitrous oxide concentrations in the past and in the future under the different scenario groups for greenhouse gas emissions. (A after IPCC 2001; B after IPCC 2001; Meure et al. 2006.)

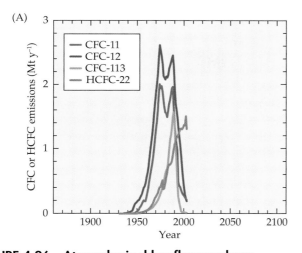

(A)

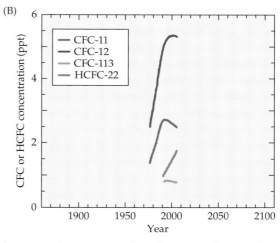

(B)

FIGURE 4.26 Atmospheric chlorofluorocarbons (CFCs) and hydrochlorofluorocarbons (HCFCs) (A) Total emissions (Mt y^{-1}) of major CFCs and HCFCs. (B) Atmospheric concentrations (parts per trillion, ppt) of CFCs and HCFCs. (A after Alternative Fluorocarbons Environmental Acceptability Study 2006; B after Butler et al. 1999; Global Monitoring Division 2007.)

leveled off. Imposing severe limits on CFC emissions is having the desired effect in that atmospheric CFC concentrations have begun to wane (**Figure 4.26B**). All in all, the influence of these compounds should diminish from about a 7% contribution to global warming to less than a 4% contribution by 2050 (Hansen and Sato 2001). This provides a clear example of how restricting emissions of greenhouse gases can reverse their negative influence.

Testing Global Climate Models

If we are given values for the past, present, and projected future of greenhouse gases, we can derive the total of all climatic forcing factors (**Figure 4.27**), and run global climate models (GCMs) to calculate the past, present, and likely future of Earth's climate. Verifying a GCM involves simulating the climate for some time past and comparing the model's output with what actually occurred. Discrepancies between a GCM and the climate record indicates problems with the model's assumptions, algorithms, initial conditions (u_0, v_0, w_0, ρ_0, p_0, and T_0), coordination among various submodels, or even the climate record. Tweaking a GCM to correct for such problems is a fine art because climate involves interactions among many variables that do not exhibit linear behavior.

Linear behavior dictates that changes in output be proportional to changes in input. For example, a spring exhibits linear behavior in that doubling the weight on a spring doubles its extension (**Figure 4.28**). In nonlinear behavior, small changes in conditions may have large consequences, or conversely, large changes may have no perceptible consequences. An example of nonlinear behavior is that the vapor pressure of water increases exponentially with temperature (see Figure 2.43), and so vapor pressure changes less with a 10°C temperature differential from 10°C to 20°C than with one from 30°C to 40°C. Another example is that lowering the temperature of an air mass has only slight

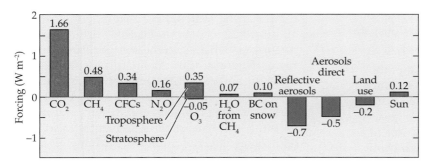

FIGURE 4.27 Changes in global forcing in W m^{-2} (Watts per square meter) by different factors from 1750 to 2005 Red bars indicate warming influences and blue bars, cooling. "BC" stands for black carbon soot. (After Hansen et al. 2005; Forster et al. 2007.)

FIGURE 4.28   Linear behavior Extension of the spring in a handheld fish scale is proportional to the gravitational force on the fish load on the scale in these manipulated photographs.

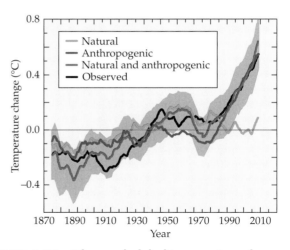

FIGURE 4.29   Observed global temperature changes versus simulations via the HadGEM1 GCM assuming only natural climate forcing factors, anthropogenic forcing factors, or the sum of the two. The gray band indicates the uncertainty of the calculations for all forcing factors. The baseline (0.0) is the mean temperature for the entire period. Plotted are the 10-year running averages. (After Stott et al. 2006.)

effects on its transmission of electromagnetic radiation until water vapor in the air mass reaches saturation, a cloud forms, and, suddenly, albedo skyrockets.

The Intergovernmental Panel on Climate Change (IPCC), in its Fourth Assessment Report released in 2007 (IPCC 2007a), presents the results from 16 or more GCMs that simulated Earth's climate under a range of forcing factors, both natural (e.g., orbital variations) and anthropogenic (e.g., human activities that emit additional greenhouse gases). The models were run three times: once with only natural climate forcing, the second time with only anthropogenic forcing, and, finally, with both natural and anthropogenic forcing (**Figure 4.29**). The best match for reconstructing global temperatures over the past 125 years occurred when the models included both types of forcing. In particular, anthropogenic forcing appears responsible for most of the 0.5°C rise in global temperatures experienced from 1970 until 2000.

Anthropogenic Forcing Factors

Several other lines of evidence support the conclusion from GCM simulations that human-generated greenhouse gases, rather than natural climatic variations or urbanization near land-based weather stations, are largely responsible for recent global warming:

- A number of recent analyses indicate that the rise in the global average surface temperature since 1987 is predominantly associated with changes in anthropogenic greenhouse gases and aerosols and that the contribution of changes in solar electromagnetic radiation to this temperature trend is negligible (Lean

and Rind 2008; Lockwood 2008). Changes in solar radiation from 1750 to 2005 account for a warming force of only 0.12 (ranging from 0.06 to 0.30) watts per square meter, whereas human activities account for 1.7 (ranging from 0.6 to 2.4) watts per square meter (see Figure 4.27) (Forster et al. 2007; IPCC 2007c; Rapp 2008).

- Global nighttime minimum temperatures show a greater increase than daytime maximum temperatures (**Figure 4.30A**). One would expect the reverse if changes in solar electromagnetic radiation were driving global warming. Higher concentrations of greenhouse gases enhance back radiation of electromagnetic energy from the atmosphere to the surface (see Figure 3.38) during the night.

- Temperatures in the lower stratosphere—the region of the atmosphere that extends from 10 km to 22 km above Earth's surface—are declining, while surface temperatures are rising (**Figure 4.30B**). Again, this is consistent with higher concentrations of greenhouse gases retaining more energy near Earth's surface. Stratospheric and surface temperatures would rise at similar rates if changes in solar electromagnetic radiation were driving global warming.

(A)

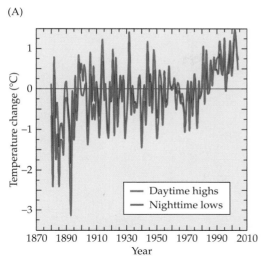

(B)

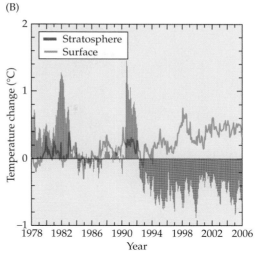

(C)

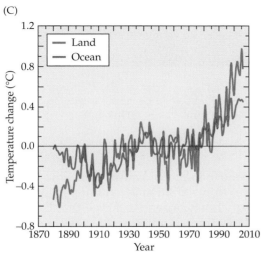

FIGURE 4.30   Historical trends in temperatures
(A) Observed changes in the global daytime maximum and nighttime minimum temperatures on land in relation to the respective means for the entire period. (B) Changes in monthly temperatures for Earth's stratosphere with respect to the mean for the entire period and Earth's surface (green) in relation to the 1961–1990 mean. Red and blue bars indicate above- and below-average temperatures. Stratospheric data is from the TIROS-N satellite. (C) Changes in annual temperatures for Earth's landmasses and oceans in relation to the means for the entire period. (A after National Climatic Data Center 2007a; B after Brohan et al. 2006, Global Hydrology and Climate Center 2007; C after National Climatic Data Center 2007b.)

- A portion of the measured rise in global temperatures may derive from an "urban heat island" effect, whereby many weather stations that were once in a rural setting are now surrounded by asphalt. Nonetheless, measurements from sea-based weather stations indicate that Earth's oceans have warmed during the last century in a manner that parallels the warming of landmasses (**Figure 4.30C**). The global warming observed during the past 50 years amounts to about 0.9°C over land and 0.4°C over oceans (IPCC 2007c), whereas urbanization near weather stations can account for about a 0.03°C temperature rise over land and a negligible temperature rise over the oceans during this period. Therefore, an urban heat island effect around weather stations does not seem to be commensurate with the global warming that has occurred.

- Jet airplanes flying at high altitudes generate contrails (**Figure 4.31**). These behave like clouds in that they reflect incoming solar radiation and decrease daytime temperature maximums but absorb long-wave radiation from Earth's surface and increase nighttime temperature minimums (see Chapter 3); they therefore have the potential to diminish temperature differentials between day and night. Commercial air traffic was shut down in the United States for three days after the destruction of the World Trade Center in New York City on September 11, 2001. During that interval, day–night temperature differentials

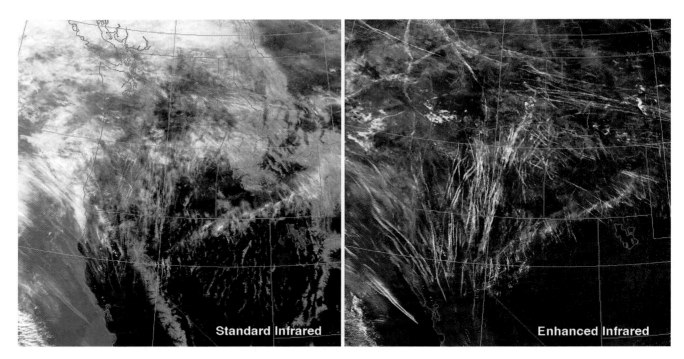

FIGURE 4.31   Images of the Pacific Northwest United States taken by the NOAA-17 satellite during the morning of March 20, 2004 in which what appears to be regular cloud cover in the standard infrared image on the left is, in fact, shown to be mostly spreading contrails in the enhanced infrared image on the right.

in the continental United States jumped by 1.8°C above those during the 3-day periods immediately before or after this episode (Travis et al. 2002). This change was significantly greater than any observed for similar intervals during the preceding 30 years. Moreover, the change in day–night differentials during the shutdown was twice as large in the regions of the United States where contrails are normally most abundant (**Figure 4.32**). These results highlight the extent to which human activities can affect regional surface temperatures.

These various pieces of evidence dispel many, if not most, of the issues raised by skeptics who do not believe in global warming or who believe that it derives from variation in natural forcing factors. Over 75% of climate scientists agree that Earth is significantly warmer today than it has been in many centuries and that human-generated greenhouse gases have been the major driving force for this warming (Bray and von Storch 2008; Lichter 2008; Doran and Zimmerman 2009). Although uncertainties remain as to the extent of warming and the relative contributions of each factor, the accumulation of additional data is steadily diminishing these uncertainties and is pointing directly at human activities.

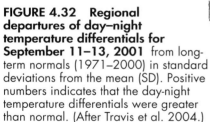

FIGURE 4.32   Regional departures of day–night temperature differentials for September 11–13, 2001 from long-term normals (1971–2000) in standard deviations from the mean (SD). Positive numbers indicates that the day-night temperature differentials were greater than normal. (After Travis et al. 2004.)

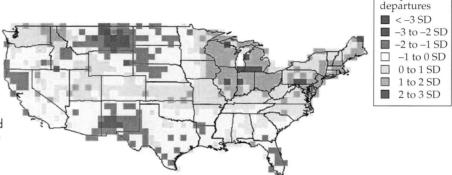

Temperature departures
- < –3 SD
- –3 to –2 SD
- –2 to –1 SD
- –1 to 0 SD
- 0 to 1 SD
- 1 to 2 SD
- 2 to 3 SD

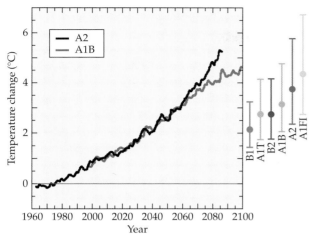

FIGURE 4.33   Global temperature projections from a GCM (the HadGEM1) under the A1B and A2 scenarios Plotted are annual means with respect to the mean from 1961–1990. At the right are temperature projections for 2100 under various emission scenarios (mean and range from 16 or more GCMs). Projections from the HadGEM1 GCM are higher than from most models. (After Stott et al. 2006, IPCC 2007c.)

Future World

Global climate models predict that the changes in climate observed over the last few decades will be small in comparison to those that will occur before the end of this century. These include changes in temperature, albedo, sea level, quantity of precipitation, ocean pH, intensity of major storms, and frequency and severity of forest fires. In the following sections, we examine each of these changes.

Temperatures and albedo

Average global temperatures are anticipated to warm by somewhere between 1.5°C and 6.8°C from 2000 to 2100, depending on human activities (**Figure 4.33**). These changes will not be uniform, although no regions are expected to cool (**Figure 4.34**). Landmasses will warm more than the oceans, and Arctic regions will warm more than other parts of the world, perhaps by as much as 8°C or 9°C above 1975 temperatures. This will melt more of the northern ice and snow pack,

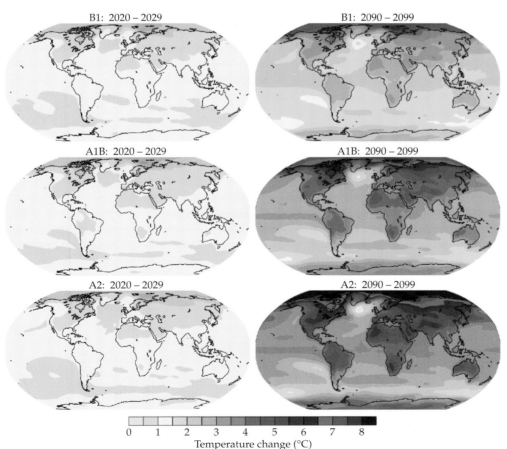

FIGURE 4.34   Projected changes in surface temperatures comparing data collected between 1961–1990 and 2020–2029, or between 1961–1990 and 2090–2099, under emissions scenarios B1, A1B, and A2. (From IPCC 2007c.)

(A)

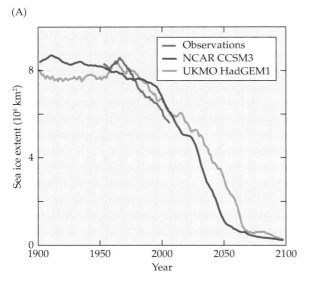

(B)

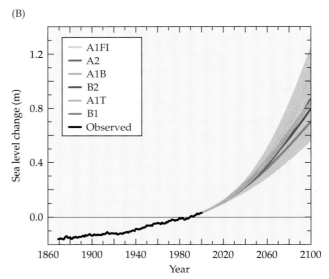

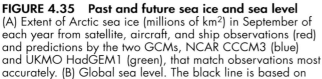

FIGURE 4.35 Past and future sea ice and sea level (A) Extent of Arctic sea ice (millions of km²) in September of each year from satellite, aircraft, and ship observations (red) and predictions by the two GCMs, NCAR CCCM3 (blue) and UKMO HadGEM1 (green), that match observations most accurately. (B) Global sea level. The black line is based on an empirical fit of data with temperature changes. Colored lines represent the sea-level changes resulting from the anticipated mean temperatures under the different emission scenarios. The gray band represents the uncertainty under the range of temperatures expected. (A after Stoeve et al. 2007; B after Rahmstorf 2007.)

thereby decrease the albedo of these regions, and further accelerate global warming. Unfortunately, changes in albedo calculated via 17 to 20 different GCMs fail to match closely those measured via satellite (Bender et al. 2006; Wang et al. 2006), and so detailed projections about Earth's albedo are not yet possible.

Sea ice and sea level

Sea ice in the polar oceans is melting and may disappear entirely during summer months (**Figure 4.35A**). This will shorten sea routes, say, from Europe and the east coast of the United States to China (see Chapter 1). It will also make more accessible some natural resources such as oil deposits in the Arctic seabed (see Chapter 11).

Sea level is predicted to rise as higher temperatures expand the volume of ocean waters and melt the snow and ice situated on landmasses, particularly Greenland and Antarctica. Over 20 different GCMs predict that the sea level will rise by 0.2 m to 0.5 m from 2000 to 2100 (Bindoff et al. 2007). These models usually assume that glaciers will keep flowing (moving) at their current speed. Recent observations in Greenland have shown that glacial flows are accelerating (Cazenave and Nerem 2004; Church and White 2006; Miller and Douglas 2006; Woodworth 2006). The likely mechanism is that as temperatures increase, the ice along the edges of a glacier melts. This water makes its way to the bottom of the glacier and acts as a lubricant that accelerates the ice flow. An alternative forecast, which takes this phenomenon into account, predicts an even greater change in sea level: the sea level in 2100 will be 0.5 m to 1.2 m higher than in 2000 (**Figure 4.35B**).

A rising sea level will put coastal regions of the world at great risk. Cities with large populations, such as those along the Gulf and East coasts of the United States, will be inundated (**Figure 4.36**). Bangladesh, one of the world's poorest yet most populous countries, is perhaps the most vulnerable to sea-level changes (**Figure 4.37**): about 80% of this country is below 10 m in elevation, and so a 0.5 m to 1.0 m rise in sea level will permanently flood between 6% and 10% of its land area and displace between 3.4 million and 17 million people (Ali 1996; Ericson et al. 2006). In addition to worldwide flooding, the rising sea level may contaminate the freshwater supplies of many coastal regions with seawater.

Precipitation

Global warming will promote greater evaporation of water from Earth's surface (see Figure 2.43). What goes up (water vapor) must come down (precipitation); therefore, global precipitation should increase on average. In particular, polar and equatorial regions are likely to become wetter (**Figure 4.38**). The mid-latitudes, however, will become drier, especially areas that are already deserts. Higher temperatures and

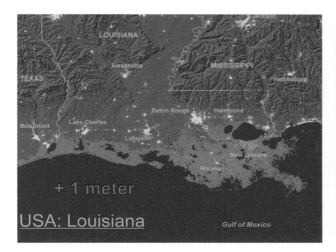

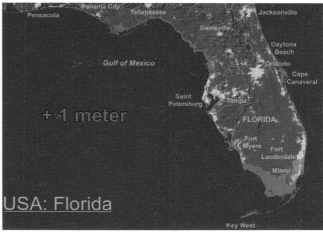

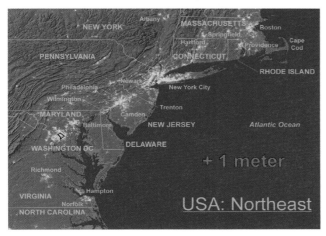

FIGURE 4.36 Coastal areas of Louisiana, Florida, and the northeast United States that are likely to be inundated if the sea level rises 1 m above current levels (red). Predictions are based on elevation and adjacency to the sea. Cities are indicated in white.

lower precipitation at mid-latitudes will decrease the snowpack in many mountain ranges. For example, the Sierra Nevada Mountains in California may receive from 30% to 90% less snow by the year 2100 (Hayhoe et al. 2004). Also, only about one-third of the ski resorts in Europe will have reliable snow by 2100 (Agrawala 2007). Finally, even in areas where precipitation does not change significantly, the amount of water available to living organisms, lakes, and rivers will decrease because at higher temperatures, more water will be lost to evaporation.

Ocean acidification

As CO_2 concentrations in the atmosphere increase by 50% to 300% during this century (see Figure 4.18), more CO_2 will dissolve in the oceans and form carbonic acid (see Equation 2.5). This carbonic acid dissociates in water to release a proton and bicarbonate

$$H_2CO_3 \rightarrow H^+ + HCO_3^- \qquad (4.12)$$

FIGURE 4.37 Floods in Dhaka, the capital of Bangladesh, in 2004 Victims gather at a relief office to collect survival assistance.

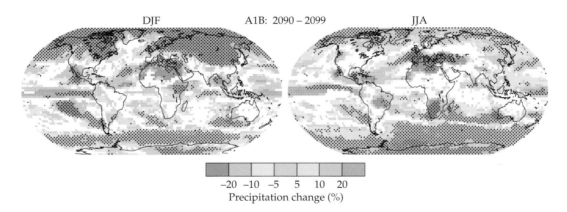

DJF   A1B: 2090 – 2099   JJA

−20 −10 −5 5 10 20
Precipitation change (%)

FIGURE 4.38 Projections of 21 GCMs about changes over the present century in precipitation during December, January, and February (DJF) or June, July, and August (JJA) under the A1B emissions scenario. White areas are where less than 66% of the models agree in the direction of the change, and stippled areas are where more than 90% of the models agree in the direction of the change.

and then again to release another proton and carbonate:

$$HCO_3^- \rightarrow H^+ + CO_3^{2-} \qquad (4.13)$$

The exact balance among dissolved CO_2, carbonic acid (H_2CO_3), bicarbonate (HCO_3^-), and carbonate (CO_3^{2-}) depends on the concentration of protons in the water (pH), its salinity, and its temperature. Waters near the poles are more alkaline because of their cooler temperatures. Higher atmospheric CO_2 concentrations and warmer temperatures increase proton concentrations in the oceans (i.e., lower the pH), and oceans are thereby becoming more acidic.

Simulations on GCMs affirm this trend (**Figure 4.39**). They indicate that the oceans in preindustrial times were more alkaline, with a pH of 8.25 (a pH of 7.0 is neutral); they dropped to pH 8.15 by around 1994; and they are expected to reach pH 7.85 by 2100 (Jacobson 2005; Orr et al. 2005). The pH scale is logarithmic, and so a shift in ocean pH from 8.25 to 7.85 means a 250% increase in proton concentrations, a change that will have a strong effect on sea life (see Chapter 6).

Stormy weather

Warmer sea surface temperatures and cooler stratospheric temperatures accentuate the temperature gradients that empower major storms (see Chapter 3). Consequently, major storms are increasing in intensity. This, together with the rising sea level, exacerbates storm surges and threatens coasts adjacent to warm ocean currents. Projections about the frequencies of major storms, however, remain uncertain because of the complex nature of storm formation and will require additional data and fine-resolution modeling (Curry et al. 2006; Witze 2006).

Forest fires

Hotter and drier weather at the mid-latitudes will foster more frequent (**Figure 4.40**) and more severe forest fires (Westerling et al. 2006). Forests in the western United States contain from 20% to 40% of the carbon

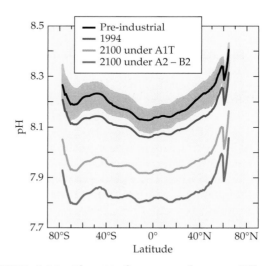

FIGURE 4.39 The pH of ocean surfaces at different latitudes Reconstruction of the pH in preindustrial times (gray band is uncertainty around the mean) was derived by subtracting anthropogenic contributions from current data. Data from 1994 and projections from three GCMs for 2100 under a scenario similar to A1T and another intermediate to A2 and B2. (After Orr et al. 2005.)

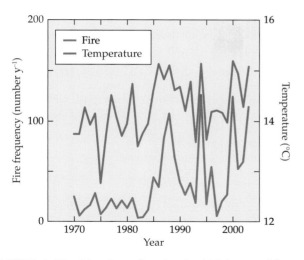

FIGURE 4.40 Number of major (> 400 hectare) forest fires per year in the western United States and mean summer temperatures (March–August) for this region. (After Westerling et al. 2006.)

FIGURE 4.41 A public service announcement of the U.S. government.

sequestered in the vegetation of the entire country. Fires might turn these forests into sources of additional CO_2 that further accelerate global warming, rather than sinks that mitigate it. Clearly, more intensive management of these lands will become necessary.

Warming Momentum

Earth will continue to warm for many decades after we curtail greenhouse gas emissions. Several factors are responsible for this momentum.

1. Several of the greenhouse gases are long-lived (lasting more than 100 years) in the atmosphere (see Table 3.1), and so their concentrations will remain high even without further additions.

2. Oceans serve as large reservoirs of CO_2 that have a slow exchange with the atmosphere, less than 0.3% per year (see Figure 4.13). This means that CO_2 released from human activities accumulated gradually over time in the oceans and will dissipate only slowly once emissions abate.

3. Positive feedback relationships among global warming, ice melting, and declining surface albedo or among global warming, decomposition of organic carbon in the soil, and release of this sequestered CO_2 tend to accelerate warming and will take time to reverse.

In physics, altering momentum requires work (see Equation 4.1). So it will be with altering the momentum of Earth's climate. Diminishing greenhouse gas emissions to avoid undesirable climate change will require work.

When faced with a direct but precarious path, some people will forge ahead, albeit cautiously, and sink ever deeper until they place themselves and those around them in mortal danger (**Figure 4.41**). Turning around may prove inconvenient, embarrassing, or even expensive, but often is the only viable option. Clearly, the time has come to turn around greenhouse gas emissions. Chapters 7, 8, and 9 introduce a broad range of mitigation strategies.

Summary

Mathematical models have the ability to examine the influence of many complex factors on climate. GCMs (General Circulation Models or Global Climate Models) are based on dividing Earth and its atmosphere into a grid system and following exchanges in energy and matter among grids at discrete time steps. Mathematical equations depict how these exchanges obey the classical laws of physics, and computers calculate the solutions to these equations via finite differences or spectral methods. To verify a GCM, scientists simulate climates in the past and compare the output from the model with the actual climate data. The models have become more accurate as computers have become faster and use finer spatial resolutions for their grid systems and calculate changes over shorter time steps. Results from GCMs and other evidence support that human-generated greenhouse gases are the dominant factor responsible for the observed rapid changes in global climate.

Predictions of climate require assumptions about how human emissions of greenhouse gases will change during this century. If human emissions continue to accelerate at current rates until the end of the century, carbon dioxide concentrations in the atmosphere are likely to increase 300%, global average temperatures are likely to warm by 6.8°C, sea level is likely to rise by 1.2 m and displace millions of people, oceans are likely to become more acidic by 0.3 pH units, major storms are likely to become more fierce, and major forest fires are likely to erupt more often. Limiting human emissions of greenhouse gases will diminish the extent but not quickly reverse the direction of these anticipated changes.

Review Questions

1. Predictions about Earth's climate during the coming decades are based primarily on
   (a) direct experimentation.
   (b) observations of other planets.
   (c) political posturing.
   (d) extrapolation from past millennia.
   (e) computer models.

2. Scenarios A1, B1, A2, and B2 for future greenhouse gas emissions do *not* differ in
   (a) rates of economic development.
   (b) adoption of technological advances.
   (c) amounts of international cooperation.
   (d) dependence on fossil fuels.
   (e) rotational orbit of Earth around the Sun.

3. Which of the following statements about Global Climate Models (GCMs) is *false*?
   (a) GCMs usually have submodels to treat land, oceans, and atmosphere.
   (b) GCMs usually divide the Earth into layers, sections, or grids for computational ease.
   (c) All GCMs require as input assumptions about how human activities will influence greenhouse gas emissions.
   (d) All GCMs must simulate past temperatures relatively accurately.

(e) All GCMs predict the same rise in global temperatures during this century.

4. Global Climate Models (GCMs) are based on the fundamental equations of fluid mechanics that depend on all but one of the following principles. Which one is not involved?
   (a) Ideal gas law
   (b) Montreal protocol
   (c) Conservation of momentum
   (d) Conservation of mass
   (e) Conservation of energy

5. Global Climate (General Circulation) Models have *not* proved useful in simulating
   (a) concentrations of atmospheric greenhouse gases.
   (b) average global temperatures.
   (c) major volcanic eruptions.
   (d) surface sea temperatures.
   (e) precipitation patterns.

6. Carbon dioxide emissions are rising most rapidly in
   (a) the United States.
   (b) China.
   (c) Japan.
   (d) Germany.
   (e) Africa.

7. Preindustrial concentration of carbon dioxide in Earth's atmosphere was about _____ ppm.
   (a) 190
   (b) 270
   (c) 385
   (d) 550
   (e) 970

8. Current atmospheric concentration of carbon dioxide is about _____ ppm.
   (a) 190
   (b) 270
   (c) 385
   (d) 550
   (e) 970

9. By the end of the century, atmospheric concentration of carbon dioxide is anticipated to be between _____ ppm and _____ ppm. (Pick two concentrations.)
   (a) 190
   (b) 270
   (c) 385
   (d) 550
   (e) 970

10. Greenhouse gases that are released in large quantities by human activities include (select one or several):
    (a) Argon (Ar)
    (b) Dinitrogen (N_2)
    (c) Hydrochlorofluorocarbons (HCFCs)
    (d) Nitrous Oxide (N_2O)
    (e) Methane (CH_4)

11. List from highest to lowest, the warming influence of the following factors on average global temperatures:
    (a) Variations in Earth's orbit around the Sun.
    (b) Rising atmospheric CO_2 concentrations.
    (c) Rising atmospheric CFC concentrations.
    (d) Rising atmospheric concentrations of aerosols.
    (e) Land use changes.

12. Global Climate Models (GCMs) indicate that
    (a) Human activities have the largest influence on current climate trends.
    (b) Natural climatic variation is responsible for most of current climate trends.
    (c) Current climate trends do not fit any predictable pattern.
    (d) Current climate trends are mainly the consequence of urbanization around weather stations.
    (e) Current climate trends can be reversed in about a decade.

13. Mark all of the global changes that are anticipated by 2100.
    (a) Severity of cyclonic storms will increase.
    (b) Average global temperatures will rise between 1.5°C and 6.8°C.
    (c) Sea level will rise between 0.5 and 1.2 meters.
    (d) Oceans will be become more acidic by 0.2 to 0.3 pH units.
    (e) Winter snow pack in mid-latitude mountain ranges like the Sierra Nevada and Himalayas will decline.

Suggested Readings

Bader, D. C. and 7 others. 2008. *Climate Models: An Assessment of Strengths and Limitations*, A Report by the U.S. Climate Change Science Program and the Subcommittee on Global Change Research. U.S. Department of Energy, Washington, D.C., *http://www.climatescience. gov/Library/sap/sap3-1/final-report/sap3-1-final-all.pdf.*

This 124-page report describes the global climate models and their ability to simulate current climate and predict future climate.

Hansen, J. E. 2006. Can we still avoid dangerous human-made climate change? *Social Research* 73: 949–971.

A general review about global climate change from a well-known NASA scientist who defied major political pressure to present this material.

Karl, T. R., J. M. Melillo, and T. C. Peterson (eds.). 2009. *Global Climate Change Impacts in the United States.* Cambridge University Press, New York.

An overview of the widespread consequences that changing climate is likely to have on the United States.

Randall, D. A. and 12 others. 2007. Climate models and their evaluation. In: *Climate Change 2007: The Physical Science Basis. Contribution of Working Group I to the Fourth Assessment Report of the Intergovernmental Panel on Climate Change*, Solomon, S., D. Qin, M. Manning, Z. Chen, M. Marquis, K. B. Averyt, M. Tignor, and H. L. Miller, eds. Cambridge University Press, Cambridge. pp. 589–662.

This section of the Fourth IPCC report describes global climate models and their use.

5

BIOLOGICAL IMPACTS OF HIGHER CARBON DIOXIDE CONCENTRATIONS

In the previous chapters of this book, we introduced the intimate relationship that exists between living organisms and climate. Organisms as major sources and sinks of greenhouse gases are important factors that force climate change. Conversely, changes in climate often bring about the success or demise of organisms. It is in our own best interest, therefore, to elucidate how anticipated changes in climate will influence the biosphere (the living portion of Earth) in general and humans in particular.

This chapter examines the direct effects of higher atmospheric carbon dioxide (CO_2) concentrations on organisms, especially plants. It explains aspects of plant environmental physiology in detail (an emphasis that reflects the research interests of the author). It then discusses CO_2 sensing in organisms ranging from plants and fungi to insects. Responses of organisms to changing temperature, precipitation, salinity, and pH as well as overall changes in the population size of certain species are the focus of the next chapter.

Rising Carbon Dioxide and Declining Oxygen

Atmospheric concentration of CO_2 has risen from about an average of 0.027% (270 ppm) in preindustrial times to 0.039% at present and are anticipated to reach between 0.05% and 0.10% by the end of the century (see Figure 4.18). These concentrations exceed any that have occurred during the last 20 million years (see Figure 2.25). Direct effects of such changes on organisms can be profound.

Do not despair! You are not in danger of suffocating from rising CO_2 levels. The atmospheric CO_2 concentration anticipated during this century is only a small fraction of the concentrations that cause respiratory distress (Table 5.1). Human beings suffer loss of mental acuity at a CO_2 concentration between 2% and 7.5%, loss of consciousness at between 5% and 10%, and loss of life at between 20% and 30%. The National Research Council recommends that long-term exposure to CO_2 in submarines be kept below 0.8% (National Research Council 2007a). During the trouble-plagued Apollo 13 lunar mission, CO_2 toxicity became a major threat when levels in the spacecraft rose to about 4% before the astronauts managed to rig a makeshift system of CO_2 scrubbers that brought levels below 0.5% (**Figure 5.1**).

Processes that alter atmospheric CO_2 concentration generally produce nearly equal but opposite changes in atmospheric oxygen (O_2) concentration. The actual ratios of CO_2 to O_2 vary among processes.

- When plants conduct **photosynthesis**, they release between 1.0 molecule and 1.2 molecules of O_2 for every molecule of CO_2 that they consume, depending on the extent to which they are synthesizing sugars versus converting **nitrate** (a form of inorganic nitrogen) into proteins (Rachmilevitch et al. 2004).

FIGURE 5.1 CO_2 and human respiration Astronauts from Apollo 13 rig a CO_2 scrubber, the black box with blue-gray duct tape, to diminish CO_2 concentration in the spacecraft.

- Combustion of natural gas, crude oil, and solid carbon consume 1.95 molecules, 1.44 molecules, and 1.17 molecules of O_2, respectively, for every molecule of CO_2 released (Bender et al. 2005; Battle et al. 2006).

- A popular motto of my generation was "You are what you eat." Someone who studies animal **respiration** might add the proviso "You are what you eat, and we can tell what you've eaten by the gases you give off." That is because if you ingest large quantities of fats (e.g., an Atkins Diet), you will inhale 1.4 molecules of O_2 for every molecule of CO_2 that you exhale; on the other hand, if you ingest mostly carbohydrates (e.g., an anti-Atkins Diet), you will inhale 1.0 molecule of O_2 for every molecule of CO_2 you exhale.

- When plants respire, they consume 1.1 molecules of O_2 for every molecule of CO_2 released if they are using **ammonium** (another form of inorganic nitrogen) as a nitrogen source and 1.4 molecules of O_2 for every molecule of CO_2 released if they are using nitrate (Bloom et al. 1989; Bloom et al. 1992).

Exceptions to these relatively fixed ratios between O_2 and CO_2 fluxes occur at the air–sea interface, where substantial O_2 exchange may be independent of CO_2 exchange and vice versa. In the summer, atmospheric O_2 concentration rises as enhanced algal photosynthesis releases more O_2 to the air. Deep waters tend to be slightly deficient in O_2 because respiration and other chemical reactions

TABLE 5.1	Atmospheric CO_2 concentrations and human responses to CO_2
Time or response	CO_2 concentration (%)
Ice-age levels	0.018–0.022
Preindustrial levels	0.026–0.028
Current levels	0.038–0.040
Predicted 2100 levels	0.05–0.10
Exhaled from lungs	5.3–5.9
Loss of mental acuity	2.0–7.5
Loss of consciousness	5.0–10.0
Loss of life	20.0–30.0

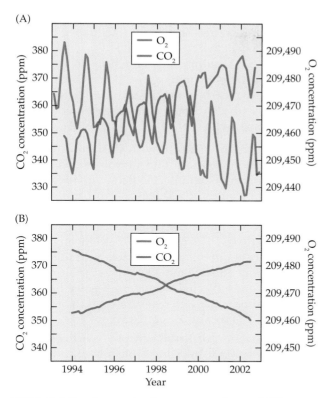

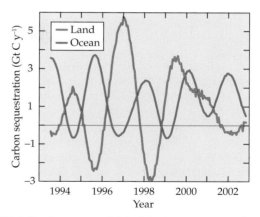

FIGURE 5.3   Rates at which CO$_2$ is sequestered in biomass on land or dissolved in oceans. Units are in 10^9 metric tons of carbon equivalents per year (carbon equivalent is the carbon content of CO$_2$ where 1 g carbon equivalent = 3.67 g CO$_2$). Based on changes in CO$_2$ and O$_2$ concentrations at Barrow, American Samoa, and Cape Grim. (After Bender et al. 2005.)

FIGURE 5.2   Atmospheric concentrations of CO$_2$ and O$_2$ over time The tick marks on the x-axis indicate January of the designated year. (A) Seasonal variations at Barrow, Alaska (northern tip of the United States). (B) Yearly averages (running means of 6 months before and after a given date to smooth out seasonal variation) at three sites: Barrow, American Samoa (the South Pacific island), and Cape Grim (the northwestern tip of Tasmania). (After Bender et al. 2005.)

consume O$_2$, while limited mixing with the atmosphere and limited photosynthesis because of low light levels fail to replenish it. When these deep waters, low in O$_2$, upwell to the surface in the wintertime, more atmospheric O$_2$ dissolves into the oceans, and the atmospheric O$_2$ concentration falls (**Figure 5.2**).

Fluctuations in atmospheric concentration of CO$_2$ are smaller in amplitude than those of O$_2$ for two reasons: CO$_2$ is more soluble in water than O$_2$, and oceans can absorb large amounts of CO$_2$ as bicarbonate (HCO$_3^-$) and carbonate (CO$_3^{2-}$) ions (see Equations 2.9, 4.12, and 4.13). As a consequence, terrestrial CO$_2$ exchanges, but not oceanic ones, are proportional to atmospheric O$_2$ fluctuations (Keeling and Garcia 2002; Bender et al. 2005). A comparison between CO$_2$ and O$_2$ fluctuations, therefore, provides an estimate of the rela-

tive amounts of CO$_2$ sequestered in the ocean versus that sequestered on land (**Figure 5.3**).

Computer models (GCMs) predict, based on the anticipated changes in atmospheric CO$_2$ concentration and the ratios between O$_2$ and CO$_2$ fluxes, that atmospheric O$_2$ concentration will decrease in the future—but this will not leave people gasping for breath. Seasonal O$_2$ oscillations are a mere 0.003% (30 ppm) on a background of 20.946%. The decline in average global O$_2$ concentrations from 1993 to 2003 of about 25 ppm (see Figure 5.2) is smaller than the one a person experiences in an elevator when it ascends one or two floors. On top of Mt. Everest, where everyone gasps for breath, O$_2$ concentration is about 7%.

Direct Effects of Carbon Dioxide on Plants

A brief introduction to plant physiology should prove helpful in understanding the responses of plants to higher CO$_2$ concentrations.

Resource triangle

Resources vital to plants and other living organisms fall into three categories: energy (carbon or light), water, and nutrients (**Figure 5.4**). Plants may use up stores of one resource to acquire additional amounts of another resource if this second resource limits their growth (Bloom et al.1985). For example, plants will

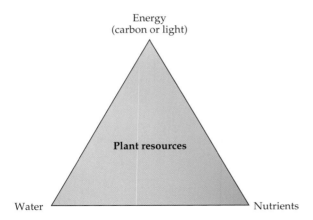

FIGURE 5.4 Plant resource triangle Three categories of natural resources that plants and other organisms must obtain from their environment.

expend more carbohydrates (energy) on root growth when water or nutrients are scarce. The following sections detail how a higher atmospheric CO_2 concentration directly enhances the availability of energy and water but inhibits the acquisition of nutrients, and how this mechanism influences plant responses to CO_2.

Energy (carbon or light) and plants

Photosynthesis is the conversion of electromagnetic radiation from the sun into biochemical energy that supports nearly all life on Earth. The initial chemical reactions, which occur only when plants or other photosynthetic organisms are exposed to light, produce the biochemical compounds ATP and NADPH. These high-energy substances, like volatile acquaintances who may break down if perturbed, are not viable in the long term. More desirable are associates who are energetic and sweet, yet stable. So it is with photosynthetic organisms: They convert volatile ATP and NADPH into stable sugars, compounds that they can easily transport across compartments, can store for long periods (like sugar in the bowl on your table), or can readily break down to generate energy whenever the plant or photosynthetic organism needs it.

All photosynthetic organisms follow the same biochemical pathway to convert ATP and NADPH into sugars. This is the Calvin–Benson cycle, also known as the **C_3 carbon fixation** pathway (see Equation 2.1). In higher plants (versus cyanobacteria), this pathway occurs in a special compartment called the **chloroplast**. Rubisco (*ribu*lose-1,5-*bis*phosphate *c*arboxylase/*o*xygenase)—the **enzyme** that comprises up to 50% of all protein in leaves and is thereby the most prevalent

protein on the planet—**catalyzes** (facilitates) the first reaction of this pathway (**Figure 5.5**).

Rubisco exhibits opposing tendencies in that it catalyzes two different chemical reactions: One reaction combines a five-carbon sugar RuBP (ribulose-1,5-bisphosphate) with CO_2, and the other reaction combines this same sugar with O_2.

- The reaction of RuBP with CO_2 produces a six-carbon compound (5 carbons + 1 carbon → 6 carbons) that intermediately divides into two molecules of a three-carbon compound PGA (3-phosphoglycerate) (6 carbons ÷ 2 → 3 carbons), hence the name C_3 *carbon fixation*. Six of these PGA molecules pass through an elaborate chemical cycle that expends the energy of 18 ATP and 12 NADPH molecules, forms one molecule of fructose-6-phosphate, a six-carbon sugar, and regenerates six molecules of RuBP (Taiz and Zeiger 2006).

- The reaction of RuBP with O_2 oxidizes the RuBP, splits it into one molecule of a three-carbon PGA and one molecule of a two-carbon

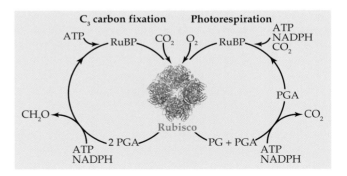

FIGURE 5.5 C_3 carbon fixation and photorespiratory pathways The enzyme rubisco catalyzes the reactions between RuBP (ribulose-1,5-bisphosphate) and either CO_2 or O_2. The first stable products are, respectively, two molecules of PGA (3-phosphoglycerate) or one molecule each of PGA and PG (2-phosphoglycolate). Driving these reactions are the high-energy compounds ATP and NADPH generated from the light-dependent reactions of photosynthesis. Carbon fixation, the reaction with CO_2, produces sugars (CH_2O), whereas photorespiration, the reaction with O_2, releases CO_2 and reincorporates it during regeneration of RuBP. Generation of NADPH releases O_2 (not shown). The relative rates of the two reactions depend on the relative concentrations of CO_2 or O_2 at the active site of the enzyme; therefore, elevated CO_2 or low O_2 atmospheres inhibit photorespiration. In the center is a depiction of the structure of rubisco with its large subunits in red and small subunits in blue.

PG (2-phosphoglycolate) (5 carbons → 3 carbons + 2 carbons), and subsequently releases CO_2, hence the names C_2 *pathway* or, more commonly, **photorespiration**. In total, photorespiration consumes five ATP and three NADPH per RuBP oxidized and regenerated but does not result in any net production of sugar (Tolbert 1994; Foyer et al. 2009). Thus, photorespiration has been viewed as a wasteful process, a vestige of the high CO_2 atmospheres under which plants evolved (Wingler et al. 2000).

The balance between C_3 carbon fixation and photorespiration depends on the relative amounts of CO_2 and O_2 entering the **active site** of rubisco (i.e., the portion of the enzyme involved in the primary chemical reactions) and the **affinity** of the enzyme for each gas (i.e., the degree to which it attracts CO_2 or O_2). Atmospheric concentrations of CO_2 and O_2 are currently 0.039% and 20.946%, respectively, yielding a CO_2:O_2 ratio of 0.0019. Gaseous CO_2, however, is much more soluble in water than O_2, and so the CO_2:O_2 ratio near the chloroplast, the part of a cell where these reactions occur, is about 0.026 at 25°C. Rubisco has about a 50-fold (in cyanobacteria) to 100-fold (in higher plants) greater affinity for CO_2 than O_2 (Galmes et al. 2005). Altogether, because of the relative concentrations and affinities, rubisco catalyzes about two to three cycles of C_3 carbon fixation for every cycle of photorespiration (Sharkey 1988) (see Figure 5.5).

Temperature influences the balance between C_3 carbon fixation and photorespiration in two ways. The solubility of CO_2 in water decreases more with temperature than the solubility of O_2, resulting in a lower CO_2:O_2 ratio at higher temperatures. In addition, the enzymatic properties of rubisco shift with temperature, stimulating the reaction with O_2 to a greater degree than the one with CO_2. Higher temperatures, therefore, favor photorespiration over C_3 carbon fixation, and photosynthetic conversion of absorbed light into sugars becomes less efficient (**Figure 5.6**).

With rising atmospheric CO_2 concentration and falling O_2 concentration (see Figure 5.2), the CO_2:O_2 ratio at the active site of rubisco increases and thereby favors RuBP carboxylation (C_3 carbon fixation) over oxidation (photorespiration). Under such conditions, plants that rely solely on C_3 carbon fixation (C_3 plants) enjoy higher efficiency in converting absorbed light into sugars (see Figure 5.6A). Conversely, when atmospheric CO_2 concentration diminished to one-half of current levels during past ice ages (see Figure 2.29), RuBP carboxylation decreased and oxygenation increased. These conditions spawned a CO_2 pumping mechanism known as **C_4 carbon fixation** (Sage 1999).

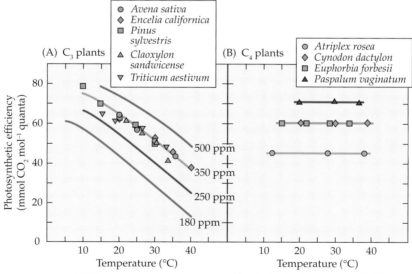

FIGURE 5.6 Temperature and CO_2 responses of photosynthetic efficiency in representative plants Photosynthetic efficiency is in terms of the net amount of CO_2 fixed into sugars per particle waves of light absorbed (quantum yield). (A) Responses of most plants (plants that conduct C_3 carbon fixation). The red, green, blue, and brown lines are theoretical responses at the designated atmospheric CO_2 concentration. The symbols are data for the indicated species at 350 ppm CO_2, which demonstrates that different C_3 species have similar responses. (B) Responses of plants that have a CO_2 pumping mechanism (C_4 carbon fixation as discussed in the following section). Symbols are data for the indicated species at 350 ppm CO_2. Photosynthetic efficiencies (quantum yields) in C_4 plants are neither sensitive to temperature (as shown) nor CO_2 concentration (not shown). (After Ehleringer et al.1997.)

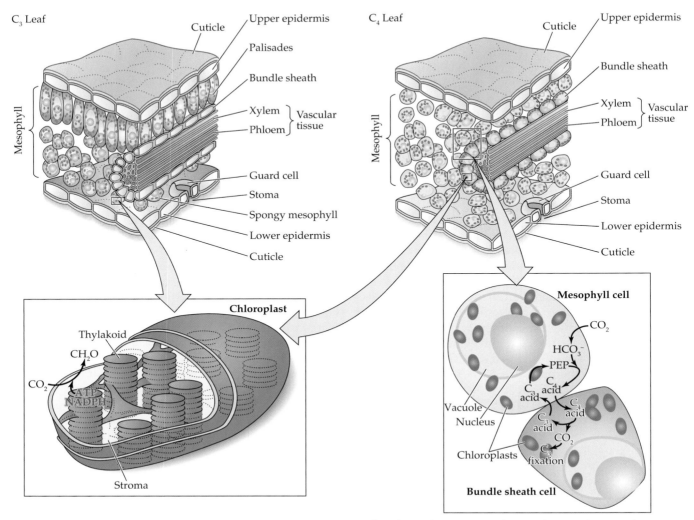

FIGURE 5.7 Cross sections of leaves in C$_3$ and C$_4$ plants Insets depict C$_3$ carbon fixation in a chloroplast (at lower left) and the biochemical pathway for a CO$_2$ pumping mechanism called C$_4$ carbon fixation (at lower right). Most plants are C$_3$ plants. In mesophyll cells of these plants, C$_3$ carbon fixation occurs in every chloroplast. In C$_4$ plants, CO$_2$ dissolves in the cytoplasm of a mesophyll cell and forms bicarbonate (HCO$_3^-$), which then reacts with PEP (phosphoenolpyruvate), a three-carbon acid, to produce a four-carbon acid. The four-carbon acid is transported to the bundle sheath, where it breaks down to release a three-carbon acid and CO$_2$, which then enters a chloroplast and participates in C$_3$ carbon fixation. The three-carbon acid returns to the mesophyll cell and is reconverted into PEP in a chloroplast. The net result of C$_4$ fixation is to pump CO$_2$ from the mesophyll cell to the bundle-sheath cell.

C$_4$ carbon fixation

Some plants, including the commercially important crops of maize (corn), sorghum, sugarcane, millet, and amaranth, conduct an additional mechanism, C$_4$ carbon fixation. In the leaf of a C$_4$ plant (**Figure 5.7**), CO$_2$ enters a **mesophyll** cell, dissolves in H$_2$O, and forms HCO$_3^-$ (bicarbonate). (See the chemical reactions outlined in Equations 2.9 and 4.12.) The enzyme PEP carboxylase catalyzes the reaction of this HCO$_3^-$ with a three-carbon acid PEP (phosphoenolpyruvate) to produce a four-carbon acid (oxaloacetate, or OAA) (1 carbon + 3 carbons → 4 carbons), hence the name *C$_4$ carbon* *fixation*. Atmospheric O$_2$ concentration has no effect on this reaction.

Subsequently, OAA is converted into another four-carbon acid (malate or aspartate, depending on species), which diffuses to the bundle sheath, a structure near the middle of a leaf (see Figure 5.7). In the bundle sheath, malate or aspartate splits into CO$_2$ and a three-carbon acid (pyruvate or alanine, respectively). This release of CO$_2$ drives bundle sheath concentration to above 0.1% CO$_2$, and standard C$_3$ carbon fixation via rubisco proceeds in bundle sheath chloroplasts with negligible photorespiration. To complete the cycle,

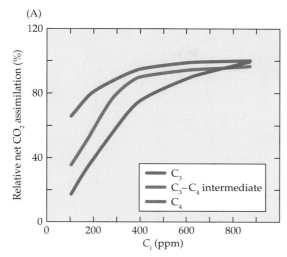

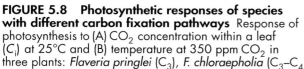

FIGURE 5.8   Photosynthetic responses of species with different carbon fixation pathways Response of photosynthesis to (A) CO_2 concentration within a leaf (C_i) at 25°C and (B) temperature at 350 ppm CO_2 in three plants: *Flaveria pringlei* (C_3), *F. chloraepholia* (C_3–C_4 intermediate), and *F. bidentis* (C_4). Plotted are net CO_2 assimilation (photosynthetic CO_2 consumption minus respiratory CO_2 release) relative to the maximum rate of the C_4 species. (A after Rachmilevitch and Bloom, unpublished; B after Ku et al. 1991.)

pyruvate or alanine returns to the mesophyll cells, and the mesophyll chloroplast regenerates PEP from this pyruvate or alanine.

Costs versus benefits of C_4 carbon fixation depend on environmental conditions. One cost is that the additional chemical reactions in this cycle expend two ATP per CO_2 molecule pumped from the mesophyll to the bundle sheath. Moreover, C_4 carbon fixation requires special anatomical features such as tightly packed bundle sheath cells to maintain large differences in CO_2 concentrations within various parts of the leaf.

The benefits of C_4 fixation are that low atmospheric CO_2 concentration or high temperature (**Figure 5.8**) does not inhibit photosynthetic sugar production as severely as in C_3 plants. At the current atmospheric CO_2 concentration of 386 ppm, the crossover point at which C_3 and C_4 plants are equally efficient in converting light into chemical energy (quantum yield) falls between 20°C and 30°C (see Figure 5.6). Plants in which C_4 carbon fixation is not fully developed, C_3–C_4 intermediate species, have CO_2 and temperature responses that are intermediate to C_3 and C_4 species in the same genus.

C_3 versus C_4: Why should you care?

The physiology of photosynthesis, outlined in the preceding sections, determines where your food is grown today and where it may be grown tomorrow as optimum growing locations shift with climate change. In particular, the physiological differences between the two carbon fixation pathways prescribe that low atmo-

spheric CO_2 concentration and high daytime growing temperature favors C_4 plants over C_3 plants (**Figure 5.9**). Geographical distributions of C_3 and C_4 plants reflect these physiological differences. For example, temperature optima of the two pathways determine the locations of the spring wheat and temperate corn

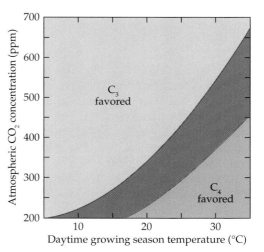

FIGURE 5.9   Conditions favoring C_3 versus C_4 plants are based on the light efficiencies of photosynthesis (quantum yield) for both pathways. High CO_2 concentration and low temperature (light blue) favor C_3 plants, whereas low CO_2 concentration and high temperature (pink) favor C_4 plants. Results under intermediate conditions (purple) vary among different types of C_4 plants. (After Ehleringer et al. 1997.)

FIGURE 5.10 Geographical distributions of spring wheat, a C_3 plant, and temperate maize (corn), a C_4 plant. (A) Actual distributions in 1970 and (B) potential distributions in 2050 based on the physiological responses of the two species and assuming non-irrigated agriculture and a doubling of atmospheric CO_2 concentration. (After Leemans and Solomon 1993.)

(A) 1970

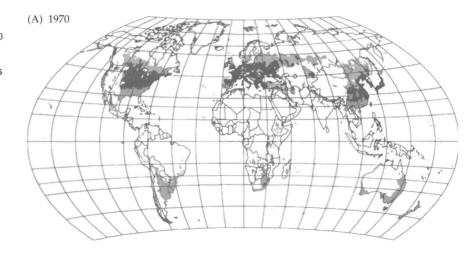

(B) 2050

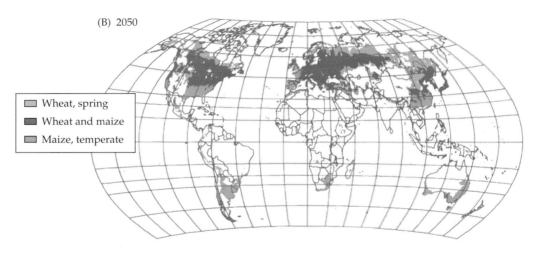

Wheat, spring
Wheat and maize
Maize, temperate

belts in the Great Plains of North America, Europe, and Asia; the colder, northern latitudes are more suitable for wheat production, and the warmer, southern latitudes are more suitable for maize (corn) production (**Figure 5.10A**). Distributions of native C_3 and C_4 grasses in the Great Plains are also consistent with such considerations (Ehleringer et al. 1997).

Global climate models predict that both atmospheric CO_2 concentration and temperature will continue to increase (see Chapter 4). Such changes will exert opposing pressures on the distribution of C_3 and C_4 plants (see Figure 5.9): Higher CO_2 concentration will favor C_3 plants, whereas higher temperature will favor C_4 plants. Nonetheless, the warming of Arctic regions, which is anticipated to be more extreme than in other parts of the world (see Figure 4.34), will push wheat and corn belts farther northward (**Figure 5.10B**).

Global changes in the availability of water (H_2O) and nutrients will also influence the distributions of

C_3 and C_4 plants. The following sections examine the interaction of these resources and plants.

Water and plants

Terrestrial plants face a dilemma. They must expose their internal organs to the atmosphere for CO_2 to enter their leaves and participate in photosynthetic carbon fixation. Yet such exposure permits H_2O vapor to escape from leaves and desiccates them.

To address this dilemma, plants precisely regulate gas exchange between leaves and the atmosphere. Leaf surfaces are covered with a waxy layer called the cuticle (see Figure 5.7) that inhibits inward diffusion of CO_2 into leaves as well as outward movement of H_2O from them. Atmospheric CO_2 enters leaves and H_2O vapor exits them through openings called **stomata** (singular, *stoma*). Plants control the extent to which they open their stomata based on a number of environmen-

(A)

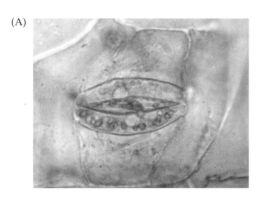

(B)

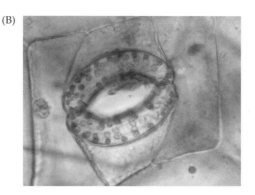

FIGURE 5.11 A stoma of the *Tradescantia zebrina*, C$_3$ plant (A) At night, the stoma closes. (B) During the day, the stoma opens.

tal cues—relative humidity, light level (**Figure 5.11**), CO$_2$ concentration, and soil H$_2$O availability—to minimize the amount of H$_2$O lost per CO$_2$ assimilated into carbohydrate (Cowan and Farquhar 1977).

Fick's first law of diffusion dictates that a gas moves from regions of high concentration to regions of low concentration at a rate that is proportional to the concentration gradient (i.e., the difference in concentration over distance). Unfortunately for plants, the concentration gradient driving H$_2$O vapor out of a leaf is usually much larger than the gradient driving CO$_2$ into the leaf. Air spaces within a leaf are saturated with H$_2$O vapor because cells that line the air spaces must be permeable to H$_2$O in order for gases to enter or exit them. At 25°C, the **partial pressure** of H$_2$O vapor within a leaf is 3.12% (see Figure 2.43). When, for example, the relative humidity of air around a leaf is 50%, the concentration of H$_2$O vapor at 25°C is one-half of 3.12%, or 1.56%, and so the gradient driving water vapor out of the leaf is 3.12% − 1.56% = 1.56%.

Compare this with the gradient for CO$_2$. The partial pressure of CO$_2$ outside a leaf is 0.0386%, and the lowest it can go within a leaf is about 0.0005%. The concentration gradient for CO$_2$, therefore, is 0.0381%, and at 25°C and 50% relative humidity, the concentration gradient driving H$_2$O vapor out of a leaf is about 40 times that driving CO$_2$ into it.

To make matters worse, the diffusion rate of a substance is proportional to its density. Water is a smaller, denser molecule than CO$_2$. Consequently, a molecule of H$_2$O vapor diffuses out of a leaf 1.6 times faster than a molecule of CO$_2$ diffuses into a leaf.

Worse yet, the escape route for H$_2$O is much shorter than the path that CO$_2$ must travel from the atmosphere to the active site of rubisco. Water vapor needs merely exit through a stoma, whereas CO$_2$ must enter through the stoma, pass through the membrane of a mesophyll cell, and pass through the double membrane of a chloroplast (see Figure 5.7). As you can imagine, gas transport through biological membranes is relatively slow.

All told, C$_3$ plants lose from 500 molecules to 1000 molecules of H$_2$O for every molecule of CO$_2$ they assimilate. In contrast, C$_4$ plants are more efficient; they lose only 200 to 300 molecules of H$_2$O per molecule of CO$_2$ assimilated because they operate at lower internal CO$_2$ concentrations (see Figure 5.8A) and have smaller stomatal apertures than C$_3$ plants. Still more efficient are plants such as cacti that lose only about 50 H$_2$O molecules per CO$_2$ assimilated using **crassulacean acid metabolism** (**CAM**), a third type of carbon fixation (Bloom and Troughton 1979).

Crassulacean acid metabolism

The nature of crassulacean acid metabolism (CAM) evokes the following joke (I consulted with my attorneys, and they are of the opinion that a joke in a textbook, while unprecedented, is not prohibited).

Astronauts from all over the world were attending a workshop. After dinner and drinks, they began to boast about their countries space programs. Russians bragged about their plans to send humans to Mars by the year 2018, Americans talked about establishing a permanent moon base by 2024, and so on until all known planets and moons in the solar system were claimed. Finally, an astronaut, who had heretofore remained silent, blurted out that his country would be sending people to the sun by the year 2100. The other astronauts gasped, "The sun! You'll burn up!" The astronaut replied, "We'll go at night."

When challenged to acquire both sunlight and CO$_2$ for photosynthesis, yet avoid excessive H$_2$O losses under hot, dry conditions, crassulacean acid metabolism (CAM) plants replied, "We'll go at night." Plants cannot store sunlight, nor can they amass large quanti-

(A) Night: Stomata opened

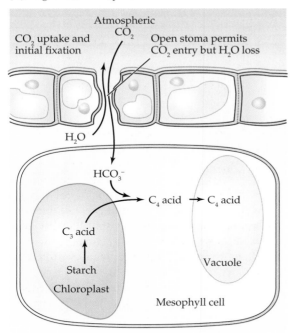

(B) Day: Stomata closed

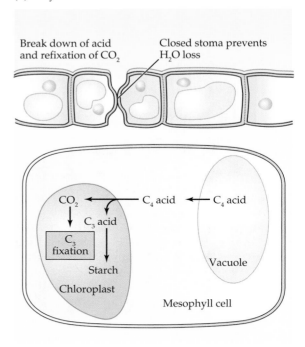

FIGURE 5.12 CAM carbon fixation (A) At night, stomata are open; CO_2 enters leaves and generates a four-carbon (C_4) acid that is stored in the vacuole, a large storage compartment in a mesophyll cell. (B) During the day, stomata remain closed, and the C_4 acid is broken down to release CO_2 internally. This CO_2 participates in C_3 carbon fixation. Notice that in CAM plants the two chemical reactions with CO_2 are separated in time (day vs. night), whereas in C_4 plants (see Figure 5.7), the two reactions are separated in space (mesophyll cell vs. bundle-sheath cell).

ties of the unstable compounds ATP and NADPH that they generate from sunlight. Instead, CAM plants have effectively developed a method of stockpiling CO_2; they open their stomata at night, convert the CO_2 that enters their leaves into an organic acid (malate), and hoard large quantities of this organic acid in vacuoles, storage compartments within mesophyll cells (**Figure 5.12A**). During the day, when H_2O loss would be highest, they close their stomata, break down the organic acid to release CO_2 internally, and fix this CO_2 via the standard C_3 pathway (**Figure 5.12B**). The downside of CAM is that the requirement for a large amount of storage seems to interfere with rapid photosynthesis under well-watered conditions.

Water-use efficiency

Regardless of which carbon fixation pathway (C_3, C_4, or CAM) a plant employs, rising atmospheric CO_2 concentrations will improve its water-use efficiency (**Figure 5.13**). That is, rising atmospheric CO_2 concentration will increase the amount of CO_2 assimilated into carbohydrate per H_2O lost. This is a consequence

of decreasing the disparity among the concentration gradients for H_2O and CO_2 between leaves and atmosphere. For the previous example at 25°C and 50% relative humidity, a doubling of CO_2 concentration diminishes the ratio of the concentration gradient driving H_2O vapor out of a leaf to that driving CO_2 into it from about 40 to 20.

Plants might respond to these more favorable conditions in either of two ways.

Strategy 1: They might decrease their stomatal apertures to conserve water while maintaining a similar CO_2 concentration within their leaves, thereby assimilating CO_2 at similar rates.

Strategy 2: They might open their stomata to the same extent and lose water at similar rates but increase the CO_2 concentration within leaves, thereby accelerating CO_2 assimilation (see Figure 5.8).

So what do plants actually do? Before revealing the secret life of plants, we will digress to some experimental methodology.

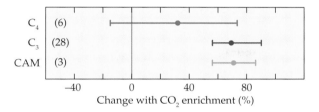

FIGURE 5.13   Water-use efficiency in response to atmospheric CO_2 levels Water use efficiency is the amount of CO_2 assimilated per amount of water lost through transpiration. Shown are the changes in water use efficiency from leaves of plants grown at ambient CO_2 concentration ($\approx$365 ppm) to those grown at elevated CO_2 concentration ($\approx$550 ppm). A change of 100% would mean that growth at elevated CO_2 doubled water use efficiency. Experiments on C_3 and C_4 plants were conducted in free air CO_2 enrichment (FACE) plots, and those on CAM plants were conducted in controlled environment or open-top chambers. In parentheses are the number of studies in each analysis. Circles and error bars designate means and confidence intervals, respectively. (After Drennan and Nobel 2000; Ainsworth and Long 2005; Leakey et al. 2006; Ainsworth and Rogers 2007.)

Growth systems and statistical analysis

In studies on plant responses to elevated CO_2 concentration, plants are grown in controlled environment chambers, open-top chambers, or *free air* CO_2 *enrichment* (FACE) plots.

Controlled environment chambers (**Figure 5.14A**) maintain light, temperature, humidity, and atmospheric CO_2 concentrations at designated levels. Such chambers consume large amounts of electricity to power lamps as well as to cool a brightly lit, confined area. Energy constraints thus limit light levels in nearly all chambers to a fraction of full sunlight. Chambers may approach normal amounts of total photosynthetically active radiation per day, however, by extending the number of hours per day that its lamps are illuminated. Plants receive water and nutrients from a solid medium in flowerpots or from hydroponic solutions in tanks.

Open-top chambers (**Figure 5.14B**) have sides of transparent plastic film and large fans that blow a

(A)

(B)

(C)

FIGURE 5.14   Methods for growing plants under different CO_2 atmospheres (A) Lesley Randall at the University of California, Davis, works in a controlled environment chamber that permits precise control of light, temperature, humidity, and CO_2 concentration. Sugar maple saplings in the tub (foreground) suffer from receiving nitrate nutrition under elevated CO_2. (B) Open-top chambers. Here soybeans are growing under different CO_2 atmospheres in North Carolina. (C) Free air CO_2 enrichment (FACE) plots. Aspen are growing under different CO_2 atmospheres in Rhinelander, WI. A control system releases pure CO_2 out of the ring of tall, white vertical standpipes, based on measurements of wind speed, wind direction, and CO_2 concentration within the ring.

specified CO_2 concentration into the chamber as well as keep temperatures inside and outside the chambers more similar. Carbon dioxide is denser than air (1.799 kg m^{-3} vs. 1.184 kg m^{-3} at sea level and 25°C), and so any CO_2 added will remain near the ground until diminished through plant CO_2 assimilation. Unfortunately, environmental conditions near the chamber sides differ in light, temperature, and humidity from those near the middle of the chamber.

A FACE plot consists of a ring of 24 to 32 vertical standing vent pipes set up in a grassland, agricultural field, or stand of trees (**Figure 5.14C**). Wind speed and direction and CO_2 concentration in the middle of the ring determine how much CO_2 each standpipe releases. For example, when the measured CO_2 concentration is lower than desired, the upwind standpipes release CO_2 in amounts proportional to their angle to the wind (i.e., a standpipe perpendicular to the wind releases the most CO_2) until the desired CO_2 concentration is achieved. This technique has proved effective in rings that range from 2 m in diameter with standpipes 1 m in height in a California grassland (Jasper Ridge Global Change Experiment 2005) to rings 30 m in diameter with standpipes 16 m in height in a North Carolina pine forest (Duke University FACE Experiment 2008). Unfortunately, such operations are very expensive. Supplying just the CO_2 for one large ring costs over \$20,000 per month. For this reason, most of the elevated-CO_2 experiments in FACE plots control CO_2 at about 570 ppm and sometimes only during the daytime.

No matter what system is chosen for exposing plants to elevated CO_2 atmospheres, it becomes a major undertaking, one that can strain facilities, funds, and personnel. Compromises must be made that invariably limit the number of times an experiment can be replicated. Because most studies on plant responses to elevated CO_2 have few replicates, a statistical approach called **meta-analysis** has proved valuable.

Meta-analysis combines results from many independent studies to increase experimental sample size and thus enhance the strength of statistical tests. Commonly, a meta-analysis presents weighted means, in which larger studies or those with greater experimental consistency (smaller variation) have a stronger influence on the final value than smaller studies or those with less consistency. Here, we review the results of meta-analyses that surveyed hundreds of independent studies.

Plant responses to elevated CO_2 atmospheres

What do plants actually do with respect to stomatal behavior when exposed to elevated CO_2 atmospheres? In general, plants behave in manner intermediate to Strategy 1 and Strategy 2, described earlier: They close their stomata by an average of 22% (**Figure 5.15A**) and

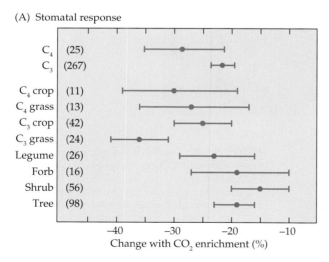

(A) Stomatal response

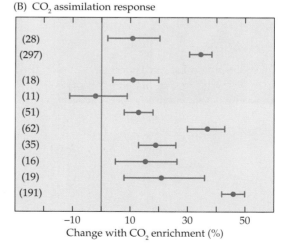

(B) CO_2 assimilation response

FIGURE 5.15 Plant responses to elevated CO_2 levels (A) Stomatal conductance to H_2O, a measure of the degree to which stomata are open and (B) maximum rates of CO_2 assimilation in response to atmospheric CO_2 levels. Shown are the changes between plants exposed to elevated CO_2 concentration (≈567 ppm) and those exposed to ambient CO_2 concentration (≈366 ppm). A change of 100% would mean that exposure to elevated CO_2 doubled stomatal conductance or CO_2 assimilation. In parentheses are the number of studies in each analysis. Circles and error bars designate means and confidence intervals, respectively, for C_4 species (red), C_3 species (blue), and different functional groups (all C_3). (After Ainsworth and Long 2005; Ainsworth and Rogers 2007.)

thereby conserve water, but they also increase CO_2 concentrations within their leaves and thereby achieve, on average, 32% faster CO_2 assimilation (**Figure 5.15B**). As a group, C_4 plants close stomata and conserve H_2O (Strategy 1) to a greater extent than C_3 plants. This reflects the fact that carbon fixation in C_4 plants is less sensitive to CO_2 concentration than in C_3 plants (see Figure 5.8). Trees tend to conserve less H_2O and assimilate CO_2 to a greater extent (Strategy 2) than grasses, legumes, forbs (other non-woody plants), or shrubs. The disproportionately large number of studies on trees in the meta-analysis biases the value for the average response of C_3 plants.

This highlights a major problem with meta-analysis: "publication bias." Meta-analyses compile the results of studies published in readily accessible sources, but such studies may not evenly represent all types of species, nor all valid studies on a question. Authors are more likely to conduct research on model species (species for which many previous results are available) and submit for publication studies with positive results rather than those with null (negative or inconclusive) results. Readily accessible journals are more likely to accept for publication studies on model species with positive results. Publication bias has become such a problem in evaluating drug efficacy that several prominent medical journals will no longer publish results of pharmaceutical research unless the research project was registered in a public database when it was initiated. In this way, negative results on drugs will not just disappear. Unfortunately, no equivalent system exists to track studies about the influence of global climate change on organisms.

CO_2 acclimation

Net CO_2 assimilation is the difference between the CO_2 consumed during photosynthetic carbon fixation and that released during respiration. Initial exposure of C_3 plants to elevated CO_2 concentration strongly stimulates net CO_2 assimilation (see Figure 5.8). Yet days, weeks, or months of exposure to an elevated CO_2 concentration slow down net CO_2 assimilation (**Figure 5.16**) and aboveground growth (**Figure 5.17**). Consequently, in the long-term, growth rates at an elevated CO_2 concentration are only 8% faster than those at an ambient (normal) CO_2 concentration (Poorter and Navas 2003). This decline in the stimulation of net CO_2 assimilation and growth by CO_2 enrichment is known as **CO_2 acclimation**.

CO_2 acclimation presents several experimental challenges. For instance, the response of even a single species within a single experiment changes with time of exposure to elevated CO_2 concentration. Such changes

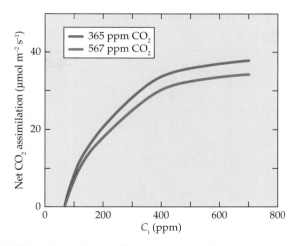

FIGURE 5.16 CO_2 acclimation Net CO_2 assimilation as a function of short-term exposures to different CO_2 concentrations within a leaf (C_i) for C_3 plants grown at either ambient (≈365 ppm) or elevated (≈567 ppm) atmospheric CO_2 concentration. Net CO_2 assimilation is in units of 10^{-6} moles of CO_2 assimilated per square meter of leaf surface area per second. Short-term exposure to elevated CO_2 stimulates net CO_2 assimilation, but this stimulation is less in plants grown under elevated CO_2 atmospheres. (After Ainsworth and Rogers 2007.)

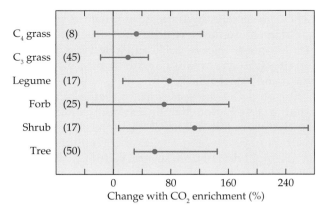

FIGURE 5.17 Aboveground plant biomass in response to atmospheric CO_2 levels Shown are the changes in biomass from plants grown at ambient CO_2 concentration (≈366 ppm) to those grown at elevated CO_2 concentration (≈567 ppm). In parentheses are the number of studies in each analysis. Circles and error bars designate means and confidence intervals, respectively. (After Nowak et al. 2004.)

TABLE 5.2 Relative change in yields for selected grains at elevated versus ambient CO_2 concentrations

Crop	Range of values	Mean chamber[a]	Mean FACE[b]	Number of studies	References
Wheat	–20 to +80	+10	+7	50	Amthor 2001
Soybean	–20 to +100	+24	+15	58	Ainsworth et al. 2002; Morgan et al. 2005
Rice	+4 to +71	+15	+12	6	Baker et al. 1992; Ziska et al. 1997; De Costa et al. 2003; Baker 2004; Bannayan et al. 2005; Weerakoon et al. 2005
Maize	–35 to +93	+29	≈ 0	57	Rogers and Dahlman 1993; Leakey et al. 2004
Sorghum	–4 to +31	+31	+6	2	Chaudhuri et al. 1986; Ottman et al. 2001

[a]Mean value for environmental and open-top chamber experiments

[b]Mean value for free air CO_2 enrichment (FACE) experiments

contribute to the wide variation in results for crop yields (Table 5.2) and aboveground biomass (Figure 5.18).

Of the explanations advanced for CO_2 acclimation, one of the most popular involves insufficient sink strength. According to this hypothesis, plants exposed to elevated CO_2 atmospheres are unable to consume all the carbohydrates that they generate via CO_2 assimilation because of genetic constraints, high temperature limitations, developmental changes, or nitrogen deficiency. These carbohydrates accumulate as sugars and starch in chloroplasts, which cause structural changes in chloroplasts (Figure 5.19) and a decrease in rubisco activity, CO_2 assimilation, growth, and yield (Ainsworth and Rogers 2007).

Elevated CO_2 increases grain yields to a smaller extent in FACE plots than in environmental or open-top chambers (see Table 5.2) (Long et al. 2006). This discrepancy between FACE and chamber experiments might derive from several factors, including the following.

- FACE experiments are so expensive and involve so many investigators that nearly all results are published, even null results, whereas chamber experiments with null results are less likely to be published.
- FACE experiments often apply modest amounts of fertilizer to plants growing in

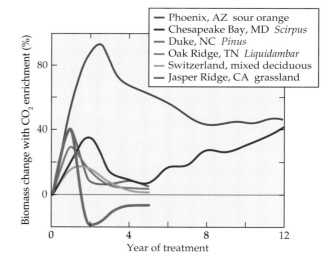

FIGURE 5.18 Differences in biomass after years of exposure to CO_2 enrichment Six different studies in which the designated types of plants were exposed in experimental plots to elevated (≈567 ppm) or ambient (≈365 ppm) atmospheric CO_2 concentration for 5 years or longer. Shown are percent differences in plant biomass between elevated-CO_2 and ambient-CO_2 plots as a function of treatment year. (After Dukes et al. 2005; Rasse et al. 2005; Korner 2006; Kimball et al. 2007.)

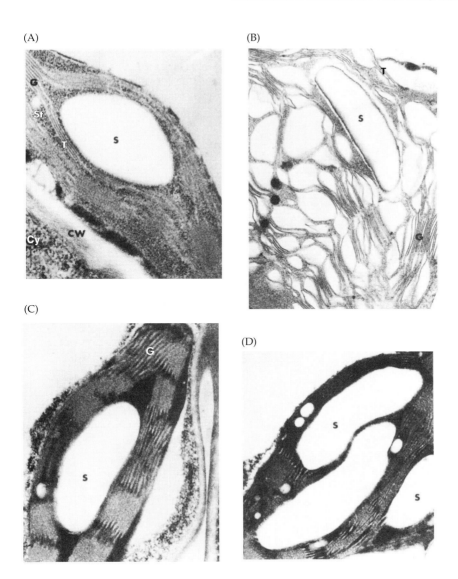

FIGURE 5.19 Chloroplasts of tomato (*Lycopersicon esculentum*) (A, B) and a wild relative (*L.chmielewski*) (C, D) grown at 330 ppm CO_2 (A, C) and 900 ppm CO_2 (B, D) for 10 weeks. S, starch; T, thylakoid; G, granum stack of thylakoids; St, stroma; CW, cell wall; Cy, cytoplasm. (From Yelle et al. 1989.)

soil, whereas chamber experiments usually apply higher levels of fertilizer to hydroponic solutions, pots, or soils. The amount of nitrogen fertilizer, as discussed in the following sections, is a major determinant of plant responses to elevated CO_2 atmospheres.

Nitrogen nutrition of plants

"Sugar and spice and everything nice; that's what little boys and girls are made of." Boys and girls obtain their sugar and spice from plants, but where do plants obtain their constituents? The most plentiful elements—oxygen, carbon, and hydrogen (Table 5.3)—derive from water or atmospheric gases. Plants "mine" all of the remaining elements from soil. These "mineral" ele-

ments constitute one corner of the plant resource triangle (see Figure 5.4).

Nitrogen (N) is the mineral element that organisms require in the greatest amounts. It constitutes a vital part of many organic compounds including proteins such as rubisco and nucleic acids such as DNA. Application of nitrogen fertilizers stimulates nearly all agricultural and natural ecosystems (Figure 5.20). Consequently, nitrogen availability is considered to be a major factor that limits plant productivity (Epstein and Bloom 2005).

You should now be able to answer this riddle: What is scarce although it literally surrounds us? (Hint: nitrogen.) Earth's atmosphere contains vast quantities of molecular nitrogen (N_2), about 78% by volume (see Figure 3.33). Most organisms, however, cannot directly

(A)

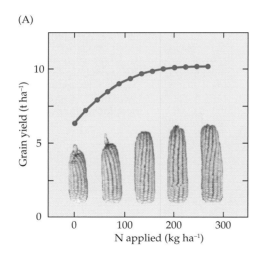

(B)

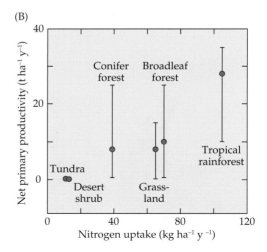

FIGURE 5.20 Influence of nitrogen on yield or productivity (A) Maize yield (metric tons per hectare) as a function of nitrogen fertilizer applied (kg ha⁻¹) for 30 Missouri farms, 1995–1998. Ears of corn, grown in Illinois at the designated fertilizer levels, show a similar response. (B) Relationship for various ecosystems between net primary productivity (metric tons of plant biomass per hectare per year as a function of plant nitrogen uptake. (A after Scharf and Wiebold 2001, corn photos courtesy of Below and Brandau 2001; B after Whittaker 1975; Haynes 1986a; Attiwill and Adams 1993.)

access this huge reservoir because of the exceptionally stable triple covalent bond between the two nitrogen atoms (N≡N) that renders it an inert, or "noble," gas. Breaking this triple bond to generate more reac-

tive forms, such as ammonia (NH_3) or nitrate (NO_3^-), requires an enormous input of energy. Chemical reactions of such energetic proportions do occur during the industrial or natural process known as nitrogen fixation (see Figure 4.23).

In industrial nitrogen fixation, molecular nitrogen (N_2) and hydrogen—when subjected to elevated temperatures (400°C to 650°C) and high pressures (200 atmospheres to 400 atmospheres) and a metal catalyst (usually iron)—form ammonia (NH_3).

$$N_2 + 3\ H_2 + energy \rightarrow 2\ NH_3 \qquad (5.1)$$

This is called the Haber-Bosch process. Before World War I, the manufacture of gunpowder and other explosives in Europe largely depended on the import of saltpeter (sodium nitrate) mined in Chile. The German government realized at the onset of World War I that it was likely to be cut off from this supply of saltpeter and sponsored an intensive research program to develop an alternative source of nitrate or ammonia. The Haber-Bosch process was the result of this program.

Nowadays, most industrial nitrogen fixation is directed toward agricultural fertilizer production (we have "beaten our swords into plowshares"), which exceeds 100×10^{12} grams of nitrogen per year (FAOSTAT 2009). Agriculture in the United States consumes about 14% of the world's total, but its fertilizer use has remained relatively constant over the last decade. Fertilizer costs, because of the high-energy demands of nitrogen fixation, are strongly correlated with power costs (**Figure 5.21**).

TABLE 5.3 Elemental composition of plants and people

| | Percent dry weight | |
Element	Maize (Zea mays)	Human (Homo sapiens)
O	44.43	14.62
C	43.57	55.99
H	6.24	7.46
N	1.46	9.33
Si	1.17	0.005
K	0.92	1.09
Ca	0.23	4.67
P	0.20	3.11
Mg	0.18	0.16
S	0.17	0.78
Cl	0.14	0.47
Al	0.11	—
Fe	0.08	0.012
Mn	0.04	—
Na	—	0.47
Zn	—	0.010
Rb	—	0.005

Source: Epstein and Bloom 2005.

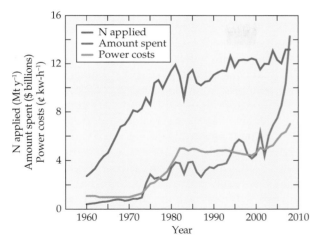

FIGURE 5.21 Nitrogen fertilizers applied in the United States (million metric tons per year) and cost of these fertilizers ($ billion) in comparison to costs of industrial electric power (¢ per kilowatt-hour). (After Economic Research Service 2009 and Energy Information Administration 2009a.)

FIGURE 5.22 Carnivorous plant in the London stage production of the musical "Little Shop of Horrors."

Certain bacteria and blue-green algae also convert N_2 into NH_3, a process known as biological nitrogen fixation. Biological fixation, in contrast to industrial fixation, accomplishes this feat at ambient temperatures and pressures. Overall, the biological reaction is

$$N_2 + 8\ e^- + 8\ H^+ + 16\ ATP \rightarrow 2\ NH_3 +$$
$$H_2 + 16\ ADP + 16\ P_i \qquad (5.2)$$

Notice that production of two NH_3 expends the equivalent of 16 ATP, an enormous amount of energy considering that most other biochemical reactions require energy expenditures of only one or, at most two, ATP.

Nitrogen-fixing microorganisms are usually free-living in the soil. A few of them, however, form symbiotic ("living together") associations with higher plants, in which the microorganism directly supplies the host plant with fixed nitrogen in exchange for other nutrients and carbohydrates. The most common type of symbiosis is between members of the pea family (Fabaceae or Leguminosae) and soil bacteria collectively called rhizobia. Another common type of nitrogen-fixing symbiosis involves several woody plant species and soil bacteria of the genus *Frankia*.

Nitrogen, once fixed into NH_3, dissolves into water to form ammonium (NH_4^+), according to

$$NH_3 + H_2O \rightarrow NH_4^+ + OH^- \qquad (5.3)$$

This NH_4^+ enters a biogeochemical cycle and passes through several organic or inorganic forms before it eventually returns to atmospheric N_2 (see Figure 4.23). Amino acids, NH_4^+, and nitrate (NO_3^-), which are pro-

duced through biological fixation or released through decomposition of soil organic matter, become the object of intense competition among plants and microorganisms. To remain competitive, plants have developed diverse mechanisms for scavenging these nitrogen forms as quickly as possible. These mechanisms range from routine to bizarre—from standard root absorption to carnivory of small animals.

Carnivorous plants have taken center stage on Broadway, in Hollywood, and at the local plant nursery (**Figure 5.22**). Nearly 600 species scattered throughout the plant kingdom are carnivorous (Ellison and Gotelli 2001). Native to nitrogen-poor sites, these plants kill small animals that have the misfortune to cross their path and digest them to acquire their nitrogen (Darwin 1875; Adamec 1997). For example, sundews (*Drosera sp.*), when fed insects labeled with the heavy isotope of nitrogen ^{15}N, developed a ratio of $^{15}N/^{14}N$ in their tissues, indicating that the plants obtained between 15% and 87% of their nitrogen through carnivory (Dixon et al.1980; Schulze and Schulze 1990; Schulze et al. 1991). Carnivorous plants, although intriguing, are relatively rare.

Most plants, even nitrogen-fixing and carnivorous ones, acquire the majority of their mineral nitrogen through root absorption of NH_4^+ and NO_3^- dissolved in the **soil solution**, the solution that fills the spaces between soil particles. The availability of NH_4^+ generally shows much less variation, both seasonally and spatially, than the availability of other nitrogen forms, especially, NO_3^-. Consequently, NH_4^+ may be the major form of nitrogen available from the soil at some times and places.

Availability of NO_3^- from the soil sometimes exceeds that of NH_4^+ by more than a factor of 10. Microbes in well-aerated, temperate soils rapidly convert NH_4^+ into NO_3^- (see Figure 4.23). Being a negatively charged ion, NO_3^- moves much more freely through the soil than NH_4^+ and other positively charged ions because NO_3^- usually does not bind to soil particles, which have negatively charged surfaces (Nye and Tinker 1977). For these reasons, most crops in temperate zones rely on NO_3^- to meet the major portion of their nitrogen requirements (Haynes 1986b). This does not mean that plants depend on NO_3^- as their sole nitrogen source; rather, it means that that NO_3^- is often their primary nitrogen source.

Once roots absorb NH_4^+ and NO_3^- from the soil solution, the two forms of inorganic nitrogen suffer distinct fates within a plant. High levels of NH_4^+ are toxic to all organisms because they interfere with energy **metabolism** (biochemical processes that generate energy) within cells. To warn of dangerous levels of NH_4^+, animals have developed a strong aversion to its smell: The active ingredient in smelling salts is ammonium carbonate, and NH_4^+ endows dirty diapers with their characteristic aroma. Plants also attempt to prevent high accumulations of NH_4^+ in their tissues. They assimilate NH_4^+ into amino acids near the site of absorption or generation and rapidly place any excess into their vacuoles, large storage compartments within plant cells. The conversion of NH_4^+ into amino acids in roots entails several reactions. In summary:

$$NH_4^+ + \text{organic acid} + 2\,ATP + e^- \rightarrow \text{amino}$$
$$\text{acid} + 2\,ADP + 2\,P_i + H_2O \qquad (5.4)$$

In contrast to NH_4^+, plants may store NO_3^- at high levels or transport it from tissue to tissue without deleterious effect. If, however, livestock or humans consume plant material that is high in NO_3^-, they may suffer methemoglobinemia. In this disease, the liver converts NO_3^- to NO_2^-, which may combine with hemoglobin and render it unable to bind oxygen. Human and other animals also may convert NO_3^- into nitrosamines, which are potent carcinogens. Some countries, therefore, limit the NO_3^- content in plant materials sold for human consumption.

Plants assimilate NO_3^- absorbed from the soil into amino acids in either their roots or their shoots via a series of energy-intensive biochemical reactions. In summary:

$$NO_3^- + 10\,ATP + 10\,H^+ + 8\,e^- \rightarrow NH_4^+ +$$
$$10\,ADP + 10\,P_i + 3\,H_2O \qquad (5.5)$$

The NH_4^+ generated in this reaction is rapidly assimilated into amino acids via Equation 5.4.

Typically, organic nitrogen compounds constitutes only about 1.5% of the dry weight of plants (see Table 5.3), but the reactions involved in NH_4^+ and NO_3^- assimilation into organic nitrogen compounds are so energy intensive that plants expend up to 25% of their energy on these reactions (Bloom et al. 1992). Even more impressive is what happens when some of these reactions run in reverse—say, from NH_4NO_3 to N_2. Kaboom! Nearly all conventional explosives (e.g., TNT, nitroglycerin, dynamite, and gunpowder) are based on the rapid breakdown of nitrogen compounds. Airport bomb detectors search for high concentrations of nitrogen compounds in luggage. Never attempt to carry nitrogen fertilizer on a commercial airline flight.

Plants vary in their relative dependence on NH_4^+ and NO_3^- as sources of nitrogen from the soil. When nitrogen is scarce, plants incorporate whatever form they can find, and so the balance between NH_4^+ and NO_3^- usage varies with relative availability in the soil. If both nitrogen forms are present in ample amounts, most plants prefer NH_4^+ (Bloom 1997). This is because NH_4^+ assimilation into amino acids requires 2 ATP (Equation 5.4), whereas NO_3^- assimilation first into NH_4^+ and then into amino acids requires 12 ATP (Equation 5.4 and 5.5). The preference for NH_4^+ becomes more pronounced in plant communities where light energy is limiting, such as forests with dense leaf canopies.

Nitrogen and plant responses to elevated CO_2

With rising atmospheric CO_2 concentration, the availability of energy (carbon) and water resources increases and stimulates plant growth (see Figure 5.17). The availability of nutrient resources, however, may not keep pace. In particular, nitrogen resources tend to lag behind. Consider the effects of atmospheric CO_2 concentration on the major sources and sinks of mineral nitrogen, NH_4^+ and NO_3^-, in soils.

Major sources of mineral nitrogen in soils are biological nitrogen fixation and nitrogen **mineralization**, microbial breakdown of soil organic matter into mineral nitrogen. These processes are generally independent of atmospheric CO_2 concentration, but nitrogen fixation may decrease under CO_2 enrichment if soils do not receive additional nutrients (**Figure 5.23**). Plants under CO_2 enrichment initially grow faster and release more carbohydrates to soils (Reich et al. 2006). In response to the input of additional energy, soil microorganisms become more active, respiring on average 18% faster than under ambient atmospheres. As soil microorganisms become more active, they remove more nutrients

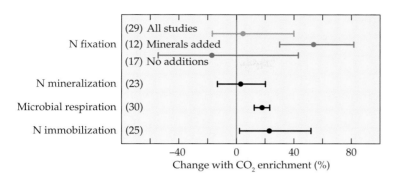

FIGURE 5.23 Soil processes in response to atmospheric CO$_2$ levels Shown are the changes in nitrogen fixation, net nitrogen mineralization, microbial respiration, and gross nitrogen immobilization from plots exposed to ambient CO$_2$ concentration ($\approx$370 ppm) to those exposed to elevated CO$_2$ concentration (430–750 ppm). A change of 0% would mean that exposure to elevated CO$_2$ had no influence on the rates of a process. Some nitrogen fixation studies added nutrients ("Minerals added"), whereas others did not ("No additions"). In parentheses are the number of studies in each analysis. Circles and error bars designate means and confidence intervals, respectively. (After de Graaff et al. 2006.)

from the soil. Iron and molybdenum, both of which are required for nitrogen fixation, may become scarce, and nitrogen fixation may decline (Hungate et al. 2004; van Groenigen et al. 2006).

A major sink for soil mineral nitrogen is nitrogen **immobilization**, the process through which soil microorganisms incorporate mineral nitrogen and thereby diminish its availability to other organisms. As soil microorganism become more active under CO$_2$ enrichment, nitrogen immobilization increases by an average of 22% (see Figure 5.23).

Because additions of mineral nitrogen to soils do not increase under CO$_2$ enrichment, while extraction from soils by microorganisms increases, nitrogen availability to plants decreases under CO$_2$ enrichment. The availability of soil mineral nitrogen, therefore, determines plant responses to elevated CO$_2$ concentration more than any other environmental factor (Poorter and Perez-Soba 2001; Ellsworth et al. 2004). Plants receiving high amounts of nitrogen fertilization respond much more vigorously to CO$_2$ enrichment than those exposed to low

amounts (**Figure 5.24**). Moreover, stimulation by nitrogen fertilization becomes more pronounced with time of exposure to elevated CO$_2$ concentration (**Figure 5.25**).

Plant nitrogen status plummets under CO$_2$ enrichment. Shoot nitrogen and protein concentrations in C$_3$ plants fall by 17% (Cotrufo et al. 1998) and 49% (Poorter et al. 1997), respectively, at elevated CO$_2$ concentration. These declines are similar in plants receiving different levels of nitrogen fertilization (see Figure 5.24) and are more than double what would result if this were simply a case of **dilution through growth** (i.e., plants absorb a constant amount of nitrogen, and this amount just becomes more diluted because the plants grow bigger). Were it dilution through growth, the additional biomass that C$_3$ plants accumulate under CO$_2$ enrichment would decrease shoot nitrogen or protein by only 8% (Poorter and Navas 2003; de Graaff et al. 2006), not the 17% to 49% that is observed. Exposure to elevated CO$_2$ seems to inhibit some aspect of plant nitrogen metabolism.

Activity of rubisco and other enzymes in the C$_3$ cycle also declines at elevated CO$_2$ concentration (Moore et

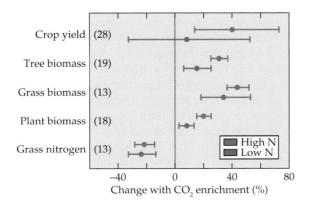

FIGURE 5.24 Influence of nitrogen fertilization on plant responses to atmospheric CO$_2$ levels Shown are the changes in crop yield, aboveground biomass, and grass total nitrogen concentrations from plants grown at ambient CO$_2$ concentration ($\approx$366 ppm) to those grown at elevated CO$_2$ concentration ($\approx$567 ppm). "High N" (red) and "Low N" (blue) designate, respectively, plants receiving greater amounts of and lesser amounts of nitrogen fertilization. In parentheses are the number of studies in each analysis. Circles and error bars designate means and confidence intervals, respectively, for crops, trees, C$_3$ grasses, and all plant species. (After Ainsworth and Long 2005; Curtis and Wang 1998; Wand et al. 1999; de Graaff et al. 2006.)

FIGURE 5.25 Influences of time and nitrogen fertilization on plant biomass responses to atmospheric CO_2 levels Shown are the changes in biomass from grasses grown at ambient CO_2 concentration (360–385 ppm) to those grown at elevated concentration of CO_2 (560–680 ppm) over years of treatment. "High N" (solid line) and "Low N" (dotted line) designate, respectively, grasses receiving heavier nitrogen and lighter nitrogen fertilization. The Swiss study measured aboveground biomass, whereas the other two studies estimated total biomass. (After Schneider et al. 2004; Dukes et al. 2005; Reich et al. 2006.)

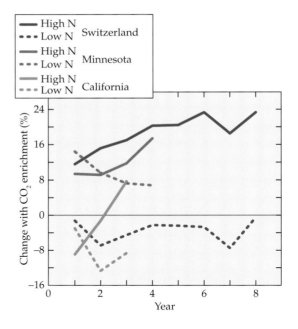

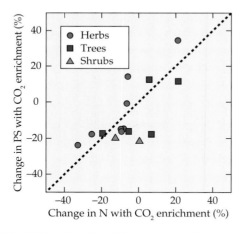

FIGURE 5.26 Relationship between changes in nitrogen and photosynthesis in response to CO_2 levels Shown are the changes in total leaf nitrogen concentrations (N) versus changes in leaf capacity for photosynthesis (PS) in terms of maximum net CO_2 assimilation from plants grown at ambient CO_2 concentration (360–380 ppm) to those grown at elevated CO_2 concentration (495–571 ppm). Each symbol designates the mean value for a species, and the dotted line designates a one-to-one relationship. The data generally follows a one-to-one relationship, indicating that the response of nitrogen content to CO_2 enrichment is important in determining the response of photosynthesis. (After Ellsworth et al. 2004.)

al. 1998). These changes are not limited to the enzymes in the C_3 cycle; rather, they follow the overall trends of shoot nitrogen and protein concentrations (Makino and Mae 1999; Ainsworth and Long 2005). The machinery for C_3 carbon fixation may be more efficient under CO_2 enrichment, but leaves produce less of this C_3 carbon fixation machinery (as well as all other enzyme machinery) because of nitrogen limitations. As a result, the influence of CO_2 enrichment on photosynthesis depends strongly on its influence on plant nitrogen status (**Figure 5.26**).

Bloom's hypothesis

We recently discovered a key interaction between carbon and nitrogen metabolism in plants that accounts for many of their responses to CO_2 enrichment (Bloom et al. 2002; Searles and Bloom 2003; Cousins and Bloom 2004; Rachmilevitch et al. 2004). We hypothesize that, contrary to popular opinion, photorespiration is not just a process that consumes sugars and generates waste heat. It is not just the unfortunate vestige of plants evolving at a time when atmospheric CO_2 concentration was much higher than it is today. In a wide range of C_3 species, conditions that inhibit photorespiration, namely an elevated CO_2 atmosphere or a low O_2 atmosphere, also inhibit shoot NO_3^- assimilation into amino acids (**Figures 5.27 and 5.28**). Photorespiration appears to generate NADPH that, via a four-carbon acid (malate), provides some of the energy required to convert nitrate (NO_3^-) to ammonium (NH_4^+). This coupling of photorespiration to NO_3^- assimilation makes

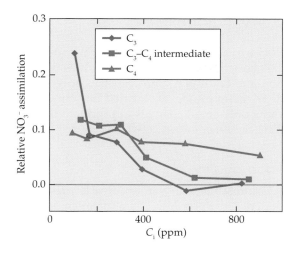

FIGURE 5.27 Response of nitrate assimilation as a function of CO_2 concentration within a leaf (C_i) in plants having different carbon fixation pathways. Relative nitrate (NO_3^-) assimilation was assessed from the change in $CO_2{:}O_2$ fluxes with a shift from NH_4^+ to NO_3^- nutrition. C_3 species included six species, C_4 species included three species, and C_3–C_4 intermediates included two species. In C_3 plants and, to a lesser extent, intermediates, nitrate assimilation declined rapidly with rising CO_2 concentration. (Bloom unpublished data.)

C_3 species more energy-efficient than C_4 or CAM species under most conditions and explains why more than 95% of Earth's flora is composed of C_3, not C_4 or CAM, species.

In C_4 plants, NO_3^- assimilation in the shoot proves to be relatively insensitive to CO_2 levels (see Figure 5.27). C_4 plants normally produce high levels of malate in mesophyll cells (see Figure 5.7), and all their shoot NO_3^- assimilation occurs in the mesophyll. Therefore,

both the C_4 pathway and photorespiration have the same previously unrecognized function: They generate the organic acid malate to power NO_3^- assimilation.

This discovery provides a simple explanation for the variation in plant responses to CO_2 enrichment. Plants differ in their relative dependence upon NH_4^+ and NO_3^- as nitrogen sources (Bloom 1997). Variation in the response of C_3 plants to elevated CO_2 concentration derives from variation in nitrogen sources. Wheat

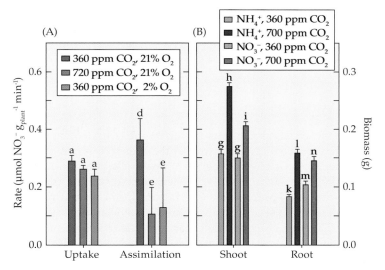

FIGURE 5.28 Wheat nitrate and growth in response to CO_2 levels (A) Nitrate (NO_3^-) uptake and assimilation (conversion into amino acids) in wheat plants exposed to atmospheres containing 360 ppm or 720 ppm CO_2 and 21% or 2% O_2. Uptake was assessed as the rate (10^{-6} moles of NO_3^- per gram plant per minute) at which plants depleted NO_3^- from a hydroponic solution and assimilation as the difference between the rates of NO_3^- uptake and net accumulation of free NO_3^- in plant tissues. Shown are the mean and confidence limits. Treatments labeled with different letters differ significantly ($P \leq 0.05$). Elevated CO_2

or low O_2, conditions that inhibit photorespiration did not influence NO_3^- uptake but did inhibit NO_3^- assimilation. (B) Shoots and root biomass (g dry mass) of wheat seedlings grown for 14 days in controlled environment chambers at 360 or 700 ppm CO_2 and under NH_4^+ or NO_3^- nutrition. Shown are mean and confidence limits for four replicate experiments, each with eight to ten plants per treatment. Treatments labeled with different letters differ significantly ($P \leq$ 0.05). Elevated CO_2 stimulated growth under NH_4^+ nutrition to a greater extent than under NO_3^- nutrition. (A after Rachmilevitch et al. 2004; B after Bloom et al. 2002.)

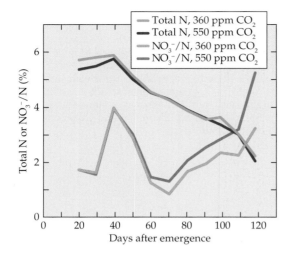

FIGURE 5.29 Wheat total shoot nitrogen (% dry matter) and the proportion of this nitrogen that is unassimilated nitrate (%) in response to CO_2 levels. The plants were grown in a FACE (free air CO_2 enrichment) experiment conducted in Maricopa, AZ. Plotted are the averages for two growing seasons. Ambient CO_2 treatment was 360 ppm (light blue and orange), and the elevated CO_2 treatment was 550 ppm (dark blue and red). CO_2 treatment did not influence the accumulation of nitrogen in the wheat, but much more of this nitrogen remained as unassimilated NO_3^- in the plants grown under elevated CO_2. (After Burger, Bloom, and Kimball, unpublished data.)

plants at ambient CO_2 concentration grew similarly under NH_4^+ and NO_3^- nutrition, but those at elevated CO_2 concentration grew much slower when their sole nitrogen source was NO_3^- than when it was NH_4^+ (see Figure 5.28). The phenomenon of CO_2 acclimation derives from the inhibition of photorespiration at elevated CO_2 concentration and thereby the decline of NO_3^- assimilation and synthesis of plant proteins required for photosynthesis and growth.

Elevated CO_2 concentration has no consistent effect on plant absorption of mineral nitrogen from soils (BassiriRad 2000). Wheat plants grown in environmental chambers absorbed NO_3^- from nutrient solutions at only slightly slower rates under CO_2 enrichment (see Figure 5.28). In a FACE plot, wheat receiving over 350 kg ha^{-1} of NH_4NO_3 fertilizer (about five times the normal application rate) absorbed similar amounts of nitrogen under ambient and elevated CO_2 concentration (**Figure 5.29**). As the season progressed, the proportion of this nitrogen that remained as unassimilated NO_3^- was much higher in the elevated CO_2 treatment than in the ambient treatment. Consequently, we can conclude that CO_2 enrichment inhibits NO_3^- assimilation directly.

The different effects of CO_2 on NH_4^+ and NO_3^- assimilation explain ecosystem responses to CO_2 enrichment. Ecosystems in which NH_4^+ is the predominant source

of nitrogen from soils, such as rice paddies (Bannayan et al. 2005), forests (DeLucia et al. 2005), and marshes (Rasse et al. 2005), show consistent stimulation under CO_2 enrichment. Ecosystems in which NO_3^- is the predominant nitrogen source, such as grasslands, become deprived of protein at elevated CO_2 concentration and show only small changes in productivity (Shaw et al. 2002; Schneider et al. 2004).

Food quality

As atmospheric CO_2 concentration rises and plants dependent upon NO_3^- as a nitrogen source become deprived of protein, the nutritive value of plant material will suffer. Wheat, rice, and potato provide 21%, 14%, and 2%, respectively, of the protein in the human diet (FAOSTAT 2009). Wheat plants receiving normal levels of nitrogen fertilizer had 10% less grain protein, 10% smaller bread loaf volume, and 10% lower scores on quality indices under CO_2 enrichment (Fangmeier et al. 1999; Kimball et al. 2001). Similarly, grain protein in rice (Terao et al. 2005) and tuber nitrogen in potato (Fangmeier et al. 2002) declined by about 10% in elevated CO_2 atmospheres.

Changes in plant nitrogen levels also affect other **herbivores** (plant eaters) such as insects (Reich et al. 2006). Not only do plant protein concentrations decline at elevated CO_2 concentration, but green tissues accumulate higher amounts of phenolics, compounds that interfere with animal digestion (Zvereva and Kozlov 2006). Herbivores in CO_2-enriched treatments may increase consumption of foliage to compensate for its diminished nutritional quality and palatability, but the herbivores still experience slower growth rates, lower fecundity, and higher mortality (Agrell et al. 2000; Zvereva and Kozlov 2006). These factors lead to an overall decline in herbivore populations under CO_2 enrichment (Knepp et al. 2005).

How will plants and the organisms dependent upon them for food adjust to rising atmospheric CO_2 concentration? If history is any guide, significant changes are imminent. Major shifts in Earth's flora and fauna (see Figure 2.26) have accompanied past changes in atmospheric CO_2 levels (see Figure 2.25).

For example, CO_2 levels were relatively low 8 million to 6 million years ago (Cerling et al. 1997). Low CO_2 levels favor the proliferation of plant species with C_4 carbon fixation (see Figure 5.9). C_4 plants tend to be more nitrogen efficient than C_3 plants and have an average carbon to nitrogen ratio (C:N) of about 50 versus an average C:N of about 30 for C_3 plants (Ehleringer et al. 2002). In C_4 plants, much of the chlorophyll and leaf protein is centralized in bundle-sheath cells that have heavy walls to prevent the escape of CO_2 released during the breakdown of C_4 organic acids (see Figure 5.7).

BOX 5.1 Isotopes provide evidence of a shift in Earth's flora and fauna

Some additional background information on isotopes is required to complete this story. As introduced in Chapter 2, a chemical element may have several isotopes that differ in the number of neutrons in their nucleus. For example, the most common form of carbon on Earth is ^{12}C, which has six neutrons in its nucleus. About 1% of the carbon on Earth is ^{13}C, which has seven neutrons in its nucleus and thus is heavier, but still stable. The carbon fixation reaction via rubisco in C_3 plants proceeds much faster with light $^{12}CO_2$ than heavy $^{13}CO_2$, whereas the reaction via PEP carboxylase in C_4 plants proceeds only slightly faster with light $H^{12}CO_3^-$ than heavy $H^{13}CO_3^-$. Therefore, C_3 plants contain a much smaller proportion of the heavy carbon isotope than C_4 plants ($\delta^{13}C$, the ratio of ^{13}C to ^{12}C, averages about $-8‰$ for Earth's atmosphere, $-27‰$ for C_3 plants, and $-12‰$ for C_4 plants). Digestive enzymes in herbivores also break down organic compounds that contain the lighter ^{12}C carbon faster than those that contain the heavier ^{13}C. Herbivores thereby preferentially release $^{12}CO_2$ during respiration; as a result, the carbon that remains in a herbivore to form tooth enamel is more enriched in the heavy isotope ^{13}C than carbon in the diet ($\delta^{13}C$ increases about $+14‰$). Putting together the discrimination of plants with that of herbivore digestion, the ratio of ^{13}C to ^{12}C ($\delta^{13}C$) in tooth enamel of herbivores eating C_3 plants averages $-13‰$ ($= -27 + 14$), whereas that of herbivores eating C_4 plants averages $+2‰$ ($= -12 + 14$).

Equipped with this background information, we can deduce from stable isotope data that lower atmospheric CO_2 concentration 8 million years ago to 6 million years ago were responsible for a major turnover in mammalian herbivores (Cerling et al. 1997). The shift in diet from C_3 to C_4 plants is evident from the abrupt rise in $\delta^{13}C$ values of fossil teeth from this period (Figure A). The shift in herbivores is evident from the fossil record: Teeth from this period had higher crowns, which are better suited for the extensive grinding required to extract nutrients from the bundle sheaths of C_4 plants (Figure B).

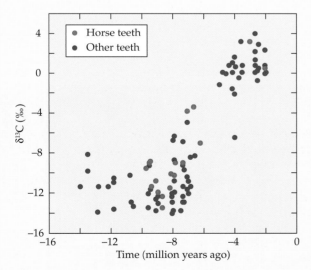

Figure A Carbon isotope ratio ($\delta^{13}C$) in fossil tooth enamel of horses and other animals from the Siwalik hills of Pakistan. (After Cerling et al. 1998.)

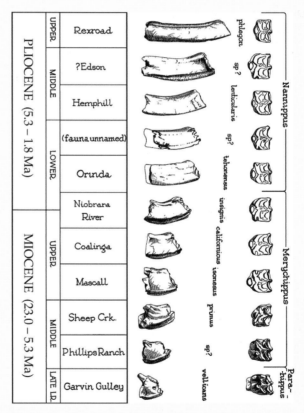

Figure B Cheek teeth from fossil horses from the Miocene (23.0–5.3 million years ago) and Pliocene (5.3–1.8 million years ago) epochs. Crown height increased steadily to accommodate wear from the grinding of tougher plant material.

These heavy cell walls also protect bundle sheaths from herbivory. In summary, C_4 plants have less nitrogen, and this nitrogen is less accessible to herbivores. The proliferation of C_4 plants 8 million to 6 million years ago thus shifted the nutritional value of the foliage and thereby produced a shift in the herbivores dependent upon this foliage to species that were more efficient in extracting plant nitrogen (**Box 5.1**).

TABLE 5.4 CO_2 sensing in various organisms

Organism	Purpose	Response	Lower limits of CO_2 (ppm) sensing
Plants	Water conservation, carbon assimilation	Stomatal closure	10
Bacteria	Virulence	Protein excretion, encapsulation	10,000
Fungi	Virulence	Filament formation, encapsulation	5,000
Basidiomycetes	Spore dispersal (mushrooms)	Stalk elongation, cap expansion	500
Social insects	Temperature control, ventilation	Hive fanning	5,000
Insect herbivores	Food location	Locomotion	0.5
Blood-sucking insects	Food location	Locomotion	10

Source: Sage 2002.

Carbon Dioxide Sensing

A wide range of organisms measure the CO_2 concentration of their surroundings (Table 5.4) and are likely to alter their behavior as atmospheric CO_2 concentration increases. Consider the following examples.

- As discussed in the preceding sections, plants adjust stomatal apertures to minimize water loss per carbohydrate gain based on CO_2 concentration. Exposure to elevated CO_2 concentration diminishes stomatal apertures (see Figure 5.15A).

- The anthrax bacterium *Bacillus anthracis* (**Figure 5.30**) determines from high external CO_2 concentration (about 5%) that it has arrived in the lungs or bloodstream of its mammalian host. It then prepares for dispersal by producing toxic proteins to kill the host and synthesizing a protective capsule around itself to form spores (Fouet and Mock 2006).

- Pathogenic fungi also employ CO_2 concentration as a means for detecting that they have arrived in the appropriate host tissue. *Candida albicans* and *Cryptococcus neoformans*, which cause life-threatening systemic infections in immunocompromised patients, synthesize protective capsules and scavenging filaments in response to CO_2 levels of around 0.5% (Bahn and Muhlschlegel 2006).

- Mushrooms elevate their fruiting bodies (caps) to an optimal height for spore dispersal based on measurements of atmospheric CO_2 concentration (Sage 2002). Root and microbial respiration drives CO_2 concentration near the soil surface to above 0.1%. Exposure of mushrooms to such CO_2 levels promotes stalk elongation instead of cap expansion. Only when the fungus reaches heights above the soil, where CO_2 concentration drops below 0.05%, does cap expansion take precedence (Kues and Liu 2000).

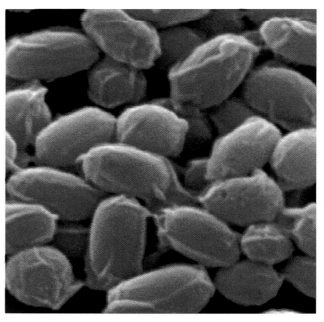

FIGURE 5.30 Spores of *Bacillus anthracis*, the bacterium responsible for anthrax disease.

FIGURE 5.31 Larvae of *Cactoblastis cactorum* feeding on prickly pear cactus.

- Social insects such as ants, bees, and termites detect the location and activity of hives via gradients of CO_2 concentration. Bees also determine whether ventilation in a hive is adequate from CO_2 concentration: If concentration in a beehive rises above 0.5%, young worker bees orient in a single direction and beat their wings to ventilate (fan) the hive.

- Insect herbivores may monitor CO_2 concentration to identify the most delectable plant tissues: The tissues that are most rapidly depleting atmospheric CO_2 are synthesizing the most carbohydrates (Stange and Stowe 1999). For example, a moth aptly named *Cactoblastis cactorum*, which feeds on prickly pear cactus

(*Opuntia* sp.) (**Box 5.2**), resolves differences as small as 0.00005% (0.5 ppm) CO_2 against the normal atmospheric background of 0.0386% (386 ppm) to locate in the dark the cactus pads that are most active in CAM carbon fixation (see Figure 5.12).

- Haematophagous insects such as mosquitoes, ticks, and tsetse flies find their next blood meal by following the trail of CO_2 to their prey (**Figure 5.32**). These bloodsuckers activate flight or searching behavior at a relatively low CO_2 concentration gradient, 0.001% to 0.030% (10 ppm to 300 ppm), but begin to fly in the direction of their prey only at a higher

FIGURE 5.32 Yellow-fever mosquito (*Aedes aegypti*), the primary vector of dengue fever. Here the insect obtains a blood meal through its tubular fascicle, which had penetrated the host's skin and was reddening in color. As the mosquito's abdomen filled with blood, it distended, allowing the collecting blood to become visible as an enlarging intra-abdominal red mass.

BOX 5.2 CO_2-sensing moth and biological control

Cactoblastis cactorum is a legend in the annals of biological pest control. Prickly pear cactus was introduced into Australia in the 19th century for use as a natural agricultural fence and in an attempt to establish a cochineal dye industry. Cochineal insects, which feed on this type of cactus, are the source of a red dye that at the time was important to European clothing industries. For example, this dye was used to color the British soldiers' red coats.

The cactus, once introduced, became a major weed in Australia, spreading from 4 million hectares in 1900 to 24 million hectares in 1920. Mechanical or chemical control measures proved inadequate, and the "pear" was advancing nearly 0.5 million hectares per year when authorities conducted a large-scale release of cactoblastis moth eggs in 1926. By 1932, the moths had destroyed most of the thick stands of cactus.

(A)

(B)

Prickly-pear cactus in Australia A property inundated with prickly pear cactus in Queensland, Australia (A) before and (B) after exposure to *Cactoblastis cactorum*.

concentration gradient of about 0.1% (1000 ppm) (Stange 1996). Many insect repellants, including DEET (meta-N,N-diethyl toluamide), work by blocking insect CO_2 receptors.

All of these organisms seem to rely on similar biochemical mechanisms for CO_2 sensing (Bahn and Muhlschlegel 2006; Jones et al. 2007; Nilson and Assmann 2007). Will rising CO_2 concentration in the atmosphere interfere with any of these processes? The anticipated changes of 0.05% to 0.10% CO_2 could present difficulties for insect herbivores and haematophagous insects that will be forced to detect a CO_2 signal against a higher background level. Other climate changes, however, may expand the ranges of these insects. The following chapter will discuss this possibility.

Summary

Seasonal variations in photosynthesis and respiration and the burning of fossil fuels drive atmospheric concentrations of CO_2 and O_2 in opposite directions. The magnitude of these changes is far below those that might cause respiratory distress in humans. Plants, however, and a few other organisms are highly sensitive to changes of this magnitude. In most plants (C_3 plants), CO_2 enrichment and O_2 depletion initially enhances CO_2 assimilation into carbohydrates. Upon longer exposures, however, this enhancement of CO_2 assimilation diminishes for reasons still under investigation. C_4 plants such as maize have a CO_2 pumping mechanism and so are far less sensitive to atmospheric CO_2 or O_2 concentration. The physiological differences between C_3 and C_4 plants determine their distributions worldwide and how these distributions will shift with warming temperatures and rising atmospheric CO_2 concentration. Rising CO_2 concentration will improve the water use efficiency of all types of plants but will decrease both the availability of mineral nitrogen in the environment and the capacity of plants to convert nitrate, one form of mineral nitrogen, into organic compounds including proteins and nucleic acids. Rising CO_2 concentration may also interfere with the ability of microorganisms and insects to use CO_2 as a signal.

Review Questions

1. In 2100, carbon dioxide concentrations in the atmosphere are anticipated to
   (a) exceed the concentrations during the last ice age.
   (b) exceed the concentrations exhaled from our lungs.
   (c) lead to loss of mental acuity.
   (d) lead to loss of consciousness.
   (e) lead to loss of life.

2. As carbon dioxide concentrations in the atmosphere increase, oxygen concentrations
   (a) do not change.
   (b) increase at the same rate.
   (c) decrease at the same rate.
   (d) increase slightly faster.
   (e) decrease slightly faster.

3. All higher plants do *not*
   (a) convert carbon dioxide into carbohydrates via the same biochemical pathway.
   (b) use the enzyme rubisco to catalyze the first reaction with carbon dioxide.
   (c) photorespire at the same rate as they conduct carbon fixation.
   (d) conduct carbon fixation in chloroplasts.
   (e) convert solar radiation into the high energy compounds ATP and NADPH.

4. Most terrestrial plants (select all that apply)
   (a) lose substantial amounts of water in allowing carbon dioxide to enter their internal spaces.
   (b) have openings in their photosynthetic organs called stomata.
   (c) have their internal spaces saturated with water vapor.
   (d) can open and close the openings in their photosynthetic organs.
   (e) have their above ground organs covered with a waxy layer called the cuticle.

5. At elevated concentrations of atmospheric carbon dioxide, most plants
   (a) quickly shrivel and die.
   (b) do not alter their internal concentrations of carbon dioxide.
   (c) require less water to achieve a certain amount of growth.
   (d) do not open or close the openings in their photosynthetic organs.
   (e) convert solar radiation directly into ethanol.

6. Carbon dioxide acclimation is the process in which plants
   (a) initially show a dramatic stimulation in growth and photosynthesis with exposure to elevated carbon dioxide, but this stimulation diminishes over time.
   (b) initially show a dramatic inhibition in growth and photosynthesis with exposure to elevated carbon dioxide, but this inhibition diminishes over time.

   (c) conducting C_3 carbon fixation start to conduct C_4 carbon fixation upon exposure to elevated carbon dioxide.
   (d) conducting CAM carbon fixation start to conduct C_4 carbon fixation upon exposure to elevated carbon dioxide.
   (e) conducting C_4 carbon fixation start to conduct CAM carbon fixation upon exposure to elevated carbon dioxide.

7. Which of the following statements about the element nitrogen is *false*?
   (a) Nitrogen is one of the top five elements in the amounts found in plants and animals.
   (b) Nitrogen is plentiful in Earth's atmosphere in a relatively inert form.
   (c) Nitrogen is converted into forms more useful to plants through industrial and biological nitrogen fixation.
   (d) Industrial and biological nitrogen fixation require the input of substantial amounts of energy.
   (e) Nitrogen in organisms is only found in nucleic acids such as DNA and RNA.

8. Which of the following statements about nitrogen in plants is *false*?
   (a) Plant nitrogen contents decrease with exposure to elevated carbon dioxide.
   (b) Most plants obtain most of their nitrogen from absorbing ammonium and nitrate dissolved in the soils around their roots.
   (c) Conversion of nitrate into organic forms of nitrogen is highly energy intensive.
   (d) Plant growth and primary productivity increase as plant nitrogen contents increase.
   (e) Nitrogen in plants is likely to explode.

9. Variation in atmospheric levels of carbon dioxide has changed in the past and in the future may change the (select the *false* statement)
   (a) amount of protein in food.
   (b) accessibility of protein to animals.
   (c) balance between aerobic and anaerobic respiration in animals.
   (d) size and shape of teeth.
   (e) relative numbers of plants with C_4 carbon fixation versus C_3 carbon fixation.

10. Which of the following statements about carbon sensing in organisms is *true*?
    (a) Animals cannot detect the changes in atmospheric carbon dioxide concentrations that have occurred during the last century.
    (b) Fungi cannot sense carbon dioxide concentrations.
    (c) Encapsulation of pathogenic organisms is not as sensitive to carbon dioxide concentrations as locomotion of organisms locating food sources.
    (d) Carbon dioxide concentrations do not change near plants.
    (e) Carbon dioxide concentrations do not change near animals.

11. Which of the following statements explaining why organisms sense small differences in atmospheric CO_2 is *false*?
    (a) To locate a source of blood to provide nourishment for eggs
    (b) To monitor temperature changes
    (c) To distinguish which parts of a plant are producing the most carbohydrate
    (d) To determine when the organism has migrated to the lungs of its host
    (e) To maximize water use efficiency

Suggested Readings

Bloom, A. J. 2009. As carbon dioxide rises, food quality will decline without careful nitrogen management. *California Agriculture* 63: 67–72.

This article discusses the relationship between photorespiration and nitrate assimilation and its implication to our food supply.

Cerling, T. E., J. R. Ehleringer, and J. M. Harris. 1998. Carbon dioxide starvation, the development of C_4 ecosystems and mammalian evolution. *Philosophical Transactions of the Royal Society of London B Biological Sciences* 353: 159–171.

This article describes the rise of C_4 plants and its influence on herbivores.

Sage, R. F. 2002. How terrestrial organisms sense, signal, and respond to carbon dioxide. *Integrative and Comparative Biology* 42: 469–480.

This article discusses CO_2 sensing.

Taiz, L. and E. Zeiger. 2006. *Plant Physiology*, 4th Ed. Sinauer Associates, Sunderland, MA.

This general textbook on plant physiology presents an introduction to many topics, including photosynthesis, plant-water relations, and plant nutrition.

6

CLIMATE CHANGE AND THE BIOSPHERE

Chapter 5 focused on how rising carbon dioxide (CO_2) levels in the atmosphere influence living organisms, especially plants, but living organisms are also highly sensitive to other environmental conditions. Of these, temperature, precipitation, salinity, and pH have undergone global-scale changes that have distinctly altered species distributions. This chapter describes biological responses to these coanditions and anticipated shifts in the **biosphere** (the part of Earth that supports life) during the coming decades.

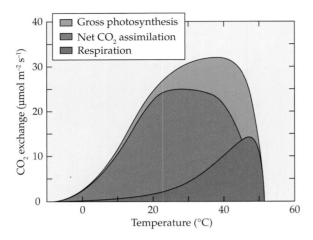

FIGURE 6.1 Typical temperature response of gross photosynthesis, net CO$_2$ assimilation, and respiration in a C$_3$ plant in terms of CO$_2$ fluxes (10^{-6} moles of CO$_2$ per square meter of leaf area per second). These complex processes involve a balance between reactions that synthesize and break down biochemicals in the plant. Gross (total) photosynthesis increases exponentially with temperature until about 10°C, whereas respiration increases exponentially until about 35°C. Net CO$_2$ assimilation, the difference between gross photosynthesis and respiration, is relatively insensitive to temperatures between about 20°C to 37°C. As a result, plant growth is relatively constant over this range.

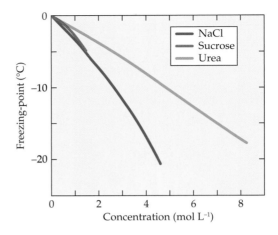

FIGURE 6.2 Freezing point (°C) of an aqueous solution as a function of concentration (moles per liter) of table salt (NaCl), sugar (sucrose), or urea (NH$_2$CONH$_2$). Shown are the range of concentrations over which these substances remain soluble.

Temperature

Average global temperatures have warmed about 0.6°C over the past 150 years and are likely to warm an additional 2°C to 6°C by the end of this century (see Figure 4.33). The temperature of an object indicates how rapidly molecules within it are moving (see Chapter 3). Biochemical reactions depend on such movements, and a 10°C increase in temperature speeds up these reactions by two-fold to three-fold. Once temperatures exceed a certain threshold, however, conformations of proteins and other organic compounds become distorted, membranes separating cells and their compartments become leaky, and reactions dependent on these compounds and compartments begin to fail. Consequently, most biochemical processes show a typical temperature response: They proceed slowly at temperatures near freezing, accelerate exponentially with rising temperatures until they level off somewhere between 15°C and 35°C, and finally crash at temperatures higher than 40°C.

For example, photosynthesis and respiration in plants follow such a pattern (**Figure 6.1**). The critical temperatures of these processes are offset from one another, with the rate of respiration reaching a peak at a higher temperature than that of photosynthesis. Consequently, the difference between them, the net CO$_2$ assimilation, remains relatively constant over the range of midday temperatures that normally occurs during the growing season. This broad temperature optimum buffers plant growth against some of the vagaries of weather.

Extreme temperatures, both low and high, threaten living organisms. At the one end, the freezing point of water (0°C for pure water at 1 atmosphere pressure) presents a formidable barrier because life on Earth entails biochemical reactions in an aqueous medium, and these reactions cease when water freezes. Organisms that are active at cold temperatures either raise their internal temperatures above 0°C through metabolic generation of heat or lower their freezing point through accumulation of dissolved solutes (**Figure 6.2**). For example, microbes continue to grow, albeit very slowly, in ice at temperatures as low as –20°C by accumulating sufficient salts from their surroundings to maintain an aqueous environment within their cells (Price and Sowers 2004).

Freezing temperatures are not the only peril at this end of the temperature scale. Low temperatures slow growth and development and thereby increase the probability that an organism will encounter other adverse environmental conditions or pests. For example, plant breeders have developed early maturing grains to insure that crops in cold climes will be ready

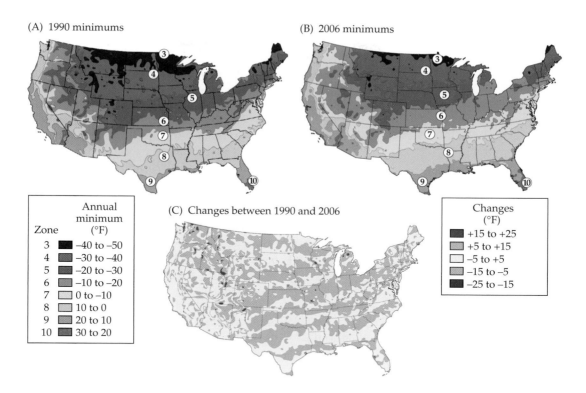

(A) 1990 minimums

(B) 2006 minimums

(C) Changes between 1990 and 2006

Zone	Annual minimum (°F)
3	–40 to –50
4	–30 to –40
5	–20 to –30
6	–10 to –20
7	0 to –10
8	10 to 0
9	20 to 10
10	30 to 20

Changes (°F)
+15 to +25
+5 to +15
–5 to +5
–15 to –5
–25 to –15

FIGURE 6.3 Changes in plant hardiness zones These maps divide the contiguous United States into eight zones based on a 10°F differences in the average annual minimum temperature. Gardening books indicate the appropriate zone for most tree and perennial plant species. (A) 1990 minimums. (B) 2006 minimums. (C) This map shows the change in average annual minimum temperature from 1990 to 2006. The banding is in part the results of round-off errors.

for harvest before some calamity strikes. Moreover, certain species of tropical origin suffer chilling injury when exposed to nonfreezing temperatures between 1°C and 12°C.

The cultivated tomato, *Lycopersicon esculentum* Mill., is a classic example of a chilling-sensitive species. Temperatures below 10°C severely inhibit tomato growth and development at all life stages, and those below 6°C inflict so much damage that commercial farmers generally cannot harvest their crop and home gardeners must again reconcile themselves to eating hard tomatoes from foreign lands (Geisenberg and Stewart 1986). Chilling injury in tomato and other crops derives in part from water stress that develops when low temperatures impede water movement from roots to shoots (Bloom et al. 2004).

Prospects for a longer tomato growing season are improving. Changes in climate have altered global temperature patterns. In particular, global warming has increased nighttime minimum temperatures to a greater extent than daytime maximums (see Figure 4.30A). Nighttime minimums, especially frosts or chilling temperatures, define the seasons at any given location (**Figure 6.3**): The last frost or chill brings the promise of spring, whereas the first frost or chill serves as the harbinger of winter.

Recent changes to the timing of the seasons have had broad biological consequences because organisms respond more to season length than to warmer temperatures alone (Bradshaw and Holzapfel 2006). On average, spring events (e.g., arrival times of migrant species, peak flight date, nesting, egg-laying, breaking of plant buds, and flowering) are occurring an average of 2.3 days to 5.1 days earlier each decade (**Figure 6.4**) (Parmesan and Yohe 2003). Autumn events (e.g., leaf fall, departure times of migrant species, and hibernation) have been delayed by 0.3 days to 1.6 days per decade (Sherry et al. 2007).

Daytime maximum temperatures also play a role. To escape the heat and drought at mid-latitudes (see Figures 2.40C and 2.45) and to take advantage of expanded growing seasons at higher altitudes and higher latitudes, species have migrated to 6.1 m higher in elevation and 6.1 km closer to the poles each decade

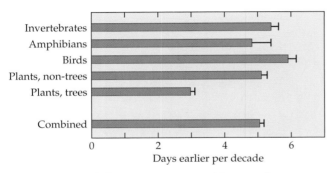

FIGURE 6.4   Shift in spring events (days earlier per decade) such as arrival, flowering, and egg laying for various groups of species. Depicted are the mean and an error bar indicating the variation in values around the mean of 694 species over the past 50 years. (After Root et al. 2003.)

(Parmesan and Yohe 2003). These shifts in geographical distributions are putting many species at risk.

A common assumption is that global warming is more disruptive to species having narrow geographical distributions (Thomas et al. 2004; Schwartz et al. 2006). Survival of a species would therefore vary with the geographical area of suitable habitats. Based on this assumption, 20% to 30% of all species will face extinction during this century (Table 6.1) (IPCC 2007). Other researchers question the applicability of species–area

relationships for predicting extinctions, especially in response to climate change (Lewis 2006).

Nonetheless, migration of temperate organisms to higher altitudes and latitudes and loss of snowpack or ice caps endangers a number of alpine and arctic species. These include several **charismatic megafauna**, large animals with popular appeal (Smetacek and Nicol 2005). Most familiar are polar bears (*Ursa maritimus*), Adélie penguins (*Pygoscelis adeliae*), emperor penguins (*Aptenodytes forsteri*), and baleen whales.

Polar bears

The polar bear (**Figure 6.5A**) is one of the world's largest carnivores. It inhabits the ice-covered waters of the circumpolar Arctic, particularly annual ice over the continental shelf where biological productivity is highest (**Figure 6.5B**). For most of the year, polar bears hunt seals from sea ice (Derocher et al. 2004). Bears optimize foraging for seals by staking out seal breathing holes and haul-out sites. Rarely do they capture seals in open water.

During ice-free periods, female polar bears swim to shore from the southern edge of the pack ice and give birth on land, in dens. Recession of the Arctic ice cap (see Figure 1.11) as a result of warmer Arctic temperatures (see Figure 2.40C) have extended the distance between denning and feeding sites. This increases the vulnerability of mothers and cubs.

TABLE 6.1   Extinctions predicted by habitat loss

Biome	Undistsurbed	World surface area (%)		
		1990	Area lost	Species expected to go extinct
Cropland	0.0	10.9	0.0	0.0
Pasture	0.0	23.1	0.0	0.0
Ice	1.7	1.7	0.0	0.0
Tundra	4.8	4.6	0.2	1.0
Wooded tundra	2.0	1.9	0.1	1.1
Boreal forest	13.0	12.5	0.5	0.9
Cool conifer forest	2.7	2.1	0.6	6.1
Temperate mixed forest	5.2	2.2	3.0	19.2
Temperate deciduous forest	4.5	1.5	3.0	24.2
Warm mixed forest	4.7	1.9	2.8	20.3
Grassland/steppe	13.7	6.9	6.8	15.7
Hot desert	14.9	11.8	3.1	5.6
Scrubland	7.3	1.9	5.4	28.9
Savannah	11.9	6.2	5.7	15.1
Tropical woodland	6.1	4.4	1.7	8.0
Tropical forest	7.6	6.4	1.1	4.0

Source: Thomas et al. 2004.

(A)

(B)

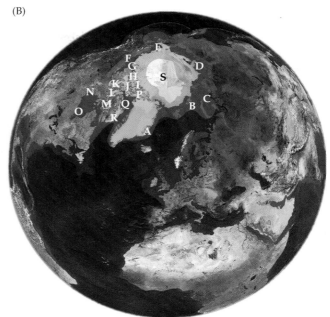

FIGURE 6.5 Polar bears (A) Polar bear mother and cub. (B) Circumpolar distribution of polar bears (pink shading). Letters refer to bear populations described in Table 6.2.

TABLE 6.2 Polar bear populations

		Aerial survey/mark–recapture			Additional/alternative analysis			
Major populations[a]		Year	Size	Range	Year	Size	Range	Method[b]
A[c]	East Greenland							
B	Barents Sea	2004	2997	2299–4116				
C	Kara Sea							
D	Laptev Sea	1993		800–1200				
E	Chukchi Sea				1993	2000		Dens
F	Southern Beaufort Sea	2006	1500	1000–2000				
G	Northern Beaufort Sea	1986	1200	133–2097				
H	Viscount Melville Sound	1992	161	121–201	1996	215	99–331	Sim
I	Norwegian Bay	1998	190	102–278				
J	Lancaster Sound	1998	2541	1759–3323				
K	M'Clintock Channel	2000	284	166–402				
L	Gulf of Boothia	2000	1523	953–2093				
M	Foxe Basin	1994	2197	1677–2717	2004	2300	1780–2820	Sim, Tek
N	Western Hudson Bay	2004	935	794–1076				
O	Southern Hudson Bay	1988	2000	684–1116				
P	Kane Basin	1998	164	94–234				
Q	Baffin Bay	1988	2074	1544–2604	2004	1546	690–2402	Sim
R	Davis Strait				2004	1650	1000–2300	Sim, Tek
S	Arctic Basin							

Source: Aars et al. 2006.
[a]Estimates are missing for East Greenland, Kara Sea, and Arctic Basin.
[b]Analyses are "Sim" for simulation, "Dens" for density, and "Tek" for traditional ecological knowledge.
[c]Letters relate to locations on Figure 6.5B.

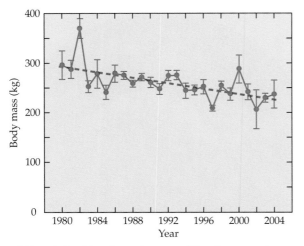

FIGURE 6.6 Yearly estimates of body mass (mean with error bars indicating the variation in values around the mean) for polar bears in western Hudson Bay. Dashed blue line highlights the overall linear trend over time. (After Stirling and Parkinson 2006.)

Polar bears, when on land, mostly scavenge along the shore for food other than seals. A longer ice-free period means more time on land and intensified competition for this food. Because females are walking and swimming longer distances and spending more time on land, their body weights are decreasing (**Figure 6.6**). If this trend continues, the average weight of females during the next 20 to 30 years in western Hudson Bay will drop below 190 kg. Females weighing less than 190 kg seldom bear cubs (Stirling and Parkinson 2006).

A longer time on land also increases encounters between polar bears and humans. Such encounters usually mean more bad news for the bears. If they damage property or threaten human safety, as they often do when scavenging for food, they are likely to be shot. Furthermore, repeated sightings of polar bears inflate estimates of their population sizes and have justified higher quotas for hunting by aboriginal people (Stirling and Parkinson 2006).

Despite widespread interest about the plight of polar bears, information about their numbers is meager. Estimates of population size in any region are based on one or at most two scientifically conducted censuses (see Table 6.2). Standard census methods such as "capture, mark, release, and recapture" may involve tracking bears from helicopters, shooting them with darts containing sedatives, tagging or collaring them, and then following them. These procedures may unduly stress the bears, change their behavior, or even compromise their survival. A novel, less invasive approach for tracking polar bears uses dogs to locate polar bear scat (feces), characterizes the age and food source of the scat, and identifies individual bears by DNA in the scat (Revkin 2007).

Penguins

At the other pole, several species of Antarctic penguins that depend on sea ice are experiencing population changes (Parmesan 2006) (**Figure 6.7**). Colonies of Adélie penguins are declining at warmer sites, such as Anvers and King George Island, but expanding at colder sites such as Cape Royds on the Ross Sea (**Figure 6.8A,B**). This indicates that populations of Adélie penguin are shifting location (Croxall et al. 2002). Emperor penguins also depend on sea ice, and their colony at Point Geologie is declining.

(A)

(B)

FIGURE 6.7 Antartic penguins that are being influenced by climate change (A) An adult Adélie penguin stands about 0.6 m in height and weighs about 4.5 kg. (B) Adult emperor penguins stand about 1.3 m in height and weigh about 45 kg.

(A)

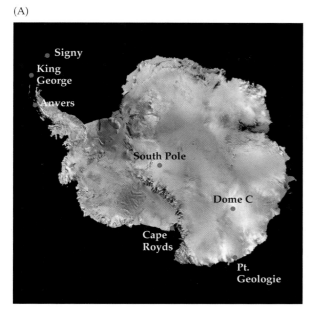

(B)

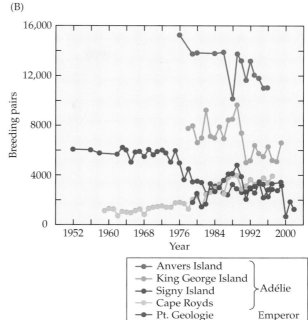

FIGURE 6.8 Penguins in Antarctica (A) Locations of Adélie and emperor penguin colonies on Antarctica for which multiyear censuses of population size are available. (B) Number of breeding pairs of Adélie penguins on Anvers Island and King George Island are declining, whereas the colony on Signy Island is highly variable in size, and that on Cape Royds is growing. The colony of emperor penguins on Point Geologie is shrinking. (C) Seasonal events for Adélie and emperor penguins nesting near Point Geologie on Antarctica. Shown are the days of the year when the birds first arrive at the rookery and when they lay their first egg. Note that the two species arrive about 163 days apart, that oscillations in the species are independent of one another, that arrivals show no trend, and that egg layings are delayed slightly. "Day 0" is January 1. (B after Croxall et al. 2002; C after Barbraud and Weimerskirch 2006.)

(C)

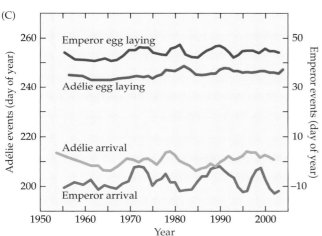

Both Adélie penguins and emperor penguins nest near Point Geologie, but at different times of the year (**Figure 6.8C**). Temperatures at this site have remained relatively constant over the past 50 years, while the extent of sea ice has diminished 66% (Barbraud and Weimerskirch 2006). The date of the penguins' arrival at the site has remained relatively constant for the two species, whereas the date on which they begin laying eggs is now delayed by a few days. Apparently, the decreased amount of sea ice has reduced the quantity and accessibility of food supplies such as krill (as we present in a following section), and these penguins need more time to build up the fat reserves necessary for breeding (Barbraud and Weimerskirch 2006).

Baleen whales

Baleen whales are a prominent group of sea mammals composed of 14 species. These range in size from the blue whale (*Balaenoptera musculus*), the largest animal ever found on Earth at 33.5 m in length and 117 metric tons (one metric ton = 2205 lbs), to the pygmy right whale (*Caperea marginata*), which extends only 4 m to 6.5 m in length and weighs a mere 3 to 3.5 metric tons. Other species include the Atlantic northern right whale (*Eubalaena glacialis*), 18 m in length and 100 metric tons (**Figure 6.9A**), and the humpback whale (*Megaptera novaeangliae*), 12 m to 16 m in length and 36 metric tons (**Figure 6.9B**).

(A)

(B)

FIGURE 6.9 Baleen whales (A) Northern right whales (*Eubalaena glacialis*), mother and calf, swimming off the coast of Florida in 2005. (B) Humpback whales (*Megaptera novaeangliae*) feeding along the coast of Alaska.

These behemoths, although impressive in stature, lack teeth (**Figure 6.10**). They filter-feed through fine, comb-like structures called baleen (**Figure 6.11**). During filter-feeding, a whale engulfs large volumes of sea water, closes its mouth, creates internal pressure by raising its tongue toward the top of the palate, pushes the water out through the slots in the baleen, traps small aquatic animals, especially krill, against the baleen, and swallows the trapped animals.

Krill are shrimp-like marine **invertebrates**, animals lacking a backbone (**Figure 6.12**). Adults of most krill species reach about 1 cm to 2 cm in a year or two, although some species grow to as large as 15 cm and mature in 5 to 7 years. Krill, like the baleen whales that eat them, are filter feeders that comb through seawater for **plankton**, microscopic algae (**phytoplankton**) or animals (**zooplankton**) suspended in the water column.

Together, these organisms form a relatively simple food web (Smetacek and Nicol 2005). Phytoplankton support zooplankton. Plankton support krill. Krill support many marine organisms, including baleen whales. Baleen whales are part of the diet for the Inuit, the indigenous people of the Arctic.

During the nineteenth and early twentieth centuries, whalers of European ancestry hunted baleen whales to near extinction. Often they harvested only the oil and large baleen plates to which the baleen filter hairs are attached (see Figure 6.11) and left the rest of the carcass to rot. Baleen plates, though generally called whalebone, are composed of keratin, the substance of human

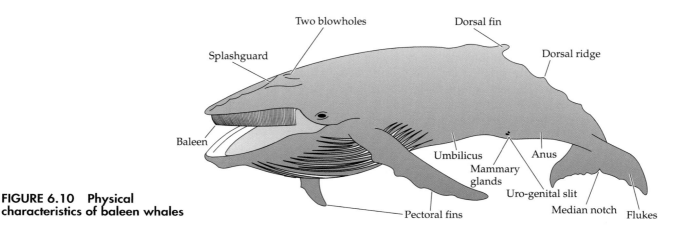

FIGURE 6.10 Physical characteristics of baleen whales

Two blowholes

Dorsal fin

Splashguard

Dorsal ridge

Baleen

Umbilicus

Anus

Mammary glands

Uro-genital slit

Pectoral fins

Median notch

Flukes

FIGURE 6.11 **Baleen filter hairs** attached to a baleen plate (whalebone).

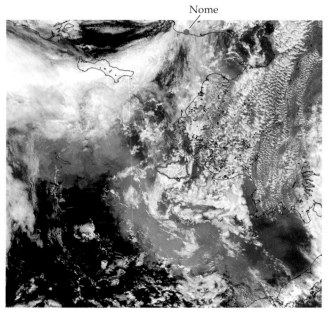

FIGURE 6.13 **Algal bloom in the Bering Sea** Single-celled phytoplankton change the color of the water from deep blue to milky blue-green. Designated are Nome, Alaska (red dot) and Nunivak Island (red outline), about 100 km wide, to provide scale and orientation.

hair and nails. These plates, because they are flexible and retain the shapes imposed on them when heated, served as the primary material for springs in vehicles and stays in ladies' corsets before the development of steel alloys and plastics. Therefore, baleen whales have supported humans in several ways.

Oceans are generally nutrient poor. When marine organisms eliminate wastes or die, the nutrients in the wastes or remains eventually sink and collect in deep waters. Ocean currents at high latitudes bring deep waters (see Figure 3.43) and their nutrients to the surface. These **upwellings** promote algal **blooms**, noticeable increases in algal populations (**Figure 6.13**) that sup-

port large swarms of krill (**Figure 6.14**). Baleen whales spend extended periods in the high-latitude oceans feeding on these swarms.

Global warming, as mentioned several times previously, is decreasing the extent of sea ice near conti-

FIGURE 6.12 **Northern krill (*Meganyctiphanes norvegica*),** a major food source for baleen whales, penguins, and seals. Narrow slots on the front legs filter plankton from the surrounding seawater. This example is about 3 cm in length.

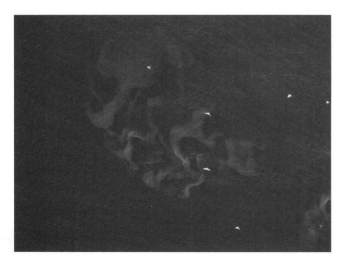

FIGURE 6.14 **Aerial view of krill in the waters near Portland, Maine** Swirls of red in the water are krill forming a backdrop for white seabirds feeding upon them.

nents. Phytoplankton living on or near sea ice are a critical food source for krill. Sea ice also shields krill from predators. Accordingly, annual mean density of krill varies with the duration of sea ice during the preceding winter (**Figure 6.15**). This dependency is complex, however, because fluctuations in krill density derive from both changes in overall abundance and movements to other locations (Murphy et al. 2007). Moreover, exceptions to the correlation between krill density and sea ice duration occur from time to time and from region to region (Smetacek and Nicol 2005).

Changes in the density of krill, given its key position in the food chain, should reverberate throughout the polar ecosystem, including baleen whales (Atkinson et al. 2004; Smetacek and Nicol 2005; Murphy et al. 2007). Nonetheless, determining the relationships among declining sea ice, krill, and whales has proved elusive, given the impact of human activities such as hunting.

Commercial whaling in the 1800s decimated baleen whale populations. Finally, in 1946, an international agreement was forged to save the whales. Each country was allotted a maximum number of catches for different whale species. Some species have begun to recover in response to more limited hunting. For example, the number of southern right whales (*Eubalaena australis*) has risen from a minimum of about 60 females in the 1920s to over 7000 today (**Figure 6.16**).

Unfortunately, illegal hunting has hampered recovery efforts. For instance, the whaling fleet of the Soviet Union caught about 48,477 humpback whales from 1948 until 1973 but reported a mere 2,710 of them. All told, the Soviet fleet admitted to only 185,000 of the 260,000 whales that it killed during this period. The extent of this fraudulent accounting was revealed only in 1994 after the dissolution of the Soviet Union (Yablokov 1994).

Restoration efforts for whales face other difficulties, including uncertainties about their population sizes and their low reproductive rates. Baleen whales travel

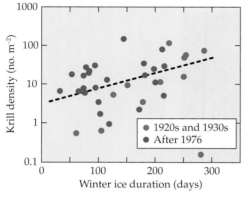

FIGURE 6.15 **Annual mean density of krill** across the southwest Atlantic near Antarctica. Plotted are density of krill versus number of days frozen coastal sea ice was observed during the previous winter. Note that the density scale is logarithmic. (After Atkinson et al. 2004.)

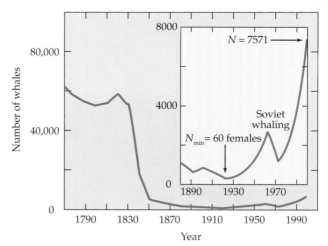

FIGURE 6.16 **Number of southern right whales (*Eubalaena australis*)** estimated from a computer model based on the natural history of the whales and counts of whale catches. The whale population fell to as few as 60 adult females (N_{min}) in 1922. The drop during the 1960s resulted from illegal hunting by the Soviet Union. Currently, this species has about 7600 individuals. (After Baker and Clapham 2004.)

Cosmos: White, star-like lights fill the dark, night-like sky across the two flukes.

Denali: Left fluke shows a snow-covered mountain top on the lower half with the rays of the northern lights above.

Dune: A dune-like formation is found on the left fluke with sand blowing across the flukes.

Fizz: White specks across the two flukes resemble the fizz in a dark-colored soda.

14: The number 14 is near the center of the left fluke.

FIGURE 6.17 Mugshots for identifying humpback whales via markings on flukes.

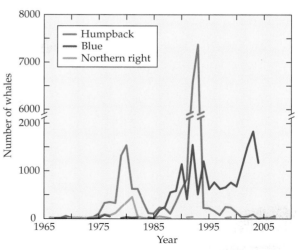

FIGURE 6.18 Estimated whale counts Number of humpback whales (*Megaptera novaeangliae*), blue whales (*Balaenoptera musculus*), and northern right whales (*Eubalaena glacialis*) based on sightings in a given year. The *y*-axis is split to accommodate the large number of humpback sightings in 1992 and 1993. Some years lacked any sightings. These data indicate the difficulty in estimating whale population sizes. (After Read et al. 2007.)

Other Organisms

Charismatic polar megafauna, the focus of the previous sections, are not the only organisms that are sensitive to the current warming trend. The following sections briefly introduce the anticipated responses of other organisms to global temperature changes. The coverage of organisms here is not comprehensive; rather, it seeks to highlight the range of responses via representation from a wide spectrum of organisms. Included are birds, grapes, corals, amphibians, pathogens, soil microbes, and humans.

Birds

Birds, particularly those that fly, have a high metabolic rate that heightens their sensitivity to their immediate surroundings. As such, they have long served as bellwethers of environmental change. For example, canaries were brought into coal mines in the United Kingdom as recently as 1986 to monitor carbon monoxide and methane because the birds gave clear indications of distress—they stopped singing and fell off their perches—if these noxious gases began to accumulate (Figure 6.19).

Seasonal temperature variations influence the proportion of short- and long-distance migratory species in European bird communities (Lemoine et al. 2007). Bird

far, wide, and deep. They are long-lived (some over 150 years), require 4 to 15 years to reach sexual maturity, and have only one or two calves every few years. Photo identification of markings on tail flukes (**Figure 6.17**) has aided the tracking of individuals (Wheelock College 2007), but sightings of various baleen whales remain intermittent (**Figure 6.18**). Estimates of whale population sizes, therefore, must rely heavily on theoretical models (Stevick et al. 2003; Baker and Clapham 2004; Branch et al. 2004).

In summary, food sources for baleen whales may have declined with the loss of sea ice that has accompanied global climate change. Nonetheless, it is not yet possible to distinguish the influence of climate change on the size of whale populations from the influence of other factors such as whaling.

FIGURE 6.19 **Canary in a cage** testing for noxious gases in the Hollinger Mine, Ontario, Canada in 1928.

species that migrate only short distances must endure winter and its meager resources, whereas long-distance migratory birds escape to warmer climes. Both short- and long-distance migrants benefit from the expanded resources that result from warm and early springs. In a study of birds at 21 sites across Europe, spring temperatures and the number of long-distance migratory species have increased significantly, while winter temperatures and the number of short-distance migratory species have not (Lemoine et al. 2007). A computer model of bird populations, which incorporates the United Kingdom's Hadley Centre HadCM3 climate model, predicts that the warming anticipated over the next half-century will favor long-distance migratory bird species to the detriment of short-distance migratory species (**Figure 6.20**).

Wine grapes

Wine grape (*Vitis vinifera*) is one of the more charismatic plant species. High temperatures during fruit ripening adversely affect wine quality and thus its market value. For example, premium Petit Verdot wine grapes from the Napa Valley of California (**Figure 6.21**) fetch an average of $5.95 per kg, whereas mixed red wine grapes from only 150 km away in the hotter San Joaquin Valley garner only an average of $0.24 per kg (California Department of Food and Agriculture 2007).

As a consequence of global warming, locations suitable for growing premium wine grapes will shift to higher latitudes and from inland to coastal areas. According to the forecasts of two climate models (the United States Department of Energy's Parallel Climate Model and the United Kingdom's Hadley Centre HadCM3 model), wine grapes in all of California

FIGURE 6.20 **Predicted changes (%) in the proportion of long- and short-distance migratory bird species in Europe** Shown are predictions for the time period 2051–2080 in relation to 1961–1990 under two different greenhouse gas emission scenarios. One scenario assumes rapid economic development based on fossil fuels (A1FI), and the other assumes large-scale adoption of clean and resource-efficient technologies (B1). (After Lemoine et al. 2007.)

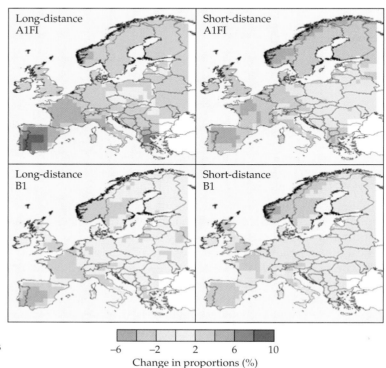

FIGURE 6.21   California The Sacramento Valley and San Joaquin Valley form the Central Valley.

FIGURE 6.23   Cavernous star coral (*Montastrea cavernosa*) in the Florida Keys National Marine Sanctuary.

except along the coast will ripen 1 to 2 months earlier and at higher temperatures by the year 2100 (Hayhoe et al. 2004). Total premium wine grape production in the United States might decline by up to 81% (**Figure 6.22**).

Coral reefs

Colonies of corals, marine invertebrate animals a few millimeters in diameter (**Figure 6.23**), develop a symbiotic relationship with coralline algae in the shallow, nutrient-poor waters of the tropics. Algae supply the coral invertebrates with carbohydrates, and their photosynthetic pigments impart color to the corals. In return, coral invertebrates provide the algae with a stable environment, carbon dioxide, and nutrients.

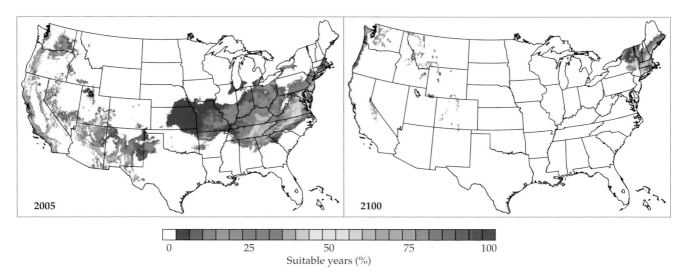

FIGURE 6.22   Wine country on the move Current (left panel) and predicted future (right panel) regions suitable for growing premium wine grapes in the 48 contiguous United States. Colors designate the proportion of years that a location would have optimal conditions. Predictions are based on a regional climate model and assume the greenhouse gas emissions scenario A2. (After White et al. 2006.)

FIGURE 6.24 **The Great Barrier Reef** extends for 2000 km along the northeastern coast of Australia. This satellite image shows a 380 km–wide portion, 200 km north from the coast of central Queensland.

FIGURE 6.25 **Bleached elkhorn coral** near St. Croix, U.S. Virgin Islands, in October 2005.

Together, they secrete calcium carbonate ($CaCO_3$) that forms elaborate reefs (**Figure 6.24**), which provide habitat for a diverse array of animal species.

Many corals live at close to their upper thermal tolerances (Harley et al. 2006). Warming of the seas by as little as 1°C or 2°C above historical averages tends to stress corals so they expel algae. Corals without algae turn white, a process appropriately called coral bleaching (**Figure 6.25**). If bleaching persists for several months, corals die. Global warming has generally exacerbated the incidence of coral bleaching (**Figure 6.26**). Coral bleaching in conjunction with other fac-

tors, including the decrease in seawater pH (discussed at the end of this chapter), may cause irreparable damage to 40% of the reefs during the next few decades (Knowlton 2001).

Amphibians

Amphibians have not fared very well in recent years (**Figure 6.27**). Since 1980, 32.3% of amphibious species (1856 out of 5743 known species) have declined in population size, 7.4% sit on the brink of extinction, and between 0.2% and 2.1% have already disappeared (Stuart et al. 2004). This contrasts with birds and mammals for which, respectively, 12% and 23% have declined over the same period, 1.8% and 3.8% are near extinction, and 0.05% and 0% have already disappeared.

Most of the threatened amphibian species belong to four families: Bufonidae (true toads), Hylidae (tree frogs), Leptodactylidae (neotropical frogs), and Ranidae (true frogs) (Stuart et al. 2004). Loss of suitable habitat has been a critical factor in the decline of all four families (**Figure 6.28**). In addition, 21 species in the Ranidae family are threatened by overexploitation from excessive harvesting for human consumption, especially in Asia. At first, the decline in populations of 80 species in the Bufonidae (**Figure 6.29**), 38 species in the Hylidae, and 47 species in the Leptodactylidae seemed puzzling because it was occurring even

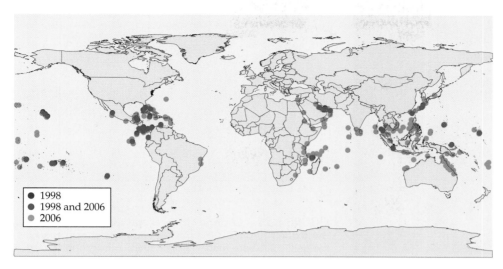

FIGURE 6.26 Locations of coral reefs that exhibited severe bleaching in 1998, 2006, and at both times. (After Marshall and Schuttenberg 2006.)

- 1998
- 1998 and 2006
- 2006

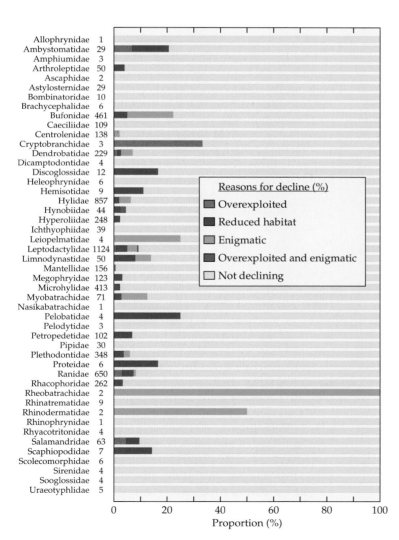

FIGURE 6.27 Amphibian families Name and number of species in a family. Plotted is the proportion of these species that are declining for various reasons. (After Stuart et al. 2004.)

Reasons for decline (%)
- Overexploited
- Reduced habitat
- Enigmatic
- Overexploited and enigmatic
- Not declining

in areas protected from habitat loss or human exploitation. Recent evidence, however, implicates global warming (Pounds et al. 2006).

A pathogenic fungus *Batrachochytrium dendrobatidis* infects many amphibians. Within 6 months after *B. dendrobatidis* initially arrived at a site in Panama, approximately 80% of the individuals from over half of the amphibian species had died from the resulting infections (Lips et al. 2006). Temperatures at many tropical highland localities are warming, and as they approach the growth optimum of the fungus, outbreaks occur and extinctions increase.

Human pathogens

The fungal epidemic in amphibians demonstrates that global warming can be a major factor in the spread of pathogens (Harvell et al. 2002; Patz et al. 2005). Of particular concern are human pathogens that spend part of their life cycle in **host vectors** (organisms that transmit diseases) that cannot regulate their temperature, such as bloodsucking insects (Table 6.3). Pathogenic viruses, bacteria, protozoans (single-celled, eukaryotic organisms), and worms need to complete certain life stages within the

FIGURE 6.28 Golden Toad, *Bufo periglenes* (family Bufonidae), in the cloud forest above Monteverde, Costa Rica. Adults, like the one in the photograph, reached 5 cm in length. This species is presumed extinct because it has not been seen since 1989.

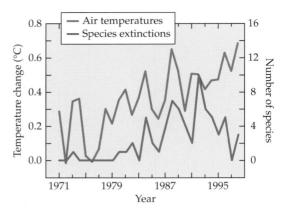

FIGURE 6.29 Changes in average air temperatures of the tropics (from 30°N to 30°S) and number of species in the genus *Atelopus* (neotropical harlequin frogs in the toad family Bufonidae) that became extinct. Temperatures are relative to the average from 1856 to 1895. (After Pounds et al. 2006.)

lifespan of their host vector, and warmer temperatures speed their development. On the host vector side, warmer temperatures not only extend the range of cold-sensitive bloodsucking insects, but also increase their egg production and thus the frequency of their blood feeds. All of these factors serve to expand the range and to accelerate the spread of epidemics.

TABLE 6.3	Vector-borne diseases that may be sensitive to global warming
Vector	*Examples of major diseases*
Blackflies	Onchocerciasis (river blindness)
Mosquitoes	Malaria, filariasis (elephantiansis), dengue fever, yellow fever, West Nile fever
Sandflies	Leishmaniasis (Orient Boils, black fever, Dum-Dum fever)
Ticks	Lyme disease, tick-borne meningoencephalitis
Triatomines (kissing bugs)	Chagas disease (American trypanosomiasis)
Tsetse flies	African trypanosomiasis (sleeping sickness)

Source: Haines et al. 2006.

Many cities were established at latitudes or altitudes that were previously too cold for such diseases to be **endemic** (common to the area). With global warming, minimum temperatures are increasing even faster than maximum temperatures (see Figure 4.30A). As a consequence, some of these locations are becoming more vulnerable to these diseases.

For example, dengue fever is considered the most important vector-borne viral disease in the world (World Health Organization 2002). It is limited to tropical and subtropical regions because freezing temperatures kill overwintering larvae and eggs of the one species of mosquito (*Aedes aegypti*) that transmits it. Computer models of *Ae. aegypti* populations suggest that climate-induced variations in population size are highly correlated with the number of reported dengue cases (Hopp and Foley 2003). Extrapolating such models to the warmer world of the future predicts that 50% to 60% of the human population will be at risk of dengue infection, versus 35% today (Hales et al. 2002). For instance, towns such as Broome and Katherine, Australia will lie squarely in the endemic region where today they are generally dengue-free (**Figure 6.30**).

Malaria, perhaps the most debilitating disease of all, afflicts 350 million to 500 million people each year. The epidemiology of malaria is complex. Parasitic protozoans from four or five species in the genus *Plasmodium* cause this disease, and mosquitoes from 30 to 40 species in the genus *Anopheles* transmit it. Some studies predict that global warming will promote outbreaks at higher altitudes (Tanser et al. 2003), whereas others find no such trend (Small et al. 2003; Reiter et al. 2004).

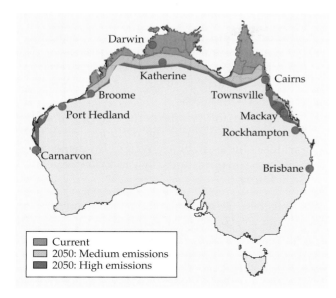

FIGURE 6.30 **Potential spread of dengue fever in Australia** A model of the virus–mosquito vector relationship predicts that rising temperatures under medium and high scenarios of greenhouse gas emissions will expand the endemic region to higher latitudes. (After McMichael et al. 2006.)

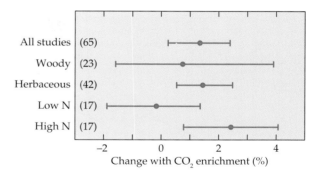

FIGURE 6.31 **Soil carbon responses to atmospheric CO₂ levels** Shown are the changes in soil carbon content from plots exposed to ambient concentrations (≈370 ppm) to those exposed to elevated CO_2 concentrations (ranging from 430 ppm to 750 ppm). Studies focused on woody versus herbaceous plant species or applied low versus high levels of nitrogen fertilizer. In parentheses are the numbers of studies in each analysis. Circles and error bars designate means and confidence intervals, respectively. (After de Graaff et al. 2006.)

Indeed, factors such as greater pathogen resistance to antimalarial drugs and failed vector-control programs may be just as important as global warming in driving recent expansions of malaria (Harvell et al. 2002).

Soil microorganisms and biogeochemistry

The vast majority of carbon on Earth is located underground (see Figure 4.13). It resides in inorganic forms such as the carbonates in sedimentary rock, mostly calcite ($CaCO_3$), and in organic forms such as deposits of coal, oil, and natural gas and accumulations of soil organic matter. The ground just beneath our feet is teeming with biological activity, and each year, soil microorganisms, in a process known as **decomposition**, break down about 2.5% of soil organic matter, converting the organic carbon to carbon dioxide (CO_2) that is subsequently released to the atmosphere. Roughly an equivalent amount of organic carbon returns to soils through additions of animal and especially plant material, a process known as soil organic carbon **sequestration**.

The balance between carbon decomposition and sequestration has a direct effect on atmospheric CO_2 levels. Evidence for this includes the prominent seasonal cycles of atmospheric CO_2 concentrations (see Figure 1.9). Atmospheric CO_2 concentration decreases during the summer, when photosynthesis and carbon sequestration exceed respiratory decomposition, and increases during the winter, when the balance reverses. Will global climate change influence the relative rates of carbon sequestration and decomposition? If so, will this exacerbate or mitigate rising CO_2 levels?

An elevated atmospheric CO_2 concentration stimulates photosynthesis and thereby plant biomass production, particularly under heavy nitrogen fertilization (see Figure 5.24). This additional plant material enhances soil organic carbon sequestration. Soil microorganisms proliferate under these amendments and accelerate respiratory decomposition (see Figure 5.23). At more natural levels of nitrogen fertilization (low N), faster soil carbon decomposition balances faster sequestration, and so soil carbon contents generally do not change significantly under CO_2 enrichment (**Figure 6.31**).

Global warming may disturb this balance. Net photosynthesis is relatively insensitive to temperature (see Figure 6.1), but a longer growing season (see Figure 6.4) should enhance plant productivity and thereby accelerate soil carbon sequestration. Soil carbon sequestration, however, may not keep pace with soil carbon decomposition (Davidson and Janssens 2006) because the activity of soil microorganisms responsible for decomposition increases exponentially with temperature (**Figure 6.32**).

Predicting the response of soil decomposition to global warming has proved difficult for several reasons (Davidson and Janssens 2006; Kirschbaum 2006).

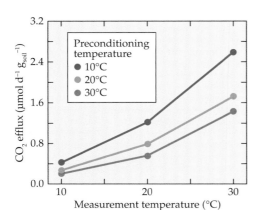

FIGURE 6.32 Temperature dependence of CO_2 released (10^{-6} moles per day per gram of soil) during microbial decomposition of soil. (After Kirschbaum 2006.)

First of all, soils vary greatly from place to place. Next, even a single soil type from one place contains thousands of different organic compounds. About 5% of these compounds are **labile**, meaning they decompose rapidly (e.g., sugars and proteins). Another 60% to 85% of these compounds are **intermediate**, meaning they decompose in years to decades (e.g., cellulose). The final 10% to 40% are **recalcitrant**, meaning they may persist in soils for millennia (e.g., lignin and tannins). Differences among soil constituents are responsible for large variations in decomposition rates over time.

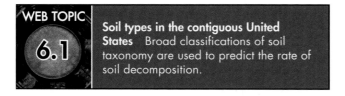

WEB TOPIC 6.1 **Soil types in the contiguous United States** Broad classifications of soil taxonomy are used to predict the rate of soil decomposition.

In regard to global warming, respiratory CO_2 efflux from soils declines exponentially with time of exposure to elevated temperatures (**Figure 6.33**). Initially the labile compounds in a soil decompose and release a substantial amount of CO_2. As labile compounds break down, intermediate and recalcitrant compounds constitute a larger portion of the remaining soil organic carbon, and CO_2 efflux slows down (see Figure 6.32).

Environmental factors other than temperature influence decomposition (Six and Jastrow 2002; Davidson and Janssens 2006).

- Soil particles form **aggregates** (masses or clods) of different sizes. Size matters because organic carbon compounds inside larger soil aggregates are more protected from water, oxygen, and microorganisms and their enzymes and, thus, decompose more slowly.

- Oxygen concentrations fluctuate widely in the disjointed air- and water-filled pores of soils. Under anaerobic conditions, soil microorganisms are limited to a few, generally slow biochemical pathways for decomposing organic compounds.

- Water availability affects the activity of soil microorganisms and the diffusion of their water-soluble enzymes. Water also displaces air from soil pores and can lead to low oxygen conditions.

Despite these uncertainties about soil decomposition in a warmer world, release of additional CO_2 from soils is anticipated. Wetlands and permafrost soils (soils frozen year round) at high latitudes contain large amounts of organic soil carbon. These high latitudes are likely to experience more extreme warming than the rest of the planet (see Figure 4.34) and, thus, the labile pool of organic carbon in their soils will be subject to faster decomposition. Consequently, global climate models predict that by the latter half of the twenty-first century, landmasses will release more CO_2 than they absorb (see Figure 4.17).

Humans and heat

As global maximum temperatures rise, heat waves are becoming more pronounced (see Figure 4.38). Europe experienced the hottest summer on record in 2003, with average temperatures 3.5°C above normal (Patz et al. 2005). During a 2-week period in August 2003, between 30,899 and 49,004 Europeans died from heat-related

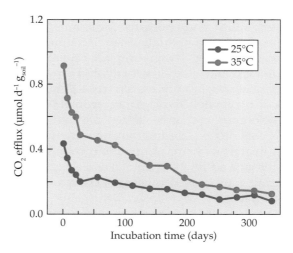

FIGURE 6.33 Time course of CO_2 released (10^{-6} moles per day per gram of soil) during microbial decomposition of soil at 25°C and 35°C. (After Conant et al. 2007.)

TABLE 6.4 Heat-related deaths in various European countries during August, 2003

Country	Excess mortality
Belgium	1,205
France	14,802
Germany	1,410
Great Britain	2,091
Italy	3,134–19,780
Netherlands	1,400–2,200
Portugal	1,854
Spain	3,574–4,687
Switzerland	975

Source: Kosatsky 2005.

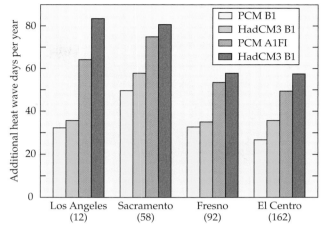

FIGURE 6.34 Predicted heat waves in California Shown are the additional number of days per year with heat wave conditions in four Californian cities over the next 100 years. Number in parentheses under city name is average number of heat waves per year during the period 1961–1990. Simulations were conducted on two global climate models (PCM and HadCM3) under two greenhouse gas emissions scenarios (B1 and A1FI). A heat wave is defined as three or more consecutive days with temperatures above 32°C. (After Hayhoe et al. 2004.)

causes (Table 6.4). Mortality was highest among the old, chronically ill, and isolated (Kosatsky 2005).

Was this a natural disaster, independent from human activities? Simulations run on the HadGCM3 global climate model indicate that human-induced climate change more than doubled the risk of such an event in 2003 (Stott et al. 2004). Under the warming anticipated during the next 40 years, similar heat waves will be 100 times more likely to occur.

Climate models also predict that California will experience more severe heat waves (Figure 6.34). From 1961 to 1990, an average of 165 people in Los Angeles died each year from heat-related causes. A computer model that relates mortality to temperature suggests that Los Angeles will suffer two times to eight times more heat-related deaths annually (319 to 1429) by the end of the century (Hayhoe et al. 2004).

Most of the studies on heat waves and human health have been conducted in developed countries located in temperate zones. One exception (Gouveia et al. 2003) studied São Paulo, Brazil, a developing city in the subtropics, and found that mortality increased at extreme temperatures, hot or cold, in a manner similar to that observed in North American and European cities (Figure 6.35). This finding suggests that the anticipated upsurge in heat-related deaths from global warming will know few boundaries.

Human bodies cool themselves via perspiration. We feel less comfortable when high temperature combines with high humidity because of our inability to lose heat when the concentration gradients that drive evaporation from our bodies are small. As the song says, "Too tough to tame... sweating in the rain" ("Fly on the Wall," AC/DC). Consequently, the apparent temperature or heat index—a measure of how hot a person feels—is a function of relative humidity as well as air temperature (Figure 6.36). Regional changes in relative humidity and precipitation will interact with global warming to influence the impact of future heat waves.

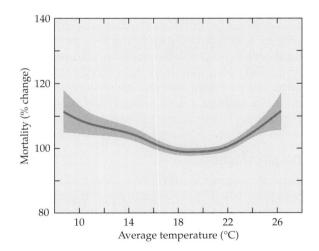

FIGURE 6.35 Mortality (% change from the average daily rate) in response to temperature (°C) averaged over the current and previous day in São Paulo, Brazil, 1991–1994. Mortality was adjusted for season, day of week, pollution, and humidity. Shown are means (red) and confidence intervals (gray). Mortality increased when temperatures become colder or hotter than 20°C. (After Gouveia et al. 2003.)

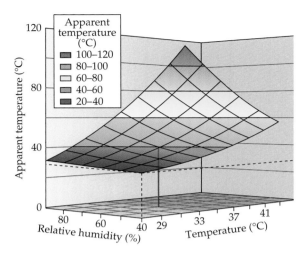

FIGURE 6.36   Apparent temperature (°C) as a function of relative humidity (%) and actual air temperature (°C) For example, when relative humidity is 60% and actual temperature is 33°C, apparent temperature is 40°C (purple line). Below 27°C and 40% relative humidity, apparent and actual temperatures are nearly equal.

Precipitation

Life on our planet—based as it is on biochemical reactions in aqueous media—is highly sensitive to water availability. Over the past quarter century, tropical and polar regions have become wetter, whereas mid-latitudes have become drier (see Figure 2.45). These trends are likely to continue into the foreseeable future (see Figure 4.38). Species are shifting their geographical distributions in response to changing patterns of temperature and precipitation, as introduced earlier in the chapter. The following sections focus on water issues associated with human well being, specifically, those of food production, irrigation, and salinity.

Food production

Worldwide acreage under cultivation has reached a plateau (**Figure 6.37A**) because much of the arable land on Earth is already in use. Agriculture, to meet the food and fiber demands of a growing human population without additional land area to expand, has become more efficient in its land use. Farmers today require only about half the land to supply a person with food and fiber than they did 50 years ago. Improved water management is, in part, responsible for this greater agricultural efficiency.

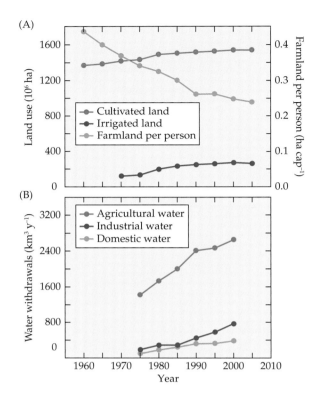

FIGURE 6.37   Water resource use in agriculture (A) Global land area (millions of hectares) of crops under persistent or intermittent cultivation or irrigation and hectares of farmland per person. (B) Volume of water (km³ per year) used for agriculture, industry, and domiciles. (After Aquastat 2007.)

Timing crop plantings to take advantage of seasonal water availability, although still widely practiced, is no longer sufficient to feed the world. More and more farmers supplement available soil water through irrigation using water collected from different locations or at different times of the year. Irrigated acreage and water withdrawals for agriculture have doubled since 1975. Over the same period, water withdrawals for industrial and domestic purposes have increased four-fold, but they still account for less than half of the amount withdrawn for agricultural purposes (**Figure 6.37B**).

The majority of the world's best farmland, in terms of climate and soils, is located in the temperate zones that lie between the tropics and the polar circles (**Figure 6.38**). A large percentage of this land is now irrigated (**Figure 6.39**). Water for irrigation in several major agricultural regions depends on snow that accumulates in nearby mountains.

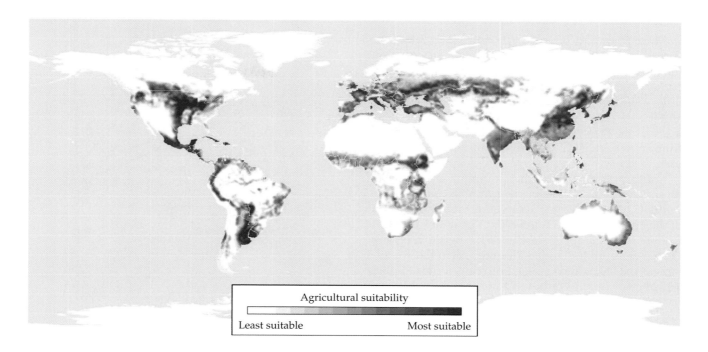

FIGURE 6.38 Agricultural suitability of different locations based on crop-growth degree days, relative humidity, density of soil organic carbon, and soil pH. Crop-growth degree days are the sum over a year of the difference between daily average temperatures and 5°C, which is the minimum temperature that allows the growth of most crops. (From Ramankutty et al. 2002.)

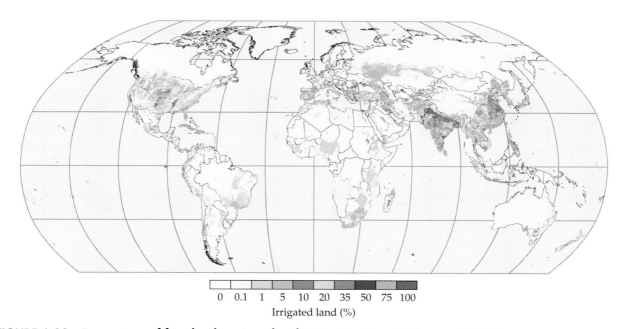

FIGURE 6.39 Percentage of farmland equipped with irrigation (From Siebert et al. 2005.)

More than half of the fruits and vegetables consumed in the United States and Canada are grown in the Central Valley of California (see Figure 6.21). This highly productive agricultural system receives precious little rainfall in spring, summer, and fall (see Figure 1.3) and thus depends on irrigation during the prime growing season. The water for irrigation derives primarily from snowpack in the Sierra Nevada Mountains to the east.

Californian agriculture is not unique in depending on snowpack. Every major river in Asia, including the Indus, Ganges, Mekong, Yangtze, and Yellow, flows out of the Himalayas. The Yellow River provides water for China's wheat crop, the largest in the world; the Yangtze River supplies China's rice crop, also the world's largest; the Indus and Ganges both contribute to India's wheat crop, second only to China's; and the Mekong irrigates the rice crop of Vietnam, the world's third largest exporter. Altogether, these agricultural areas in Asia feed half of the human population (Brown 2006). Both the extent and timing of water flow in these rivers varies with the snowpack in the Himalayas.

Global warming will diminish the quantity of water stored in snowpacks of mountain ranges at mid-latitudes such as the Sierra Nevadas and Himalayas. In such ranges, most of the precipitation falls at mid-elevations when moist air, blown in from the oceans, rises and cools as it approaches the mountains. Cooler air has a lower capacity to hold water vapor (see Figure 2.43), and any excess water vapor precipitates. If this precipitation falls as snow, the resulting snowpack provides convenient storage until it melts during warmer, drier seasons. If this precipitation falls as rain, this run-off may engorge rivers, overflow reservoirs, and be discharged directly back to the oceans. Warmer climates will shift snowline to higher altitudes and thereby decrease snowpack and increase rainfall.

In the Sierra Nevadas of California, the snowpack at lower elevations (1000 m to 2000 m) currently holds about 30% of the total water in snow and provides about 40% of the water flowing into reservoirs during the spring. Global climate models predict that most of this low-elevation snow will disappear by the end of the century. Moreover, the climate models predict that mid elevations (2000 m to 3000 m), which receive 52% of the total snowfall, will receive only 25% of the snow that they do now (**Figure 6.40A**). Water flows into reservoirs from snowmelt during the spring will halve (**Figure 6.40B**), and maximum flows from snowmelt will occur 14 to 32 days sooner in the year (Hayhoe et al. 2004).

In many parts of Asia, water from local precipitation and snowmelt in the Himalayas no longer meet the

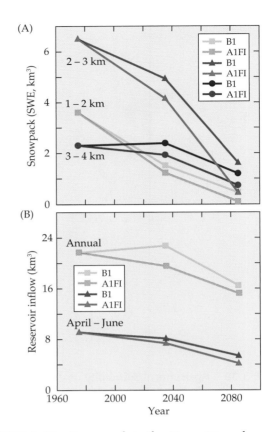

FIGURE 6.40 Snowpack in the Sierra Nevada Mountains (A) Average snowpack, in snow water equivalents (SWE, km³) and (B) reservoir inflow (km³) measured (1961–1990) or predicted (2020–2049 and 2070–2099) with a global climate model under the B1 or A1FI scenarios for greenhouse gas emissions. Snowpack is shown at three different elevation ranges. Reservoir inflow is over the entire year or from April–June snowmelt. (After Hayhoe et al. 2004.)

demands of irrigated agriculture (Brown 2006). To meet this water deficit, farmers are pumping groundwater from wells as deep as 1000 m. Not only are energy costs for pumping from such depths becoming prohibitive, but deep wells tap ancient aquifers that are likely to be depleted in the near future because they replenish only at a very slow pace.

In China, acreage of irrigated farmland has peaked for lack of water, and grain production has leveled off or even declined. China, however, has experienced steady population growth during its rapid economic expansion of the past decade (**Figure 6.41**). China receives only 5% of the world's total precipitation but is home to 21% of the human population. To meet its food needs, China now imports grain and other agricultural commodities.

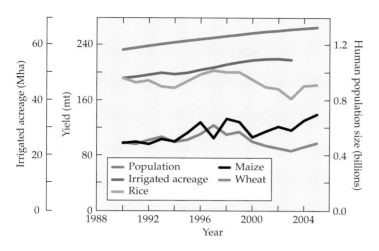

FIGURE 6.41   Trends in China, 1990–2005 Human population size (billions of people), irrigated acreage (millions of hectares), and yields (millions of metric tons) of rice, maize, and wheat. (After FAOSTAT 2009.)

FIGURE 6.42   Lake Chad in 2004 The lake is observable from space only during the wet season, when wetlands surrounding the lake appear as a green "L."

In December 2006, the Chinese Ministry of Science and Technology released its first National Assessment of Climate Change (Yihui et al. 2007). It predicts that average precipitation in China will increase by 2% to 3% by the year 2020 and by 5% to 7% by the year 2050 (see Figure 4.48). Such increases are unlikely to alleviate water shortages in many areas because warmer temperatures—from 1.3°C to 2.1°C by 2020 (see Figure 4.44) and from 2.3°C to 3.3°C by 2050—will cause greater evaporation. During the second half of the century, water availability will limit production of major crops such as rice, maize, and wheat by as much as 37%. The assessment concludes that climate change poses a great threat to China's food security.

Another example of competition for water is Lake Chad, located at the intersection of four countries (Chad, Niger, Nigeria, and Cameroon) in West Africa. The lake once served as a landmark for astronauts but now is difficult to spot from space at most times of the year (**Figure 6.42**). Between 1983 and 1994, the volume of water diverted from the lake for irrigated agriculture quadrupled. Moreover, rainfall in the region has declined significantly from the 1960s. As a result, Lake Chad is currently about 5% of its previous size (**Figure 6.43**).

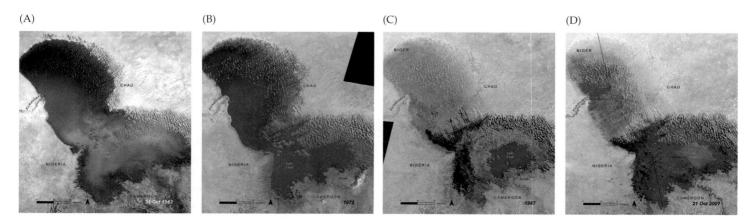

FIGURE 6.43   Satellite photographs of Lake Chad in (A) 1963, (B) 1972, (C) 1987, and (D) 2001.

TABLE 6.5	Concentrations in red blood cells (RBC), plasma, and seawater		
Component	RBC (mmol kg⁻¹ solids)	Plasma (mmol kg⁻¹ fluids)	Seawater (mmol kg⁻¹ fluids)
Sodium	39.6 ± 5.4	141.3 ± 1.5	480.7
Potassium	266.9 ± 13.9	4.0 ± 0.3	10.5
Chloride	152.5 ± 12.0	101.2 ± 3.5	559.4
Bicarbonate	—	29.0 ± 3.5	2.1
Calcium	1.2	—	10.5
Magnesium	11.1	—	54.5
Phosphate	57.9 ± 10.2	—	—
Protein	65	36	—

Source: Keitel et al. 1955; Pilson 1998.

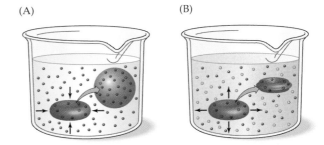

(A)   (B)

FIGURE 6.44 Responses of red blood cells to pure water and saline solutions The membrane of a red blood cell is permeable to water but impermeable to many salts. (A) A red blood cell placed into a beaker of pure water absorbs water and expands until it bursts because the concentration of water is smaller within the cell than outside the cell. (B) A red blood cell placed into a beaker of saline water loses water and shrinks until the concentration of salt inside and outside are equal. Blue spheres represent water molecules and green spheres represent salt molecules such as sodium chloride.

Salinity

The common saying "blood is thicker than water" is also true in terms of physical chemistry. What flows through our veins are red blood cells (erythrocytes, constituting 45% of our blood volume), white blood cells (leukocytes, 1%), platelets (thrombocytes, less than 1%), and plasma (serum, greater than 53%). The blood cells contain complex solutions of biochemical salts and inorganic salts (**Table 6.5**). When blood is exposed to pure water, the concentration of water inside the cells becomes less than the concentration outside because the water inside is diluted with biochemical and inorganic salts; this concentration gradient for water drives it into cells until they burst (**Figure 6.44A**). Conversely, when blood is exposed to a more saline medium (e.g., seawater), the concentration of water inside the blood cells becomes higher than the concentration outside, and this concentration gradient for water expels it from the cells until they collapse (**Figure 6.44B**). Blood cells have an elaborate system of pumps, channels, and carriers that compensates for moderate salt imbalances, yet the salt composition of plasma and fluids delivered intravenously must remain within strict tolerances to avoid damage to blood cells.

Most organisms respond to the salinity of their environment in a manner analogous to blood cells. Exposures to concentrations either too high or too low are problematic. Plant roots, for example, must maintain salt concentrations higher than their surroundings in order to extract water from the soil. Plants cells avoid bursting under such conditions because they are surrounded by a strong cell wall. Plants experience drought stress in soils containing high concentrations of salt because extracting water is more difficult. In contrast, plants may suffer nutrient deficiencies when salt concentrations in the soil solution are low (Epstein and Bloom 2005).

Agricultural systems that depend on irrigation inevitably increase soil salinity. Natural sources of water contain some dissolved salts. **Transpiration** (loss of water vapor) from plants and evaporation from soils, lakes, and rivers act as distillation processes in which pure water vaporizes and the salts left behind become more concentrated in the smaller volume of liquid that remains. Unless large amounts of water are periodically flushed through soils and discarded, irrigated farmlands (see Figure 6.40) eventually become salt-afflicted (**Figure 6.45**).

Mesopotamia is located at confluence of the Tigris and Euphrates Rivers in what is currently Iraq. It is considered the "cradle of civilization" because societies that formed there sometime between 4000 B.C. and 3500 B.C. developed sophisticated writing systems and used them in the context of complex governmental administration. Supporting these societies were many agricultural advances including plant breeding, animal domestication, crop rotations, water storage, and irrigation. Centuries of irrigation with inadequate drainage, however, progressively added salt to the soil (Algaze 2001). Eventually, farmers were forced to abandon large acreage, and prosperity declined. Much of this area still

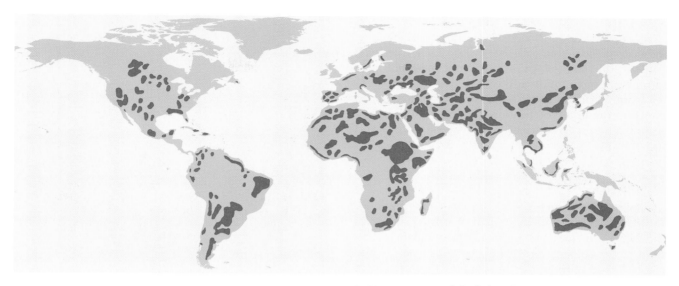

FIGURE 6.45 Global distribution of salt-inflicted areas (After Pessarakli and Szabolcs 1999.)

suffers from salinity problems as the result of ancient and more recent agricultural mismanagement.

The Salinas Valley of California, affectionately known as America's Salad Bowl, is the primary source of the salad greens, celery, and broccoli consumed in the United States and Canada. Water for irrigation primarily comes from pumping of groundwater. This and, to a smaller extent, rising sea level (see Figure 2.42) have led to progressively greater intrusion of seawater into the local aquifers (**Figure 6.46**).

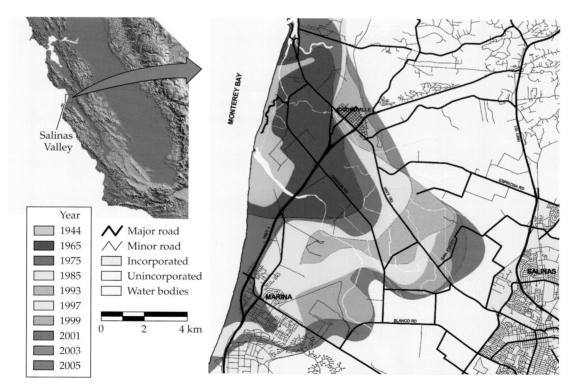

FIGURE 6.46 Areas afflicted with seawater intrusion into the aquifer at 60-m depth near Salinas, California. (After Monterey County Water Resource Agency 2006.)

Global warming will increase transpiration of plants and evaporation from soils (see Figure 2.43). Increased transpiration and evaporation, if not matched with additional precipitation and proper disposal of drainage water, will accelerate soil salinization. Rising sea level and depletion or contamination of aquifers will diminish the availability of fresh water for crops and humans alike, particularly in coastal regions.

pH: Acid–Base

The balance between acids and bases in a medium has multiple effects on the biological processes within it. These effects include the conformation of biochemicals, the storage of biochemical energy, and the stability and availability of inorganic compounds. A brief review of water chemistry follows.

Hydrogen is the lightest element. The common isotope of hydrogen consists of a single negatively charged electron spinning around the nucleus of a single positively charged proton. When hydrogen ionizes and loses its electron, only the proton (H^+) remains.

Water (H_2O) is a polar solvent; that is, it has domains with equal but opposite charges. The oxygen atom (O) in a water molecule strongly attracts the electrons of the hydrogen atoms (H_2), resulting in a net positive charge on the hydrogen atoms and a net negative charge on the oxygen (**Figure 6.47**). The uneven distribution of electrical charges across a water molecule fosters interactions with other water molecules and with other charged molecules. This uneven distribution also facilitates the dissociation of water into protons (H^+) and hydroxyl ions (OH^-):

$$H_2O \leftrightarrow H^+ + OH^- \tag{6.1}$$

The activity of protons in a solution determines its pH, a measure of acidity or alkalinity (baseness). In precise terms,

$$pH = -\log_{10} \alpha_{H^+} \tag{6.2}$$

where $\log_{10}$ is the logarithm to base 10 and α is the activity of protons. In dilute solutions, the activity level of protons is nearly equal to their concentration (in concentrated solutions, interactions among protons diminishes their activity). An aqueous solution is considered neutral—neither acid nor base—when the pH equals 7.0 because at this value the concentrations of protons (H^+) and hydroxyl ions (OH^-) are the same, 10^{-7} moles per liter.

Conformation of biochemicals

Many biochemicals have electrical charges on their surface. For example, the "A" in DNA stands for "acid,"

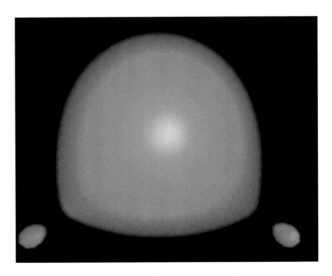

FIGURE 6.47 Relative electron densities in a water molecule The red object depicts an oxygen atom, and light blue ones depict hydrogen atoms. The electron densities around the oxygen are much higher than those around the hydrogens, resulting in a polar molecule. (After Trindade and Fiolhais 1998.)

meaning that in a neutral solution, DNA molecules tend to lose protons and assume a negative charge. Proteins are constructed from linking together a string of amino acids. An amino acid, as the name implies, has one end that tends to lose protons and assumes a negative charge. Less obvious is that the other end of an amino acid is a base that tends to gain protons and assumes a positive charge.

These tendencies to assume positive or negative charges depend on the pH of the solution in which the biochemicals are dissolved. If the solution is acidic, the DNA and amino acids are less likely to lose a proton and are more electrically neutral. If the solution is alkaline, the biochemicals are more likely to lose protons and assume a negative charge.

Because opposite charges attract and like charges repel, pH-dependent changes in the surface charges of biochemicals such as DNA and proteins influence their conformation in aqueous solutions (**Figure 6.48**). Conformation of these molecules has a large effect on their functioning. Organisms have sophisticated mechanisms that maintain cellular pH within narrow limits to prevent conformation changes that might interfere with the proper functioning of their biochemicals.

Energy storage

Biological compartments are surrounded with a **membrane** composed of two layers of non-charged molecules (i.e., lipid bilayer). Ions such as protons pass through biological membranes only slowly if unaided

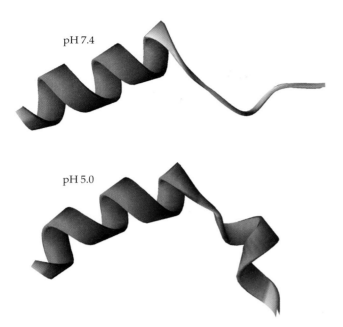

pH 7.4

pH 5.0

FIGURE 6.48 Herpes simplex virus Shown here are pH-dependent (top, pH 7.4; bottom, pH 5.0) changes in the conformation of a protein from the herpes simplex virus. These changes enable the protein to move from one cellular compartment to another and may affect the virulence of the virus.

by specialized protein structures that transverse these layers. Cells of most organisms exploit this low **mem-**

brane permeability of protons to store energy. When surplus energy is available, integral membrane proteins called proton pumps transport protons across a membrane and establish a proton concentration gradient, or pH gradient. To retrieve this energy, cells let these protons flow back across the membrane through integral membrane proteins called H^+-ATPases. The flow of protons through a H^+-ATPase produces the high-energy compound ATP (adenosine triphosphate) (**Box 6.1**).

An acidic or alkaline environment may interfere with establishing or maintaining pH gradients across biological membranes and thus disrupt cellular energy production. Organisms can compensate for short-term disruptions, but long-term disruptions may prove toxic. This is yet another reason that organisms buffer themselves against pH extremes.

Inorganic compounds and pH

The pH of a solution determines the **solubility** of several elements that are essential for life (**Figure 6.49**). For example, in an aerated solution with a pH above 8, ferric iron (Fe^{3+}) precipitates as the extremely insoluble ferric hydroxide, $Fe(OH)_3$, with the consequence that iron may not be available for absorption by organisms. Difficulty in maintaining iron dissolved in an absorbable form impairs treatment of iron-deficiency anemia in humans and of iron deficiencies in plants. The solu-

---

BOX 6.1   Using a pH gradient to generate ATP

The interplay between the structure and function of H^+-ATPases is fascinating. As protons flow from a compartment with a high concentration of protons to one with a low concentration, they pass through a channel in the protein that causes several subunits to rotate. Some of these subunits form a complex with ADP (adenosine diphosphate) and a phosphate group P_i, and as they rotate, conformational changes in the subunits squeeze the ADP and P_i together and generate ATP. This process is analogous to placing a waterwheel in a flowing stream and using the resultant rotational motion of the wheel to run a press that makes high-energy fuel bars.

H^+-ATPase   Shown is an F-type ATPase protein found in the membranes of mitochondria and chloroplasts. Passage of protons (H^+) through the protein causes the rotor parts of it to rotate. This squeezes ADP and phosphate (P_i) together, thus generating ATP. The stator parts of the protein remain immobile.

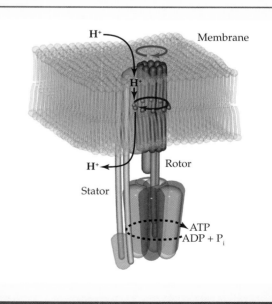

H+ ... Membrane

H+

H+ ... Rotor

Stator

ATP
ADP + P_i

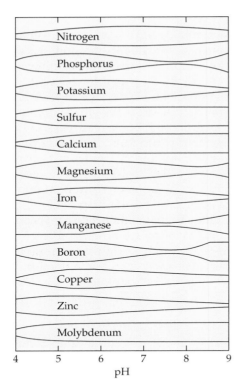

FIGURE 6.49   Influence of pH on the solubility of different elements essential to organisms Width of the beige shaded area reflects the relative solubilities. For example, nitrogen ions are most soluble at a pH of about 6.7. (After Lucas and Davis 1961.)

bilities of many other positively charged ions are also strongly pH sensitive.

Furthermore, chemical reactions between water and phosphorus (P) or carbon dioxide (CO_2) depend on pH (Table 6.6). The **acid dissociation constant** (pK_a) is

defined as the pH at which the protonated and deprotonated forms are present in equal amounts. For example, the concentrations of H_3PO_4 and $H_2PO_4^-$ in a dilute solution will be equal at pH 2.12, and therefore the pK_a of this conversion equals 2.12. Organisms often modify the pH of their surroundings to enhance acquisition of the essential nutrient phosphorus because complexes of monobasic phosphate ($H_2PO_4^-$) with aluminum or iron limit phosphorus availability under acid conditions, whereas complexes of dibasic phosphate (HPO_4^{2-}) with calcium or magnesium limit it under slightly alkaline conditions.

The pH alters the structural integrity of many organisms through its effect on calcium salts. Tooth and bone of vertebrate animals are composed primarily of calcium phosphate in hydroxylapatite crystals ($Ca_{10}(PO_4)_6(OH)_2$). Shells of animals with exoskeletons are composed primarily of calcium carbonate crystals ($CaCO_3$). These calcium compounds become more soluble at acid pHs (Larsen and Jensen 1989; Brandao-Burch et al. 2005; Morse et al. 2007). One reason toothpaste, blood, and seawater have alkaline pH is to maintain the structural integrity of calcium compounds.

Global change and pH

Rising atmospheric CO_2 concentrations have increased dissolved inorganic carbon (DIC = CO_2 + HCO_3^- + CO_3^{2-}), which in turn has increased water acidity. If atmospheric CO_2 concentrations climb from 350 ppm to 1000 ppm, as they well may do (see Figure 4.18), DIC in a glass of pure water sitting on your table will go from 14.1 µmol L^{-1} to 37.4 µmol L^{-1}, and its pH will decrease from 5.65 to 5.42. Such a pH shift may seem nominal, but remember that pH is based on a logarithmic scale, and so the corresponding proton concentra-

TABLE 6.6   Reactions with phosphorus (P) and carbon dioxide (CO_2) and their acid dissociation constants (pK_a) in pure water at 25°C

P		CO_2	
Reaction	*pK_a*	*Reaction*	*pK_a*
$H_3PO_4 \leftrightarrow H_2PO_4^- + H^+$	2.12	$CO_2 + H_2O \leftrightarrow H_2CO_3$	2.77
$H_2PO_4^- \leftrightarrow HPO_4^{2-} + H^+$	7.21	$H_2CO_3 \leftrightarrow HCO_3^- + H^+$	3.60
$HPO_4^{2-} \leftrightarrow PO_4^{3-} + H^+$	12.67	$CO_2 + H_2O \leftrightarrow HCO_3^- + H^+$	6.36
		$HCO_3^- \leftrightarrow CO_3^{2-} + H^+$	10.25
$10\ Ca^{2+} + 6\ PO_4^{3-} + 2\ OH^- \leftrightarrow Ca_{10}(PO_4)_6(OH)_2$		$Ca^{2+} + CO_3^{2-} \leftrightarrow CaCO_3$ and	
		$Ca^{2+} + 2\ HCO_3^- \leftrightarrow CaCO_3 + CO_2 + H_2O$	8.30–8.48

tions in the glass will increase 70%, from 2.2 µmol L^{-1} to 3.8 µmol L^{-1}.

Seawater, because of the salts dissolved in it (see Table 6.5), is significantly more alkaline than pure water. The pH of seawater now averages 8.1 (see Figure 4.49), which translates to a proton concentration of 0.008 µmol L^{-1}. Given this low starting point, proton concentrations in seawater may increase 250% as atmospheric CO_2 concentrations rise. Sea life will be subjected to pH-induced changes in conformation of biochemicals, storage of biochemical energy, and the stability and availability of inorganic compounds.

Of particular concern is that many forms of sea life have exoskeletons of calcium carbonate ($CaCO_3$) that will dissolve into, rather than precipitate out of, seawater. The balance between these two processes depends on pH and CO_2 (see Table 6.6) as well as on temperature, pressure, and salinity. Aragonite is the form of $CaCO_3$ found in the exoskeletons of macroalgae (seaweeds), corals, and mollusks. According to some predictions, oceans in the southern hemisphere will become undersaturated for aragonite by the end of the century (**Figure 6.50**). This means that aragonite will be more likely to dissolve into seawater; exoskeletons will begin to thin; and macroalgae, corals, and mollusks will be put at risk (Orr et al. 2005; The Royal Society 2005).

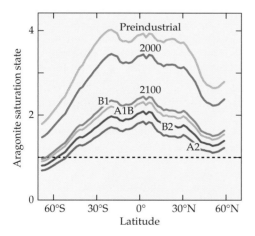

FIGURE 6.50 Solubility of calcium carbonate in seawater Shown is the stability of aragonite, a form of $CaCO_3$, in the oceans at various latitudes at three different times: preindustrial; in the year 2000; and in the year 2100 under four different scenarios for greenhouse gas emissions (B1, A1B, B2, and A2). The black dashed line depicts where $CaCO_3$ dissolves and precipitates at equal rates. This predicts that shells of marine animals may dissolve in the southern oceans at high latitudes by the year 2100. (After Caldeira and Wickett 2005.)

Perspective

This chapter outlines factors associated with global climate change that may influence biological organisms. It emphasizes the decline of certain species and possible threats to the human condition. Unfortunately, data are sparse on the number and distribution of species in many regions of the world, population estimates for most species are based on inexact and infrequent censuses, and reports of human resource use are often unreliable. These gaps in our knowledge obscure present circumstances and heighten anxiety about the future. We are not entirely "flying blind" with respect to potential conflicts between climate and the biosphere, but the poor resolution of our information make it more difficult to navigate an appropriate route through the various options for mitigation. Chapters 7 through 9 will continue on the theme of mitigation.

Summary

All creatures, big and small, respond to environmental conditions such as temperature, water availability, salinity, and pH. Global climate changes are anticipated to influence all of these conditions and, thus, most biological organisms. Warming of polar regions has been more rapid than other areas on Earth, and polar organisms including polar bears, penguins, whales, and krill are showing signs of distress. Population sizes of these species, however, are difficult to assess.

Organisms from temperate and tropical zones are also sensitive to Earth's warming trend. For example, bird migration patterns may become disrupted, quality of wine grapes may diminish, coral reefs may bleach, amphibian species may disappear, insect-borne diseases may spread to higher latitudes and altitudes, soil microorganisms may decompose more soil carbon, and human heat-related fatalities may increase.

Water availability around the world is likely to change during this century, and this will affect human food production and salinization of soils. To meet human needs, agriculture will continue to intensify and depend more heavily on irrigation. Warmer temperatures will increase evaporation, decrease snowfall in temperate mountain ranges, and accelerate salinization of soils. These are all factors that may strain water resources in major agricultural areas.

Higher CO_2 concentrations in the atmosphere will acidify bodies of water. This acidification, in turn, influences organisms through altered conformations of biochemicals; inhibition of energy metabolism; and less solubility of some nutrients and greater solubility of others. Of particular concern is that the exoskeletons of marine organisms begin to dissolve as the seas acidify.

Review Questions

1. Select all of the following environmental factors that directly influence the biosphere.
   (a) pH
   (b) Precipitation
   (c) Temperature
   (d) Atmospheric concentrations of CFCs
   (e) Salinity

2. Select all of the following phenomena that increase with temperature.
   (a) Motion of molecules
   (b) Wavelengths of electromagnetic radiation from a perfect blackbody
   (c) Leakage through biological membranes
   (d) Deformation of proteins
   (e) Rate of chemical reactions

3. Select all the reasons that most biological processes cease at 0°C.
   (a) Molecules cease to move.
   (b) Saltwater becomes less concentrated.
   (c) Water changes phase from a liquid to a solid.
   (d) Water is at its maximum density.
   (e) Molecular motion in ice is too limited.

4. Over the past few decades, most biological organisms have (select one)
   (a) altered the timing of their development.
   (b) have returned to lower elevations.
   (c) come close to extinction.
   (d) migrated closer to the equator.
   (e) emitted more greenhouse gases.

5. Select all the reasons why polar organisms may be at greater risk than many other organisms.
   (a) Polar organisms are now experiencing greater competition from temperate organisms.
   (b) Hurricanes that originate in polar regions are becoming stronger.
   (c) Recovery from perturbations is faster in cold regions.
   (d) Human development of natural resources in polar regions has ceased.
   (e) The poles are warming faster than other regions.

6. Polar bears (select one)
   (a) eat almost anything.
   (b) readily catch seals in open water.
   (c) bear their young on ice flows.
   (d) are difficult to count.
   (e) migrate to the mountains in summer.

7. Penguins (select one)
   (a) of all species are declining.
   (b) have been subjected to extensive hunting.
   (c) breed in large colonies.
   (d) now encounter more polar bears during breeding.
   (e) migrate to the mountains in summer.

8. Baleen whales (select one)
   (a) feed on krill.
   (b) mature in a year.
   (c) chew their food slowly.
   (d) remain year-round in a given location.
   (e) have nothing to fear from modern whalers.

9. Phytoplankton (select one)
   (a) migrate from pole to pole.
   (b) are at the bottom of the polar food chain.
   (c) chew their food slowly.
   (d) bloom only in tropical waters.
   (e) avoid upwellings of nutrients in the oceans.

10. Birds (select one)
    (a) all migrate to the tropics in winter.
    (b) are at the bottom of the tropical food chain.
    (c) have relatively high metabolic rates.
    (d) are placed in gilded cages because of their reaction to heavy metals.
    (e) avoid upwellings of nutrients in the oceans.

11. Wine grapes (select one)
    (a) vary more than tenfold in value.
    (b) improve in quality under heat stress.
    (c) may be made into prunes.
    (d) ferment slowly at warmer temperatures.
    (e) are native to cold climates.

12. Corals (select one)
    (a) bleach to rid themselves of pathogens.
    (b) are a symbiotic relationship between phytoplankton and zooplankton.
    (c) cover the oceans depths.
    (d) are native to polar regions.
    (e) are sensitive to heat stress and changes in ocean pH.

13. Insect vectors of pathogens (select one)
    (a) all point to the magnetic North Pole.
    (b) are spreading to higher latitudes and altitudes.
    (c) often generate metabolic heat to avoid freezing.
    (d) do not transfer human diseases.
    (e) are the state bird of Minnesota.

14. Select all of the following statements about soil carbon decomposition that are true.
    (a) It will exceed carbon sequestration by the end of the century.
    (b) It is significant only under elevated CO_2 conditions.
    (c) It decreases exponentially with increasing temperature.
    (d) It depends on soil particle size.
    (e) It occurs at a similar rate for all organic compounds.

15. Select all the changes in water resources that are anticipated by the end of the century.
    (a) Precipitation will be limited to higher latitudes and altitudes.
    (b) More water will be diverted into irrigated agriculture.
    (c) Snowpacks in mid-latitude mountains will decline.

(d) Wet places will become wetter and drier places will become drier.

(e) Extensive pumping of ground water will be the best long-term solution.

16. Select all of the statements about salinity that are *true*.
    (a) It influences the movement of water in and out of biological organisms and cells.
    (b) It tends to increase in agricultural soils that are being irrigated.
    (c) In water vapor, it is high.
    (d) It will probably increase in coastal aquifers if sea level rises.
    (e) It influences the freezing point of a solution.

17. Select all of the following statements that explain why a pH-balanced shampoo is desirable.
    (a) pH provides hair with greater energy.
    (b) Hair is composed of proteins.
    (c) pH influences the conformation of proteins.
    (d) pH determines the color of hair.
    (e) Extreme pH changes can cause protein breakage.

18. Select all the likely consequences of global climate change.
    (a) It will increase the acidity of the oceans.
    (b) It will increase the alkalinity of the oceans.
    (c) It will have no major influence on either ocean pH or solubility of calcium carbonates in the oceans.
    (d) It will decrease the solubility of calcium carbonates in the oceans.
    (e) It will increase the solubility of calcium carbonates in the oceans.

Suggested Readings

Constible, J., L. Sandro, and R. E. Lee, Jr. 2008. *Climate Change From Pole to Pole: Biology Investigations*. NSTA Press, Arlington, VA.

This textbook for high-school students and non-science undergraduates presents biology-based case studies and background information on climate change.

National Academy of Sciences. 2009. *Ecological Impacts of Climate Change*. National Academies, Washington, D.C., *http://dels.nas.edu/dels/rpt_briefs/ecological_impacts.pdf*.

This is a recent report by a panel of distinguished experts.

World Wildlife Federation International. 2009. *Polar Bear—Diet, Habitat, Threats, Pictures, Status. http://www.panda.org/what_we_do/where_we_work/arctic/area/species/polarbear/*, accessed June 9, 2009.

This web site provides information on polar bears, including live tracking, and other megafauna.

7

MITIGATION STRATEGIES: TRANSPORTATION

The first step in most rehabilitation programs is to admit that one has a serious problem, but as the the cartoon character Bugs Bunny warned, "Look out for that first step; it's a lulu." Such has been the case for global climate change. Uncertainty and, in a few instances, outright denial that global climate change is a serious, human-caused problem have stalled many attempts at intervention.

An expanding body of evidence has now convinced most scientists that the world is warming at an unprecedented rate, that emissions of greenhouse gases from human activities are responsible for most of this warming, and that the effects of this warming may prove detrimental to life as we know it. Governmental bodies at regional, national, and international levels seem poised to act. The next step is to develop a reasonable plan for diminishing the severity of the problem: a **mitigation** strategy.

(A)

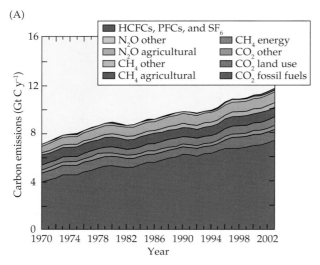

(B)

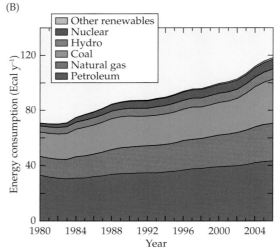

FIGURE 7.1 Worldwide greenhouse gas emissions and energy consumption, from various sources (A) Greenhouse gas emissions in carbon equivalents (10^9 metric tons of carbon per year). "PFC" designates perfluorocarbons; "SF_6" designates sulfur hexafluoride. The categories "CH_4 energy" and "CO_2 fossil fuels" relate

to emissions during fossil-fuel extraction and burning. (B) Energy consumption (10^{18} calories per year). Note that 1 calorie = 4.186 joules = 1.163×10^{-3} watt-hours = 3.968×10^{-3} Btu. (After Olivier et al. 2005; Energy Information Administration 2009b.)

Extraction and combustion of **fossil fuels** (i.e., petroleum, coal, and natural gas) account for 69% of the greenhouse gas emissions from human activities (**Figure 7.1A**). Abstinence, a successful approach for some dependencies, is not a viable option for fossil fuels. They provide 86% of the world's total energy production (**Figure 7.1B**) and virtually all of the energy for transportation (**Figure 7.2**). Banning the use of fossil fuels—going "cold turkey"—would literally stop the world.

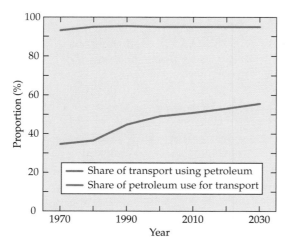

FIGURE 7.2 Past and predicted petroleum dependence Percentage of transportation that depends on petroleum as a fuel and percentage of petroleum consumed for transportation. (After International Energy Agency 2004b.)

Many other mitigation strategies are possible, ranging from relatively simple ones such as changing a lightbulb to some that are literally out of this world, such as a giant space umbrella. This chapter sorts through a number of strategies for the transportation sector and categorizes them according to their energy conservation potential. It also discusses alternative fuel use. Chapter 8 addresses strategies for electric power generation, and Chapter 9 examines emissions from agricultural, forestry, industrial, commercial, and residential sources. All three chapters present an engineering perspective, first outlining the technologies required for a mitigation strategy, then discussing the potential efficacies, and, finally, describing the collateral damages to the environment. Detailed economic analyses will be presented in Chapter 10.

Energy Conservation

World energy consumption has increased by an average of about 2% per year over the past few decades (see Figure 7.1B). Much of this increase derives from population growth and economic development. On a per capita basis, worldwide energy consumption remains relatively constant (**Figure 7.3A**): that is, each person uses only slightly more energy than he or she did in previous decades. Moreover, less energy is required to stimulate a given amount of economic activity: calories per GDP (*gross domestic product*) have declined (**Figure 7.3B**). This means that we now get more bucks for our bang.

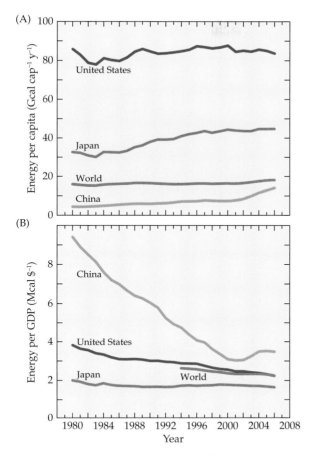

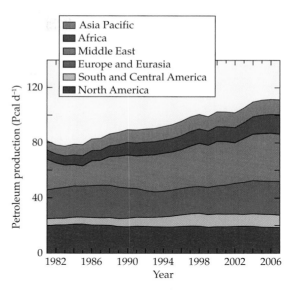

FIGURE 7.4   Petroleum output (10^{15} calories day^{-1}) from different regions This includes crude oil, shale oil, oil sands, and natural gas liquids (the liquid content of natural gas, where this is recovered separately from oil). (After BP 2008.)

FIGURE 7.3   Energy usage (A) Annual energy consumption per capita per year (10^9 calories cap^{-1} y^{-1}) and (B) energy per gross domestic product (10^6 calories per U.S. dollar) adjusted for inflation with respect to year 2000 and for purchasing power of the various currencies. Shown are the world average, and values for United States, Japan, and China. (After Energy Information Administration 2008d,e.)

Countries vary widely in their energy usage (see Chapter 4). For example, each person in the United States and Japan uses more energy than the world's average but requires equal or less energy to generate a dollar's worth of economic activity (see Figure 7.3). In contrast, each person in China uses less energy but requires more energy to generate a dollar's worth of economic activity.

Slowing global energy consumption would abate greenhouse gas emissions but would require a departure from current economic activities. Some people argue that such changes might unduly hamper economic development and threaten human prosperity, while others argue that such changes might even improve quality of life. The subsequent chapter on economics (Chapter 10) examines these arguments.

Slowing global energy consumption would conserve nonrenewable natural resources. Finite resources include supplies of fossil fuels such as petroleum, coal, and natural gas; deposits of radioactive materials such as uranium; productive sites for hydro, wind, solar, geothermal, tidal, and ocean-wave electricity generation; and arable land with sufficient water for **biofuels** (fuels derived directly from living organisms) production. In specific, the amount of petroleum extracted per year in North America, Europe, and Eurasia has already peaked (**Figure 7.4**). If fuel consumption continues at its 2006 rate, known reserves of petroleum will last approximately 41 more years; coal, 147 more years; natural gas, 63 more years; and uranium, 70 more years (BP 2007). Of course, these reserves will deplete more quickly if fuel consumption continues to accelerate (see Figure 7.1B).

Fossil fuels obey the law of supply and demand, albeit very loosely (see Chapter 10): petroleum prices rise when supply cannot keep pace with demand. Demand exceeds supply today, and so prices of petroleum have returned to historic highs (**Figure 7.5**). These high prices have stimulated exploration for new oil fields as well as broadened the range of economically feasible options for petroleum extraction from established fields. Expanded exploration and extraction extends the world's petroleum supplies (**Figure 7.6**).

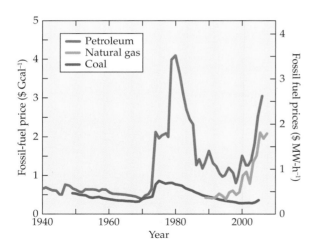

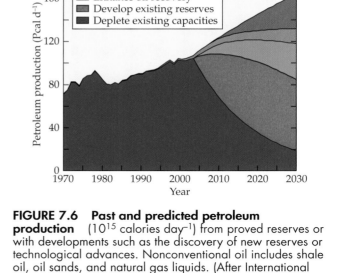

FIGURE 7.5 Prices of fossil fuels in constant 2006 U.S. dollars per gigacalories (10⁹) or megawatt-hours (10⁶) Petroleum prices are for the international market, whereas natural gas and coal prices are for the United States. (After BP 2008; Energy Information Administration 2008a,b.)

FIGURE 7.6 Past and predicted petroleum production (10^{15} calories day^{-1}) from proved reserves or with developments such as the discovery of new reserves or technological advances. Nonconventional oil includes shale oil, oil sands, and natural gas liquids. (After International Energy Agency 2004b.)

As petroleum becomes more expensive, so do other fossil fuels (see Figure 7.5) and other energy sources. Higher energy prices promote conservation. The following section examines the potential for energy conservation in the transportation sector.

Transportation

Globalization of the world's economy requires that humans and their goods move long distances. Con-

sequently, passenger and freight travel are likely to double during the next half century (**Figure 7.7**). This expanded travel derives both from human population growth (see Chapter 12) and from everyone traveling a greater distance each year (**Figure 7.8A**). In addition, more people possess their own light-duty vehicle (passenger car or small truck) and are using them for travel (**Figure 7.8B**).

Petroleum serves as the energy source for over 95% of transportation (see Figure 7.2). The transportation

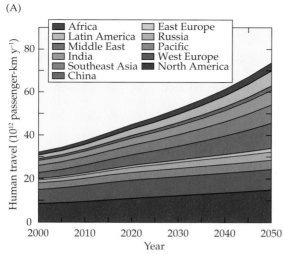

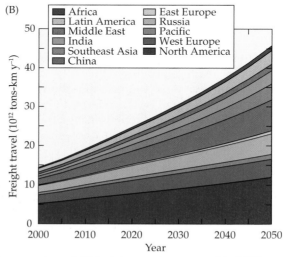

FIGURE 7.7 Past and predicted annual volume of passenger travel and freight travel in different regions (A) Passenger travel, measured in 10^{15} passenger-

km year^{-1}, and (B) freight travel, measured in 10^{15} metric tons-km year^{-1}. (After World Business Council for Sustainable Development 2004.)

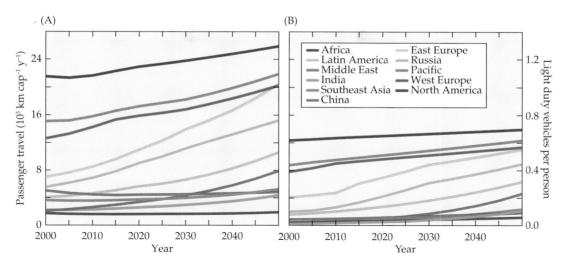

FIGURE 7.8 Past and predicted annual travel and vehicle ownership per person in different regions (A) Travel per person in 10³ km per person per year. (B) Light duty vehicles per person. (After World Business Council for Sustainable Development 2004.)

sector consumes half of the world's petroleum output, a proportion that is likely to increase in the future. A straightforward strategy for prolonging petroleum supplies is to invoke additional conservation measures in the transportation sector.

The Middle East supplies most of the world's petroleum (see Figure 7.4) and has the world's largest proved oil and natural gas reserves (i.e., reserves that have a reasonable certainty of being extracted). For this reason, among others, the political volatility of this region has global consequences. In October 1973, when the Yom Kippur War erupted between Israel and its Arab neighbors, Arab petroleum exporters initiated an oil embargo against Israel's supporters including the United States, Western Europe, and Japan that they maintained until March 1974. Supplies of petroleum dropped, prices shot up (see Figure 7.5), and motorists in these countries had to wait in long lines at filling stations. Five years after this embargo ended, the Iranian Revolution precipitated another oil crisis that again doubled petroleum prices and generated long lines at filling stations (**Figure 7.9**). Wars in Iraq from 1990 to 1991 and from 2003 to 2009 have also influenced oil supplies and prices. Conservation in the transportation sector would provide greater independence from Middle East oil supplies.

Conservation measures in the transportation sector also would reduce greenhouse gas emissions (**Figure**

FIGURE 7.9 Cars lining up for fuel at a gas station in the United States on June 15, 1979.

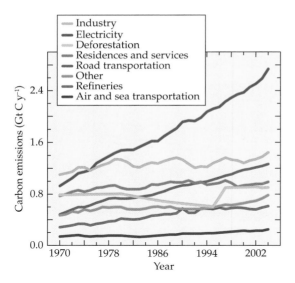

FIGURE 7.10 Greenhouse gas emissions in carbon equivalents (10⁹ metric tons C per year) from different human activities "Deforestation" includes burning of firewood. "Industry" excludes cement production. "Other" includes domestic surface transport, non-energy use of fuels, and cement production. (After IPCC 2007c.)

Light-duty vehicles account for only 10% of human greenhouse gas emissions worldwide and 19% in the United States (Davis and Diegel 2007). They are, however, one of the few sources of greenhouse gases over which individuals have direct control and thus have high symbolic value. Mitigation strategies that involve light-duty vehicles are often contentious because of the important role that these vehicles play in modern life. One indication of this importance is that many people spend more on vehicles than they do on food: Americans allocate 33% of their disposable income to housing, 17% to vehicles, and 13% to food (U.S. Bureau of Labor Statistics 2007); the British allocate 12% of their disposable income to vehicles, 10% to food, and 10% to housing (National Statistics 2007).

7.10). About 21% of global emissions currently derive from transportation (33% in the United States, see Figure 4.15A; 10% in China, see Figure 4.15C), and this fraction is expected to reach 23% by 2030 (International Energy Agency 2004b). Emissions from all modes of transportation except buses are climbing at about 2% per year. Light-duty vehicles (passenger cars and small trucks) contribute the largest share of emissions (**Figure 7.11**); therefore, an obvious strategy for greenhouse gas abatement is to improve the fuel efficiency of light-duty vehicles.

Efficiency of light-duty vehicles

Average fuel efficiencies of light-duty vehicles vary by 60% among industrialized countries. During the past 15 years, efficiencies have improved steadily in Western Europe but have remained relatively unchanged in Japan, Australia, and the United States (**Figure 7.12**).

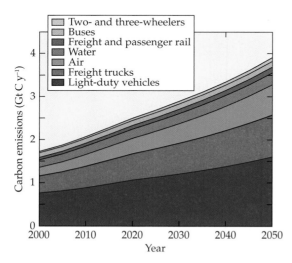

FIGURE 7.11 Past and predicted greenhouse gas emissions in carbon equivalents (10⁹ metric tons C per year) from various modes of transportation (After World Business Council for Sustainable Development 2004.)

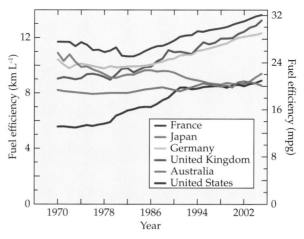

FIGURE 7.12 Average fuel efficiencies (km L⁻¹ or mpg) of all registered light-duty vehicles in selected countries (After Australian Bureau of Statistics 2006; Schipper 2007.)

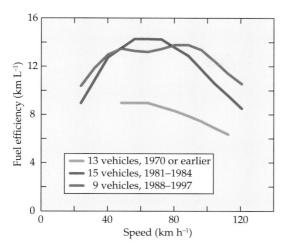

FIGURE 7.13 Fuel efficiencies (km L⁻¹) of representative vehicles in the United States as a function of speed (km h⁻¹) in studies conducted in 1973, 1984, and 1997. (After Davis and Diegel 2007.)

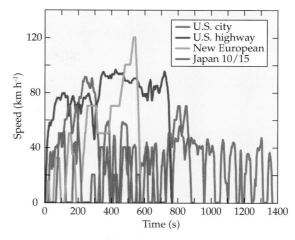

FIGURE 7.14 Models of driving behavior used in standardized tests of fuel efficiencies: the U.S. EPA city and highway standard, New European standard, and Japanese 10/15 standard. (After U.S. Environmental Protection Agency 2007a.)

Major factors that influence fuel efficiency include driving conditions; taxes on petroleum and vehicles; consumer preferences; use of diesel-powered vehicles; and agreements with automobile manufacturers.

DRIVING CONDITIONS The wind resistance of a vehicle is proportional to the square of its speed, and so fuel efficiencies decline at speeds above 50 mph (80 kmph) (**Figure 7.13**). Newer vehicles have better aerodynamics and thus are more efficient at higher speeds than older vehicles. Nonetheless, vehicles moving at 75 mph (121 kmph), the speed limit on many rural U.S. highways, are still 20% less efficient than those moving at 55 mph (89 kmph).

After the 1973 oil crisis, the U.S. Congress capped the speed limit on all roads at 55 mph to promote higher fuel efficiencies. This restriction was so unpopular that Congress raised the limit to 65 mph in 1987 and repealed the law entirely in 1995. Local authorities now set speed limits; the limit is 75 mph on many rural highways, and two remote stretches of road in Texas have a speed limit of 80 mph (129 kmph).

Speed, although important, is second to driving behavior in determining fuel efficiency. Quick acceleration and heavy braking diminish fuel economy by more than 30% (Reed and Hudson 2005). To mimic "typical" driving conditions in the United States, Europe, or Japan, standardized tests of fuel efficiency follow proscribed patterns of accelerations and decelerations (**Figure 7.14**). The stop-and-go nature of driving in Japanese cit-

ies such as Tokyo (**Figure 7.15**) compromises fuel efficiency more severely than the other patterns, cutting in half the efficiency obtained at the constant, moderate speeds of U.S. highways (**Table 7.1**).

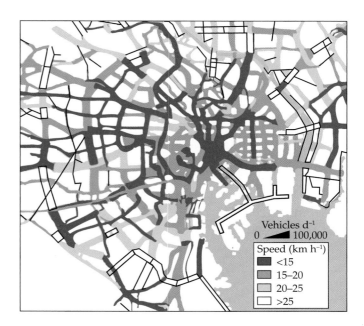

FIGURE 7.15 Rush hour traffic in Tokyo, Japan Line width indicates the average number of vehicles over a day on different roads. Color designates the average vehicle speed on a road. In the city center, traffic congestion limits average speeds on most roads to less than 15 km h⁻¹. (After World Business Council for Sustainable Development 2004.)

TABLE 7.1 Fuel economies of 1995 midsize vehicles[a]

Driving cycle	Percent time stopped or decelerating	Average speed (km h $^{-1}$)	Maximum acceleration (km h $^{-1}$ s $^{-1}$)	Fuel economy	
				(km L $^{-1}$)	(mpg)
U.S. EPA City	42.3	31.4	5.3	8.4	19.8
U.S. EPA Highway	9.3	77.6	5.3	13.7	32.1
New European	24.9	33.6	3.9	9.4	22.0
Japanese 10/15 Mode	52.3	23.8	2.9	7.4	17.5

Source: Davis and Diegel 2007.
[a]Evaluated using United States, European, and Japanese driving cycles. The vehicles were a Chevrolet Lumina, Chrysler Concord, and Ford Taurus.

Japan is not alone in having traffic conditions that often interfere with the free flow of vehicles and thereby compromise fuel efficiencies. Indeed, the terms "traffic jam" and "gridlock" are universally understood. Worldwide, the number of vehicles and the distances they travel (see Figure 7.8) are many times greater than they were 2 decades ago, while the construction of new roads has lagged far behind (**Figure 7.16**). Such disparities have led to greater traffic congestion.

The United States, for example, paved a large proportion of its roads during the past 50 years but extended the total length of roads by only 11% (**Figure 7.17**). Over this same period, the number of vehicles on these roads tripled. Not surprisingly, traffic congestion worsened. A 10-minute drive across a U.S. city during slack times took an average of 11 minutes during rush hours 20 years ago, but now takes nearly 13 minutes.

The causes of traffic congestion fall into two categories: recurring and nonrecurring. Recurring causes are those such as rush hour traffic, holiday traffic, and routine road maintenance, events that are somewhat predictable. Nonrecurring causes are unplanned events such as traffic incidents (accidents or mechanical breakdowns) and poor weather (snow, fog, or heavy rains). Recurring and nonrecurring causes are responsible for travel delays in almost equal measure (**Figure 7.18**).

Many strategies can relieve recurring traffic congestion. Constructing entirely new roads is an obvious but

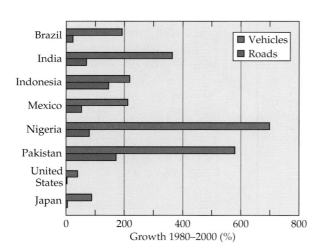

FIGURE 7.16 Change in number of vehicles and length of roads in various countries (After Bureau of Transportation Statistics 2002; Universal Traffic Management Society of Japan 2004; Davis 2007.)

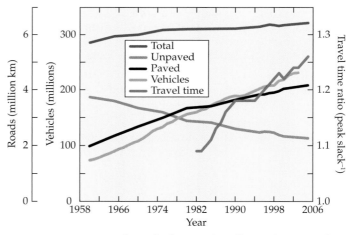

FIGURE 7.17 Roads, vehicles, and traffic in the United States: Length (millions of km) of paved, unpaved, or total roads in the United States; number of vehicles in the United States (millions); and ratio of travel time required during peak vs. slack hours in 437 U.S. cities of various sizes. (After Bureau of Transportation Statistics 2002; Central Intelligence Agency 2007; Schrank and Lomax 2007.)

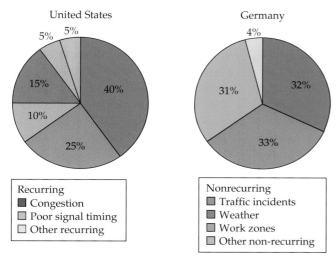

FIGURE 7.18 Proportion of traffic congestion in the United States and Germany attributable to various recurring and nonrecurring causes. Traffic incidents include accidents and disabled vehicles. Other nonrecurring causes include special events. (After Joint Transport Research Centre 2007.)

TABLE 7.2 Strategies for alleviating recurring and non-recurring traffic congestion on existing roads

Recurring

Encourage flexible work hours and stagger school opening times.
Control the timing of traffic lights and vary speed limits during rush hours.
Ban turns of vehicles across oncoming traffic at intersections during rush hours.
Meter entry onto highways at on-ramps.
Collect tolls electronically or only in one direction.
Split lanes to allow motorcycles or scooters to ride between automobiles.
Reverse travel direction of certain lanes to match asymmetric demand.
Open emergency lanes or shoulders of roads to flowing traffic during rush hours.
Convert center lanes of highways into express lanes with limited access to on- and off-ramps.
Charge additional fees to drive into city centers (London, Singapore, Stockholm).
Restrict entry into the city centers on alternate days of the week to odd or even license plate numbers (Athens, Mexico City, São Paulo).
Schedule road work for off-peak times and off-peak seasons.

Nonrecurring

Respond rapidly to traffic incidents and remove disabled vehicles quickly.
Disseminate information about traffic and weather conditions via radio, variable traffic signs, and internet and suggest alternative routes.
Provide emergency crossover gates between opposing lanes of traffic that permit safety personnel more ready access to an incident site.
Establish accident investigation areas where motorists and safety personnel can exchange information about an accident away from the roadway.
Construct tall lane barriers that block the view of opposing lanes of traffic to diminish rubbernecking, slowdowns by curious onlookers in opposing lanes.
Cut grooves into highway surfaces to reduce hydroplaning.

expensive approach. Municipalities usually find it more cost-effective to eliminate "choke points" (narrow passages) on existing roads by building new intersections or even new lanes between intersections. Even within the confines of existing roads and lanes, a number of other strategies may diminish recurring and nonrecurring traffic congestion (**Table 7.2**).

Implementing strategies to reduce traffic congestion not only improve fuel efficiencies, but advance public safety as well. Traffic accidents are the ninth leading cause of death and disability in the world and are the second leading cause of death after HIV/AIDS for people 15 to 29 years of age (Kapp 2003). The highest absolute numbers of road deaths occur in Asia, and the highest rates occur in Africa (Hyder and Peden 2003). These same regions are experiencing rapid increases in numbers of vehicles (see Figures 7.8 and 7.16), which are only making matters worse. On the bright side, although vehicle traffic has doubled over the past 30 years in the United States, Australia, and Finland, safety measures such as mandatory use of seat belts, motorcycle helmets, and child car seats have halved traffic fatalities over this period (Kapp 2003).

Benjamin Franklin said "In this world nothing can be said to be certain, except death and taxes." This chapter now turns to the topic of taxes.

TAXES ON FUEL AND VEHICLES The average cost of extracting, transporting, and refining gasoline and diesel fuel in 2006 was $0.53 per liter ($2.01 per gal) for gas and $0.59 per liter ($2.23 per gal) for diesel (Metschies 2007). Average prices at the pump, however, ranged from $0.02 to $1.90 per liter ($0.08 to $7.19 per gal) in different countries, depending on fuel subsidies or taxes (**Figure 7.19**). Countries subsidizing fuel are, in general, major oil exporters. Most other countries tax fuels, often at rates higher than other goods.

Taxes on fuels are an attractive source of revenue. They amass vast sums, amounting to $1.2 trillion in 2004, or 10.8% of all revenues collected by governments (**Figure 7.20**).

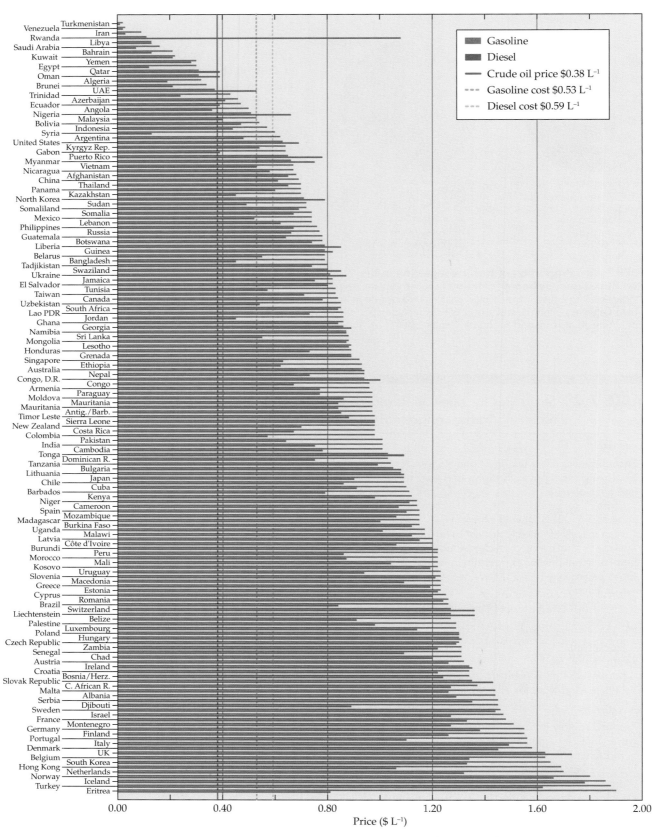

FIGURE 7.19 Gasoline and diesel fuel costs ($U.S. per liter) in 2006 Without taxes, a liter of gasoline and diesel would cost about $0.53 and $0.59, respectively. (After Metschies 2007.)

FIGURE 7.20
Percentage of total state revenues that derives from taxes on fuels Negative percentages (red) indicate subsidies for fuels. (After Metschies 2005.)

They are easier to collect than other taxes because they are assessed at refineries or large wholesalers, which are few in number. People are somewhat less adverse to fuel taxes than other taxes because a major portion of these revenues is spent on highly visible infrastructure projects. Road signs in construction zones along U.S. highways proudly announce "Your Tax Dollars at Work." The public also recognizes that, in addition to high fuel prices, high fuel taxes discourage superfluous driving as well as encourage the purchase of more fuel-efficient and less polluting vehicles.

Taxes on fuels are only the beginning of revenue collection. Supplemental taxes on vehicles are nearly universal, although there is little uniformity among countries or sometimes even among regions within a single country (Table 7.3). Most governments assess a value-added tax or sales tax at the time of purchase that is based on a percentage of the purchase price. Most also assess a license or registration fee at the time of purchase; this tax often varies with the value of the vehicle and some measure of its fuel consumption. Finally, most assess an annual circulation or road use fee; this tax is often keyed to some measure of its fuel efficiency.

CONSUMER PREFERENCES   Economics and regulations aside, consumers prefer larger vehicles with faster acceleration. This requires engines that are more powerful. Since the mid-1980s, when petroleum prices began to decline after the oil crises of the previous decade, vehicles in the United States and the European Union have gained weight, more powerful engines, and faster acceleration (Figure 7.21). Only through technological advances (discussed in the following section) have vehicle fuel efficiencies in the United States, Europe, and Australia held steady or increased slightly (see Figure 7.12). In Japan, the number of "miniature" vehicles (having **engine displacement** of less than 661 cubic centimeters) and "standard" vehicles (engines of greater than 2000 cc) have grown at the expense of "small" vehicles (having engine displacement between 667 cc and 2000 cc), and now annual sales of all three sizes are nearly equal (Figure 7.22).

Reviews of new vehicles often focus on size and acceleration. For example, the front page of a 2007 newspaper poses the question, "Redesigned Toyota SUV gets a little bigger, but is it big enough?" (Healey 2007). Automobile manufacturers complain, with some justification, that they have had difficulty in selling

TABLE 7.3 Taxes on the average personal light-duty vehicle in various countries

Country	VAT or sales tax (%)[a]	Additional taxes			
		Purchase[b]	Basis[c]	Annual	Basis
Argentina	21.0				
Australia	10.0	2%–5%	V	$183	W, C
Austria	20.0	$1911	V, F	$292	H
Belgium	21.0	$682	H	$431	H
Bosnia	17.0				
Bulgaria	20.0	$360	H	$28	H
Canada	6.0–14.0	$1023	F	$60	W
Chile	19.0				
China	17.0				
Colombia	16.0				
Croatia	22.0			$160	D, A
Cyprus	15.0			$146	D
Czech Republic	19.0				
Denmark	25.0	$21,559	V	$637	F
Dominican Republic	6.0,				
Ecuador	12.0				
Egypt	10.0				
El Salvador	13.0				
Estonia	18.0	$680	D		
Finland	22.0	$10,332	V	$257	A
France	19.6				G
Germany	19.0			$256	D
Greece	19.0	$3672	V	$183	D
Guyana	16.0				
Hungary	20.0			$1500	W
Iceland	24.5	30%–45%	D		
India	12.5				
Indonesia	10.0				
Ireland	21.0	$6090	V, D	$424	D
Israel	15.5				
Italy	20.0	$368	H	$41	H
Japan	5.0	3%–5%	V	$425	D
Jersey	3.0				
Jordan	16.0				
Kazakhstan	14.0				
Kosovo	15.0				
Latvia	18.0	$154		$93	W
Lebanon	10.0				
Lithuania	18.0	5%	V		
Luxembourg	15.0				

the small, fuel-efficient cars that they offer (European Automobile Manufacturers Association 2007). Consequently, manufacturers canceled many of these models and supersized others.

For example, I purchased a Toyota Scion xB in 2006. This model, which comfortably seats three hockey players and their equipment, had by one accounting the lowest total energy cost per mile of any car in the United States, including gas-electric hybrids (CNW Marketing Research 2007). Hybrids may have higher total energy costs than some non-hybrids if one considers the manufacture, replacement, and disposal of such items as batteries and electric motors. The xB model introduced a year later, in 2007, is 8% longer and 4% wider, yet has little more useable interior space. It is 29% heavier than the previous model, its

TABLE 7.3 Taxes on the average personal light-duty vehicle in various countries (*continued*)

Country	VAT or sales tax (%)[a]	Additional taxes			
		Purchase[b]	Basis[c]	Annual	Basis
Macedonia	18.0				
Malaysia	5.0				
Malta	18.0				
Mexico	15.0	5%	V	3%–10%	
Moldova	20.0				
Montenegro	17.0			$75	
Netherlands	19.0	$6368	V	$604	
New Zealand	12.5			$36	
Norway	25.0				
Pakistan	7.5				
Panama	5.0				
Paraguay	10.0				
Peru	19.0				
Philippines	12.0				
Poland	22.0				
Portugal	21.0	$2700	D	$171	
Romania	19.0	3%–12%	D, G	$8	
Russia	18.0				
Serbia	18.0			$23	
Singapore	7.0				
Slovakia	19.0				
Slovenia	20.0				
South Africa	14.0				
South Korea	10.0	2%	V	$468	D
Spain	16.0	7%–12%	D		
Sri Lanka	15.0				
Sweden	25.0			$344	
Switzerland	7.6			$355	
Thailand	7.0				
Trinidad and Tobago	15.0				
Turkey	18.0			$101	
United Kingdom	17.5			$383	
Ukraine	20.0				
Uruguay	23.0				
United States	0–9.0	$1300	F	$139	
Venezuela	9.0				
Vietnam	10.0				

Source: European Commission 2002; Idaho Transportation Department 2006; Organisation for Economic Co-operation and Development 2007.
[a]Value-added tax (VAT) or sales tax is placed on most purchases.
[b]In many countries, an additional tax is placed on the purchase of vehicles based on age (A); no. of cylinders (C); engine displacement (D); fuel consumption (F); greenhouse gas emissions (G); horsepower (H); vehicle value (V); and weight (W).
[c]Many countries also assess an annual road use or circulation tax based on these factors.

engine displacement is 60% larger, its acceleration is 19% faster, and its fuel efficiency in city driving has decreased by more than 30%. Nevertheless, the new model with its larger size and side airbags should be safer than the previous one (Insurance Institute for Highway Safety 2006). Toyota obviously believes that these changes will make the Scion xB appeal to a wider audience.

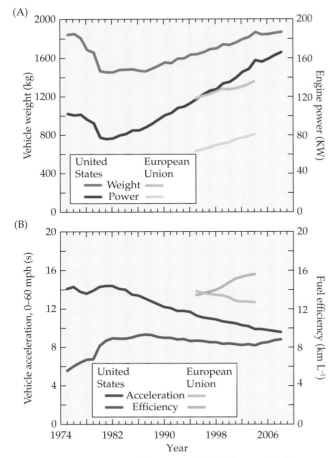

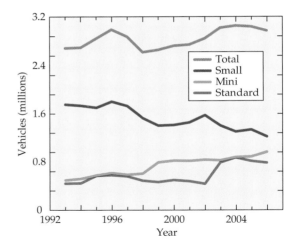

FIGURE 7.22 Size of new cars registered in Japan each year from 1993 to 2006 Mini cars have engine sizes below 661 cc; small cars, below 2000 cc; and standard cars, greater than 2000 cc. (After Japanese Automobile Manufacturers Association 2007.)

FIGURE 7.21 Data about new light-duty vehicles (passenger cars and small trucks) in the United States and European Union (A) Average vehicle weight and engine power (1 kW = 1.3 hp) and (B) acceleration, measured in seconds to accelerate from a stop to 60 mph (97 kmph), and fuel efficiency (1 km L⁻¹ = 2.35 mpg). (After Greene 2005; European Automobile Manufacturers Association and European Commission Services 2006; U.S. Environmental Protection Agency 2008.)

DIESEL-POWERED VEHICLES Diesel fuel is denser than gasoline (petrol) and contains about 11% more energy per volume, yet both fuels cost about the same per volume to extract and refine. Diesel engines are also inherently more efficient than gasoline engines in converting the energy in the fuel into mechanical work because they operate at higher pressures (higher compression ratios) and temperatures. Altogether, diesel engines obtain roughly 40% higher fuel efficiency per volume of fuel than gasoline engines of the same power (Heavenrich 2006). Large trucks and agricultural vehicles have diesel engines to take advantage of this higher efficiency.

Many countries, in consideration of the role of such vehicles in their economy, do not tax diesel fuel as heavily as gasoline (see Figure 7.19).

Still, small diesel engines have been plagued with performance issues such as being noisier, generating more vibrations, being more difficult to start, emitting thick black smoke in their exhaust, and exhibiting slower acceleration than their gasoline counterparts. Technical advances during the last few decades—principally, computer-controlled electronic ignition and turbocharged direct fuel injection—have largely overcome these performance issues. In response to these improvements and the rising costs of petroleum fuels, the market share of diesel-powered light-duty vehicles in the European Union has doubled during the past decade and now accounts for half of all new vehicles (**Figure 7.23**).

Comparing the greenhouse gas emissions of diesel-powered and gasoline-powered light-duty vehicles is tricky. Diesel fuel contains 11% more energy per volume of fuel combusted but also releases 15% more CO_2 per volume of fuel combusted (U.S. Environmental Protection Agency 2005a). Diesel engines weigh more than gasoline engines of the same power because diesels require a larger engine displacement (volume swept by the pistons) for complete combustion and heavier components to withstand higher pressures and temperatures (**Table 7.4**). In small vehicles, the engine contributes a greater percentage of the total vehicle weight, and the weight penalty of a diesel engine diminishes

TABLE 7.4   Fuel efficiencies and curb weights for car models with various types of powertrains[a]

Model		Gas	Hybrid	Diesel	CNG	H_2 fuel cell
Honda Civic	City/hwy	26/36	40/45	36/55	28/39	62/51
	Curb weight	2690	2875	2960	2904	3700
Toyota Camry	City/hwy	21/31	33/34	NA[b]	NA	NA
	Curb weight	3307	3680			
Toyota Highlander	City/hwy	19/25	28/25	NA	NA	64
	Curb weight	3880	4245		4100	
Ford Focus	City/hwy	22/31	NA	48	NA	26/37
	Curb weight	2677		3234		4255
Ford Escape	City/hwy	22/28	34/30	NA	NA	NA
	Curb weight	3313	3638			
Saturn Aura	City/hwy	18/29	24/32	NA	NA	NA
	Curb weight	3578	3529			
Mercedes A-Class	City/hwy	33/51	NA	40/64	NA	50
	Curb weight	2695		2915		3000

Source: Alliance Technical Services 2007; U.S. Department of Energy 2007.
[a]Fuel efficiency (mpg) in the U.S. EPA city vs. highway test (see Figure 7.14); curb weight (lbs) 1 mpg = 0.43 km L^{-1}; 1 lb = 0.45 kg.
[b]NA, not available.

its advantage in fuel efficiency from 40% to approximately 20%. Moreover, people tend to purchase diesel-powered vehicles with a larger engine (compared to the size of a gasoline-powered engine) because of diesel's higher fuel efficiency (Schipper et al. 2002). The net result is that the higher fuel efficiency of diesel engines in light-duty vehicles sometimes barely compensates for their higher CO_2 emissions, heavier weights, and larger cylinder displacements: In total, therefore, diesel-powered light-duty vehicles emit around 5% to 30% less greenhouse gases per distance traveled than their gasoline equivalents (Table 7.5).

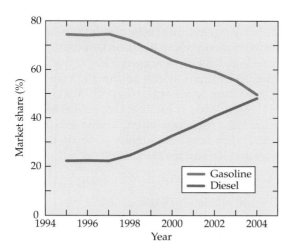

FIGURE 7.23   Percent market share of new gasoline- and diesel-powered light-duty vehicles in the European Union The remaining few percent are powered by electricity, natural gas, or unidentified sources. (After Commission of the European Communities 2006.)

TABLE 7.5   Greenhouse gas emissions from new cars in the United Kingdom[a]

	CO_2 emissions (g C km^{-1})	
Model	Gasoline	Diesel
Volkswagon Jetta	45–52	37–44
Toyota Yaris	35–46	32
Ford Focus	43–46	34–41

Source: Vehicle Certification Agency 2007.
[a]Driving conditions are the combined European driving cycles. The range of values covers the different engine sizes available for the model.

Diesel and gasoline vehicles have larger differences in their emissions of other atmospheric pollutants (ECOtravel 2007).

- Diesel engines, in comparison with gasoline engines, emit over 20 times more nitrogen oxides (NO_X). Nitrous oxides include N_2O, a greenhouse gas, and NO_2, which produces photochemical smog. **Catalytic converters** can remove NO_X from diesel exhaust as they do in modern gasoline vehicles, but many diesel fuels contain sulfur at concentrations that foul catalytic converters. Regulations in Europe and the United States now require refineries to remove most of this sulfur. This will enable most diesel vehicles to be equipped with catalytic converters.

- Diesels also release potentially harmful particles of carbon soot. Newer diesel vehicles, which have electronic ignition systems that compensate for cold starts and have self-cleaning particle filters, are much better in this regard than older models.

- Diesel vehicles emit less carbon monoxide (CO) than gasoline vehicles.

Technical advances in both gasoline-powered and diesel-powered vehicles should soon minimize differences between their engine emissions.

Gasoline and diesel engines also differ in their ability to accommodate alternative fuels such as natural gas (methane) and biofuels. Biofuels, which include bioethanol, biomethane, and biodiesel, are derived directly from living organisms. Below, we discuss the conversion of engines to use multiple fuels. Biofuels themselves are described in greater detail later in this chapter.

Gasoline engines built after 1980 can use mixtures of gasoline and up to 10% ethanol without modification. With modifications (i.e., fabricating the engine from materials that are resistant to ethanol or methanol corrosion and installing sensors that detect the fuel mixture and allow the electronic ignition control unit to adjust engine timing), gasoline vehicles can use mixtures that contain up to 85% ethanol or methanol and 15% gasoline. Converting a gasoline engine to use compressed natural gas (CNG) entails the addition of a pressure regulator and an electronic multi-point gas injection system similar to a gasoline injection system.

Diesel engines are more flexible in their fuel mixtures. Rudolf Diesel, the inventor of the device that bears his name, ran his engines on vegetable oils and became an advocate for independence from petroleum fuels.

Today's diesel engines, with minor modifications (larger fuel lines and more frequent replacement of the fuel filter), can switch freely between petroleum-based

Rudolf Diesel's repudiation of the petroleum industry has spawned a number of conspiracy theories because of the mysterious circumstances surrounding his death. He disappeared on September 29, 1913 from a steamer crossing the English Channel. Ten days later, Dutch fishermen found a body floating at sea, removed the personal effects, but left the decaying corpse in the water. The personal effects later identified the body as Rudolf Diesel.

diesel fuel (petrodiesel) and mixtures of petrodiesel and biodiesel. Biodiesel, the product of the reaction between vegetable oils or animal fats with ethanol or methanol, has a viscosity similar to petrodiesel. Some, and perhaps many, diesel vehicles can operate on unadulterated biodiesel.

AGREEMENTS WITH AUTOMOBILE MANUFACTURERS As discussed previously, higher vehicle fuel efficiencies would conserve limited natural resources, reduce dependence on unreliable sources of petroleum, improve the balance of trade for many countries, and mitigate air pollution and greenhouse gas emissions. Most people agree with these goals, yet economic market forces such as taxes on fuel and vehicles have proved insufficient to achieve them (National Research Council 2002). Consequently, governments have pressured automobile manufacturers to improve the fuel efficiency of their light-duty vehicles.

In the aftermath of the 1973 oil embargo, the U.S. Congress enacted the Corporate Average Fuel Economy (CAFE) regulations. These regulations established the average fuel economy of passenger cars in a manufacturer's U.S. fleet at 18 mpg (7.7 km L^{-1} or km/L) in 1978 and 27.5 mpg (11.7 km L^{-1}) in 1985 and of small trucks at 17.2 mpg (7.3 km L^{-1}) in 1979 and 21.6 mpg (9.2 km L^{-1}) in 1985. These regulations have been controversial from their inception. They have generated sharp disagreements about their effects on the following (National Research Council 2002):

1. Mix of vehicles in the United States;

2. Overall efficiency of this mix;

3. Safety of vehicles;

4. Costs to consumers; and

5. Health of the domestic automobile industry

The following section describes each of these issues.

Mix of vehicles The CAFE regulations have several loopholes. One, which proved large enough to drive

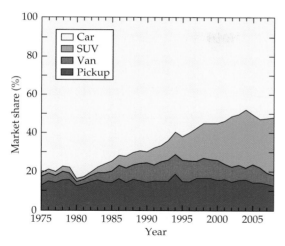

FIGURE 7.24 Percent market share of new cars, SUVs, vans, and light-duty (pickup) trucks in the United States as a function of model year (After U.S. Environmental Protection Agency 2008.)

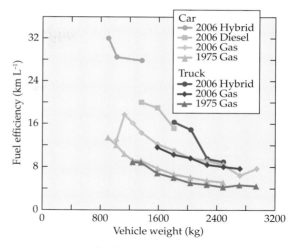

FIGURE 7.25 Fuel efficiency (1 km L^{-1} = 2.35 mpg) in the United States as a function of vehicle weight (1 kg = 2.2 lb) for new cars and trucks having different types of engines or introduced in 1975 or 2006. The symbols denote the fuel efficiencies of individual models based on laboratory measurements. (After Heavenrich 2006.)

a truck through, is that sport utility vehicles (SUVs), light-duty pickup trucks, and vans fall under the more lax standards of small trucks and thus require less-costly technology for compliance. With their lower prices and their desirable larger sizes, the public purchased more and more of these vehicles as passenger vehicles. The market share of SUVs, small trucks, and vans in the United States grew from less than 20% of new vehicles sold in 1975 to nearly 50% of new vehicles sold today (**Figure 7.24**).

Another loophole is that, in the calculation of the fleet fuel-economy averages for a manufacturer, CAFE regulations hold alternative-fuel or dual-fuel vehicles to a standard different from that for other vehicles. Fuel efficiencies of dual-fuel vehicles are multiplied by their percentage of alternative fuel use times 6.67. For instance, a 15-mpg vehicle supposedly dedicated exclusively to alternative fuels would rate as if it were a 100-mpg vehicle (15 × 100% × 6.67). In reality, most dual-fuel vehicles in the United States never burn anything but gasoline, and most owners of a duel-fuel vehicle do not even know that they are driving one (Energy Efficiency and Renewable Energy 2009). Thus, this loophole inflates fleet fuel-economy averages.

Efficiency of the mix The CAFE regulations did spur the development of more fuel-efficient technologies (**Table 7.6**), and automobile manufacturers have adopted many of them. Cars and small trucks, despite differences in the CAFE regulations, have similar fuel efficiencies at equivalent weights (**Figure 7.25**). Also, fuel efficiencies of cars and small trucks at a given

weight improved about 70% between 1975 and 2006. Indeed, vehicles in the United States would be much more fuel efficient today if they were the same size as yesteryear, but vehicle weight and engine power increased 30% and 115%, respectively, from 1981 to 2006. As a result, fuel efficiencies of both new and all vehicles in the United States have remained stagnant for over a decade (see Figures 7.12 and 7.21B).

Safety The laws of physics dictate that when a small and a large vehicle collide, the small one suffers more damage. Accordingly, accident fatalities (**Figure 7.26**) decrease as vehicle weights increase. Cars and vans, however, have lower fatality rates than pickup trucks and SUVs of equivalent weights. Pickup trucks and SUVs attach a body shell to a rigid chassis, and the shell tends to collapse upon the occupants during a collision, whereas passenger cars integrate the body shell with a flexible chassis to provide better protection for the occupants. In addition, SUVs have high centers of gravity and short wheelbases that are more prone to rollovers. Nonetheless, safety records vary greatly among models within every class of vehicle (Insurance Institute for Highway Safety 2007).

Costs There are two general approaches for improving fuel efficiencies (see Table 7.6). One focuses on the design of engines or transmissions to extract more energy out of the fuel combusted. The other focuses on vehicle weight, aerodynamics, or rolling resistance

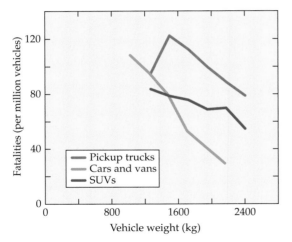

FIGURE 7.26 Fatalities per million registered vehicles by vehicle weight of 2000 to 2004 models during 2001 through 2005. (After Insurance Institute for Highway Safety 2006.)

to diminish the effort required to propel the vehicle. Additional costs for these technological improvements range from negligible for simple body aerodynamics to many thousands of dollars for gasoline-electric hybrid systems (see Table 7.6). Many of these technologies also enhance attributes such as reliability or acceleration. Details about technologies to increase vehicle fuel efficiencies are available in several reports (ACIL Consulting 1999; National Research Council 2002; World Business Council for Sustainable Development 2004; Heavenrich 2006; IPCC 2007b).

Health of the industry The economic well being of industries with respect to climate change mitigation is discussed in Chapter 10.

The CAFE fuel efficiency regulations have resisted major revisions over the years. The state of California passed legislation in 2006 requiring higher vehicle fuel

TABLE 7.6 Available and emerging technologies for improving the fuel efficiency of light-duty vehicles

Area	Gasoline-powered vehicles[a]	Diesel-powered vehicles	Approximate cost ($U.S.)
Engine	Reduced engine friction losses	Reduced engine friction losses	35–140
	Four valves per cylinder	Four valves per cylinder	105–140
		Piezo injectors	
	DI/homogeneous charge (stoichiometric)		125–175
	DI/stratified charge (lean burn)		320–480
	Downsizing with turbocharging	Downsizing with turbocharging	350–560
	Variable valve timing		35–140
	Variable valve control		70–210
	Cylinder deactivation	Cylinder deactivation	112–252
	Variable compression ratio		210–490
	Advanced cooling and electric water pump	Advanced cooling and electric water pump	35–120
		Exhaust heat recovery	45
Transmission	Six-speed manual/automatic gearbox	Six-speed manual/automatic gearbox	140–280
	Piloted gearbox	Piloted gearbox	300–400
	Continuous variable transmission	Continuous variable transmission	140–840
	Dual-clutch	Dual-clutch	450–675
Hybrid	Idle start-stop function	Idle start-stop function	210–350
	Regenerative braking	Regenerative braking	515–685
	Mild hybrid (motor assist)	Mild hybrid (motor assist)	900–2000
	Full hybrid (electric drive)	Full hybrid (electric drive)	2100–4200
Body	Improved aerodynamic efficiency	Improved aerodynamic efficiency	0–140
	Weight reduction	Weight reduction	210–350
Other	Low rolling resistance tires	Low rolling resistance tires	14–56
	Electrically assisted steering (EPS, EPHS)	Electrically assisted steering (EPS, EPHS)	105–150
	Improved air conditioning	Improved air conditioning	
	DeNOx catalyst	DeNOx catalyst	
		Particulate trap/filter	

Source: National Research Council 2002; Smokers et al. 2006.

[a]DI, direct injection; EPS, electronic power steering; EPHS, electrically powered hydraulic steering.

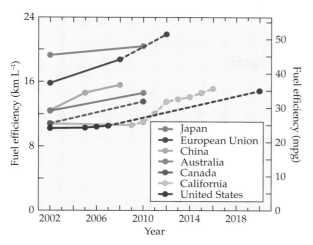

FIGURE 7.27 Current (solid) and pending (dashed) governmental regulations on fuel efficiency standards for passenger vehicles (After An and Sauer 2004.)

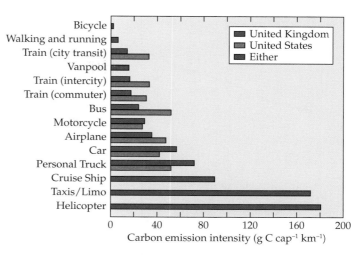

FIGURE 7.28 Greenhouse gas emissions for various modes of transportation, in grams of carbon per person per kilometer Differences between the United States and the United Kingdom often hinge on the estimated passenger occupancy rate for the various modes; for example, the U.S. estimate for personal trucks (which includes SUVs and "king cabs") assumes 1.72 occupants, whereas the U.K. estimate assumes 1.25 occupants. Only 40% to 85% of the fuel consumption on a cruise ship serves for propulsion; the rest serves for passenger amenities such as electricity and hot water. (After Defra 2007; Davis et al. 2008.)

efficiencies, but these are being contested in the courts. In 2007, the U.S. Congress seemed poised to raise the standards gradually to about 35 mpg (14.9 km L^{-1}) for passenger cars as well as set higher standards for SUVs and small trucks based on weight or volume, but as of June 2009, passage of these provisions were still under debate.

Many other countries also have standards for vehicle fuel efficiencies (**Figure 7.27**). Some emphasize distance traveled per volume of fuel (km L^{-1} or mpg) and others compare the amount of CO$_2$ emitted per distance traveled. Testing protocols also differ among countries (see Figure 7.14). If one converts the various standards to common units and common testing protocols, the standards in many countries will soon exceed 15 km L^{-1} (35.3 mpg).

Public Transport and Energy Conservation

Light-duty vehicles are only one of many modes of personal transportation. Greenhouse gas emissions from these modes vary widely. A person on a bicycle emits about 2.3 grams of carbon per kilometer, and a person walking or running emits 6.3 grams per kilometer (**Figure 7.28**); these modes are the ultimate in biofuel conversion. At the other extreme, a commuter helicopter emits carbon at a rate of 180.5 grams per kilometer per passenger.

A central factor in transportation efficiency is passenger occupancy. Except for cycling, walking, or run-

ning, the incremental energy required per additional passenger is small for most modes of transportation. Therefore, doubling passenger occupancy nearly halves the effective greenhouse gas emissions per distance traveled.

The vast majority of travel in the United States and Europe is via private vehicles, which include cars, small trucks, SUVs, and motorcycles (**Figure 7.29**). Attempts to encourage carpooling through special highway lanes and toll-free passage over bridges have met with only marginal success. In fact, the percentage of people who carpool to work in the United States has not changed significantly over the past 18 years (**Figure 7.30**). The percentage of those who walk or cycle to work has also not changed.

By contrast, ridership on public ground transport has expanded significantly from a low point in the 1970s (**Figure 7.31**). This includes a 50% increase in travel via rail and other forms of public ground transport except for buses (**Figure 7.32**). Public air travel—except for the period after the airplane crashes on September 11, 2001—has also risen steadily. Nevertheless, travel in vehicles carrying only one passenger is still gaining ground on all other forms of transportation (see Figures 7.29 and 7.30).

This chapter has focused almost entirely on personal transportation, although freight transportation

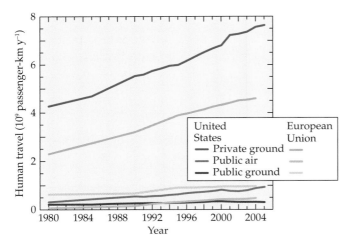

FIGURE 7.29 Private and public modes of transportation Travel (10^{12} passenger-km) in the United States and Europe via private ground vehicles, public air carriers, or public ground transport (buses, railroads, vans, or ferries). (After Eurostat 2003, 2006; American Public Transportation Association 2007; Bureau of Transportation Statistics 2009.)

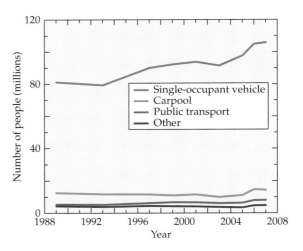

FIGURE 7.30 Number of people (millions) in the United States who travel to work via different modes of transportation "Public transport" denotes ground transport. "Other" includes those who walk and bicycle. (After Bureau of Transportation Statistics 2009.)

is increasing faster than that of people (see Figure 7.7). Freight is more sensitive to economic market forces than personal transportation. For example, no one cares whether their packages arrive in an elegant limousine or a big, brown, boxy van as long as they are delivered quickly and intact. The shipping industry (air, sea, road, and rail) thus responds more directly to rising petroleum costs than does personal transportation. For instance, the shipping industry quickly adopted

technologies that improve fuel efficiency such as diesel, hybrid diesel-electric, compressed natural gas, and aerodynamic cowlings for truck cabs.

Alternative Fuels for Transportation

Many countries are vigorously developing alternatives to gasoline and petrodiesel in order to achieve greater energy independence, attain a more favorable balance

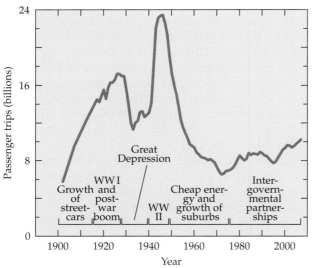

FIGURE 7.31 Ridership on public ground transport (10^9 passenger trips per year) in the United States during the twentieth century and major events influencing ridership. (After American Public Transportation Association 2007.)

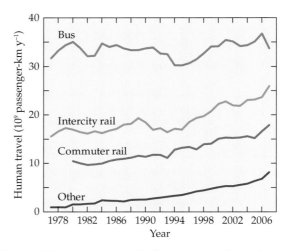

FIGURE 7.32 U.S. travel (10^9 passenger-km) via different modes of public ground transport "Other" includes vanpools, jitneys, ferries, trolleys, light rail, and non-route service for the elderly or disabled. (After American Public Transportation Association 2009.)

of trade, and diminish greenhouse gas emissions. Brazil, for example, produces enough ethanol from sugarcane to replace about 14% of its use of petroleum fuel in vehicles and has become the world's leading exporter of ethanol (International Energy Agency 2006). The U.S. Department of Energy has set a goal with the catchy title of "30-by-30" whereby the United States would replace 30% of its gasoline use with alternative fuels by 2030. Alternative energy sources for transportation include compressed and liquid natural gas, hydrogen, electric and electric-hybrid, and bioethanol and biodiesel.

Natural gas

Natural gas is extracted from oil wells, coal beds, natural gas fields, and landfills. After processing, natural gas contains a mixture of **hydrocarbons**: between 70% and 90% methane (CH_4), between 5% and 15% ethane (C_2H_6), and smaller amounts of propane (C_3H_8) and butane (C_4H_{10}). Combustion of these hydrocarbons releases carbon dioxide and water and produces chemical energy:

$$CH_4 + 2\,O_2 \rightarrow CO_2 + 2\,H_2O + \text{chemical energy} \qquad (7.1)$$

Natural gas has a very low energy content at normal atmospheric pressure (CNG at 0°C, 1 atm. in **Table 7.7**). Carrying a sufficient amount of natural gas to power a vehicle any distance would take a huge storage tank; therefore, natural gas for transportation is pressurized to several hundred atmospheres (2000 psi to 3600 psi or 13.8 MPa to 24.8 MPa). This compressed natural gas (CNG) still has a quarter or less of the energy content in gasoline and thus requires larger storage tanks at refueling stations and on the vehicles themselves. Refueling of CNG vehicles is also slower than refueling with liquid fuels, requiring about double the time to fill the tank.

On the plus side, natural gas costs less on the world market than gasoline or diesel fuel (see Figure 7.5). Some countries, such as Argentina and Pakistan, have greater reserves of natural gas than oil (BP 2007) and have promoted its use (Table 7.8). Conversion of engines from gasoline to CNG is straightforward, as mentioned previously.

Natural gas deserves its reputation as a "clean" fuel because it produces fewer particulates, non-methyl hydrocarbons, and NO_X (a contributor to photochemical smog) during combustion than gasoline or petrodiesel (U.S. Environmental Protection Agency 1999). Buses in areas suffering from air pollution are often powered by CNG and may be distinguished by the large storage tank on their roofs (**Figure 7.33**). Most urban areas in the United States have CNG refueling stations; for example, California now has more than 200 CNG refueling stations (**Figure 7.34**).

TABLE 7.7 Energy content and greenhouse gas emissions[a] during combustion of various fuels

Fuel[b]	Energy content (MJ L^{-1})[c]	Carbon emissions (g C MJ^{-1})
Asphalt and road oil	44.0	19.5
Aviation gas	33.5	17.9
Biodiesel	35.2	19.0
Butane	28.7	16.1
Cellulose	26.9	
CNG (0°C, 1 atm)	0.041	13.7
CNG (2,000 psi)	4.9	13.7
CNG (3,600 psi)	8.8	13.7
Coal (anthracite)	38.0	24.7
Coal (bituminous/lignite)	32.9	24.3
Crude petroleum	38.5	19.2
Ethanol	23.6	17.3
Fuel oil (distilled)	38.7	18.9
Fuel oil (residual)	41.7	20.4
Gasohol	33.7	18.0
Gasoline	34.8	18.3
Hydrogen (0°C, 1 atm)	0.013	0
Hydrogen (10,000 psi)	5.3	0
Hydrogen (liquid)	10.0	0
Jet fuel (kerosene)	37.6	18.7
Jet fuel (naphtha)	35.5	18.3
Lignin	32.5	
LNG	25.3	13.7
Lubricants	40.2	19.2
Methanol	18.0	16.4
Petrodiesel	38.7	18.9
Petroleum coke	40.0	26.4
Propane	25.4	16.1
Starch	22.4	
Sucrose	27.0	
Waxes	36.7	18.8

Source: Davis and Diegel 2007; Tiax LLC 2007.
[a]Gram carbon equivalents per 10^6 joules energy.
[b]CNG, compressed natural gas; LNG, liquefied natural gas.
[c]One joule (J) = 0.239 calories. Energy contents assume the Higher Heating Value and 100% combustion. 1 psi (pound per square inch) = 0.0069 MPa.

More importantly, from the perspective of this book, combustion of natural gas emits smaller amounts of greenhouse gases than any fuel except for hydrogen (see Table 7.7). Leakage of natural gas during extrac-

FIGURE 7.33 Buses from the current shuttle bus fleet in Grand Canyon National Park A bus powered by compressed natural gas (CNG) is on the left, a bus powered by liquid natural gas (LNG) is in the middle, and an electric-powered bus is on the right.

FIGURE 7.34 Locations of public compressed natural gas refueling stations in California (After Consumer Energy Center 2006.)

tion, refining, distribution, and combustion is an issue because its main constituent, CH_4, is a greenhouse gas with 23 times the global warming potential of CO_2 (see Table 3.1). Such leakage is small, however, and CNG

| TABLE 7.8 | Number of compressed natural gas vehicles and refueling stations in the top 10 countries, and the percentage of vehicles that use CNG |

Country	Vehicles (1000s)	Refueling stations	Percent of all vehicles
Argentina	1650	1640	25.8
Brazil	1357	1410	6.4
Pakistan	1300	1230	64.0
Italy	410	558	1.1
India	335	321	1.7
Iran	292	203	14.8
United States	147	1340	0.1
China	127	355	0.5
Columbia	100	90	4.1
Bangladesh	80	129	37.9

Source: International Association of Natural Gas Vehicles 2007; United Nations Statistics Division 2007.

vehicles typically emit about 12% less greenhouse gases than do gasoline-powered vehicles (Science Applications International Corporation 2002).

Natural gas will condense into a liquid when cooled to between -120° and -170°C, depending on the relative proportions of the various constituent hydrocarbons. Liquefied natural gas (LNG) has an energy content similar to other liquid fuels (see Table 7.7) and thus can be stored in smaller tanks and transferred more quickly than CNG. Such storage tanks, however, require heavy-duty thermal insolation to keep the LNG from boiling. Because of the difficulties of working with a **cryogenic liquid** (i.e., liquid that must be stored at a very low temperature), use of LNG has been generally limited to long-distance transport of natural gas when CNG pipelines are unavailable. For example, cargo ships carry LNG for transport across oceans.

Hydrogen

George W. Bush, the 43rd president of the United States, declared in his State of the Union address of January 28, 2003, "A simple chemical reaction between hydrogen and oxygen generates energy, which can be used to power a car producing only water, not exhaust fumes. With a new national commitment, our scientists and engineers will overcome obstacles to taking these cars from laboratory to showroom so that the first car driven by a child born today could be powered by hydrogen,

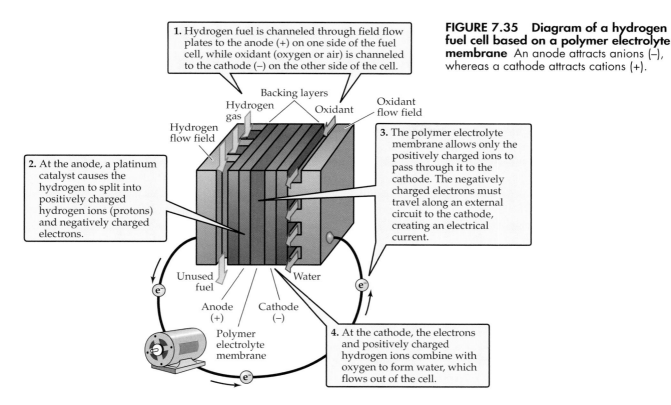

1. Hydrogen fuel is channeled through field flow plates to the anode (+) on one side of the fuel cell, while oxidant (oxygen or air) is channeled to the cathode (–) on the other side of the cell.

FIGURE 7.35 Diagram of a hydrogen fuel cell based on a polymer electrolyte membrane An anode attracts anions (–), whereas a cathode attracts cations (+).

Backing layers
Hydrogen gas
Oxidant
Oxidant flow field
Hydrogen flow field

2. At the anode, a platinum catalyst causes the hydrogen to split into positively charged hydrogen ions (protons) and negatively charged electrons.

3. The polymer electrolyte membrane allows only the positively charged ions to pass through it to the cathode. The negatively charged electrons must travel along an external circuit to the cathode, creating an electrical current.

Unused fuel
Water
Anode (+)
Cathode (–)
Polymer electrolyte membrane

4. At the cathode, the electrons and positively charged hydrogen ions combine with oxygen to form water, which flows out of the cell.

and pollution-free." Today, most experts agree that this assessment was overly optimistic given the technical and logistical problems that remain unresolved (National Research Council 2004). Fifty years, not fifteen, would be a more reasonable timeline for some general use of hydrogen as a fuel for transportation.

Internal combustion engines with some modifications can burn hydrogen, but the thermodynamics of such engines limit their efficiency to about 23%, somewhat independently of the fuel being combusted (Malte 2001; von Helmolt and Eberle 2007). This means that less than a quarter of the energy in the fuel is converted into propelling the vehicle. Fuel cells, in which hydrogen reacts with oxygen to form water and generates electricity for an electric motor that moves a vehicle (**Figure 7.35**), may be more efficient, theoretically above 36% (Malte 2001; von Helmolt and Eberle 2007). This approach, however, depends on advances in hydrogen storage, fuel cell costs, fuel cell reliability, and hydrogen production and distribution.

Hydrogen storage Vehicles running on alternative fuels need a range of travel comparable to that of gasoline vehicles, something greater than 300 km (186 miles), before they will be adopted for general use. Unfortunately, hydrogen has the lowest energy content per unit volume of any fuel (see Table 7.7). To carry enough

hydrogen for an adequate travel range, a vehicle must accommodate either compressed hydrogen gas, liquefied hydrogen gas, or hydrogen gas absorbed onto solid particles.

Compressed gas To match energy contents with natural gas, hydrogen must be compressed to a much higher pressure: A tank filled with 10,000 psi of hydrogen (680 times the normal atmospheric pressure) holds the same amount of energy as one filled with 2000 psi of CNG (see Table 7.7). Under such pressures, hydrogen is such a small, energetic molecule that it diffuses through most materials that line the walls of compressed gas tanks. Once hydrogen penetrates the liners, it reacts with many wall materials (e.g., steel, aluminum, and titanium), and weakens them, a process called **embrittlement**. Lightweight storage tanks that can withstand hydrogen at 10,000 psi have liners of high molecular weight polymers, walls of filament-wound carbon fibers, and shapes that are nearly spherical. Not only do such tanks cost roughly $3,600 if purchased in large quantities (Chalk and Miller 2006), but their shape does not fit easily into vehicles (round peg, square hole). Finally, the energy required to compress hydrogen to 10,000 psi equals about 15% of the energy value of the hydrogen in the tank (von Helmolt and Eberle 2007).

Liquefied hydrogen   Hydrogen condenses into a liquid at −253°C, about 90°C lower than the temperature required for liquid natural gas (LNG). Liquid hydrogen contains nearly double the energy content per volume of hydrogen gas at 10,000 psi, but only 40% of the energy content per volume of LNG (see Table 7.7). Liquid hydrogen has been used in the space program for decades (the main engine of the space shuttle burns liquid hydrogen and liquid oxygen), and storage systems for liquid hydrogen might be less expensive to mass produce than those for compressed hydrogen gas, on the order of $1200 (Chalk and Miller 2006). The process of liquefying hydrogen, however, expends about 30% of the energy contained in the tank (von Helmolt and Eberle 2007). Even greater energy losses occur during storage, when some of the liquid heats up and boils off, and during refueling, when transfers between the service station storage tank and the vehicle requires precooling tubing and valves.

Solid absorbers   Hydrogen will react with metals such as aluminum, boron, lithium, magnesium, or sodium to form chemical complexes known as **hydrides**. When subjected to heat or pressure, metal hydrides will release the hydrogen absorbed. Hydrides can achieve hydrogen densities that surpass liquid hydrogen, but they have serious drawbacks. Some of them can be dangerous, exploding upon exposure to moist air. Some are highly toxic to humans. They all require a substantial input of energy, 11% to 26% of the energy contained in the material, for their generation and release (von Helmolt and Eberle 2007). The materials themselves are expensive, on the order of $1600 to $3200 per vehicle (Chalk and Miller 2006). For these reasons, the aerospace industry, although it has conducted extensive research on metal hydrides, has never actually deployed them in a vehicle.

Research is also proceeding on systems that absorb hydrogen onto tiny carbon particles (nanotubes or buckyballs). These systems usually operate at temperatures below −200°C. So far, such systems have not reached the efficiencies needed for practical applications.

FUEL CELL COSTS   Fuel cells today require some very expensive components (International Energy Agency 2005). In particular, the electrode/catalyst and polymer electrolyte membrane (see Figure 7.35) are literally worth their weight in gold. A **catalyst** is a substance that promotes a chemical reaction without itself being consumed. As such, it must remain chemically inert under the conditions necessary for the reaction. Nearly all electrode/catalysts for hydrogen fuel cells contain platinum, a chemically inert metal that costs over $35 per gram, substantially more than the price of gold. At present, a typical fuel cell vehicle (100 kilowatt = 134 horsepower) contains over $70,000 of platinum (ERDC/CERL Fuel Cell Team 2007). Much of the current research on fuel cells focuses on minimizing the platinum needed for efficient catalysis. For example, one approach uses nanotechnology to produce tiny particles of platinum alloy with higher reactivity. These developments may someday bring platinum costs for a vehicle to below $5000. Then again, large-scale deployment of hydrogen fuel cell vehicles would increase the demand for platinum and thereby increase its price on world markets.

Polymer electrolyte membranes in hydrogen fuel cells are usually constructed from a special type of Teflon called Nafion that is selectively permeable to positive electrical charges. This material is expensive, adding roughly $1000 to the cost of a vehicle. It also breaks down at temperatures above 80°C. Other materials, such as porous polyethylene film and aromatic polysulfone copolymers, may lower membrane costs to somewhere around $10 (Chalk and Miller 2006) and can operate at higher temperatures. Higher temperatures allow for more efficient use of platinum in the electrodes (International Energy Agency 2005).

RELIABILITY   Vehicles must function dependably despite being subjected to constant vibration, rapid temperature changes ranging from cold starts to high running temperatures, frequent bombardment with dirt and water, and occasional neglect or incompetence. Hydrogen fuel cell vehicles may not be ready for such abuse. Their membranes must be thin to attain sufficient permeability to gases, but this increases their fragility. Their catalysts must have a high surface area to attain sufficient reactivity, but this makes them vulnerable to contamination by dirt or poisoning by carbon monoxide.

The water and temperature management of hydrogen fuel cells is complex. The principal reaction in a cell generates not only electric current, but also water and heat.

$$2\,H_2 + O_2 \rightarrow 2\,H_2O + \text{electric current} + \text{heat} \qquad (7.2)$$

If a fuel cell floods with water, hydrogen will not reach the catalyst, and the reaction will stop. If a fuel cell has too little water or the water freezes or boils, the membrane will suffer damage and lose permeability, and the fuel cell will overheat. To sustain operation, a fuel cell must be maintained under environmental conditions that are just right: not too wet, not too dry, not too cold, and not too hot.

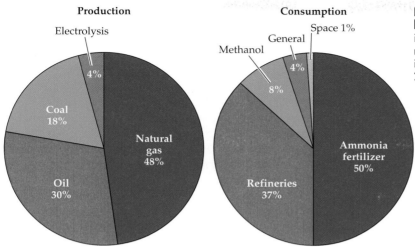

Production

Electrolysis

Coal 18%

Oil 30%

Natural gas 48%

4%

Consumption

Space 1%

General

Methanol

4%

8%

Ammonia fertilizer 50%

Refineries 37%

FIGURE 7.36 Current sources of hydrogen and uses of hydrogen "General" includes usage in the chemical, food, microchip, and metal industries. "Space" includes uses for aviation. (After Kruse et al. 2002; International Energy Agency 2007b.)

HYDROGEN PRODUCTION Aside from the technical difficulties in developing an affordable and reliable vehicle based on hydrogen fuel cells is the daunting task of expanding the infrastructure for hydrogen production and distribution. First and foremost, hydrogen is not an energy source: Earth has no recoverable deposits of hydrogen. Rather hydrogen serves a carrier, a means for transferring energy from one source to another.

Over 95% of the hydrogen generated today derives from fossil fuels (**Figure 7.36**) and is used at the site of production for synthesizing ammonia fertilizer (see Equation 5.1), converting heavier hydrocarbons in crude oil into lighter fractions that are more suitable for fuels, or producing methanol from carbon monoxide (described in the following equation).

$$CO + 2\,H_2 \rightarrow CH_3OH \qquad (7.3)$$

Several methods are available for generating hydrogen from fossil fuels (**Table 7.9**). All require high temperatures (over 700°C) and pressures (more than 3 times atmospheric pressure). They all finish with the "water-gas shift reaction," in which CO reacts with steam to produce CO_2 and H_2. Their conversion efficiencies (energy content of products divided by energy content of ingredients) are around 70%.

The only common method for producing hydrogen that does not directly require fossil fuels is electrolysis, in which an electric current passes through water and releases hydrogen and water (Equation 7.2 in reverse). There are two main types of electrolysis. One

TABLE 7.9 Hydrogen production from fossil fuels

Technology	Reaction	Enthalpy[a] (kJ mol⁻¹)	Efficiency (percent)
Steam reforming	$CH_4 + H_2O \rightarrow CO + 3\,H_2$	−206	
	$C_nH_m + nH_2O \rightarrow nCO + (n + \frac{1}{2}\,m)\,H_2$	−1,175	71–76
	$CO + H_2O \rightarrow CO_2 + H_2$	41	
Catalytic partial oxidation	$CH_4 + O_2 \rightarrow CO_2 + 2\,H_2$	38	66–76
	$CO + H_2O \rightarrow CO_2 + H_2$	41	
Autothermal reforming	$CH_4 + \frac{3}{2}\,O_2 \rightarrow CO + 2\,H_2O$	520	66–73
	$CH_4 + H_2O \rightarrow CO + 3\,H_2$	−206	
	$CO + H_2O \rightarrow CO_2 + H_2$	41	
CO_2 reforming	$CH_4 + CO_2 \rightarrow 2\,CO + 2\,H_2$	−247	> 75
	$CO + H_2O \rightarrow CO_2 + H_2$	41	

Source: International Energy Agency 2005.
[a]A negative enthalpy indicates that the reaction releases heat and a positive enthalpy indicates that it requires energy from an external source.

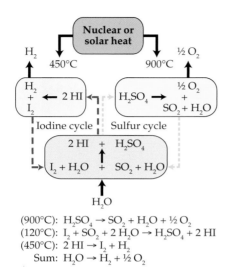

(900°C): $H_2SO_4 \rightarrow SO_2 + H_2O + \frac{1}{2} O_2$
(120°C): $I_2 + SO_2 + 2 H_2O \rightarrow H_2SO_4 + 2 HI$
(450°C): $2 HI \rightarrow I_2 + H_2$
Sum: $H_2O \rightarrow H_2 + \frac{1}{2} O_2$

FIGURE 7.37 Hydrogen production via direct water splitting of the sulfur–iodine cycle (After International Energy Agency 2005.)

employs an alkaline solution containing potassium hydroxide and electrodes of nickel alloy. The other type runs a fuel cell backwards (see Figure 7.35) and separates hydrogen across a membrane. Conducting either type of electrolysis at high temperatures or high pressures saves electricity at the expense of energy to heat or pressurize the reaction vessel. In practice, electrolysis achieves conversion efficiencies of 40% to 63% (International Energy Agency 2005).

Many other methods of hydrogen production are under investigation. One is the direct splitting of water into hydrogen and oxygen. This occurs spontaneously at temperatures above 2500°C. The sulfur–iodine cycle (**Figure 7.37**) lowers the temperature needed for direct water splitting to around 900°C. In a pilot study, this process achieved efficiencies of about 43% (International Energy Agency 2005). The eventual goal is to couple direct water splitting with heat production from nuclear reactors or solar collectors, but existing nuclear and solar facilities are not designed to operate at 900°C and would require major modifications.

Biological production of hydrogen is yet another method. Photosynthesis normally absorbs solar energy to split water, synthesizes high-energy biochemicals, and releases oxygen (see Equation 2.1). Some organisms use this biochemical energy to fix dinitrogen gas (N_2) into ammonia (NH_3), a process that releases hydrogen gas (see Equation 5.2). Other organisms—in particular, cyanobacteria and green algae deprived of sulfur—alter their photosynthesis under anaerobic conditions and release hydrogen instead of oxygen (Melis and Happe 2001). Efficiency of this hydrogen

production, however, is less than 2% (Rupprecht et al. 2006).

HYDROGEN DISTRIBUTION Hydrogen vehicles that are currently under development will require frequent refueling, at least as often as gasoline- or diesel-powered vehicles. General deployment of such vehicles thus awaits a functioning network of hydrogen refueling stations, and vice-versa. There are two strategies for supplying hydrogen to a network of refueling stations: A few large facilities might produce hydrogen and then ship it long distances, or many small facilities might produce hydrogen locally.

Large hydrogen production facilities would benefit from economies of scale. They could optimize the size of their equipment for energy efficiency. They also could be located near large power sources that could economically capture and store greenhouse gases (see Chapter 8). Consequently, large facilities should be able to produce hydrogen at much less cost than small facilities (**Figure 7.38**). The disadvantage of central facilities is distributing hydrogen, a low energy content fuel (see Table 5.7), over long distances.

One immediate solution to long-distance hydrogen transport is hydrogen gas pipelines (International Energy Agency 2005). Several thousand kilometers of such pipelines are already in use in Europe and the United States, but this is small in comparison to the 3 million kilometers of natural gas pipelines that are

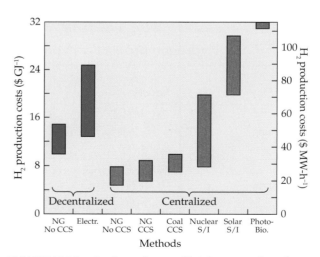

FIGURE 7.38 Projected costs ($U.S. per 10^9 joules or 10^6 watt-hours) of hydrogen produced in small, decentralized and large, centralized facilities via various methods NG denotes natural gas; CCS, carbon capture and storage; Electr., electrolysis with greenhouse gas-free electricity; S/I, the sulfur–iodine cycle; and Photo-Bio., photo-biology. (After International Energy Agency 2005.)

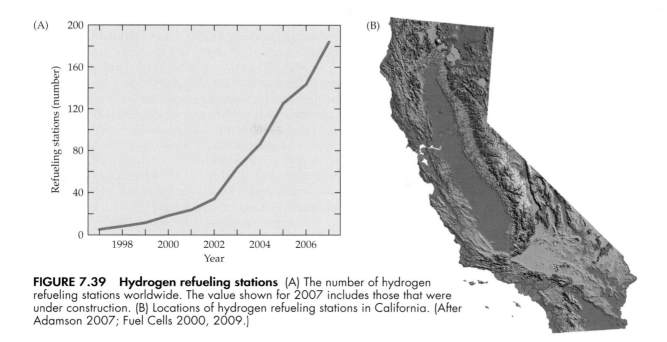

FIGURE 7.39   Hydrogen refueling stations (A) The number of hydrogen refueling stations worldwide. The value shown for 2007 includes those that were under construction. (B) Locations of hydrogen refueling stations in California. (After Adamson 2007; Fuel Cells 2000, 2009.)

currently in use. Most natural gas pipelines, however, are unsuitable for hydrogen because they are usually composed of materials that leak the smaller hydrogen molecules or become brittle by reactions with hydrogen. Moreover, hydrogen has less than a quarter of the energy content of natural gas (see Table 7.7); to deliver the same amount of energy, the hydrogen pipelines must operate at higher pressures (72 psi to 145 psi, or 5 MPa to 10 MPa) or be constructed of tubes of larger diameter than natural gas pipelines. Constructing a hydrogen pipeline network would cost over $600 billion for the United States and over $2.5 trillion worldwide (International Energy Agency 2005). Energy for pressurizing a pipeline and moving hydrogen gas 2000 km would expend 1.2 joules of energy for every 1 joule of hydrogen energy delivered (Bossel et al. 2007) and would add about $1 per GJ ($10^9$ joules) to the cost of the hydrogen, around 5% to 15% of the total cost (International Energy Agency 2005).

Other solutions for distributing hydrogen involve truck or sea transport. A large load of hydrogen gas, if compressed at high pressures (5000 psi to 10,000 psi, or 34 MPa to 69 MPa) to achieve a practical energy density, is probably too dangerous for the nation's highways or seaways. More likely is that trucks or ships with cryogenic storage capabilities would carry liquid hydrogen. Energy costs for liquefying hydrogen (–253°C) are high at about $7 per GJ, and losses during transport over a few thousand kilometers would amount to $2 per GJ (International Energy Agency 2005). These high transport costs favor the development of more decentralized hydrogen production.

The number of hydrogen refueling stations has grown exponentially, to a total of 182 worldwide in 2007 (**Figure 7.39A**). Only in certain locations, such as around San Francisco or Los Angeles, can a hydrogen vehicle travel any distance from home and expect to find a compatible refueling station (**Figure 7.39B**). Nearly all existing hydrogen vehicles use compressed gaseous hydrogen (5000 psi to 10,000 psi, or 34 MPa to 69 MPa), and so refueling stations are equipped to transfer hydrogen gas at such pressures (**Figure 7.40**). Filling up a vehicle with sufficient hydrogen to drive 300 km takes about 10 minutes (Wipke et al. 2007), many times slower than refueling a gasoline-powered vehicle.

SUMMARY OF THE HYDROGEN ECONOMY   The so-called "hydrogen economy," in which hydrogen serves as a major form of energy storage for mobile applications, has received a major share of the research effort on alternative fuels. Despite steady progress, not one piece of this economy—from fuel cell vehicles to facilities for production and distribution of hydrogen—is ready for general adoption. More promising in the short term are vehicles powered by electricity or biofuels, as discussed in following sections.

Electric and hybrid vehicles

Vehicles powered by electric motors or by a combination of gasoline engines and electric motors are becoming more commonplace. Such vehicles are highly efficient in the slow, stop-and-go traffic of cities, but are

FIGURE 7.40 Grand opening of the hydrogen refueling station built by the Total oil and gas company for fuel cell buses in Berlin Pictured from left to right are Mr. Wolf-Dietrich Kunze (Vattenfall Europe, head of heating, Berlin), Prof. Dr. Johann Köppel (vice president of research, TU Berlin), and Mr. Michel Mallet (CEO, Total Deutschland).

less advantageous for high-speed, long-distance travel. The following sections present the reasons for these differences.

ELECTRIC ONLY A new car dealership recently opened in Davis, California. What distinguishes this enterprise from the six others in the Davis Auto Mall is that it sells only electric automobiles and scooters. This dealership offers two brands of electric automobiles, each of which costs about $12,000, uses lead-acid batteries, and has a top speed of 40 mph (64 kmph), a range up to 25 miles (40 km), and a recharge time of about 8 hours. These specifications highlight the problems with electric vehicles: Battery technology limits the amount of energy that a vehicle can carry and thus its speed, range, and recharge time. Lead-acid batteries are inexpensive and reliable but are heavy and have low energy and power per size and weight (Table 7.10). One of the manufacturers of these vehicles is developing a model equipped with lithium polymer batteries that will extend its range to 100 miles (161 km), but pricing and availability have yet to be determined.

At the other end of the spectrum is a Tesla Roadster. For $101,500, one can purchase this electric-powered sports car that has a top speed of 125 mph, an acceleration from 0 mph to 60 mph in under 4 seconds, and a range of about 220 miles. This vehicle is equipped with 6831 lithium-ion cells, each about the size of an "AA" battery (Berdichevsky et al. 2007). Together, they weigh 992 pounds (450 kg), have a service life that should extend beyond 100,000 miles (161,000 km), and cost more than $30,000 retail. Recharge time is as short as 3.5 hours with a special, stationary charging unit and 8 hours or longer with a mobile unit.

Electric vehicles, despite their limited range, long recharge (refueling) times, and high costs, have advantages over other types:

- Electric motors often achieve 90% conversion efficiency. Conversion efficiency is the ratio of the input of electrical energy to the output of kinetic energy in the moving vehicle.

TABLE 7.10 Properties of various rechargeable electrical storage technologies[a]

Type[b]	Energy/ weight (kJ g^{-1})	Energy/ size (kJ cc^{-1})	Power/ weight (W kg^{-1})	Energy/ price (kJ $U.S.$^{-1}$)	Self dis-charge rate (% month^{-1})	Durability (No. of cycles)	Cell voltage (V)
NiCd	0.14–0.29	0.18–0.72	150	12	10–100	800–2000	1.2
PbSO$_4$	0.11–0.18	0.22–0.37	180–600	18–65	3–20	200–800	2.0
NiMH	0.11–0.43	0.50–1.18	220–1000	4–5	30	300–1000	1.2
Li-Ion	0.36–0.58	0.97–1.31	1800–2000	4–18	5–10	600–1200	1.2, 3.6
Li-Polymer	0.47–0.72	0.79–1.69	1000–2800	4		>600	3.7
Zn-Air	0.36–0.72	0.72–0.90	357–1000	12	rapid in air	many	1.4
Ultra-Cap	0.43	0.15–0.22	2000	1	slow	>200,000	2.5

Source: Spotnitz 2005; BatteryUniversity.com 2007; PowerStream 2007; Thompson 2007.
[a]The large range of values for some parameters reflects different sources of data and different manufacturing processes.
[b]NiCd, nickel-cadmium; PbSO$_4$, lead-acid; NiMH, nickel-metal hydride; Li-Ion, lithium-ion; LiPolymer, lithium polymer; Zn-Air, zinc-air; UltraCap, ultra-capacitor. Zinc-air is a type of fuel cell that requires non-oxidized zinc fuel. Ultra-capacitors are mechanical rather than chemical storage. Neither zinc-air nor ultra capacitors are commercially available for vehicles.

- Electric vehicles themselves emit no greenhouse gases. The power plants that generate the electricity to charge the vehicles may produce greenhouse gases, but large power plants are far more efficient than small engines and may enlist carbon capture and storage technologies (see Chapter 8).

- Electric vehicles usually recharge at night during the slack hours for electric power grids and thus, may not strain current generating capacities.

- An electric motor provides high **torque** (rotational force down a shaft) over the full range of speeds, and therefore electric vehicles do not need gears, belts, or chains between the motor and the wheels.

- All the components in an electric vehicle operate at temperatures near room temperature.

- Electric vehicles require much less maintenance than gasoline- or diesel-powered vehicles. Typically, only the tires and brakes need regular service.

- Most electric vehicles incorporate regenerative braking, in which the electric motor driving the wheels also acts as a generator, which helps stop the vehicle by converting the kinetic energy of the moving vehicle into electricity that recharges its batteries.

HYBRIDS Petroleum-electric hybrid vehicles combine the desirable properties of electric propulsion with the portability and convenient refueling of petroleum fuels. Most of the locomotives and many of the ships built during the past 50 years are diesel-electric hybrids. In such vehicles, a large diesel engine turns a generator that recharges batteries, which in turn power small electric motors connected directly to the wheels or propellers. This arrangement operates the diesel engines under the uniform loads and low revolutions per minute that optimize their fuel efficiency. It also eliminates the need for complex transmission systems between an engine and several wheels or propellers. Finally, isolating the diesel engine's vibrations from the frame of a locomotive or hull of a ship leads to much smoother and quieter operation.

Today's hybrid cars take a different approach: They connect both a gasoline-powered engine and electric motors to the wheels. In some vehicles, the electric motors assist the gasoline engine, whereas in others the gasoline engine assists the electric motors; in most, the gasoline engine shuts down when the vehicle stops. The vehicles currently on the market use nickel-metal hydride batteries and recharge the batteries with the gasoline-powered engine and regenerative braking. Unfortunately, these batteries have small capacities (e.g., a 2008 Toyota Prius can travel only 8 miles on the electric motors without a recharge), are difficult to recycle, and are guaranteed for 8 years or 100,000 miles. A replacement battery for a 2005 Toyota Prius costs $3000 retail (O'Dell 2005).

In city driving (see Figure 7.14), a hybrid vehicle takes advantage of engine shutdown, electric motors, and regenerative braking to achieve fuel efficiencies of 30% greater than the same vehicle equipped with a gasoline-powered engine alone (see Table 7.4). Efficiency gains of hybrids are far more modest under highway driving, where they depend heavily on their gasoline engines. The next generation of hybrid cars should be able to travel substantial distances (40 miles) at high speeds on electric power alone when lithium batteries, which have higher energy-storage capacities per size and weight (see Table 7.10), become more reliable and affordable.

Most automobile manufacturers have developed prototypes of diesel-electric hybrid vehicles that achieve substantial improvements in fuel efficiency. A number of them have attained 70 mpg (30 km L^{-1}) or better. The main impediment to such a propulsion system is cost: A diesel engine plus electric drive system adds over $5000 to the purchase price of the same vehicle equipped with a gasoline engine. At current gasoline prices, an owner would not recover the additional capital costs over the life of the vehicle.

Biofuels

As discussed earlier in this chapter, most modern engines have the flexibility to combust mixtures containing biofuels. Biofuels are those produced from living materials and include bioethanol, biomethane, and biodiesel. The energy in biofuels originally derives from photosynthesis:

$$6\ CO_2 + 6\ H_2O + \text{light} \rightarrow 6\ CH_2O + 6\ O_2 \quad (7.4)$$

where CH_2O denotes a carbohydrate.

Bioethanol is produced through the fermentation of a carbohydrate such as glucose ($C_6H_{12}O_6$) into ethanol (C_2H_5OH):

$$C_6H_{12}O_6 \rightarrow 2\ C_2H_5OH + 2\ CO_2 \\ + \text{chemical energy} \quad (7.5)$$

Combustion of ethanol yields:

$$2\ C_2H_5OH + 6\ O_2 \rightarrow 4\ CO_2 + 6\ H_2O \\ + \text{chemical energy} \quad (7.6)$$

Summing these three reactions together yields the simple energy conversion:

light (solar electromagnetic energy) →
$$\text{chemical energy} \qquad (7.7)$$

In theory, the chemical energy in bioethanol propels a vehicle without any net production of greenhouse gases.

Biomethane or biogas follows a similar series of reactions (see Equations 2.3 through 2.7). Carbohydrates produced via photosynthesis are converted into methane (CH_4), the main component of natural gas, and combustion of this "natural" natural gas (as opposed to natural gas extracted as a fossil fuel) releases carbon dioxide (CO_2) in amounts equal to those recently removed from the atmosphere during photosynthesis. Once again, a vehicle propelled by biomethane, at least in theory, does not contribute additional greenhouse gases to the atmosphere.

Biodiesel involves a more complex series of chemical reactions. In brief, several biochemical pathways synthesize **fats** (a class of biochemicals composed of triglycerides) from photosynthetic carbohydrates (Taiz and Zeiger 2006). Plants store some of these fats as "vegetable oils" in seeds to serve as an energy source for germinating embryos. Extracting these oils and treating them with methanol (CH_2OH) or sometimes ethanol (C_2H_5OH) in the presence of a strong base (e.g., NaOH) produces methyl or ethyl esters of fatty acids (Equation 7.8). The "R_1," "R_2," and "R_3" in the triglyceride and esters of fatty acid molecules in Equation 7.8 represent a wide variety of different long carbon chains that may occupy these positions in the molecules. Methyl esters and ethyl esters of fatty acids have combustion properties very similar to those of petrodiesel. Combustion of biodiesel, however, returns to the atmosphere CO_2 that was recently withdrawn during photosynthesis, whereas combustion of petrodiesel returns to the atmosphere CO_2 that was withdrawn millennia ago.

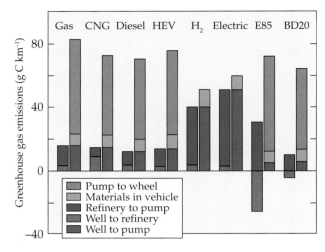

FIGURE 7.41 Greenhouse gas emissions (grams carbon equivalents per km) for light-duty vehicles with various types of powertrains Estimates are based on a computer model (Burnham et al. 2007) and vehicle weights for a Honda Civic. (See Table 7.4 for relative fuel efficiencies of the same model with different powertrains). CNG, compressed natural gas; HEV, hybrid electric vehicle; H_2, hydrogen fuel cell; E85, 85% ethanol from cornstarch; and BD20, biodiesel 20%. "Well to pump" is the sum of emissions from the "Well to refinery" and from the "Refinery to pump." (After Brinkman et al. 2005.)

cough syrup), personal care products (e.g., soap), and foods (e.g., a solvent for food colorings and vanilla flavoring). Expanded production of biodiesel has created a glut of glycerol on the world market, and its value has plunged. New commercial applications of glycerol, such as its conversion into propylene glycol or ethanol, are still under development.

Biofuels, in practice, are not even close to being neutral with respect to generation of greenhouse gases (**Figure 7.41**) because their production and use deviate significantly from 100% efficiency. Inefficiencies arise during production of **biomass**, biologically generated material; transport of biomass to a processing facility; conversion of biomass to biofuels; distribution of biofuels; and combustion of biofuels in vehicles. Chapters 5 and 6 addressed certain aspects of biomass production, and the previous sections of this chapter compared distribution of alternative fuels as well as combustion efficiencies of vehicles using various propulsion systems. The following sections present the influence of biofuels on greenhouse gas emissions during production and processing.

BIOMASS PRODUCTION Estimates of greenhouse gas emissions during the production of biomass for biofuels often differ (Farrell et al. 2006b). The greatest expense in growing plants for human use is the purchase of arable land (**Figure 7.42**); therefore, the profit-

The reactions to produce biodiesel also generate glycerol, a substance used in pharmaceuticals (e.g.,

$$\begin{array}{ccc}
& \text{O} & \\
& \| & \\
H_2C-O-C-R_1 & & H_2C-OH \\
& \text{O} & | \\
& \| & \\
HC-O-C-R_2 \ + 3\ CH_2OH \xrightarrow{\text{base}} HC-OH \ + \\
& \text{O} & | \\
& \| & \\
H_2C-O-C-R_3 & & H_2C-OH \\
\text{Triglyceride} & \text{Methanol} & \text{Glycerol}
\end{array}$$

$$\begin{array}{cc}
\text{O} & \text{O} \\
\| & \| \\
H_2C-O-C-R_1 \ + \ H_2C-O-C-R_2 \ + \\
& \text{O} \\
& \| \\
& H_2C-O-C-R_3
\end{array}$$

$$\text{Methyl esters of fatty acids} \qquad (7.8)$$

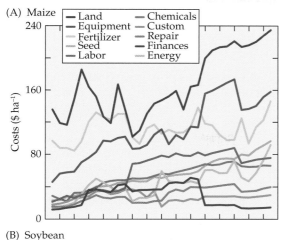

(A) Maize

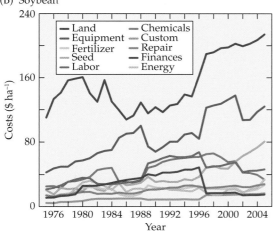

(B) Soybean

FIGURE 7.42 Costs ($U.S. per hectare) for maize (corn) and soybean production in the United States "Custom" includes custom operations such as seeding and harvesting and "Finances" includes taxes and insurance. New methods for assessing costs were adopted in 1997 and account partially for the abrupt shifts at that time. (After Economic Research Service 2007.)

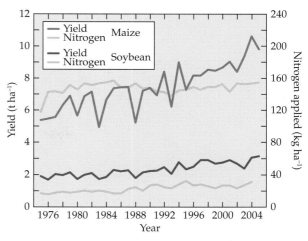

FIGURE 7.43 Yields and nitrogen fertilizer Historical trends in the United States for grain yields (metric tons per hectare) of and applications of nitrogen (N) fertilizer (kg per hectare) on maize (corn) and soybeans. (After Economic Research Service 2007.)

ability of biomass production depends on high yields, the economic productivity per unit of land area. This promotes the use of high-input agriculture that is likely to employ sophisticated farm machinery; adequate amounts of fertilizers, pesticides, and water; and new cultivars bred for high yields under these amounts of fertilizer, pesticides, and water. Grain yields in the United States have improved more than 50% during the past 30 years (**Figure 7.43**), an increase attributable in equal measure to altered agricultural practices and new genotypes (Duvick 2005).

Application of nitrogen fertilizer dramatically enhances crop yields (see Figure 5.20), but manufacture, distribution, and application of nitrogen fertilizer contributes the majority of greenhouse gases emitted during biomass production (**Figure 7.44**). A large por-

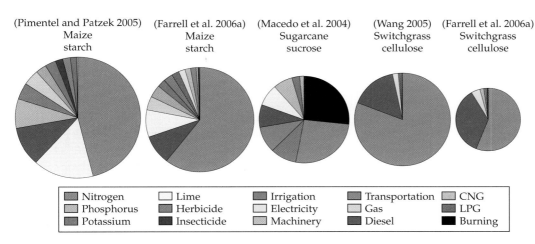

FIGURE 7.44 Five estimates of greenhouse gas emissions during cultivation of crops for bioethanol Pie sizes represent total emissions: From left to right are 0.31, 0.21, 0.10, 0.11, and 0.05 gram carbon equivalents per liter of ethanol. (After Land and Plant Nutrition Management Service 2004.)

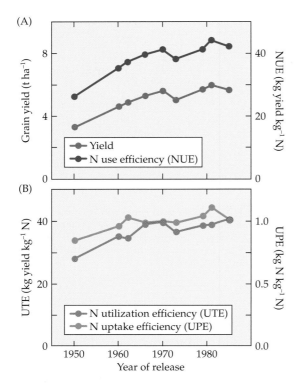

(A)

(B)

FIGURE 7.45 Nutrient efficiency Grain yield (metric tons per hectare), nutrient use efficiency (NUE, kg yield per kg nitrogen applied), utilization efficiency (UTE, kg yield per kg nitrogen absorbed), and uptake efficiency (UPE, kg nitrogen absorbed per kg nitrogen applied) for bread wheat cultivars grown in northwestern Mexico as a function of the year that the cultivar was released by CIMMYT (the International Maize and Wheat Improvement Center). The releases in 1950 and 1960 were traditional tall cultivars, whereas those released afterward were semi-dwarfs. Shown are the averages of three seasons (1987, 1988, and 1989) and three nitrogen treatments (75 kg ha^{-1}, 150 kg ha^{-1}, and 300 kg ha^{-1}). The uptake efficiency can exceed a value of 1 because the crop also absorbs residual nitrogen in the soil, nitrogen from atmospheric deposition, and nitrogen in the irrigation water. (After Ortiz-Monasterio-R et al. 1997.)

tion of these greenhouse gases derive from the power demands of industrial nitrogen fixation, the extremely energy-intensive process that makes ammoniacal nitrogen (NH_3) out of thin air (see Chapter 5). Moreover, nitrogen fertilization stimulates soil microorganisms that transform soil nitrogen into nitrous oxide (N_2O), a greenhouse gas (see Figure 4.23).

Crop nitrogen use efficiency has almost doubled over the past 50 years (**Figure 7.45A**). This improvement stems from both higher crop yields per amount of nitrogen absorbed from the soil and greater crop nitrogen absorption per amount of fertilizer applied (**Figure 7.45B**). As a result, yield gains in the United States over the past 30 years have required little additional nitrogen fertilization (see Figure 7.43) and have generated few additional greenhouse gases (see Figure 7.1). Here lies the promise of biofuels: the expanded cultivation of crops that require lower inputs of nitrogen, other chemicals, and water and less use of mechanical equipment to produce high, sustainable yields of biomass rich in organic carbon compounds that can be readily converted into fuels.

One promising source of biofuel is sugarcane (*Saccharum officinarum*). Brazil devotes about 10% of its farmland to sugarcane, and ethanol from sugarcane supplies about 14% of its fuel needs for vehicles. In prime growing regions of the tropics, fields of mature plants accumulate up to 20 kg of sucrose (table sugar) per square meter. Sugarcane in Brazil grows throughout the year and produces about 130 GJ (10^9 joules) of ethanol per hectare over 12 months, as compared to maize (corn) in the United States, which produces about 87 GJ of ethanol per hectare over 6 months (Patzek and Pimentel 2005). Sugarcane serves as a host for nitrogen-fixing bacteria within its tissues and therefore maintains high productivity under relatively low nitrogen fertilization (see Figure 7.43).

Farm management of sugarcane, like that of other biomass crops, must balance the practice of leaving stubble in the field to return vital nutrients and organic carbon to the soil against the practice of removing this material to serve as an energy source during processing. In Brazil, a large portion of the aboveground biomass is removed from the field and burned at the processing plants to provide both heat for distilling the alcohol and electricity to run the machinery. Still unknown is the long-term sustainability of such a practice (Patzek and Pimentel 2005).

A major controversy concerning cultivation of maize and sugarcane for ethanol (or soybeans for biodiesel, discussed in a subsequent section) is the diversion of food to fuels. In the past 2 years, the boom in maize-derived ethanol in the United States has contributed to a 50% increase in maize prices (**Figure 7.46**). These price increases resonate throughout the food chain, causing higher prices at the market because maize is a major feedstuff for meat and dairy cattle, and corn syrup is a major ingredient in many foods and beverages for humans.

Another potential biomass crop, one which grows on lands unsuitable for cultivation of human food crops, is switchgrass (*Panicum virgatum*). Switchgrass produces prodigious quantities of biomass that is 33% cellulose, 26% hemicellulose, and 18% lignin. One can convert

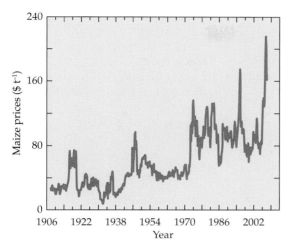

FIGURE 7.46 Average farm price ($U.S. per metric ton) received for maize (corn) in the United States during the past century. (After Economic Research Service 2009.)

TABLE 7.11 Energy density of dry fuels[a]

Fuel	Energy density (MJ kg⁻¹)
Biodiesel	37.5–40.6
Cellulose	16.1–17.9
CNG (0°C, 1 atm)	48.7–51.6
CNG (2000 psi)	50.2
Coal	24.7–28.8
Crude petroleum	44.9
Ethanol	26.8–29.9
Gasoline	44.0–47.5
Hydrogen (0°C, 1 atm)	120–142
Lignin	21.5–29.4
Maize kernels	15.0–18.8
Maize oil	38.8–39.5
Maize stover[b]	15.8–17.7
Methanol	21.1–22.7
Petrodiesel	42.4–46.1
Starch	16.7
Sucrose	16.7
Sugarcane shoots	2.1–2.4
Switchgrass shoots	8.1–8.5
Wood pellets	18.6–25.1

Source: Ragland et al.1991; Ma and Hanna 1999; Macedo et al. 2004; Patzek 2004; Demirbas 2005; Farrell et al. 2006a; Davis and Diegel 2007; Leaver 2007; Tiax LLC 2007.
[a]10^6 joules per kilogram
[b]Maize stover is mature stalks minus ears.

these constituents into ethanol with varying degrees of difficulty. Some scientists reckon that nitrogen, other chemicals, and water requirements for sustainable production of switchgrass are less than one-third those of maize. On this supposition—and a few others—hinges the future of the biofuels industry. Other scientists reckon that the input requirements for both crops are somewhat similar (see Figure 7.44).

TRANSPORT OF BIOMASS TO PROCESSING Plant biomass has a low energy density (MJ kg⁻¹) in comparison with fossil fuels (**Table 7.11**). In specific, maize kernels for starch, sugarcane shoots for sucrose, and switchgrass shoots for cellulose contain less than one-third of the energy in the equivalent weight of gasoline. It is economically unviable to transport materials such as biomass over long distances. Consequently, biofuel processing facilities are located near the sites of biomass production. For example, bioethanol processing facilities in the United States are concentrated in the midwestern corn belt (**Figure 7.47**). Transport of biomass to the processing facility, even over short distances, contributes between 2% and 4% to the total greenhouse gas emissions resulting from bioethanol production (Farrell et al. 2006a).

BIOMASS PROCESSING Some procedures for converting biomass to biofuels are as old as human civilization, whereas others still await technological breakthroughs. The procedures for direct burning of biomass, fermentation of sugars to ethanol, and cellulose purification

are well established. Still under development are efficient procedures to convert cellulose into ethanol.

Biomass burning Starting a fire using wet wood is very difficult. This is because vaporizing any water in the biomass dissipates 2.27 MJ of energy per kg of H_2O vaporized (see Chapter 3). For this reason, drying biomass before combustion enhances its energy density as well as diminishes its transport weight. Unfortunately, drying biomass without the use of ovens or kilns, which themselves consume energy, proceeds slowly in many regions of the world.

Wood pellets (typically, cylinders 6 mm in diameter and 10 mm in length) are a form of biofuel that travels well. They require only simple processing, have moderate energy densities (see Table 7.11), easily conform to packaging, and are safe to store. Processing entails harvesting the aboveground portions of trees, stripping off leafs and twigs, drying the raw wood from about

FIGURE 7.47 Locations of processing facilities for bioethanol in the United States, both those existing and those under construction as of April 3, 2007. (After Renewable Fuels Association 2007.)

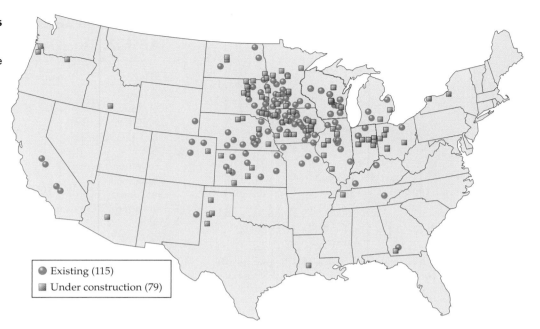

- Existing (115)
- Under construction (79)

55% water to less than 10%, and forming the pellets in a die and roller press (Leaver 2007). These processes consume, on average, 6.4 MJ per kg, or around 30% of the energy in the final product (Patzek and Pimentel 2005).

The burning of leaves, twigs, underbrush, and other so-called "waste" biomass can offset some of the energy requirements of pellet processing. Removal of such material for energy purposes, however, does not compensate for the loss of nutrients and carbon from the soils from which the biomass sprang. Long-term sustainability of biomass plantations is doubtful without periodic restoration of the soils (Patzek and Pimentel 2005).

Conversion of sugar to ethanol Fermentation of sugars to ethanol has been a common human festivity since hominids first ingested overripe berries. Today, a vintner creating the next great cabernet, a moonshiner making whiskey, and an engineer processing biomass employ brewer's yeast, the fungus *Saccharomyces cerevisiaea*, to convert glucose or fructose in plant juices into ethanol (see Equation 7.5). This reaction proceeds until ethanol concentrations reach between 14.0% and 15.5% by volume, the upper limit of tolerance for the yeast. After fermentation, the vintner, moonshiner, or engineer distills the liquid to remove water and produce nearly pure ethanol.

Distillation separates chemicals by their differences in boiling temperatures. First, the fermentation solution (mash) is heated in a still (**Figure 7.48**) to above 78.4°C (the boiling point of ethanol) but below 100°C (the boil-

ing point of water). Ethanol vapors rise to the top of the still, where they cool and condense into a liquid again. Finally, the distillate collects into a holding tank. Stills designed for biofuels produce a distillate that is up to 95.6% ethanol and as low as 4.4% water. By contrast, stills designed for alcoholic beverages do not seek such purity; their distillates contain lower concentrations of ethanol and higher concentrations of natural flavors.

FIGURE 7.48 A still for biofuels that heats the fluid in the copper tank on the right and condenses ethanol fumes in the narrow copper column The two white fiberglass tubs at the left are for fermentation, the one in the middle is for ethanol cooling, and the one on the right is for ethanol storage. This system can generate 32 gallons per day. (Photo courtesy of Revenoor Company 2007.)

The last step of bioethanol production is dehydration. The most common method is to pass the distillate through a molecular sieve that absorbs the remaining water. A molecular sieve is a material, such as an aluminosilicate clay, with pores large enough to trap water molecules but too small to restrict ethanol molecules from passing through. Another drying method is via the addition of an organic solvent, such as benzene or cyclohexane, that forms a chemical association with the water and ethanol. This association has a boiling point of 64.9°C and can be removed by a second distillation. This method leaves traces of benzene or cyclohexane, rendering the product undrinkable (in other terms, denatures the ethanol) and not subject to liquor taxes. The end product of these dehydration methods is 99% or even purer ethanol, ready for combustion.

The processes of sugar extraction from the plant biomass, fermentation of the sugars, and distillation and dehydration of the ethanol have an overall efficiency of about 87% in terms of the energy content of the ethanol produced over that of the sugar initially present in the sugarcane (Macedo et al. 2004).

Conversion of starch to ethanol Starch is a polymer of glucose sugar, meaning that it is composed of many glucose molecules linked in a chain (**Figure 7.49**). Plants store starch instead of sugars because starch is less susceptible to pathogens. Maize kernels, for example, contain about 71% starch and only 1% to 3% simple sugars.

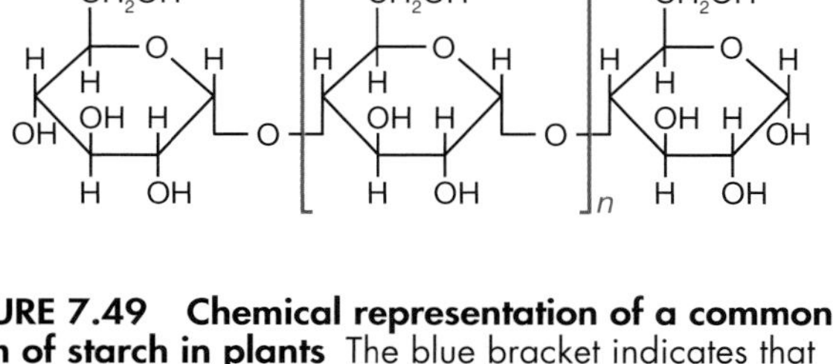

Conversion of maize kernels into ethanol begins with dry- or wet-milling. Dry-milling grinds the entire kernel. Wet-milling first removes valuable components of the kernel before ethanol conversion. The process involves steeping kernels in a dilute sulfuric acid bath at about 52°C (125°F) for a day or longer and then coarsely grinding and separating them into the outer skin (bran or hull), which contains most of the fiber; the

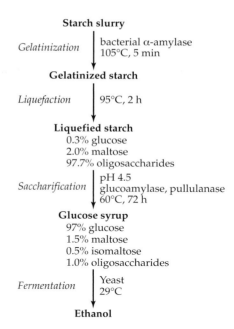

FIGURE 7.50 Stages of starch hydrolysis and the conditions under which they are usually conducted (After Chaplin 2004.)

germ, which is rich in oil; and the endosperm, which is mostly starch and protein. Centrifugation of the endosperm separates starch from protein.

Hydrolysis of starch is the next step in maize biofuels production. During hydrolysis, enzymes break down starch into its glucose subunits. This process proceeds in several stages, beginning with the conversion of raw starch into gelatinized starch that has shorter chains of glucose. This, in turn, is converted into liquefied starch, which has even shorter chains, and then into glucose syrup (**Figure 7.50**). Once hydrolysis is complete, yeast are added to ferment the glucose syrup to 10% ethanol or higher. Fermentation followed by distillation and dehydration results in a product that is purer than 99% ethanol.

Processing maize kernels to ethanol is less efficient than processing sugarcane because of the extra steps involved in converting starch to sugar. Starch extraction from the plant biomass, hydrolysis from starch to sugar, fermentation of the sugar, and distillation and dehydration of the ethanol has an efficiency of about 66% in terms of the energy in the final product (ethanol) versus the energy in the original kernel (Patzek 2004).

Conversion of cellulose to ethanol Structural integrity of plants depends on cellulose, a polymer that, like starch,

FIGURE 7.49 Chemical representation of a common form of starch in plants The blue bracket indicates that the enclosed $(1\rightarrow4)\alpha$-D-glucose subunit is repeated several hundred (n) times. Notice that each of the adjacent subunits faces the same direction.

FIGURE 7.51 Section of a cellulose molecule depicted as (A) a molecular sketch, (B) a space-filling model, and (C) a ball-and-stick model. Dotted circle encloses a (1→4)β-D-glucose subunit. Notice that adjacent subunits alternate in their orientation.

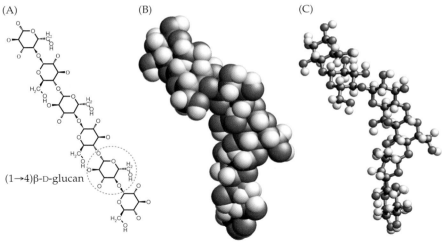

(1→4)β-D-glucan

is composed of glucose subunits (**Figure 7.51**). The glucose subunits in cellulose alternate direction, however, and this configuration promotes the formation of strong chemical bonds (hydrogen bonds) between adjacent cellulose chains. Plant cell walls contain dozens of cellulose chains, packed together into microfibrils; these microfibrils intertwine with hemicelluloses, pectins, proteins, and lignins (**Figure 7.52**). This complex construction gives cell walls the strength to withstand pressures that exceed 1.5 MPa (218 psi) and to resist pathogen invasion and herbivore attack. Unfortunately, this strength exposes the weakness of many mitigation strategies involving biofuels: Cell walls are highly recalcitrant to most physical and chemical treatments that would convert them into ethanol.

The first step in processing ethanol from cellulose is the liberation of cellulose and hemicellulose from cell walls (**Figure 7.53**). The paper manufacturing industry has centuries of experience in cellulose processing. The paper industry grinds or steam explodes plant material and then subjects it to solutions of hot, concentrated base (sodium hydroxide) or acid (sulfuric acid). Unfortunately, this process is energy intensive, expending about 30 MJ per kg of paper, more than the energy value of the paper (Patzek 2006), and generates a range of products that not only inhibit subsequent steps in converting cellulose into ethanol, but are potentially harmful to the environment. Paper manufacturing is only the twelfth largest industry in the United States according to product value (Bureau of Economic Analysis 2008), but is the fifth largest polluter after metal mining, electric utilities, chemicals, and metal smelting (U.S. Environmental Protection Agency 2007b). Therefore, processors of biofuels seek improved methods of releasing cellulose and hemicellulose from cell walls.

The next step is the decomposition of cellulose and hemicellulose into sugars via hydrolysis. Given the sta-

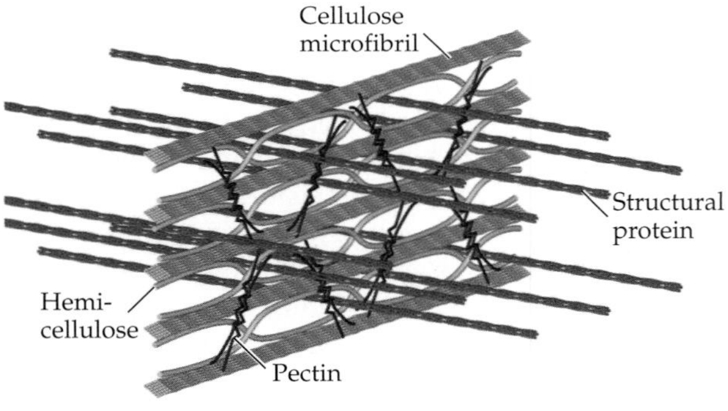

FIGURE 7.52 Simplified schematic of a primary plant cell wall Microfibrils of 36 cellulose molecules are cross-linked by hemicellulose, pectins, and structural proteins. As cells mature, they may develop walls impregnated with lignin, polymers of aromatic alcohols that provide additional strength. (After Epstein and Bloom 2005.)

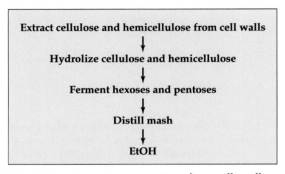

Extract cellulose and hemicellulose from cell walls
↓
Hydrolize cellulose and hemicellulose
↓
Ferment hexoses and pentoses
↓
Distill mash
↓
EtOH

FIGURE 7.53 Steps in processing plant cell walls to ethanol

bility and complexity of these compounds, no single enzymatic or chemical treatment achieves adequate yields or rates. The long periods required for hydrolysis increase the possibility of contamination by microorganisms that consume sugars without producing anything of human value. Extensive research efforts are underway to develop a fast, efficient method to hydrolyze cellulose and hemicellulose into sugars.

The step that follows hydrolysis is fermentation, in which microbes convert sugars into ethanol. Yeast are highly effective at converting hexoses (six-carbon sugars such as glucose and fructose), which are produced from sugarcane or during hydrolysis of maize starch. They are less effective at converting pentoses (five-carbon sugars such as xylose and arabinose), which are generated during the hydrolysis of hemicellulose. For instance, about 30% of the sugars produced during hydrolysis of maize stover (mature stalks minus the ears) is xylose, which yeast fermentation cannot efficiently convert to ethanol. Genetic engineering of bacteria (e.g., *Zymomonas mobilis* and *Escherichia coli*) as well as fungi (e.g., *Saccharomyces cerevisiae* and *Pichia stipitis*) for efficient ethanol production from pentoses is an active area of research (Jeffries and Jin 2004).

In summary, despite the promise of cellulosic ethanol, most of the steps in processing are still experimental, and the costs are not yet competitive with other energy sources (**Figure 7.54**).

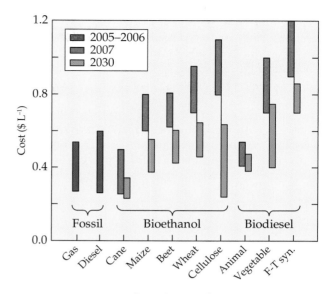

FIGURE 7.54 Cost of production for various fuels Depicted are the range of costs for gasoline and diesel from 2005–2006, for bioethanol and biodiesel from different starting materials in 2007, and for optimistic predictions regarding how technological advances might lower costs of biofuel production in 2030. Cane, sugarcane; Maize, maize starch; Beet, sugarbeet; Cellulose, ligno-cellulose; Animal, animal fat; Vegetable, vegetable oil; and F-T syn., Fischer-Tropsch synthesis. (After International Energy Agency 2007a.)

Conversion of vegetable oils to biodiesel   Crops for biodiesel contain between 4% and 62% extractable vegetable oil by weight (**Table 7.12**). Oil is extracted physically, in a mechanical press, or chemically, by the use of an

organic solvent such as hexane. Chemical extraction produces higher yields and is faster and less expensive, but it is less suitable to rural areas because it requires sophisticated equipment. Production of biodiesel involves the reaction of extracted vegetable oil with methanol or ethanol in the presence of a strong base (Equation 7.8) and then separation of the resulting biodiesel from the unreacted alcohol and base and from the glycerol that forms. The efficiency of extraction and conversion is on the order of 89% (Balat 2007), but processors must deal with the 38% to 97% of plant biomass that remains after conversion (96% to 38% of the plant biomass that is not oil plus 11% of the oil that is not converted). The economic viability of biodiesel depends on the value of this leftover biomass.

An alternative process for producing biodiesel is Fischer-Tropsch synthesis, in which gasification of carbon from natural gas, biomass, or coal with steam under heat and pressure produces carbon monoxide and hydrogen:

$$C + H_2O \rightarrow CO + H_2 \qquad (7.9)$$

This reaction requires the input of energy that usually is generated by combusting a portion of the natural gas, biomass, or coal. The products of this reaction, car-

TABLE 7.12   Extractable oil content and oil yields from various crops

Crop	Oil % (w/w)	Yield (L ha^{-1})
Castor bean	50	1413
Coconut	62	2689
Cotton seed	13	325
Maize kernel	4	172
Mustard seed	35	572
Olive fruit	30	1212
Palm oil fruit	20	5950
Palm oil kernel	36	
Peanut	48	1059
Rapeseed	37	1190
Sesame	50	696
Soybean	14	446
Sunflower seed	32	952

Source: Addison and Hiraga 2007.

bon monoxide and hydrogen, then react in the presence of a catalyst (typically, iron or cobalt) to produce hydrocarbons and water:

$$n\,CO + (2n + 1)\,H_2 \rightarrow C_nH_{(2n+2)} + n\,H_2O \quad (7.10)$$

where "n" represent a positive whole number. This reaction is allowed to proceed until it synthesizes long-chain hydrocarbons that are liquid at room temperatures. The overall result of this process is to convert natural gas, coal, or biomass into liquid fuels of high energy density that are suitable for use in vehicle engines. During World War II, petroleum-poor but coal-rich Germany produced diesel fuel for its vehicles via Fischer-Tropsch synthesis. Today, this process is relatively uncommon because its energy efficiencies are lower and costs are higher than fuel production through other means (see Figure 7.54).

OVERALL EFFICIENCY Fundamental questions remain about the extent to which biofuels could supplant fossil fuels. Different budget analyses for biofuels reach different conclusions. Estimates of greenhouse gas emissions per unit of energy obtained from biofuels vary tenfold (**Figure 7.55**). In addition, some analyses indicate that the energy in biofuels is nearly double that of the fossil fuels expended to grow the biomass and process it, whereas a few others find that the energy in biofuels does not even match that of the fossil fuels expended to produce it.

Several assumptions are responsible for these discrepancies. Some analyses assume that sustainable production of biomass will require substantial fertilization and irrigation, together with prolonged fallow periods, whereas others assume that cellulosic crops will require minimal additions of fertilizer and water. Analyses also differ in their assumptions on the proportion of energy for processing that might derive from the biomass itself. Some argue that as much biomass (such as sugarcane stubble) as possible should remain in the field to maintain soil fertility; others argue that leaving biomass in the field just lets a valuable resource rot. Some predict that energy-rich lignin (see Table 7.11), when separated from biomass, can provide much of the energy for processing; others point out that current methods for removing lignin from biomass expend large amounts of fossil fuel energy. As the biofuels industry matures, many of these discrepancies will be resolved.

How realistic is the goal of replacing 30% of oil use in the United States with alternative fuels by 2030? Given the following information,

1. The maize grain yield in the United States = 10,000 kg ha^{-1} (see Figure 7.43),

2. The ethanol yield from maize grain = 0.37 L kg^{-1} (Patzek 2006),

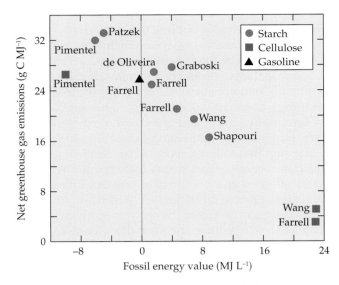

FIGURE 7.55 Net greenhouse gas emissions (gram carbon equivalents per 10^6 joules energy in the ethanol) during the growth and processing of bioethanol from maize starch or switchgrass cellulose as a function of fossil energy value. Fossil energy value (10^6 joules energy per liter, MJ L^{-1}) is the energy content of delivered fuel minus fossil energy expended during growth and processing and, for gasoline, minus fossil energy contained in the initial oil. A fossil energy value of "0" for bioethanol means that energy in the final ethanol product merely equals fossil energy expended during biomass growth and processing. For gasoline, energy content of the oil plus that expended in oil extraction and refining is 0.24 MJ L^{-1} greater than the delivered fuel. For reference, energy content of ethanol is 23.6 MJ L^{-1}. Names by data point are the first authors of the studies cited. (After Farrell et al. 2006b.)

3. Annual U.S. crude oil consumption = 1.19 × 10^{12} L y^{-1} (Energy Information Administration 2006),

4. Crude oil energy content = 38.5 MJ L^{-1} (see Table 7.7), and

5. The fossil energy value for maize ethanol = 4.6 MJ L^{-1} (see Figure 7.55),

one can calculate

$$\frac{(30\%)(1.19 \times 10^{12}\,L\,y^{-1})(38.5\,MJ\,L^{-1})}{(10{,}000\,kg\,ha^{-1})(0.37\,L\,kg^{-1})(4.6\,MJ\,L^{-1})} = 808 \times 10^6\,ha\,y^{-1}$$

To replace 30% of the U.S. oil consumption with bioethanol from maize grain would require cultivation of 808 million hectares of maize per year just for this purpose. For perspective, the total land area of the United States, including Alaska and Hawaii, is 916 million hectares, and 38 million hectares are currently in maize production. Even optimistic estimates for cellulosic ethanol—yields of 13,450 kg ha^{-1}, ethanol yields of 0.382 L kg^{-1}, and fossil energy values of 22.9 MJ L^{-1}—indicate that biomass to meet the "30-by-30" goal would cover 117 million hectares, an area larger than the states of Texas and California combined. We can only hope that there is a calculation error somewhere and that the "30-by-30" goal at least has some possibility of success.

Summary

Mitigation strategies for global climate change are plans to diminish the severity of the problem. Fossil fuels are responsible for 69% of the greenhouse gases emitted from human activities. Banning the use of fossil fuels, however, is not a viable option because nearly all of the energy for transportation depends on them.

Transportation is one of the fastest-expanding sources of greenhouse gas emissions, accounting for 33% of the emissions from the United States. The contribution of personal vehicles (cars and light-duty trucks) is increasing worldwide as public transportation loses ground to single-occupancy vehicles. The fuel efficiencies of personal vehicles depend mainly on driving conditions and engine size and type. Driving conditions are deteriorating worldwide as many more vehicles take to roads that are only slowly increasing in number. Engine sizes are growing as consumers show a preference for larger vehicles with faster acceleration. Fortunately, technological advances have helped keep vehicle fuel efficiencies constant or slightly improving.

Rising oil costs have spurred developments in the production of alternatives fuels such as natural gas, hydrogen, electric, and biofuels. For example, ethanol from sugarcane now provides about 14% of the vehicle fuel in Brazil. Difficulties abound, however, in the production, distribution, and economic viability of all of these alternatives, and none will be ready for wide-scale deployment in the next few decades.

Review Questions

1. Select all the reasons why gasoline and diesel fuel provide nearly all of the energy for transportation.
   (a) They have a high energy density.
   (b) Oil is evenly distributed worldwide.
   (c) They do not explode under most conditions.
   (d) They are liquids that are easy to transfer and carry.
   (e) They are used in internal combustion engines that generate more torque at slow speeds than any alternative.

2. Which of the following statements about high oil prices is *false*?
   (a) They encourage exploration for and extraction of new oil supplies.
   (b) They increase the price of natural gas.
   (c) They occur every other year.
   (d) They transfer money to volatile regions of the world.
   (e) They promote energy conservation.

3. Fuel efficiency of vehicles increases significantly (select one)
   (a) at speeds above 80 kmph (50 mph).
   (b) in stop-and-go traffic.
   (c) with vehicle size.
   (d) with new vehicles of the same weight as old vehicles.
   (e) with engines powered by gasoline versus compressed natural gas.

4. Select all of the following statements that apply to taxes.
   (a) Fuel taxes are relatively easy to collect.
   (b) Fuel taxes amount to more than 10% of the revenues collected by governments.
   (c) Countries vary widely in their taxation of fuels.
   (d) Most governments collect taxes when a vehicle is purchased.
   (e) Most governments collect an annual road use tax per vehicle.

5. Which of the following statements relating to diesel-powered vehicles is *true*?
   (a) Diesel fuel contains less energy per volume than gasoline.
   (b) Diesel-powered engines run at lower temperatures and pressures than gasoline engines.
   (c) Diesel-powered vehicles are inherently quieter than gasoline vehicles.
   (d) Diesel-powered vehicles emit more greenhouse gases per distance than equivalent gasoline-powered vehicles.
   (e) Diesel- and gasoline-powered vehicles have about equal market share in Europe.

6. Which of the following statements apply to vehicles powered by mixed fuels?
   (a) Modern gasoline engines can burn mixtures of up to 10% ethanol without modification.
   (b) With some modifications, gasoline engines can burn mixtures of up to 85% ethanol.
   (c) Some engines can burn either gasoline or diesel fuel.
   (d) With minor modifications, diesel engines can burn mixtures of petrodiesel and biodiesel.
   (e) Most diesel engines can directly use vegetable oils.

7. CAFE regulations in the United States (select one)
   (a) have frequently changed to accommodate new technologies.
   (b) treat cars and small trucks similarly.
   (c) treat all fuels similarly.
   (d) treat vehicles of all sizes similarly.
   (e) enforce vehicle safety standards.

8. Greenhouse gas emissions per person per distance traveled (select all that apply)
   (a) is lowest for tandem bicycles.
   (b) is highest for helicopters.
   (c) is lower for public transport than for personal vehicles.
   (d) varies with passenger occupancy.
   (e) is decreasing as more people carpool to work.

9. Natural gas is considered a clean fuel because (select one)
   (a) it produces fewer particulates and NO_x than gasoline or diesel.
   (b) it has a higher energy content than gasoline or diesel.
   (c) it does not require drilling to extract it from belowground.
   (d) it contains mostly ethanol.
   (e) in liquid form, it cleans fuel tanks.

10. Hydrogen gas (select all that apply)
    (a) reacts with oxygen to produce water.
    (b) has an energy content per volume higher than natural gas.
    (c) is extracted from deep wells.
    (d) sank the Titanic.
    (e) condenses to a liquid at temperatures a few degrees below the freezing of water.

11. Hydrogen-powered vehicles (select one)
    (a) are likely to become commonplace during this decade.
    (b) usually have a longer driving range than current gasoline-powered vehicles.
    (c) are quickly refueled at many existing refueling stations.
    (d) will be relatively inexpensive.
    (e) have received considerable attention.

12. Which of the following statements about electric-powered vehicles is false?
    (a) Their popularity is limited by battery capacity, reliability, and cost.
    (b) They usually have a longer driving range than current gasoline-powered vehicles.
    (c) They usually recharge at night during the slack hours for electric power grids and thus may not strain current generating capacities.
    (d) They require less maintenance than gasoline-powered vehicles.
    (e) They emit smaller quantities of greenhouse gases per passenger mile than gasoline-powered vehicles do.

13. Growing of biofuel crops (select all that apply)
    (a) has become more economically viable with rising oil prices.
    (b) involves some of the same agricultural practices used in the production of food.

    (c) emits no greenhouse gases.
    (d) is likely to supplant 30% of crop production in the United States by 2030.
    (e) is partly responsible for increased food prices.

14. Processing of biofuels (select all that apply)
    (a) requires the chemical breakdown of complex organic compounds.
    (b) may involve some of the same industrial practices used in the production of paper.
    (c) produces no net greenhouse gas emissions.
    (d) is conducted close to the sites of biomass production because of transportation costs.
    (e) is approaching 100% efficiency in which the energy content of the product is nearly equal to the energy content of the initial biomass.

Suggested Readings

BP 2008. *BP Statistical Review of World Energy, June 2008*. BP, London, *http://www.bp.com/liveassets/bp_internet/globalbp/globalbp_uk_english/reports_and_publications/statistical_energy_review_2008/STAGING/local_assets/downloads/pdf/statistical_review_of_world_energy_full_review_2008.pdf*.

BP, formerly British Petroleum, compiles this book of statistics about fossil fuels every year.

International Energy Agency. 2007a. *Biofuel Production, Organization for Economic Cooperation and Development*. Paris, *http://www.iea.org/Textbase/techno/essentials2.pdf*.

This is a brief fact sheet on biofuels from the International Energy Agency.

International Energy Agency. 2007b. *Hydrogen Production and Distribution*. Organization for Economic Cooperation and Development, Paris, *http://www.iea.org/Textbase/techno/essentials5.pdf*.

This is a brief fact sheet on the hydrogen economy from the International Energy Agency.

Sperling, D. and D. Gordon. 2009. *Two Billion Cars: Driving toward Sustainability*, Oxford University Press, New York.

A comprehensive treatise of the role of private vehicles in society and possible strategies for mitigating their deleterious effects.

8

MITIGATION STRATEGIES: ELECTRIC POWER GENERATION

Personal vehicles, contrary to popular belief, account for only 10% of the total greenhouse gas emissions worldwide and 19% in the United States (see Chapter 7). Larger sources of greenhouse gases include electric power generation, land-use changes, food and fiber production, manufacturing, and heating of buildings. Strategies for mitigating emissions from these sources are distinct from those for mitigating emissions from personal vehicles for several reasons.

Personal vehicles undergo more rapid technological change and faster replacement of older models by newer models than other sources of greenhouse gases (**Figure 8.1**). In specific, fuel efficiencies of personal vehicles per unit weight have improved by over 2% per year over the past 30 years (see Figure 7.25), whereas greenhouse gas emissions per unit of electricity generated from power plants have decreased by only 0.4% per year

FIGURE 8.1 Range of expected lifetimes for selected energy-related equipment and facilities "Office equipment" includes computers, printers, fax machines, and copiers; "Consumer electronics" include TVs, videos, and stereos; "Appliances" include stoves, refrigerators, washers, and dryers; "Commercial vehicles" denotes trucks, buses, tractor trailers, and tractors; "Transp. and urban development" denotes transportation and urban development. (After Philibert 2007.)

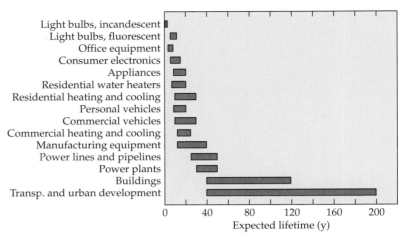

(Energy Information Administration 2009b), and greenhouse gas emissions per product produced from manufacturing plants have decreased by only 1.3% per year (Schipper 2006). Vehicles on U.S. roads have a median age of 8 years (**Figure 8.2**) as a result of both the short lifetime of current vehicles and the addition of new vehicles to an expanding population (see Figure 7.16). By comparison, electric power plants in the United States have a median age of 30 years (Energy Information Administration 2006a). Slow replacement of power plants means that, even with wide-scale adoption of new technologies, abatement of greenhouse gases may be relatively slow.

The economics of personal vehicles, in comparison with the economics of other emission sources, are less sensitive to energy prices. Brand, appearance, and seldom-used features (e.g., top speeds of 120 mph)

of personal vehicles often take precedence over fuel efficiency (see Chapter 7). In contrast, fuel efficiency is generally a primary factor in the purchase and operation of commercial vehicles. Similarly, energy prices directly influence the price of nitrogen-based fertilizers that a farmer uses. Likewise, consumers focus more on the cost of electricity reflected in their electric bill than on what type of power plant is connected to their wall socket. In the parlance of economics, commodities such as delivery trucks, fertilizers, and electric power are more responsive than personal vehicles are to market forces.

Because of these differences, Chapter 7 treated greenhouse gas emissions from transportation separately from other sources. This chapter focuses on electric power generation, the human activity that emits the largest share of greenhouse gases, producing more than double the amount of any other source (see Figure 7.10). As such, electric power generation has the greatest potential for good or evil, to diminish greenhouse gas emissions or to increase them. The emphasis here, as in the previous chapter, is on engineering and environmental aspects of mitigation. Chapter 9 addresses the topic of emissions from the agricultural, forestry, industrial, commercial, and residential sectors.

Industrial Power

Many heavy industries have relocated from developed countries such as the United States and Japan to developing countries such as China and India to benefit from the latter's lower labor costs and less stringent environmental regulations. Heavy industries are highly energy intensive (see Chapter 9). Consequently, greenhouse gas emissions from electric power generation are burgeoning in China (see Figure 4.15C) and India (**Figure 8.3**).

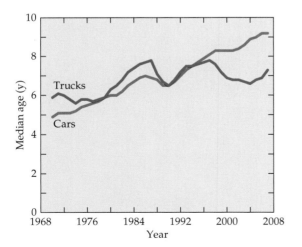

FIGURE 8.2 Median age of cars and trucks on U.S. roads (After Davis et al. 2008.)

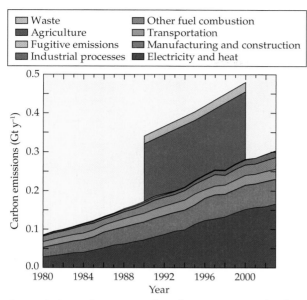

FIGURE 8.3   India's annual carbon emissions (10^9 metric tons of carbon per year) according to economic sector Data for emissions from waste and agriculture are available for only 1990–2000 and include the carbon equivalents from CH_4 and N_2O emissions. "Fugitive emissions" are emissions other than those from stacks or vents, such as from equipment leaks and evaporative processes. (After World Resources Institute 2007.)

China, in 2006 alone, added power-generating capacity that equaled the sum total of France's electricity grid (Campbell 2007). Similarly, India doubled its power-generating capacity between 1990 and 2000 (Kroeze et al. 2004). Coal-fired power plants provide over 80% of the new capacity in China and India (Kroeze et al. 2004). Coal-fired plants also generate half of the electricity in the United States (**Figure 8.4**). Such plants, in comparison with other types, have higher greenhouse gas emissions per unit of power generated but are cheaper to construct and operate (**Figure 8.5**). Coal is an attractive energy source for China, India, and the United States because these countries have much larger reserves of coal than of gas, oil, or uranium (**Table 8.1**). China plans to build 500 additional coal-fired electric power plants during the next decade, nearly one new plant per week (Campbell 2007). India plans to double its capacity over the next decade (Kroeze et al. 2004).

Power plants of all types have a life span of 30 to 50 years (see Figure 8.1). Installing sophisticated equipment to abate greenhouse gas emissions during the initial construction is far cheaper than retrofitting older plants with such equipment. Therefore, decisions made about new power plants in China or India today will influence greenhouse gas emissions for generations to come. International agreements including the

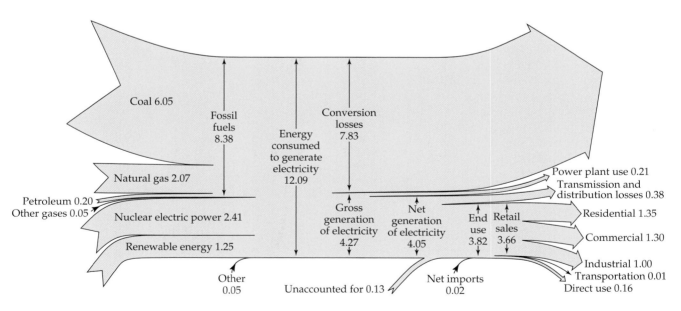

FIGURE 8.4   Sources and consumers of U.S. electricity (10^{15} Watt-hours per year) "Other gases" include those from blast furnaces, propane, and biofuels. "Other" inputs include batteries, chemicals, hydrogen, and tires. "Direct use" includes self-generated electricity. Notice that efficiency of electricity generation, if one accounts for conversion losses and power plant uses, is 4.05/12.09 = 33.5%. (After Energy Information Administration 2007b.)

FIGURE 8.5 Costs ($U.S. per 10⁶ Watt-hours) versus greenhouse gas emissions (kilograms of carbon equivalents per 10⁶ Watt-hours) for various types of electric power plants. Shown are dot (mean) and error bars (range). "CCS" denotes coal with carbon capture and storage. Costs include construction (loans factored at a 5% rate) and operation, but neither waste disposal nor insurance. In particular, nuclear power plants have low operating costs and low greenhouse gas emissions (shown) but expensive waste disposal and insurance costs (not shown). (After Nuclear Energy Agency and International Energy Agency 2005; Sims et al. 2007; Weisser 2007.)

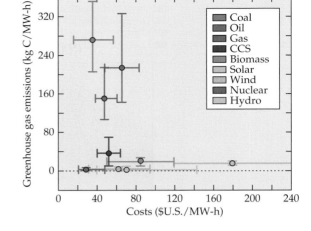

Kyoto Protocol provide economic incentives for incorporating cleaner technologies into new power plants (see Chapter 11). The following sections describe some of these technologies.

Coal-Fired Power Plants

Electric power plants commonly use a pulverized coal steam-generating system (**Box 8.1**). In such a system, metal spheres or cylinders crush pieces of coal into a fine powder. Hot air blows the coal powder into a furnace, where it combusts at high temperatures and converts water to steam in tubing that lines the walls of the furnace. The steam passes through a series of high-, mid-, and low-pressure turbines connected to a common shaft that spins a generator. This type of power plant requires high-grade coal—coal that has a low ash content (ash is the nonflammable component of coal).

Another type of coal-fueled power plant utilizes "fluidized bed combustion," a method that involves burning coal in a layer (bed) of heated particles suspended in flowing air (World Coal Institute 2004). At sufficiently high air velocity, the bed acts as a fluid that thoroughly mixes coal particles with air. This fluidizing action completely combusts the coal at relatively low temperatures. The advantage of fluidized bed combustion is that it can use almost any grade of coal. In the United States, these power plants often burn coal rejected by power plants that use pulverized coal systems.

One more type of power generation from coal is an *i*ntegrated *g*asification *c*ombined *c*ycle (IGCC) system. This method does not combust coal directly, but first converts it

TABLE 8.1 Proven reserves[a] of fuels in selected countries

Country	Coal[b]	Natural gas	Oil	Uranium
United States	27.1	3.3	2.5	10.4
Russia	17.3	26.3	6.6	4.0
China	12.6	1.3	1.3	1.1
India	10.2	0.6	0.5	1.3
Australia	8.6	1.4	0.3	22.7
South Africa	5.4	<0.1	<0.1	7.8
Ukraine	3.8	0.6	<0.1	2.0
Kazakhstan	3.4	1.7	3.3	15.6
Poland	1.5	1.6	<0.1	<0.1
Brazil	0.7	0.2	1.0	4.8
Germany	0.7	0.1	<0.1	<0.1
Columbia	0.7	0.1	0.1	<0.1
Canada	0.7	0.9	1.4	10.5
Czech Republic	0.6	<0.1	<0.1	<0.1
Mexico	0.1	0.2	1.1	<0.1
Venezuela	0.1	2.4	6.6	<0.1
Iran	<0.1	15.5	11.4	<0.1
Qatar	<0.1	14.0	1.3	<0.1
Saudi Arabia	<0.1	3.9	21.9	<0.1
United Arab Emirates	<0.1	3.3	8.1	<0.1
Nigeria	<0.1	2.9	3.0	<0.1
Iraq	<0.1	1.7	9.5	<0.1
Kuwait	<0.1	1.0	8.4	<0.1
Libya	<0.1	0.7	3.4	<0.1
Japan	<0.1	<0.1	<0.1	<0.1
Namibia	<0.1	<0.1	<0.1	5.5
Niger	<0.1	<0.1	<0.1	5.5

Source: NEA and IAEA 2006; BP 2007.
[a]Percent of world total
[b]Includes all types of coal

BOX 8.1 Pulverized coal-fired electric power plant

Railroad cars or barges transport coal from the mines to the plant, and a conveyor belt carries the coal into the plant and fills a hopper. A pulverizing fuel mill with large metal spheres or cylinders grinds the coal to a fine, powdery dust. A fan drives preheated air into the coal dust, and another fan forces the mixture of hot air and coal dust into the boiler furnace at high pressure, where it rapidly ignites. Ash from combustion falls from the boiler and collects at the bottom.

Water of high purity flows vertically up tubes lining the walls of the boiler, where it turns into steam and passes into the boiler drum. In the boiler drum, the steam separates from liquid water, rises through the roof of the boiler drum, and runs through tubing of the superheater. In the superheater, the temperature and pressure of the steam increase rapidly to 540°C and 20 MPa (197 atmospheres). A steam governor valve controls the flow of this superheated steam into a high-pressure turbine. As the steam passes through the turbine, it loses temperature and pressure. The exhaust from the turbine returns to the boiler reheater. The reheated steam passes through a mid-pressure turbine

and then a low-pressure turbine. The exhaust from these turbines enters a condenser, where it comes into thermal contact with a cool heat exchanger, condenses back to water, and creates vacuum-like conditions inside the condenser chest, thus enhancing temperature and pressure differentials. A series of pumps force this condensed water through a deaerator, heater, and economizer (denoted as "Econ." in figure) before returning it to the boiler drum.

The steam turbines have a common shaft and spin an electrical generator, which produces high voltages, typically 20 kilovolts (kV) to 25 kV. A step-up transformer boosts these to very high voltages, typically, 250 kV to 500 kV, for transmission. Cooling water, which circulated through the heat exchanger of the condenser, sprays inside a cooling tower, creating a highly visible plume of water vapor. The water collects at the bottom of the cooling tower for recirculation. A draft fan draws the exhaust gas from the boiler through an electrostatic precipitator that removes particulates and finally vents through the flue gas stack.

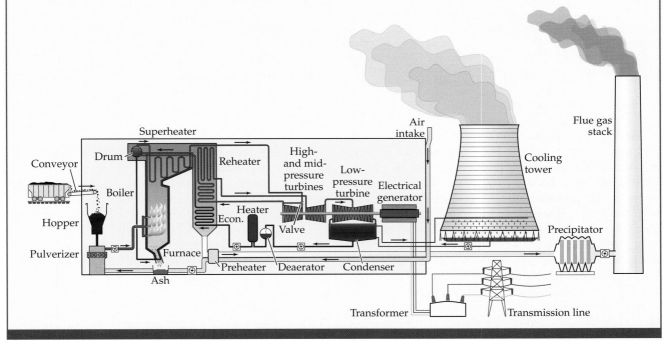

into a gas called "syngas" composed of carbon monoxide (CO) and hydrogen (H_2) (**Figure 8.6**). This process involves some of the same reactions as hydrogen production from fossil fuels (see Table 7.8) and Fischer-Tropsch synthesis (see Equation 7.5).

$$C + \tfrac{1}{2} O_2 \rightarrow CO + energy \qquad (8.1)$$

$$C + H_2O + energy \rightarrow CO + H_2 \qquad (8.2)$$

$$CO + H_2O + energy \rightarrow CO_2 + H_2 \qquad (8.3)$$

$$H_2 + \tfrac{1}{2} O_2 \rightarrow H_2O + energy \qquad (8.4)$$

FIGURE 8.6 **Integrated gasification combined cycle (IGCC) electric power plant** From left to right, railroad cars or barges transport coal from the mines to the plant, and a conveyor belt carry it into the plant and fills a hopper. Water mixes with the coal before crushing in a rod mill. Drying and chilling air to about −185°C separates O_2 and N_2. A pump carries the coal slurry to the gasifier where, in the presence of O_2 and under heat and pressure, organic carbon converts to $CO + H_2$. Impurities in the coal fall to the bottom of the gasifier or are washed out in a water scrubber. Further gas treatment removes the sulfur, leaving pure syngas (H_2 and CO). In a shift converter, CO and steam change to CO_2 and additional H_2. Injecting N_2 into the syngas minimizes NO_x emissions when syngas combusts. Combustion spins a turbine connected to a generator. Hot exhaust gases pass through a heat recovery unit and convert water to steam. This steam drives a turbine and its generator. Other components are similar to those described in Box 8.1.

Essentially, a gasification reactor introduces a limited amount of oxygen to burn some of the organic carbon material in the coal and release carbon monoxide (CO) and energy. This drives a second reaction that converts additional organic carbon material in the coal to CO and H_2 and a third reaction, in what is known as a water-gas shift converter, changes CO and steam to CO_2 and H_2. The H_2 thus produced combusts directly in a gas turbine that spins an electrical generator. Heat from this combustion also generates steam that drives a steam turbine and another electrical generator.

IGCC power plants have several advantages over other types of coal-fired power plants. They release less sulfur dioxide (SO_2) and nitrogen oxides (NO_x) pollution in the flue gas (combustion gas that exhausts to the atmosphere) because filters remove these compounds from the syngas before it combusts. They achieve high fuel efficiencies, from 38% to 56% (World Coal Institute 2004; MIT Study Group 2007). They can capture CO_2 more easily and more cheaply than other types of coal-fired plants because they generate it at high concentra-

tions in the water-gas shift converter before combustion rather than emit it, diluted, in a large volume of flue gas after combustion.

The main problem with IGCC plants is their high capital costs. Construction of an IGCC plant costs from 15% to 20% more than construction of a conventional pulverized coal-fired plant (Moore 2005; MIT Study Group 2007). Currently, IGCC plants generate only 0.1% of the world's electricity (IPCC 2005), and only two IGCC plants are operating in the United States.

Laws of thermodynamics dictate that conversion of chemical energy into mechanical energy becomes more efficient at higher temperatures and pressures. For example, diesel engines run at higher temperatures and pressures than gasoline engines and are thereby more efficient (see Chapter 7). Likewise, efficiencies of coal-fired electric power plants increase and greenhouse gas emissions per unit energy generated decrease at higher temperatures and pressures (**Figure 8.7**).

Water at temperatures above 374°C and at pressures above 22 MPa (218 atmospheres = 3191 psi) transforms

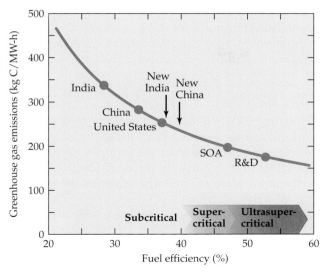

FIGURE 8.7   Fuel efficiency (100 × the ratio of electrical energy output to fuel energy input) and greenhouse gas emissions (kg carbon equivalents per 10⁶ Watt-hours) of coal-fired power plants Dots designate overall averages for the countries shown; "SOA," new state-of-the-art plants; "R&D," represents pilot plants built for research and development. Arrows denote typical values for new plants in India and China. Presented at the bottom of the graph are typical efficiencies for subcritical, supercritical, and ultrasupercritical power plants. A typical pressure and temperature for a subcritical plant is 15 MPa and 540°C, that for a supercritical plant is 25 MPa and 600°C, and that for a ultrasupercritical plant is 35 MPa and 760°C (After Moore 2005; World Energy Council 2007.)

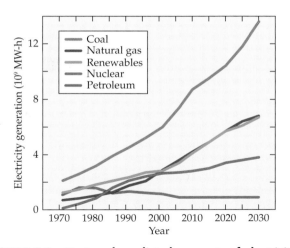

FIGURE 8.8   Past and predicted amounts of electricity in the world (10¹⁵ Watt-hours) generated from coal, natural gas, nuclear, renewables (hydro, wind, solar, geothermal, etc.), and petroleum. (After Energy Information Administration 2009a; International Energy Agency 2008.)

from a subcritical fluid that has distinct liquid and gaseous phases to a supercritical fluid that has properties of both a liquid (i.e., many soluble materials dissolve in it) and a gas (i.e., it diffuses rapidly through many materials). Moreover, supercritical water has a density that changes with temperature and pressure in a continuous manner, and so power plants designed to operate under supercritical conditions scale to higher temperatures and pressures without having to accommodate two phases of water (liquid and gas) in various locations.

The downside of supercritical operation is that higher temperatures and pressures demand stronger materials, tighter tolerances, and more complex control systems. In addition, supercritical water is highly reactive, and only components made from corrosive-resistant chrome or nickel alloys can withstand long exposures to it. These requirements add about 7% to the cost of constructing and maintaining these power plants (MIT Study Group 2007). Nevertheless, super-

critical plants are more efficient, and their lower fuel costs eventually compensate for higher construction and maintenance costs. Currently, more than 400 supercritical power plants are in service worldwide (World Coal Institute 2004), and a few ultra-supercritical plants, which operate at even higher temperatures and pressures, are beginning operations in Europe and Japan (MIT Study Group 2007).

Natural Gas

Natural gas is the second-largest primary energy source for generating electricity in the United States (**Figure 8.8**) and the world (see Figure 7.1B). Natural gas power plants have the advantage of modest capital costs, high fuel efficiency, operating flexibility, rapid deployment, and low greenhouse gas and pollutant emissions (see Chapter 7). New natural gas power plants are coming online at an annual rate of 2% worldwide, faster than any other type, and their global share of electricity production should increase from 20.1% in 2006 to 22.3% in 2030 (International Energy Agency 2008).

A power plant fueled by natural gas has many of the same components as a coal-fired IGCC plant, except that the natural gas plant needs neither a gasifier to convert coal to syngas for the combustion turbines nor gas filters to mitigate SO_2 or NO_x emissions. This is because refineries, during the processing of natural gas, remove most of the sulfur and nitrogen impurities that generate SO_2 and NO_x. A basic natural gas power plant contains only a combustor, a combustion turbine, and a generator. A higher-efficiency, "combined-cycle" plant adds a heat recovery unit, a steam turbine, a condenser, and another generator (similar to the system depicted in Figure 8.6).

Because of their simplicity, natural gas power plants are relatively inexpensive and rapid to construct, efficient to operate, and easy to maintain. Construction of a power plant that generates 1000 megawatts (MW) may take as little as 2 years from start to finish. Most of the plants in the United States use combined-cycle combustion and steam turbines and attain fuel efficiencies that average 39% (Bluestein 2006), and plants currently under construction should approach 60% efficiency (Sims et al. 2007). The economics of such plants, however, fluctuate with the price of natural gas, which is usually several times more expensive than coal (see Figure 7.5).

Carbon Capture and Storage

Hydrocarbon fuels—be they coal, natural gas, petroleum, or biomass—release carbon dioxide (CO_2) upon combustion. Carbon capture and storage (CCS) refers to the practice of collecting CO_2, concentrating it, transporting it, and storing it in a manner that prevents it from mixing freely with the atmosphere (IPCC 2005). Storage options include diversion into industrial processes, conversion into mineral carbonates, pumping to the depths of the oceans, and burial in deep geological formations (**Figure 8.9**). All phases of CCS (capture, concentration, transportation, and storage) face significant technological and economic challenges.

CO_2 capture

Capturing CO_2 is already a standard practice for several industries. For example, natural gas treatment plants and ammonia fertilizer factories separate CO_2 from their product streams, but often just vent the CO_2 to the atmosphere. Petroleum refineries, after extracting as many high value products from crude oil as possible, may burn the remaining hydrocarbons and produce compressed CO_2 for commercial sale to the food and chemical industries. Yet despite extensive industrial experience with this process, or perhaps because of it, not a single large (i.e., one generating more than 500 MW) power plant currently captures CO_2, although several facilities with this capability are under construction (IPCC 2005; MIT Study Group 2007; National Energy Technology Laboratory 2007b). The reasons for this are largely economic: Carbon capture almost doubles the construction cost of a power plant and adds 50% to the cost of electricity generation (see Figure 8.5). Only with adequate financial incentives does carbon capture becomes a viable option.

Power plants could capture CO_2 either during fuel processing, before combustion, or from the flue gas, after combustion. Capture before combustion is straightforward for integrated gasification combined

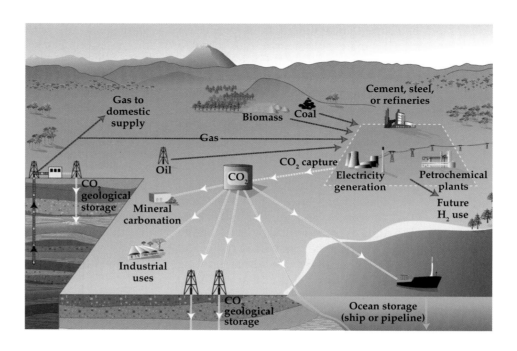

FIGURE 8.9 Carbon capture and storage (CCS) Various sources and sinks for carbon dioxide (CO_2).

TABLE 8.2 CO_2 concentrations and pressures at various locations in different types of power plants

Source	Concentration (%)	Pressure (MPa)
Atmosphere	0.038	0.1
IGCC precombustion	8–20	2–7
IGCC postcombustion	12–14	0.1
Natural gas turbine	3–4	0.1
Natural gas boiler	7–10	0.1
Coal boiler	12–14	0.1
Oxy-fuel combustion	95–98	0.1

Source: IPCC 2005.

TABLE 8.3 Boiling temperature (°C) of components in flue gases from power plants

Gas	Common name	Boiling temperature (°C)
Hg	Mercury	356.7
H_2O	Water	100.0
SO_2	Sulfur dioxide	–10.1
$CO_2{}^a$	Carbon dioxide, 5 atm	–56.6
CO_2	Carbon dioxide, 1 atm	–78.5
N_2O	Nitrous oxide	–88.5
O_3	Ozone	–111.9
CH4	Methane	–161.6
O_2	Oxygen	–182.9
Ar	Argon	–185.8
CO	Carbon monoxide	–191.5
N_2	Nitrogen	–195.8

$^a CO_2$ differs from the other gases in that it changes directly from a solid to a gas at one atmosphere pressure (1 atm = 0.1 MPa).

cycle (IGCC) electric power plants, in which the water–gas shift converter (see Figure 8.6) produces a gas stream containing CO_2 at relatively high concentrations and pressures (Table 8.2). Capture after combustion is easiest for oxy-fuel plants, which combust fuel in pure oxygen:

$$C + O_2 \rightarrow CO_2 + energy \qquad (8.5)$$

and thereby produce flue gas that is nearly pure CO_2. Oxy-fuel plants are discussed later in this chapter.

Physical separation, chemical separation, or combinations of the two may remove CO_2 from power plant gas streams. Physical methods include cryogenic separations and membrane separations. Chemical methods include liquid solvents and solid **sorbents**.

PHYSICAL METHODS Cryogenic separation of CO_2 from a gas stream is one method that is common in commercial operations. It is based on differences in the temperatures at which gases condense to a liquid or, in the case of CO_2 at normal atmospheric pressure, freeze directly to a solid "dry ice" (Table 8.3). In specific, cooling a gas stream to below –88.5°C removes mercury, water, sulfur dioxide, carbon dioxide, and nitrous oxide, yet allows nitrogen and oxygen to pass. Cooling large volumes of flue gas (after combustion) to these temperatures requires a prohibitive amount of energy, but cooling the precombustion stream to purify small volumes of oxygen and nitrogen for IGCC or oxy-fuel combustion adds only about 15% to the fuel use of a power plant (Figure 8.10).

Membrane separation, another physical method, depends on differences in partial pressures (concentration multiplied by pressure) to transfer CO_2 or H_2

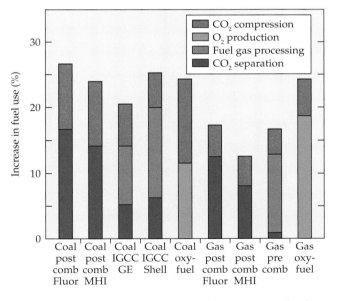

FIGURE 8.10 Increased fuel use of various methods for CO_2 capture at electric power plants Note that "comb" denotes combustion, and "IGCC," integrated gasification combined cycle. "Fluor," "MHI," "GE," and "Shell" are companies offering different liquid solvent methods: Fluor uses Econamine, MHI uses KS-1, and GE and Shell use Selexol (see Table 8.4). (After IPCC 2005; National Energy Technology Laboratory 2007b.)

across a gas-permeable membrane. For instance, if the precombustion stream (20% CO_2 concentration, 7 MPa pressure) from a gasifier in an IGCC power plant flows on one side of a membrane and normal air (0.038% CO_2 concentration, 0.1 MPa pressure) on the other, a partial pressure difference of up to 1.4 MPa [(20% × 7 MPa) – (0.038% × 0.1 MPa)] drives CO_2 out of the precombustion stream into the normal air stream. Some membranes are selectively permeable: Microporous ceramics or metals allow H_2 formed in a water-shift converter (see Equation 8.3) to pass while retaining most of the CO_2.

Membranes have problems with durability, reliability, and costs (IPCC 2005). Films of plastic polymers can achieve CO_2 recoveries of up to 57% but can only withstand temperatures up to 150°C, whereas ceramic and metal membranes can function at temperatures above 400°C but recover only about 7% of the CO_2 from the gas stream (International Energy Agency 2004a). Development of membrane materials is an active area of research.

CHEMICAL METHODS Several industrial processes, such as production of natural gas and ammonia, use chemical substances (solvents) to remove CO_2 from gas streams. These substances, once they become fully loaded with CO_2, are isolated from the main gas stream and exposed to higher temperatures or lower pressures to release the CO_2 and regenerate their capacity for CO_2 removal from gas streams. For example, the solvent monoethanolamine (MEA) reacts with CO_2 to form an acid-base pair:

$$2 \ (HO–CH_2–CH_2–NH_2) + CO_2 \rightarrow HO–CH_2–CH_2–NH–COO^- + HO–CH_2–CH_2–NH_3^+ \quad (8.6)$$

Heating the solution to above 100°C reverses this reaction, expelling the CO_2 and restoring the MEA.

A large number of such "liquid scrubbers" for CO_2 are in commercial use (Table 8.4). Those that release CO_2 via high temperatures are advantageous at low partial pressures of CO_2 such as from flue gases because their energy use is relatively insensitive to operation at low concentrations. At high CO_2 partial pressures, however, such as after IGCC gasification, scrubbers that absorb and release CO_2 via changes in pressure are more energy efficient. Some capture systems based on liquid CO_2 scrubbers have achieved above 90% recov-

---

TABLE 8.4 Liquid solvents for CO_2 scrubbing

Type[a]	Trade name	Chemical components	Reaction conditions[b]
Pressure	Rectisol	Methanol	–10°C to –70°C, > 2 MPa
	Purisol (NMP)	n-2-methyl-2-pyrolidone	–20°C to +40°C, > 2 MPa
	Selexol (DMPEG)	Dimethyl ethers of polyethyleneglycol	–40°C, 2–3 MPa
	Fluor solvent	Propylene carbonate	< 20°C, 3.1–6.9 MPa
Temperature	MEA	2,5n monoethanolamine and inhibitors	40°C, > 0.1 MPa
	Amine guard	5n monoethanolamine and inhibitors	40°C, > 0.1 MPa
	Econamine	6n diglycolamine	80°C to 120°C, 6.3 MPa
	DIPA	2-4n diisopropanolamine 2n methyldiethanolamine	35°C to 40°C, > 0.1 MPa
	MDEA	2n methyldiethanolamine	
	Flexsorb, KS–1, KS–2, KS–3	Sterically hindered amine	
	Benfield and versions	Potassium carbonate and catalysts; Lurgi and Catacarb processes with arsenic trioxide	70°C to 120°C, 2–7 MPa
Temperature and pressure	Sulfinol-D, Sulfinol-M	mix of DIPA or MDEA, water, and tertahydrothiopene (DIPAM) or diethylamine	>0.5 MPa
	Amisol	Mix of methanol and MEA, DEA, diisopropylamine (DIPAM), or diethylamine	5°C to 40°C, >1 MPa

Source: International Energy Agency 2004a.

[a] Solvents are categorized on the change in conditions (pressure, temperature, or both) that releases the CO_2 and restores the solvent.

[b] Those under which the solvent binds CO_2.

ery of CO_2 in the laboratory, yet none for electric power generation has progressed beyond the pilot plant stage (National Energy Technology Laboratory 2007a).

Solid CO_2 sorbents provide an alternative to the liquid scrubbers. One category of solid sorbent includes molecular sieves and activated carbons, materials filled with pores of the exact size that allows a CO_2 molecule to enter and adsorb on their surface but excludes larger molecules. Like the liquid scrubbers, these solid sorbents release CO_2 and regenerate when subjected to large temperature or pressure swings. Release of CO_2 via temperature swings entails long cycle times for heating up and cooling down a bed of the solid sorbents and thus has received much less attention than release via pressure swings (IPCC 2005).

Another category of solid sorbent includes metal oxides or carbonates. The metals are typically sodium (Na), potassium (K), lithium (Li), calcium (Ca), or magnesium (Mg). Using calcium oxide (CaO) and sodium carbonate (Na_2CO_3) as examples, these compounds combine with CO_2 according to the following reactions:

$$CaO + CO_2 \rightarrow CaCO_3 \qquad (8.7)$$

$$Na_2CO_3 + CO_2 + H_2O \rightarrow 2\,NaHCO_3 \qquad (8.8)$$

Exposure to high temperatures and/or pressures reverses the reactions, releasing CO_2 and regenerating the sorbent.

Solid sorbents are common in the chemical industry and may achieve 90% or higher recoveries of CO_2 from the gas streams of industrial processes. Use of solid sorbents in large electric power plants is still unrealized, however (National Energy Technology Laboratory 2007a).

OXY-FUEL POWER PLANTS   Oxy-fuel power plants combust coal, syngas, or natural gas in nearly pure O_2 rather than normal air (which is 20.94% O_2) and produce exhaust gases that are over 95% CO_2 (see Table 8.2). Under these circumstances, the challenge shifts from capturing CO_2 to extracting O_2 from air in order to provide pure O_2 for the combustion process. Current methods for O_2 extraction parallel those for CO_2 capture: separation across polymer membranes, cryogenic distillation, or adsorption on pressure-swing solid sorbents. The energy requirements of O_2 extraction and CO_2 capture are also similar (see Figure 8.10).

Concentrating CO_2

Once a power plant captures the CO_2, the next step is purification and compression. Purification may involve removal of water, sulfur compounds, nitrogen compounds, and trace metals, depending on the fuel source and the processing of the gas stream that is required pre- and post-combustion. Compressing a gas stream to greater than 0.5 MPa and chilling it to $-57°C$ selectively freezes CO_2 (see Table 8.3).

CO_2 transport

Proximity to energy consumers and fuel sources largely determines the placement of new electric power plants. Only some of these facilities are located in areas with ready access to geological formations or deep waters that might provide long-term CO_2 storage (**Figure 8.11**). Widespread adoption of CO_2 capture and storage therefore hinges on the ability to transport CO_2 from where it is produced to where it will be stored.

Several chemical properties of CO_2 influence its transport.

- CO_2, if kept dry, is noncorrosive and compatible with pipes and storage vessels made of regular steel. In the presence of water, CO_2 forms carbonic acid (see Equation 2.9), which is corrosive and must be contained in resistant alloys that are several times more expensive than regular steel.

- At standard temperature and pressure (0°C, 0.1 MPa), CO_2 is a colorless, odorless gas that is heavier than air. (CO_2 has a density of 1.98 $kg\ m^{-3}$; air has a density of 1.29 $kg\ m^{-3}$.) Thus, CO_2 that leaks out during transport tends to accumulate in depressions and may endanger people if its concentration exceeds 2% (see Table 5.1).

- Liquid CO_2 has a density of 929 $kg\ m^{-3}$ at 0°C and 3.5 MPa, and so moving CO_2 as a liquid transports 500 times more CO_2 per unit volume than as a gas. Most systems, therefore, transport liquid CO_2 despite the high pressures, low temperatures, and extra energy required for maintaining CO_2 in this state. Unfortunately, the high pressures and low temperatures of the liquid make it more dangerous than the gas.

Given the risks from leaks, facilities for CO_2 transport should be located as far as possible from densely populated areas.

Two major modes of CO_2 transport are pipelines and tanker ships. Pipelines operate at ambient temperatures and high pressures (10 MPa to 80 MPa), conditions under which CO_2 becomes a supercritical fluid, and so its density varies continuously with pressure without a distinct phase change. Tanker ships are more

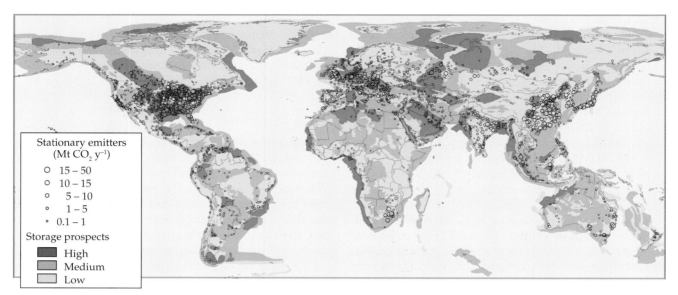

FIGURE 8.11 **Locations of large, stationary CO₂ emission sources** (dots represent megatons of carbon dioxide per year) and geological formations that are prospective CO₂ storage sites. This map illustrates the distances that might be involved in transporting CO₂ from sources to storage sites. India, for example, has many emission sources but few geological storage sites, whereas South America has many geological storage sites but few emission sources. (After IPCC 2005.)

economical than pipelines for CO_2 transport over long distances (**Figure 8.12**). Ships carry liquefied CO_2 at low temperatures (approximately -50°C) and several atmospheres pressure (approximately 0.7 MPa), using technology similar to that for transport of liquefied natural gas (LNG). Perhaps in the future, ships will transport LNG from gas fields to power plants and return with loads of liquefied CO_2 for injection into the gas wells for long-term storage.

CO₂ storage

Options for CO_2 storage include diversion into industrial processes, conversion into mineral carbonates, pumping to the ocean depths, and burial in deep geological formations. The first option, diversion into industrial processes, is unlikely to play a significant role because the current demand for CO_2 is miniscule in comparison to the potential supply. The following sections examine the other options.

CONVERSION INTO MINERAL CARBONATES   Calcium and magnesium silicon oxides, which are prevalent in certain rocks (e.g., wollastonite, olivine, or serpentine), react spontaneously with CO_2 to produce mineral carbonates such as limestone ($CaCO_3$) or magnesium carbonate ($MgCO_3$) in a process known as natural mineral carbonation.

Wollastonite:
$$CaSiO_3 + CO_2 \rightarrow CaCO_3 + SiO_2 \qquad (8.9)$$

Olivine:
$$Mg_2SiO_4 + 2\,CO_2 \rightarrow 2\,MgCO_3 + SiO_2 \qquad (8.10)$$

Serpentine:
$$Mg_3Si_2O_5(OH)_4 + 3\,CO_2 \rightarrow 3\,MgCO_3 \\ + 2\,SiO_2 + 2\,H_2O \qquad (8.11)$$

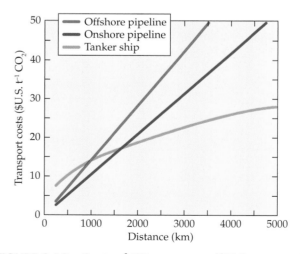

FIGURE 8.12 **Costs of CO₂ transport ($U.S. per metric ton)** via offshore pipelines, onshore pipelines, and tanker ships as a function of distance. (After IPCC 2005.)

TABLE 8.5 Estimates of CO_2 storage costs[a]

Method	Costs ($ $t^{-1}C$)
Geological formation	2–30
Ocean via pipeline	22–114
Ocean via ship	44–59
Mineral carbonation	180–370

Source: IPCC 2005.
[a] In $U.S. per metric ton carbon.

Mineral carbonates offer long-term storage because they are stable under ambient temperatures and pressures and can serve as construction materials or as fillers in abandoned mines. To fix a metric ton of CO_2 into carbonates requires between 1.6 metric tons and 3.7 metric tons of silicon oxide rock (IPCC 2005). Nonetheless, Earth's crust contains more than enough of these silicon oxides to react with all of the CO_2 released from combusting the planet's entire reserves of fossil fuels. Natural mineral carbonation, however, is prohibitively slow.

Accelerating this process to acceptable rates involves mining rocks containing silicon oxides, crushing and grinding them into a powder, and removing impurities (e.g., iron compounds, via magnetic extraction). The silicon oxide powder is then dissolved in a mild acid solution, and CO_2 is bubbled through the solution at high temperature and pressure (approximately 185°C and 15 MPa). The reaction forms particles that are smaller in size than the original silicon oxide particles. Finally, the process sieves the fine carbonate particles from the larger silicon oxide particles, disposes of the carbonates, and returns the unreacted silicon oxides to the reaction vessel.

These procedures consume 30% to 50% of the energy generated by a power plant (IPCC 2005). Together with the 10% to 30% energy penalty for CO_2 capture (see Figure 8.10), a power plant equipped CO_2 capture and storage (CCS) capability via mineral carbonation expends nearly double the energy for operations than a plant without CCS. As a result, costs for mineral carbonation CCS are several times that of other methods (**Tables 8.5 and 8.6**), explaining why mineral carbonation CCS has yet to be implemented on a commercial scale.

STORAGE IN THE DEEP BLUE SEA Earth's oceans are vast, covering over 70% of the planet's surface (see Figure 3.12) at an average depth of 3800 meters. Injecting CO_2 into the deep ocean would isolate it from the atmosphere for several centuries and halve peak atmospheric concentrations (**Figure 8.13**). Eventually, after about 2 millennia, CO_2 concentrations in the atmosphere and oceans would come back into equilibrium. Therefore, oceans can offer long-term, but not permanent, storage.

TABLE 8.6 Estimated costs of CO_2 capture and storage[a]

CCS system components	Cost range ($ $t^{-1}C$)	Remarks
Capture from a coal- or gas-fired power plant	55–275	Net costs of captured CO_2 compared to the same plant without capture
Capture from H_2 and NH_3 production or gas processing	18–202	Applies to high-purity sources requiring simple drying and compression
Capture from other industrial sources	92–422	Range reflects use of a number of different technologies and fuels
Transport	4–30	Per 250 km pipeline or shipping for mass flow rates of 1 to 11×10^{12} g C y^{-1}
Geological storage	2–30	Excludes potential revenues from EOR or ECBM[b]
Geological storage: monitoring and verification	0.4–1.1	Covers pre-injection, injection, and post-injection monitoring, and depends on the regulatory requirements
Ocean storage	18–114	Includes offshore transportation of 100–500 km, excludes monitoring and verification
Mineral carbonation	180–370	Includes additional energy use for carbonation

Source: Sims et al. 2007.
[a] In $U.S. per metric ton carbon.
[b] EOR, enhanced oil recovery; ECBM, enhanced coal-bed methane.

FIGURE 8.13 Anticipated changes in atmospheric CO₂ concentrations (parts per million) resulting from CO₂ emissions to the atmosphere or injection into oceans at 3000 m depth. This computer simulation assumes that annual CO₂ emissions reach a maximum of 25 Gt (10⁹ metric tons) of C in the year 2146 and cumulative emissions reach 5000 Gt C. The uppermost curve assumes that all of the CO₂ is released to the atmosphere. The line below assumes that all of the CO₂ is released to the atmosphere until 2050, when half of the CO₂ is injected into the oceans. The next line assumes that all of the CO₂ is released to the atmosphere until 2050, when half of the CO₂ is sequestered in permanent geological formations. The lowermost two lines assume that 100% of CO₂ is injected into the Pacific or Atlantic ocean. CO₂ mixes faster with the atmosphere in the Pacific than the Atlantic because of global current patterns (see Figure 3.44). The Atlantic also contains more natural deposits of CaCO₃ sediment than the Pacific. (After Kheshgi and Archer 2004.)

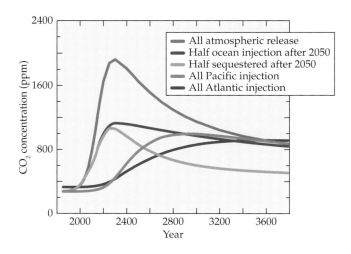

Several mitigation strategies envision that electric power plants would pump captured CO₂ down pipelines into the ocean depths (**Figure 8.14**). Alternatively, tanker ships carrying CO₂ would release it at depth or would carry the CO₂ to platforms that release it at depth. Finally, some plants would convert CO₂ into CaCO₃ (see Equation 8.7) and dissolve it in seawater for release at depth.

The fate of CO₂ in water is complex, both in its dissociation into H_2CO_3, HCO_3^-, and CO_3^{2-} (see Equations 2.9, 4.1, and 4.2) and in its transitions among gas, liquid, and solid phases (**Figure 8.15A**). In oceans, as

depth increases, hydrostatic pressure increases and temperature drops. Injected CO₂ remains a gas until about 400 m depth, at which point it becomes a supercritical fluid (i.e., its density varies continuously with pressure, without a distinct phase change).

For reasons that are not well understood, a portion of the CO₂ at these depths forms hydrates. In a hydrate, a solid, crystalline cage of water molecules surrounds each CO₂ molecule ($CO_2 \cdot 6H_2O$) (**Figure 8.15B**). Hydrates of CO₂, denser than seawater, tend to sink in the water column, whereas liquid CO₂, less dense than seawater at depths between 400 m and 2700 m, tends

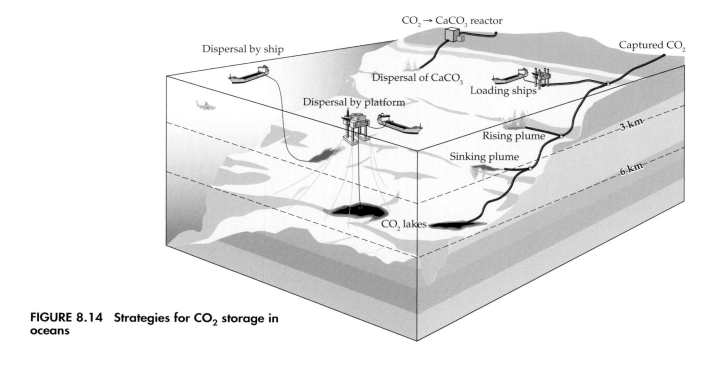

FIGURE 8.14 Strategies for CO₂ storage in oceans

(A)

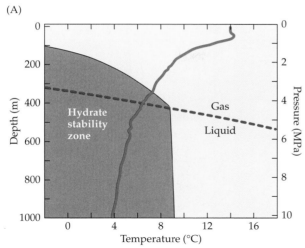

(B)

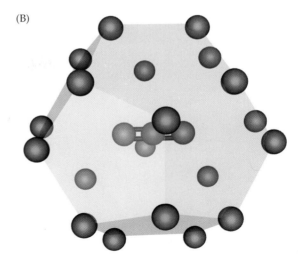

FIGURE 8.15 Behavior of CO_2 at different temperatures and depths or pressures (A) Depicted are the gas-to-liquid transition (dashed blue line), the hydrate phase conditions (gray), and the typical temperature profile for Monterey Bay, California (red line). (B) Drawing of CO_2 hydrate, where the red spheres depict the oxygen atoms of water or CO_2 and the gray sphere depicts the carbon of CO_2. (A after Brewer et al. 2004.)

to rise. Below 3000 m, liquid CO_2 becomes denser than seawater and sinks. Injection of CO_2 into deep water, therefore, may establish CO_2 lakes in depressions on the sea bottom. Such lakes would delay the dispersion of CO_2 into the surrounding environment.

Large-scale injection of captured CO_2 into the oceans would influence sea life. High CO_2 concentrations themselves can interfere with aerobic respiration (see Chapter 5). Moreover, liquid CO_2 behaves as an organic solvent that may prove toxic to certain organisms. Finally, dissolution of CO_2 into its various bicarbonate and carbonate forms would acidify the surrounding water (**Figure 8.16**), and sea life would respond to these pH changes (see Chapter 6). Nonetheless, we have such limited knowledge about life in deep oceans that we cannot anticipate all of the consequences of massive CO_2 injections (Nealson 2006).

STORAGE IN GEOLOGICAL FORMATIONS Injection of CO_2 into geological formations is currently the most practical option for long-term CO_2 storage. The oil and gas industry routinely conducts analogous procedures for several purposes, including disposal of acid gas (natural gas contaminated with large amounts of hydrogen sulfide and carbon dioxide) or liquid waste into spent wells, storage of natural gas in geological formations, and injection of CO_2 into wells to enhance oil and gas recovery (**Figure 8.17**). Large commercial facilities for geological CO_2 storage have been operating for several years at two sites, In Salah, Algeria, and Sleipner, Norway, and a number of similar facilities are under

construction (**Figure 8.18**). The cost of such storage is significantly less than that of other CO_2 storage approaches (see Tables 8.5 and 8.6).

Several types of geological formations in sedimentary basins are suitable for geological CO_2 storage. These include depleted oil or gas fields, deep coal seams, and saline formations (see Figure 8.17). The density of injected CO_2 increases with depth; it becomes a supercritical fluid at between 800 m and

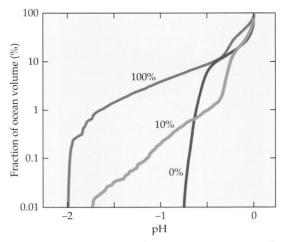

FIGURE 8.16 Fraction of ocean experiencing changes in pH as a result of injecting 100%, 10%, or 0% of 5000 Gt C (10^9 metric tons of carbon) into the oceans, versus emitting it to the atmosphere, in a computer simulation for the year 2100. Note that the y-axis is on a logarithmic scale. (After Caldeira and Wickett 2005.)

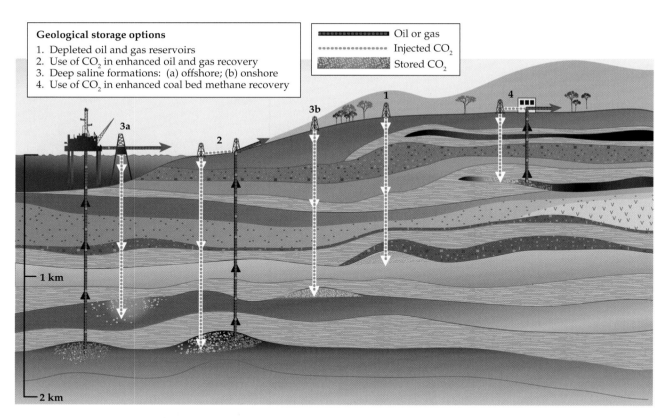

Geological storage options

1. Depleted oil and gas reservoirs
2. Use of CO_2 in enhanced oil and gas recovery
3. Deep saline formations: (a) offshore; (b) onshore
4. Use of CO_2 in enhanced coal bed methane recovery

Oil or gas
Injected CO_2
Stored CO_2

FIGURE 8.17 Strategies for CO_2 storage in geological formations

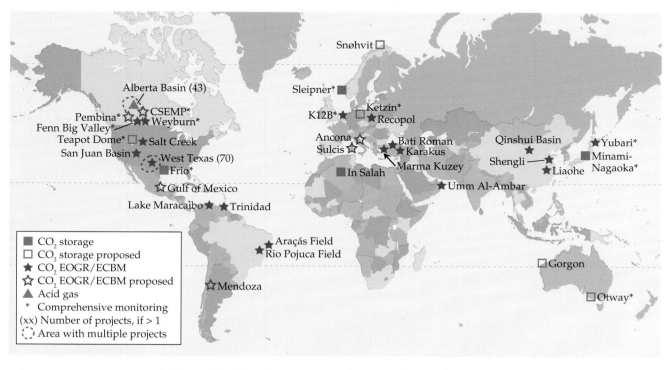

FIGURE 8.18 Current and planned facilities for CO_2 storage in geological formations "EOGR / ECBM" denotes enhance oil or gas recovery and enhanced coal-bed methane. (After IPCC 2005.)

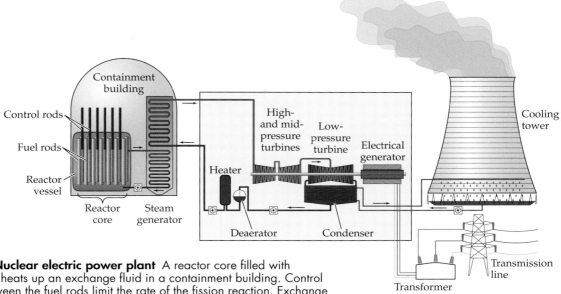

FIGURE 8.19 **Nuclear electric power plant** A reactor core filled with uranium fuel rods heats up an exchange fluid in a containment building. Control rods inserted between the fuel rods limit the rate of the fission reaction. Exchange fluid produces steam in a steam generator, and this steam leaves the containment building and drives several turbines that turn an electrical generator.

1000 m, depending on the rate at which the temperature increases with depth. (Geothermal gradients are discussed later in this chapter.) Supercritical CO_2 can remain trapped underground in porous rock formations that hold or have previously held oil, natural gas, coal, or salt water, and that are situated below a layer of impermeable, confining shale and clay rock (caprock). In the absence of deteriorating wells, open fractures, or geological faults, these rock formations should retain nearly all of the injected CO_2 for millennia.

Global capacity for geological CO_2 storage is large but highly uncertain (IPCC 2005). Depleted oil and gas reservoirs may store from 180 Gt C (10^9 metric tons of carbon to 250 Gt C), coal formations may store from 1 Gt C to 55 Gt C, and deep saline formations may store from 270 Gt C to 2700 Gt C. Potential storage sites are broadly distributed in most of the world's sedimentary basins, near many of the emission sources (see Figure 8.11). Given that the world's total cumulative emissions from fossil fuels may amount to roughly 5000 Gt C, geological CO_2 storage would be able to handle a significant proportion of those emissions.

Summary of CO_2 capture and storage methods

Many aspect of CO_2 capture and storage (CCS) are still at the preliminary stages of development, and this mitigation approach requires sophisticated equipment, consumes extra fuel, and puts additional constraints on the location of facilities. The additional costs of a CCS system are substantial (see Figure 8.5), and estimates depend on source, transport, and storage options (see Table 8.6). In most systems, CO_2 capture is the most

expensive step, but this cost should decline by 20% to 30% over the next few decades as new technologies become available.

Nuclear Power Plants

Nuclear electric power is sometimes heralded as one of the best options for mitigating greenhouse gas emissions. Two major events, however, have shaped the nuclear power industry.

- A reactor core at the Three Mile Island Nuclear Power Plant near Harrisburg, Pennsylvania melted down on March 28, 1979. Fortunately, most of the radioactivity was contained, and no one experienced severe injuries.

- On April 26, 1986, a reactor at the Chernobyl Nuclear Power Plant in the Ukraine exploded and sent a plume of highly radioactive material over large portions of Europe. Of the 237 people who suffered from acute radiation sickness, 56 died within a short period. Millions more were exposed to above-normal doses of radiation. Today, radiation levels remain so high near the plant that all residential, civil, and business activities are forbidden within a 30 km radius.

During the past 2 decades, designs and operating procedures for nuclear power plants have improved substantially (**Figure 8.19**), and so the probability of similar accidents has become exceedingly small. At present, 439 nuclear power plants operate in 30 different countries (**Figure 8.20A**) without major incident and generate about 15% of the world's electricity (**Figure 8.20B**).

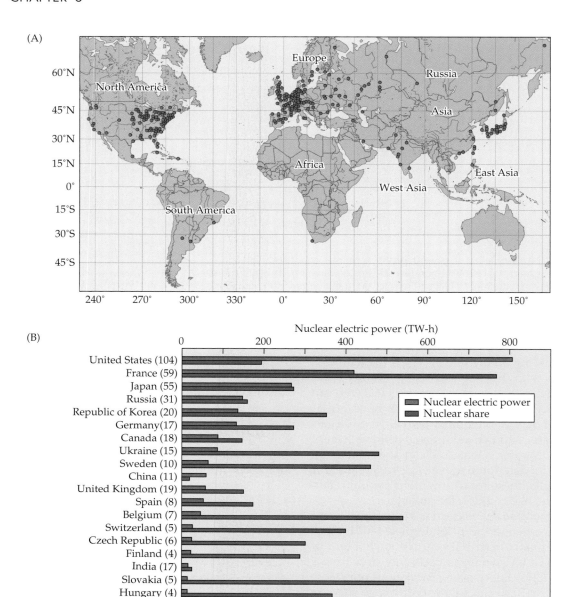

FIGURE 8.20 Location of nuclear power plants
(A) World map indicating operating nuclear power plants.
(B) Nuclear electric power in various countries. Amount
(1 TW–h =10^{12} Watt-hours) generated in 2006 (red) and
the share of the total electricity that nuclear plants contribute
(blue). In parentheses is the number of nuclear power plants
currently operating in a country. For example, the United
States has 104 nuclear power plants that generated 811
TW-h in 2006, which constituted 19% of the total electric
power. (A after International Nuclear Safety Center 2009
and B after International Atomic Energy Agency 2008.)

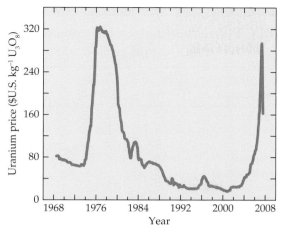

FIGURE 8.21   Price of uranium (in 2007 \$U.S. per kilogram of U₃O₈) The large number of nuclear electric power plants coming into operation contributed to the peak in 1976, whereas depletion of surplus uranium from decommissioned Soviet-era nuclear weapons contributed to the peak in 2007. (After Bureau of Labor Statistics 2007; Uranium Info 2007.)

FIGURE 8.22   A 61-kiloton nuclear bomb detonated on June 4, 1953, at the Nevada Proving Ground

Still, the accidents at Three Mile Island and Chernobyl cast a pall over the nuclear power industry. This, together with the relatively low cost of coal power, the difficulties in disposing of nuclear waste, and the volatility in uranium prices (**Figure 8.21**), have deterred construction of new facilities. Worldwide, only 33 new nuclear power plants are under construction and will begin operation during the next decade: 6 are in India, 6 in Russia, 5 in China, 3 in the Republic of Korea, 2 each in Bulgaria, Taiwan, and the Ukraine, and 1 each in Argentina, Finland, France, Iran, Japan, Pakistan, and the United States (International Atomic Energy Agency 2008). By contrast, 5 older plants in the world closed down permanently in 2007.

Nuclear power plants harness the energy released during **nuclear fission**. In this process, unstable (radioactive) isotopes of uranium (^{235}U) or plutonium (^{239}Pu) in fuel rods expel neutrons and disintegrate into ^{231}Th or ^{235}U, respectively. A tiny fraction of the mass of the original materials converts into energy, but this releases vast amounts of energy (77×10^{12} J per kg of ^{235}U) because of the high conversion value ($E = mc^2$, where E is energy in joules, m is the mass in kg converted into energy, and c^2 is the speed of light squared $= 9 \times 10^{16}$ m^2 s^{-2}). Neutrons expelled from the fuel rods may strike other uranium or plutonium atoms and induce them to expel neutrons, building a self-sustaining fission reaction.

The critical mass of fissionable material in the fuel rods is always below that necessary to start a nuclear explosion (**Figure 8.22**). Nonetheless, if the fission reaction accelerates too quickly, a reactor core may overheat and melt down, as it did at Three Mile Island and Chernobyl. To control the rate of the fission reaction and, thereby, the temperature of the core, nuclear power plants insert control rods of neutron-absorbing silver, indium, cadmium, or boron alloys between the fuel rods (see Figure 8.19). Moreover, elaborate heat-exchange systems transfer energy from the reactor core to a steam generator that converts water to high-pressure, high-temperature steam. The components outside of the containment building are common to most types of electric power plants: Steam, generated by the reactor core, drives turbines that spin electrical generators.

Nuclear power has major advantages. It is "clean" in certain respects because it neither releases significant amounts of greenhouse gases nor air pollutants such as particulates, sulfur dioxides, or nitrous oxides. Known reserves of fissionable materials (see Table 8.2) are adequate to supply nuclear power plants at the current rate for roughly 50 years (NEA and IAEA 2006). If the price of uranium were to rise above \$200 per kilogram, exploration for new sources, new nuclear reactor technologies, and fuel rod reprocessing have the potential to supply nuclear fuel for several centuries (World Nuclear Association 2007).

The downside of nuclear power is the radioactive waste produced during extraction of fissionable materials from mines, processing of these materials into fuel rods, and disposal of spent fuel rods. A typical nuclear plant generates 3 cubic meters of highly radioactive waste each year. After about 40 years, 99% of the radioactivity has decayed, but the remainder is still very dangerous and will remain so for another 10,000 years.

The danger of radioactive materials lies in the energetic ionizing radiation that these materials emit when they decay. When this radiation strikes DNA (deoxyribonucleic acid), it may cause changes in the coding sequence (resulting in genetic mutations) and interfere with DNA replication and, thereby, cell division. This often proves lethal to rapidly dividing cells such as cancer cells, bone marrow cells, and cells that form the walls of the intestines, and provides the basis of medical radiation treatments as well as radiation poisoning.

Storing highly radioactive materials for millennia is a formidable task. In the United States, a facility for long-term storage of radioactive materials is under construction at Yucca Mountain, Nevada, 100 km northeast of Las Vegas, near the site of the first nuclear bomb tests. Opposition to the facility has delayed completion for more than 2 decades, and the earliest feasible opening date is now in 2021. Debate continues about the stability of geological formations below Yucca Mountain (**Figure 8.23**) and the risks of transporting radioactive materials from all around the country to a specific location. The lack of such a facility, however, forces nuclear power plants in the United States to store their radioactive waste on site, a circumstance that further diminishes the desirability of having a nuclear power plant as your neighbor.

TABLE 8.7		World electricity generation in 2004 and projected for 2030	
Power source	2004 (TW-h)[a]	2030 (TW-h)	Increase (times)
Hydroelectric	2,810	4,903	2
Wind	82	1,440	18
Solar	4	238	60
Geothermal	56	185	3
Tide and wave	1	25	25
All sources[b]	17,408	33,750	2

Source: International Energy Agency 2006a, 2007b
[a]TW-h = 10^{12} Watt-hours
[b] Includes renewables, fossil fuels, and nuclear

Renewable Energy Sources

Hydroelectric, wind, solar, geothermal, and tide or wave power are energy sources that are naturally replenished and do not directly emit greenhouse gases. Although facilities based on such sources meet less than 3% of global energy requirements today (see Figure 7.1B), they should proliferate rapidly and generate 20% of our electricity by 2030 (**Table 8.7**).

The three most important factors for these power sources are location, location, and location. Access to natural resources (water, wind, or sunshine) as well as proximity to power consumers is critical. Although many of the most suitable locations are already in use, technological advances and rising energy prices have bolstered the economic viability of these energy

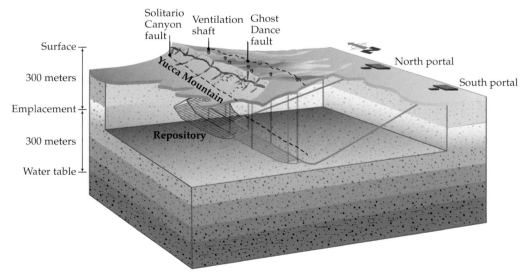

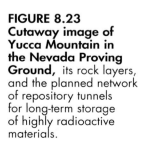

FIGURE 8.23 Cutaway image of Yucca Mountain in the Nevada Proving Ground, its rock layers, and the planned network of repository tunnels for long-term storage of highly radioactive materials.

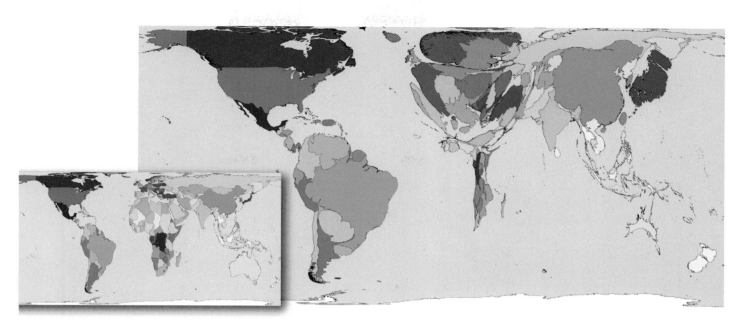

FIGURE 8.24   Hydroelectric power Size of landmasses is proportional to the hydroelectric power generated in a place. Inset at lower left shows the undistorted land area of each location. For example, Scandinavia and Western Europe generate a disproportionately large amount of hydroelectric power, whereas Africa and Australia generate a disproportionately small amount. (After SASI Group and Newman 2006.)

sources, and many additional sites are under development or receiving serious consideration.

Hydroelectric

Hydroelectric power provides the vast majority of renewable energy (see Table 8.7) and generates most of the electricity in places such as Scandinavia (**Figure 8.24, Table 8.8**). In developed countries, rivers that are suitable for generating hydroelectric power typically already have one or several dams or are located in conservation areas that prohibit dams, and so the majority of new hydroelectric projects are in developing countries (**Figure 8.25**). Hydroelectric power

TABLE 8.8   Highest and lowest amounts of hydroelectric power generation

Highest		Lowest	
Country	Amount (kW-h cap⁻¹)[a]	Country	Amount (kW-h cap⁻¹)
Norway	28,722	Bangladesh	7.7
Iceland	23,257	Tunisia	6.9
Canada	11,191	Denmark	5.9
Paraguay	8,458	Estonia	4.6
Sweden	7,475	Israel	3.3
New Zealand	6,436	Belarus	2.8
Austria	4,930	Algeria	1.8
Switzerland	4,891	Togo	0.6
Uruguay	2,805	Turkmenistan	0.6
Tajikistan	2,400	Benin	0.3

Source: International Energy Agency 2006a, 2007b.
[a]Designates 10^3 Watt-hours per person

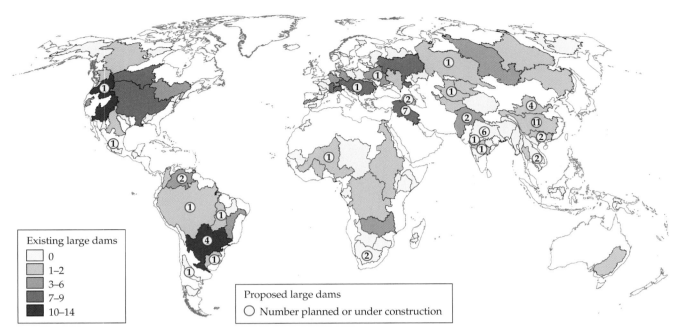

Existing large dams

☐	0
▨	1–2
▨	3–6
▨	7–9
■	10–14

Proposed large dams
○ Number planned or under construction

FIGURE 8.25   Number of existing and planned large dams in major watersheds as of 1994 A large dam meets at least one of the following criteria: height greater than 150 m, water volume greater than 15×10^6 m³, reservoir storage capacity of at least 25 km³, or generating capacity greater than 1000 megawatts. (After Revenga et al. 1998.)

derives from transfers of kinetic energy when a volume of water flows down from a high elevation source, such as a reservoir, through a generator turbine and out to the streambed below (**Figure 8.26**):

$$P = k \cdot r \cdot \Delta h \qquad (8.12)$$

where

P = power (Watts)

k = efficiency factor

r = flow rate (m³ s⁻¹)

Δh = height differential (m)

Hydroelectric, in comparison other renewable energy sources, is highly predictable and flexible and thus can readily meet daily peak loads. Operating and maintenance costs of hydroelectric facilities, after construction, are lower than costs for any other energy source. Reservoirs also serve for flood control, irriga-

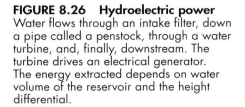

FIGURE 8.26   Hydroelectric power Water flows through an intake filter, down a pipe called a penstock, through a water turbine, and, finally, downstream. The turbine drives an electrical generator. The energy extracted depends on water volume of the reservoir and the height differential.

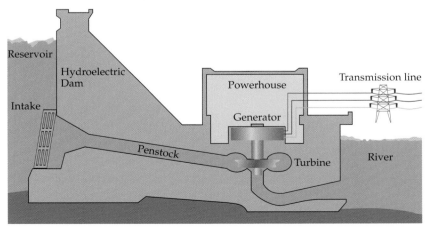

FIGURE 8.27   Abu Simbel temple, built about 1270 B.C. by Pharaoh Ramses II, was moved to higher ground to avoid submersion by the Aswan Dam.

tion, and urban water supplies. Hybrid systems are envisioned that would use wind, solar, or tide/wave power to pump water behind dams and then release the water to generate electricity when it is needed.

On the negative side, construction costs of large hydroelectric projects are enormous (see Figure 8.5), and such projects often displace communities that have developed along rivers and disrupt river ecosystems. For example, the Aswan Dam on the Nile River in Egypt, completed in 1976 at an estimated cost of $4.4 billion (in year 2000 $U.S.), required relocation of 90,000 people and of archaeological sites such as the

temple at Abu Simbel (**Figure 8.27**). The dam retains silt that the river previously deposited down river in yearly floods. This silt made the Nile Valley fertile for agriculture and fishing. Adjustments to these changes have proved difficult.

The Three Gorges Dam on the Yangtze River in China, when it becomes fully operational in 2011, will be the world's largest hydroelectric facility (**Figure 8.28**). Construction costs will total more than $23 billion (in year 2000 $U.S.), but sale of the electricity that it generates should recover these costs in about a decade. Unfortunately, the rising waters behind the

FIGURE 8.28   Three Gorges Dam on the Yangtze River, China spans 2309 m (7575 feet) and is 185 m (607 feet) high.

dam will displace more than 1.5 million people and submerge some 1300 archaeological and cultural sites. Sedimentation of silt behind the dam eventually will interfere with its operation, and lack of silt deposited below the dam will lead to erosion and sinking of land far downriver, in the delta region. The dam has slowed the flow of the Yangtze, and this reduces the ability of the river and its tributaries to flush out polluted areas. Altered fish migrations and other changes to the river ecosystem are likely to drive endangered species such as the Siberian Crane and Baiji Yangtze river dolphin to extinction.

Experiences with giant hydroelectric projects such as the Aswan and Three Gorges dams have led many governmental and nongovernmental agencies to favor smaller projects. What smaller dams sacrifice in efficiency, they compensate for in being more affordable and causing fewer societal and environmental disruptions. The amounts of electricity generated can often integrate directly into existing distribution systems. A smaller dam also presents less of a security risk: Breach of the dam from natural causes or human actions will be less catastrophic.

Wind

Wind power is the second-largest source of renewable energy that does not produce greenhouse gases, surpassed only by hydroelectric power (see Table 8.7). Denmark produces over 20% of its electricity from **wind turbines**. In California, the wind installation at Altamont Pass near San Francisco has more than 5000 wind turbines, and those at Tehachapi and Palm Springs have over 3000 each. These California sites have a range of turbines, from small units that generate as little as 100 kilowatts to some with 90 m blades that generate 1.5 megawatts. Many countries are rapidly constructing wind power facilities (Table 8.9), and within a few decades, wind power should provide about 4% of global electricity. As much as 72,000 gigawatts (GW; 10^9 Watts) of wind power might be commercially feasible worldwide, many times the current total energy demand (Archer and Jacobson 2005).

Winds derive from temperature differences between the Tropics and the Poles, between land and sea, and between Earth's surface and upper atmosphere that produce pressure differentials among air masses (see

TABLE 8.9 Installed wind power capacity and average total electricity consumption (GW) in different countries

Rank	Country	2005	2006	2007	Consumed
1	Germany	18.42	20.62	22.25	69.98
2	United States	9.15	11.58	16.82	487.21
3	Spain	10.03	11.62	15.15	30.45
4	India	4.43	6.27	8.00	79.79
5	China	1.26	2.60	6.05	285.05
6	Denmark	3.14	3.14	3.13	4.12
7	Italy	1.72	2.12	2.73	37.93
8	France	0.76	1.57	2.45	65.18
9	United Kingdom	1.33	1.96	2.39	45.43
10	Portugal	1.02	1.72	2.15	5.62
11	Canada	0.68	1.46	1.85	71.69
12	Netherlands	1.22	1.56	1.75	13.02
13	Japan	1.06	1.39	1.54	124.89
14	Austria	0.82	0.97	0.98	7.41
15	Greece	0.57	0.75	0.87	6.64
16	Australia	0.71	0.82	0.82	26.74
17	Ireland	0.50	0.75	0.81	2.96
18	Sweden	0.51	0.57	0.79	15.91
19	Norway	0.27	0.33	0.33	13.27
20	New Zealand	0.17	0.17	0.32	4.56
	World total	59.09	74.13	94.11	2081.62

Source: International Energy Agency 2007a; Global Wind Energy Council 2008.

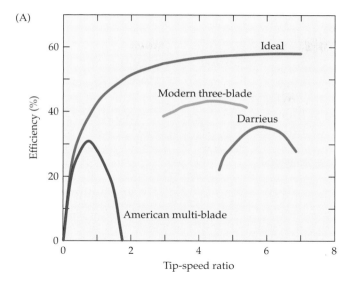

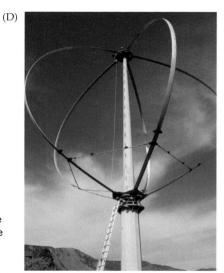

FIGURE 8.29 Wind turbines (A) Efficiency of a wind turbine as a function of the tip-speed ratio (the velocity of the blade tip divided by the wind velocity). Shown are values for an American multi-bladed windmill; a modern, three-bladed horizontal turbine; a Darrieus-type vertical turbine; and the theoretical maximum ("Ideal"). (B) American multi-bladed windmill at an old homestead site in New Mexico. (C) Wind farm of three-bladed horizontal turbines off the coast of Denmark. (D) Darrieus-type vertical wind turbine.(A after Somerton 2005.)

Figure 3.44). Electric power generated from wind is proportional to its speed (velocity) to the third power:

$$P = \tfrac{1}{2} \cdot \alpha \cdot \rho \cdot \pi \cdot r^2 \cdot v^3 \qquad (8.13)$$

where

P = power (Watts)

α = efficiency of the turbine

ρ = density of air (kg m^3)

r = radius of turbine (m)

v = wind velocity (m s^{-1})

In theory, the efficiency of a wind turbine (α) cannot exceed 59.3% (Betz' Law) because increasing the blade

pitch to extract more energy from the wind reaches a point that creates so much turbulence along the surface of the blade that the turbine stalls like an airplane attempting to climb too rapidly.

A turbine blade moving through the air, even at an appropriate angle, produces turbulence in its wake. The adjacent blade of the spinning turbine, if it passes through air that is still turbulent, will not be able to extract power effectively from the air. Therefore, a wind turbine is most efficient at a moderate tip speed ratio (i.e., the velocity of the blade tip divided by wind velocity) when the turbine spins relatively slowly, and so each blades passes through air that is no longer turbulent (**Figure 8.29A**). For example, the traditional, multi-bladed windmill (**Figure 8.29B**) is most efficient at low wind speeds, when the blade tips are spinning at about the same velocity as the wind, whereas modern, horizontal turbines (**Figure 8.29C**) and vertical turbines (**Figure 8.29D**) are most efficient at higher speeds, when their blades tips are spinning four to six times faster than the wind.

Imagine the feeling of the wind when you stick your hand out the window of a moving vehicle (the author accepts no responsibility for any injuries that may occur during this experiment).

- If your hand is parallel to the direction of motion, it allows the moving air to pass relatively undisturbed.
- If you tilt your hand at a slight angle to the direction of motion, it converts some of kinetic energy of the moving air into an up or down motion of your hand.
- If you tilt your hand at an angle that is perpendicular to the direction of motion, it blocks the moving air, and the moving air just buffets your hand (i.e., your hand creates turbulence).

Winds near ground level encounter many impediments, and their velocities approach zero. Velocities increase markedly with height above ground but decrease with surface roughness (**Figure 8.30**). Because the power that can be extracted from wind is very responsive to the wind velocity (it increases with wind velocity to the third power; see Equation 8.13),

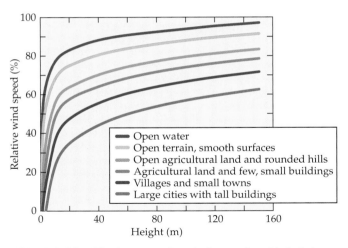

FIGURE 8.30 The increase in wind speeds with height above ground, and the decrease over rougher surfaces Open water or smooth surfaces sustain the fastest winds, and cities with tall buildings sustain the slowest. Most of the changes in wind speed occur within 20 m of the ground. (After Danish Wind Industry Association 2003.)

the hub heights of large turbines are usually placed 40 m to 80 m above the ground where wind velocity is relatively high. Evaluations of potential sites for wind turbines are based on average wind speeds at 50 m above the ground (**Figure 8.31**). Most of these sites are located just offshore, where land/sea temperature differentials drive winds and where open water presents few impediments.

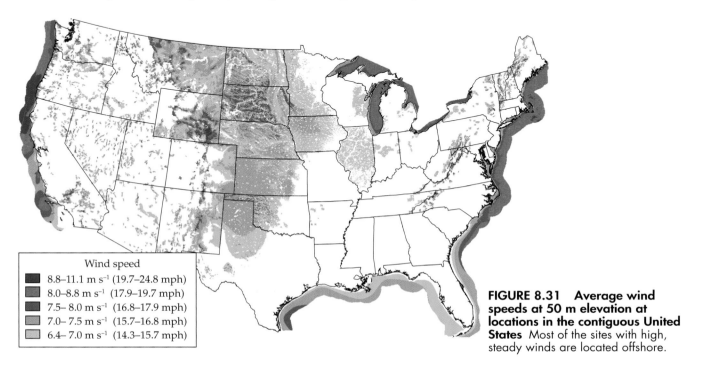

Wind speed
- 8.8–11.1 m s⁻¹ (19.7–24.8 mph)
- 8.0–8.8 m s⁻¹ (17.9–19.7 mph)
- 7.5–8.0 m s⁻¹ (16.8–17.9 mph)
- 7.0–7.5 m s⁻¹ (15.7–16.8 mph)
- 6.4–7.0 m s⁻¹ (14.3–15.7 mph)

FIGURE 8.31 Average wind speeds at 50 m elevation at locations in the contiguous United States Most of the sites with high, steady winds are located offshore.

Modern wind turbines benefit from: improved component materials, such as carbon composites and aluminum alloys that reduce weight and increase durability; more sophisticated designs, in which the angles of blades change with wind velocity; gear boxes that convert the energy of slow-spinning turbines to fast-spinning generator shafts; and dynamic loading and braking that keep the turbines from spinning too fast in high winds. Noise from a wind turbine increases dramatically with the speed of the blade tips (because noise is proportional to the fifth power of the velocity of the blade tips, a doubling of velocity increases the noise 32 times), so gearing and braking to maintain blade speed within an optimum range keeps the turbines from becoming a loud nuisance.

Horizontal turbine designs position the main turbine rotor shaft and electrical generator at the top of a support tower. They are more efficient in slower winds than are vertical designs. Nonetheless, they require precise alignment of the blades into the wind to be efficient and to avoid interference from the support tower itself as well as uneven stresses on the blades and the bearings. Small, multi-bladed windmills use a large tail vane to orient the turbine upwind, whereas large turbines use sensors and electric motors to position the blades into the wind (**Figure 8.32**). The blades also require sufficient stiffness and a position far enough in front of the support tower to prevent their tips from being blown into the tower.

Modern horizontal turbines have an odd number of blades, most commonly three. An even number of blades puts unsustainable stress on the rotor shaft because one blade is highest off the ground, where wind speeds are fastest, while another is closest to the ground, where the speeds are slowest. Three blades provide better balance because one blade is highest while two others are at intermediate heights.

The other major category of wind turbine has a vertical main rotor shaft. Typical designs, such as the Darrieus "eggbeater" (see Figure 8.29C), allow ground placement of the generator and gearbox for easy access. These turbines operate without needing to be oriented into the wind but suffer from low efficiency, particularly at slow wind speeds, because one blade of the turbine is moving into the wind while the other is moving away from the wind. Vertical turbines sometimes require electric motors to start their rotation at low wind velocities.

The intermittent nature of wind makes it an unpredictable energy source. Worse yet, demand for electricity may not coincide with periods of high winds. For instance, in the United States, the state of Texas generates the most wind power (**Table 8.10**); the hottest days in Texas, which have the peak demands for electricity to power air conditioning, tend to be those with the least wind. Thus the success of wind power depends on integration into a large energy grid with a

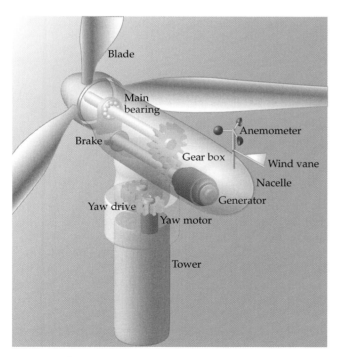

FIGURE 8.32   Modern, three-bladed vertical wind turbine Signals from the wind vane control the yaw drive that turns the turbine into the wind. "Nacelle" is the aerodynamic housing that covers the machinery.

TABLE 8.10   Wind power capacities (MW) in the United States

State	Number of turbines	
	Existing	Under construction
Texas	4,356	1,238
California	2,439	165
Minnesota	1,300	46
Iowa	1,273	117
Washington	1,163	126
Colorado	1,067	0
Oregon	885	15
Illinois	699	108
Oklahoma	689	0
New Mexico	496	0
U.S. Total[a]	16,819	3,626

Source: America Wind Energy Association 2008.
[a]Includes data for all 50 states

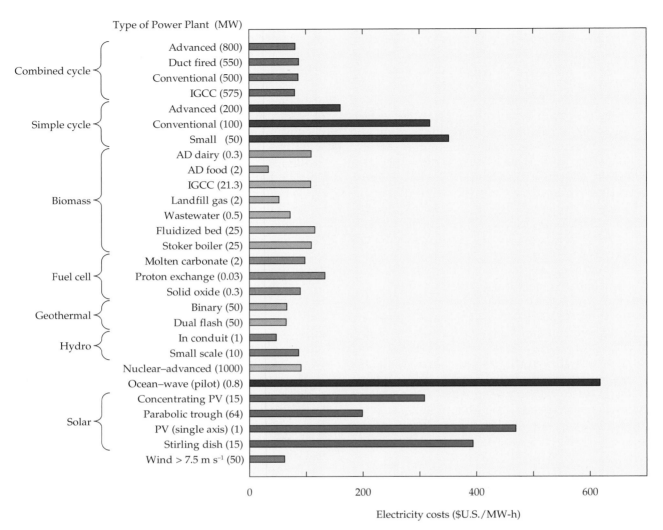

Type of Power Plant (MW)

FIGURE 8.33 Cost of generating electricity (in 2007 $U.S. per 10⁶ Watt-hours) for various types of power plants of a (MW) size. "Combined cycle" and "Simple cycle" are types of natural gas power plants described in the text. "Biomass" power plants are divided by source of biomass and type of furnace. "AD" designates anaerobic digestion. "Fuel cell" power plants react hydrogen and oxygen (see description of fuel-cell–powered vehicles in Chapter 7). "Hydro–In conduit" refers to a turbine placed in a flowing river. "PV" designates photovoltaic. (After California Energy Commission 2007.)

high capacity for energy storage such as hydroelectric power. Danish wind farms, for example, export almost half of their electricity to neighboring countries such as Sweden and Norway that have large hydroelectric power storage (Techconsult 2007).

On the positive side, the cost of wind power has diminished to where it is competitive with all other energy sources (**Figure 8.33**). Wind power, like hydroelectric, has negligible fuel costs and relatively low maintenance costs; most of the costs are for construction and transmission. European power companies, which have extensive experience in wind turbines (e.g., Iberdrola of Spain, Energias de Portugal, and Wind-

kraft Nord of Germany), have found them so profitable that they are financing two-thirds of the wind projects under construction in Texas (Krauss 2008).

Ecological concerns about wind turbines center largely on "birds and bats in a blender," injuries to animals that fly into turbine blades (National Research Council 2007b). Collisions with existing wind turbines kill 20,000 to 37,000 birds each year. By comparison, annual bird fatalities from collisions with buildings are between 97 million and 976 million; with high-tension lines, between 130 million and 1 billion; with communication towers, between 4 million and 50 million; and with vehicles, approximately 80 million. Therefore,

FIGURE 8.34   Simulated view from Craigville, Massachusetts of the proposed wind farm in Nantucket Sound

wind turbines are responsible for a minute fraction—less than 0.003%—of bird deaths from collisions with human structures. Data is sparse, but fatal encounters between bats and turbines appear to occur at rates similar to those for birds in most places in the United States with the exception of the Appalachian Mountain region, where the rates for bats are several times higher (National Research Council 2007b).

Wind farms also face local opposition if their turbines obstruct scenic views. To improve their aesthetics, new installations have more widely spaced turbines, which look less cluttered than old installations, and offshore installations are located farther out to sea. For example, a 420 MW wind farm proposed for Nantucket Sound would place 130 turbines, each 75 m tall, at distances 8 km to 15 km from the beaches of Cape Cod, Nantucket, and Martha's Vineyard. Nonetheless, the turbines would still appear prominent on the horizon (Figure 8.34), and this project has been embroiled in controversy since 2001.

Solar

Conversion of electromagnetic radiation from the sun into other forms of energy drives most processes on Earth. Plants generate chemical energy from sunlight during photosynthesis; the maximum efficiency of this process in terms of the energy recoverable from sugars versus the energy in the incident sunlight is about 2%, for sugarcane growing in the tropics (Taiz and Zeiger 2006). Solar energy is also responsible for air movements that produce wind power and for the evaporation of water from Earth's surface and its subsequent precipitation that are the bases of hydroelectric power. Solar power as a renewable energy source, however,

usually refers to the conversion of electromagnetic energy into electricity through the **photovoltaic effect** or high-temperature solar thermal collectors.

Currently, less than 0.1% of the world's electricity derives from solar power. Use of this energy source is expanding rapidly, however, and by the middle of the century, solar power should meet nearly 1% of human requirements (see Table 8.7). Solar power has the advantage that some of the times and places with the highest peak electricity demands have some of the highest incident solar radiation (Figure 8.35). Disadvantages are that solar power is not available at night and is limited during cloudy weather conditions. It thus requires energy storage and/or a complementary power system.

PHOTOVOLTAIC CELLS The photovoltaic effect occurs in semiconductors such as silicon or germanium. Semiconductors, as their name implies, are materials through which charged particles move more slowly than through conductors (e.g., metals), but faster than through insulators (e.g., plastics). Infusing small amounts of specific impurities into the crystals of semiconductors enhances their ability to move charges. Certain impurities, such as phosphorus, arsenic, or antimony, facilitate the movement of electrons and other negative charges (n-type), whereas other impurities, such as boron or aluminum, facilitate the movement of positive charges (p-type). Layering n-type and p-type semiconductors upon one another forms a depletion region (Figure 8.36). When sunlight strikes a depletion region, electrons in the region become excited and may migrate into the n-type layer while an equal number of positive charges migrate into the p-type layer. This establishes an **electrical potential** that can do work such

FIGURE 8.35 Average daily solar radiation available at various locations in the continental United States for a flat-plate collector facing south and tilted at an angle equal to the latitude of the site. This is the angle that captures the maximum amount of solar radiation throughout the year.

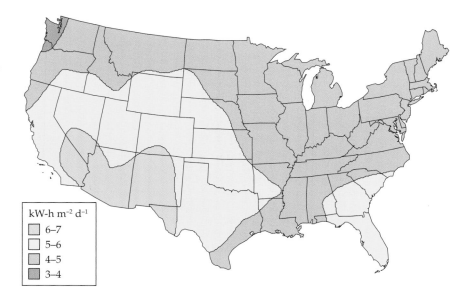

kW-h m^{-2} d^{-1}
- 6–7
- 5–6
- 4–5
- 3–4

as illuminate a lightbulb. Electrodes on the surfaces of the n-type and p-type layers collect the negative and positive charges for current to flow through an external circuit.

The efficiency with which photovoltaic cells convert solar electromagnetic energy into electricity has doubled over the past 30 years and now can exceed 40% under ideal conditions (**Figure 8.37**). Lifespans of photovoltaic cells have lengthened to over 20 years (Alsema et al. 2006). Moreover, photovoltaic cell man-

ufacturing has become more efficient, and so after about 2 years in a sunny environment, a cell generates enough electricity to recoup the energy expended in its manufacture (**Figure 8.38**). Installation costs for photovoltaic cells have declined (**Figure 8.39**), but they are still higher than for most other forms of electricity (see Figure 8.33). Operating costs for photovoltaic cells, however, are as low as any energy source, and the reliability of photovoltaic cells is unmatched under conditions as extreme as outer space. The photovoltaic cells

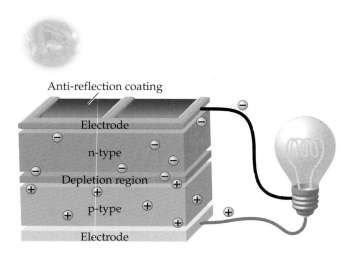

FIGURE 8.36 Photovoltaic solar cell Sunlight absorbed in the depletion region of the cell causes a separation of charges that generates an electrical potential, which enables current to flow through an external electrical circuit (lightbulb).

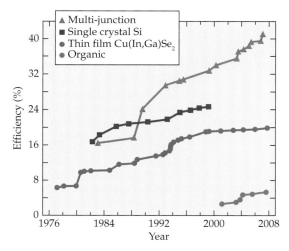

FIGURE 8.37 Highest energy efficiencies for different types of photovoltaic solar cells under ideal conditions For descriptions of the different types, see the references. (After Green et al. 2008; Kazmerski et al. 2008.)

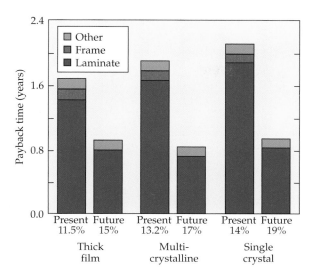

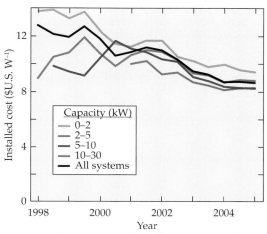

FIGURE 8.38 Time required for photovoltaic solar cells of different types to recoup the energy expended in their manufacture, given present and future efficiencies of the cells (%). "Other" includes the energy required to manufacture the inverter, support, and cable. "Laminate" includes the energy required to produce cell modules. Future modules probably will not require aluminum frames. (After Alsema et al. 2006.)

FIGURE 8.39 Total installed cost ($U.S. in 2004) per Watt for photovoltaic solar cell power systems of different capacities (After Wiser et al. 2006.)

can even be incorporated into buildings or airplanes (**Figure 8.40**).

HIGH-TEMPERATURE SOLAR POWER PLANTS Solar thermal electric power plants use mirrors or lenses to concentrate sunlight and heat a fluid (synthetic oil, molten salt, or pressurized steam) to high temperatures. A heat

engine converts the temperature difference between this hot fluid and the surrounding air into mechanical motion to spin an electrical generator. Pilot plants based on several different types of solar thermal collectors are in operation.

Most solar thermal power plants have mirrors to focus sunlight on receiver cylinders and heat them up. A parabolic trough system uses a curved mirror and a receiver cylinder that runs above the mirror (**Figure 8.41A**), whereas a linear Fresnel reflector uses an array of multiple flat mirrors focused on one receiver cylinder to achieve similar results (**Figure 8.41B**). The

(A)

(B)

FIGURE 8.40 Unusual uses of solar cells (A) Residential building in Bern, Switzerland with one side covered with solar panels. (B) Helios solar airplane flying over the Hawaiian islands of Niihau and Lehua during its first test flight on July 14, 2001.

(A)

(B)

FIGURE 8.41   Solar thermal facilities (A) These parabolic trough solar collectors at Kramer Junction near Barstow, California began operation in 1986. The facility can generate 150 MW peak output. (B) This pilot plant of linear Fresnel solar reflectors, an array that measures 30 m by 300 m near San Luis Osbispo, California, has a 5 MW capacity. A small parabolic reflector above the receiver tube further concentrates the sunlight.

mirrors are usually aligned on a north–south axis and rotate from east to west to keep sunlight focused on the receiver cylinder as the sun progresses along its daily arc. The receiver cylinder has an outer glass tube that maintains a vacuum around an inner, black-coated metal tube (**Figure 8.42**). The inner tube absorbs sunlight and transfers the thermal energy to an exchange fluid; the vacuum minimizes conductive and convective losses. The exchange fluid then flows through a heat exchanger and produces superheated steam that spins a turbine connected to a generator.

Another type of solar thermal system is a solar power tower in which an array of mirrors focus sunlight on receiver tubes in a central tower (**Figure 8.43**). At the National Solar Thermal Facility near Albuquerque, New Mexico, receiver tubes carry an exchange fluid of molten salt (60% sodium nitrate and 40% potassium nitrate) that is pumped up into the tower at 278°C, is heated to 566°C, and flows back down into an insulated hot-storage tank. The demand for electricity determines how much hot salt from this tank is pumped through a heat exchanger to produce super-

FIGURE 8.42   Solar receiver consisting of a glass outer tube with an anti-reflective coating, a black-coated steel inner tube that carries the exchange fluid, a vacuum (not shown) between the two, and a bellowed end cap that compensates for the differences in thermal expansion between the tubes.

FIGURE 8.43   Solar tower power plant (5 MW capacity) at the National Solar Thermal Test Facility near Albuquerque, New Mexico. It consists of 212 computer-controlled mirror assemblies that focus sunlight on a 60 m solar tower.

heated steam that spins a turbine connected to a generator. Advantages of this type of solar thermal system are that the tower contains all of the plumbing and that the mirrors in the array can be flat, and flat mirrors are much less expensive to manufacture than precisely curved ones. The disadvantage is that the mirrors must individually rotate around two axes to follow the sun along its daily arc; by contrast, parabolic troughs rotate in unison around a single axis.

Stirling dish solar power plants consist of reflective parabolic dishes that focus sunlight onto a central receiver (**Figure 8.44**). The receiver transfers the heat to a Stirling engine that efficiently converts temperature differences into rotary motion (**Figure 8.45**) and spins a generator. This type of power plant usually concentrates more sunlight than parabolic troughs and thus reaches higher temperatures and achieves higher efficiencies. Unfortunately, the Stirling engine and generator are part of the moving structure, which therefore requires a rigid frame and a heavy-duty dual-axis solar tracking system.

The major impediment to widespread deployment of high-temperature solar thermal power plants has

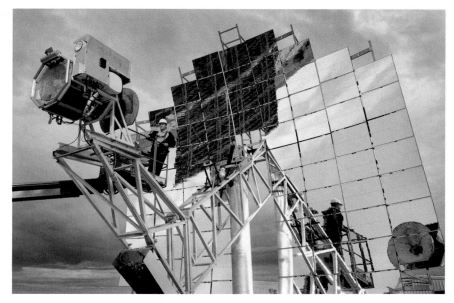

FIGURE 8.44 **Stirling dish system** at the National Solar Thermal Test Facility near Albuquerque, New Mexico. This unit has a 25 kW capacity. The researcher is making adjustments on a cloudy day to avoid sunburn.

been their high construction costs: Electricity from solar thermal is several times more expensive than from other energy sources, except for photovoltaic solar (see Figure 8.33). Nonetheless, proponents of solar thermal claim that this technology will become more cost effective as it is more widely adopted, as the price of fossil fuels soars, and as caps or taxes on greenhouse

(A) (B) (C) (D)

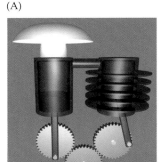  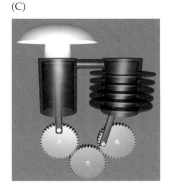 

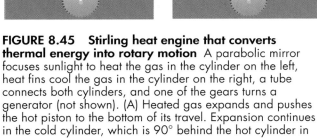

FIGURE 8.45 **Stirling heat engine that converts thermal energy into rotary motion** A parabolic mirror focuses sunlight to heat the gas in the cylinder on the left, heat fins cool the gas in the cylinder on the right, a tube connects both cylinders, and one of the gears turns a generator (not shown). (A) Heated gas expands and pushes the hot piston to the bottom of its travel. Expansion continues in the cold cylinder, which is 90° behind the hot cylinder in

its cycle and extracts additional work from the hot gas. (B) The hot piston begins to move most of the gas into the cold cylinder, where the pressure drops as it cools. (C) Nearly all the gas is in the cold cylinder, where it continues to cool. Momentum drives the cold piston to compress this gas. (D) Most of the gas is in the hot cylinder, where it expands as it heats and drives the hot piston through its power stroke.

FIGURE 8.46 Proportion of energy devoted to various uses in households in the United States in 2005 and in China in 2000. (After Levine et al. 2007.)

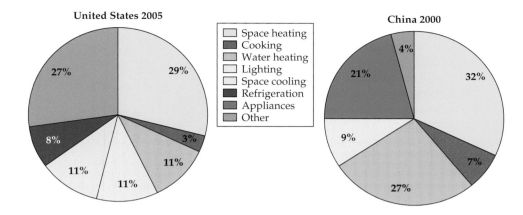

United States 2005

China 2000

Space heating
Cooking
Water heating
Lighting
Space cooling
Refrigeration
Appliances
Other

gas emissions are instituted. Indeed, commercial solar thermal power plants of each type described above are scheduled to begin operations in California during the next few years.

SOLAR HEATING Heating the interior spaces of buildings and producing hot water account for a major portion of the energy use in a typical home (**Figure 8.46**). Traditional methods for solar space heating maximize interception of sunlight in winter, yet minimize it in summer by placing awnings over windows or landscaping with deciduous trees (**Figure 8.47**). Also, sophisticated, low-temperature solar thermal systems are available to collect sunlight for space and water heating.

For example, glass-enclosed flat-plate solar collectors on rooftops have black plates that absorb sunlight and transfer the thermal energy to fluids (water or a less-corrosive liquid) passing through flow tubes lying flat against the surface of the plates (**Figure 8.48**). More efficient, but more expensive, systems use receiver tubes similar to those in parabolic trough solar systems (see Figure 8.42). The fluid heated in flow tubes or receiver tubes goes into the building to warm the interior and/or generate hot water.

Solar residential heating is becoming more commonplace. Israel—a country blessed with ample sunlight but little petroleum—leads the world in solar heating

systems per capita, and solar water heating accounts for 3.4% of the country's total energy consumption (Israeli Central Bureau of Statistics 2008). In terms of sheer number of installations, however, China dominates: The country accounts for about 75% of the global capacity added in 2006 and for nearly 65% of the world total (**Figure 8.49**). Half of the households in some Chinese cities use solar water heaters (Hodum 2007).

Geothermal

As one journeys from the surface to the center of Earth (**Figure 8.50**), temperatures warm by 17° to 30°C for every kilometer in depth, reaching 5100°C in the inner core (Alfe et al. 2007), nearly the temperature of the sun (see Chapter 3). The energy that heats Earth's interior derives from several sources. First and foremost is the decay of radioactive isotopes, especially ^{238}uranium, ^{235}uranium, ^{232}thorium, and ^{40}potassium. Second, 4.5 billion years ago, when Earth first condensed from a cloud of hot gases (see Chapter 2), planetary materials were distributed without regard to density. Friction between heavy metals (e.g., iron, nickel, and copper), as

Summer   Winter

FIGURE 8.47 Residential temperature control via window awnings and deciduous trees.

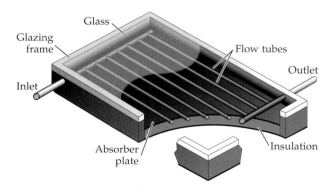

FIGURE 8.48 Glazed flat-plate solar collector for heating buildings A black absorber plate, oriented toward the sun, warms the fluid passing through the flow tubes.

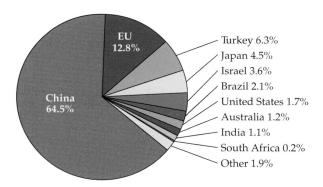

FIGURE 8.49   Distribution of solar hot water and heating capacity among countries (After REN21 2008.)

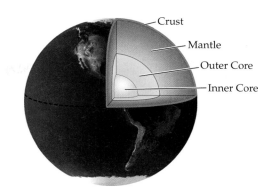

FIGURE 8.50   Cross section of Earth Thickness of the crust ranges between 5 km and 75 km, thickness of the mantle is about 2900 km, and thickness of the outer and inner cores combined is about 3400 km.

they sink toward the center of Earth and displace lighter elements (e.g., aluminum, sulfur, and silicon), contributes another 5% to 10% to Earth's thermal energy. Third, although Earth's surface has cooled during the last 4.5 billion years, its interior retains some of the energy from the original hot gases, and this accounts for an additional 5% to 10%. Net heat flow from the interior of the

planet to the surface amounts to about 30×10^{12} Watts (Anderson 2007); for perspective, this is only 0.03% of the energy that Earth's surface receives from the sun. Earth's crust varies in thickness from about 5 km to 75 km. Locations at which the crust is thinner, such as at boundaries of tectonic plates, fault lines, and volcanoes (**Figure 8.51**), receive more heat from the interior and

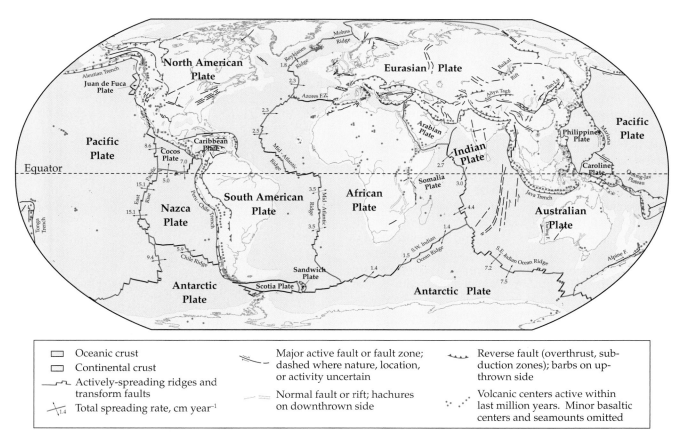

FIGURE 8.51   Global tectonic activity Movements of the major tectonic plates and centers of volcanic activity during the past million years.

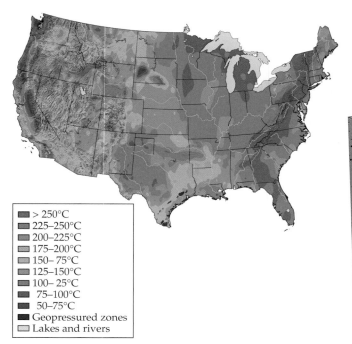

FIGURE 8.52 **Estimated temperatures at 5 kilometers depth below ground** Higher temperatures indicate locations in the contiguous United States that might be suitable sites for geothermal power. A "geopressured zone" is one where the increase in pressure with depth may be double that is normally found.

thus are promising sites for geothermal power (**Figure 8.52**). For example, California, which lies at the intersect between the Pacific and North American tectonic plates, is home to more geothermal power plants than anywhere else on Earth, and they supply the state with over 7% of its electricity (**Table 8.11**).

A geothermal power plant pumps water down a deep injection well (**Figure 8.53**). This water flows through fractures in the fiery brimstone and heats up until it escapes up a second borehole, the production well, as steam or superheated water (depending on temperature and pressure). If the outflow from the production well is steam, it may directly drive a turbine in the power plant (dry steam type). If the outflow is super-heated water, a release of pressure generates steam that may drive a turbine (flash-steam type). Alternatively, superheated water from the production well may pass through a

FIGURE 8.53 **Geothermal power plant** Production and injection wells typically descend to a depth of 2 km or 3 km, where temperatures exceed 180°C. Well liners prevent water from penetrating porous rock.

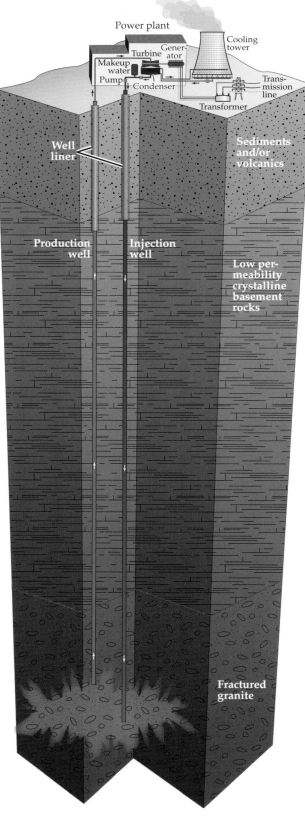

TABLE 8.11 Geothermal electric power generation in various countries and the states of California and Nevada

Country or state	Installed capacity (MW)	Running capacity (MW)	Energy produced (GW–h y^{-1})	Number of power plants	Percent of capacity	Percent of electricity
Australia	0.2	0.1	0.5	1	< 0.1	< 0.1
Austria	1.2	1.1	3.2	2	< 0.1	< 0.1
China	28	19	96	13	30% (Tibet)	30% (Tibet)
Costa Rica	163	163	1,145	5	8.4	15
El Salvador	151	119	967	5	14	22
Ethiopia	7.3	7.3	0	2	1	N/A
France	15	15	102	2	9	9
Germany	0.2	0.2	1.5	1	< 0.1	< 0.1
Guatemala	33	29	212	8	1.7	3
Iceland	202	202	14,838	19	13.7	17.2
Indonesia	797	838	6,085	15	2.2	6.7
Italy	791	699	5,340	32	1	1.9
Japan	535	530	3,467	19	0.2	0.3
Kenya	129	129	1,088	9	11.2	19.2
Mexico	953	953	6,282	36	2.2	3.1
New Zealand	435	403	2,774	33	5.5	7.1
Nicaragua	77	38	271	3	11.2	9.8
Papua New Guinea	6	6	17	1	10.9	N/A
Philippines	1,930	1,838	9,253	57	12.7	19.1
Portugal	16	13	90	5	25	N/A
Russia	79	79	85	11	< 0.1	< 0.1
Thailand	0.3	0.3	1.8	1	< 0.1	< 0.1
Turkey	20	18	105	1	< 0.1	< 0.1
United States	2,564	1,935	17,917	209	0.3	0.5
California	2,269	1,748	15,556	138	N/A	7.2
Nevada	239	239	1,943	50		6.1
World Total	8,933	8,035	56,786	490	N/A	0.3

Source: Bertani 2005.

heat exchanger that vaporizes a fluid with a lower boiling point such as butane, which then drives a turbine and returns to the heat exchanger (closed-loop binary type). A condenser and cooling tower lower the temperature of the water before it returns back down the injection well. Boreholes for the wells are lined with metal or concrete near the top to prevent water from penetrating through porous rock. A makeup tank replenishes the water lost during its passage through the system.

Today most commercial geothermal wells are shallower than 3 km (Bertani 2005). Deeper wells reach higher temperatures and thus generate more power, but drilling becomes prohibitively expensive (**Figure 8.54**). The petroleum industry is developing more economical methods for drilling oil or gas wells 6 km to 10 km deep (MIT Study Group 2006). Transfer of this technology should soon reduce the cost of constructing deeper, hotter geothermal wells (**Figure 8.55**).

Geothermal electric power plants have many advantages. They are economical; in California, geothermal plants cost less than most other types (see Figure 8.33). They have the potential for expansion; geothermal plants could provide about one-fourth of the electricity in the United States by 2030 at a reasonable price (see Table 8.9). They provide power day and night, in any weather. They discharge much lower amounts of CO_2, nitrogen or sulfur oxides, and particulates than fossil fuel power plants (**Table 8.12**).

Disadvantages of geothermal power are severalfold. Sites suitable for geothermal power are often far from population centers; for instance, the East Coast of the United States has limited possibilities for geothermal power (see Figure 8.51). Injection of water into wells could compromise the stability of the surrounding geological formations, although based on substantial evidence collected so far, the probability of

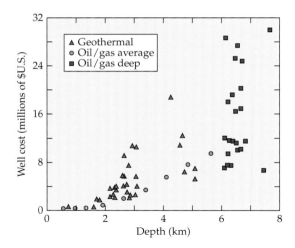

FIGURE 8.54 Cost ($U.S. in 2004) of drilling individual geothermal wells (red), industry averages for oil/gas wells (green), and individual deep oil/gas wells (blue) as a function of depth (kilometers). (After MIT Study Group 2006.)

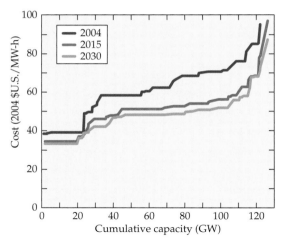

FIGURE 8.55 Probable cost of geothermal electric power (in 2004 $U.S. per 10^6 Watt-hours) as a function of cumulative United States capacity in the years 2004, 2015, and 2030. For example, generation of 50 GW of geothermal electricity in the United States would cost an average of $58 per MW-h in 2004 and $48 per MW-h in 2030, whereas adding less efficient sites to generate 50 GW more electricity, or 100 GW in total, would cost an average of $70 per MW-h in 2004 and $52 per MW-h in 2030. (After Petty and Porro 2007.)

inducing a damaging seismic event is low (MIT Study Group 2006). Overexploitation of a geothermal site may cause it to cool to the point where electricity generation is uneconomical; if left alone, however, these places will recover their lost heat as it transfers from lower depths.

For example, one of the power plants at the Geysers Geothermal Field, located 120 km north of San Francisco, extracted more heat than was sustainable. After less than a decade of operation, temperatures and pressures had dropped so precipitously that the plant was shut down and dismantled in 1989. In the intervening period, the wells have substantially recovered, and plans for a new power plant, one-third the size of the original, are proceeding (Western GeoPower Corporation 2009).

Tidal and wave

Sailors and surfers can attest to the power of the sea. Tides rise and fall from the interplay between Earth's rotation and the gravitational forces of the sun and moon. Waves develop from the pressure and friction of winds blowing over the open seas.

Humans have a long history of harnessing tidal and wave power. The ruins of a tide mill, used for grinding grain on Strangford Lough in Northern Ireland, date to A.D. 619. On a larger scale, a tidal power plant on the Rance River in northwest France has generated

TABLE 8.12 Emissions from different types of electric power plants[a]

Plant type[b]	CO_2	SO_2	NO_X	Particulates
Coal	994	4.71	1.955	1.012
Oil	758	5.44	1.814	N/A
Gas	550	0.0998	1.343	0.0635
Flash steam	27.2	0.1588	0	0
Dry steam	40.3	0.000098	0.000458	Negligible
Closed-loop binary	0	0	0	Negligible

Source: MIT Study Group 2006.
[a]In kg per MW-h
[b]Flash steam, dry steam, and closed-loop binary are variations of geothermal power.

FIGURE 8.56 Tidal power plant on the Rance River, France

600 GW annually since 1967 (**Figure 8.56**). The basic design of tidal plants has not changed in over a millennium: The power plant lowers gates when the tide is rising, allowing water to enter, and raises the gates as the tide begins to recede, trapping the water behind them. At low tide, this water escapes through a turbine that turns the millstones or an electrical generator. A wave power plant couples the rising and falling of a float with the turning of a generator.

Despite this long history, tidal and wave power are the least common forms of renewable energy (see Table 8.7). To date, only three tidal and no wave power plants are in commercial operation. Costs for these sources of electricity are not yet competitive with others (see Figure 8.33) because of the expense of equipment that works reliably under the constant buffeting of the sea and that withstands saltwater corrosion.

Nonetheless, plans for new tidal and wave facilities abound. In September 2008, for instance, Ocean Power Technologies deployed a wave-powered buoy, which generates 1.4 MW, 5 km off the coast of Spain (**Figure 8.57**). The advantage of these types of power plants is that a good site for tidal or wave power may generate nearly 50 kW per meter of shoreline (Sims et al. 2007) with negligible emissions of greenhouse gases or other pollutants. The disadvantages are their relatively high costs and the possibility of their interfering with fish and marine mammals.

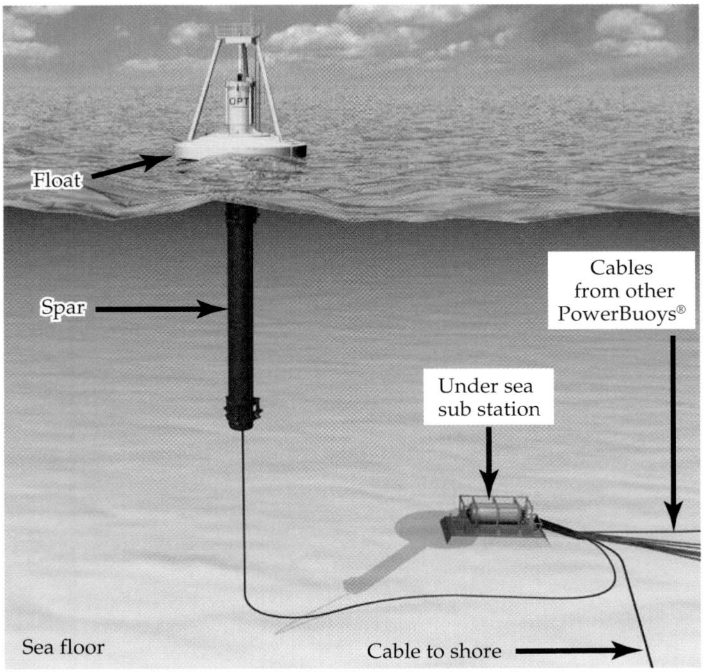

Summary

Electric power generation accounts for over 30% of the greenhouse gas emissions from human activities, more than any other activity. Coal-fired power plants generate the vast majority of electricity worldwide because they are the least expensive type of plant to construct and operate. Coal-fired power plants are the least efficient type in terms of greenhouse gases produced per electricity gener-

FIGURE 8.57 Artist's depiction of the proposed wave power plant off the coast of Spain.

ated, however. Fortunately, modern coal-fired plants operate at higher temperatures and pressures than older ones, which improve their fuel and greenhouse gas emission efficiencies. Natural gas power plants are cleaner and simpler to maintain than coal plants but incur higher fuel costs. Both coal-fired and natural gas power plants will probably incorporate carbon capture and storage (CCS) in the future as the costs of these technologies diminish. Nuclear power plants, although they have operated for more than two decades without major incident, still struggle with public perceptions of the dangers of nuclear energy and with the problem of nuclear waste disposal.

Hydroelectric, wind, solar, geothermal, and tide and wave power are renewable energy sources that will generate an ever-increasing share of electricity worldwide as their technologies improve and their costs become more competitive with fossil fuel- and nuclear-based power. As a consequence, many renewable power facilities are currently under development. There are many reasons for optimism about the future of renewable energy sources.

Review Questions

1. The human activity that emits the greatest quantity of greenhouse gases is
   (a) releasing flatus gas.
   (b) driving personal vehicles.
   (c) producing cement.
   (d) generating electricity.
   (e) heating residences.

2. Which of the following lists is sorted by lifespans from shortest to longest?
   (a) Buildings < incandescent bulbs < appliances < large trucks < power plants < fluorescent tubes
   (b) Large trucks < power plants < incandescent bulbs < fluorescent tubes < appliances < buildings
   (c) Incandescent bulbs < fluorescent tubes < appliances < large trucks < power plants < buildings
   (d) Power plants < fluorescent tubes < incandescent bulbs < appliances < large trucks < buildings
   (e) Appliances < incandescent bulbs < buildings < fluorescent tubes < large trucks < power plants

3. Many heavy industries have relocated to China and India because they (select all that apply)
   (a) benefit from lower labor costs.
   (b) take advantage of less-stringent environmental regulations.

   (c) exploit the excess generating capacity of their electric power plants.
   (d) pump their extensive oil reserves.
   (e) seek nuclear-free zones.

4. Which one of the following statements about different types of power plants is *false*?
   (a) Costs of energy from power plants should include construction, operation, maintenance, and distribution.
   (b) Coal-fired plants without carbon capture and storage are the least expensive per energy generated but emit the most greenhouse gases.
   (c) Oil- and natural gas–fueled plants are more expensive per unit of energy generated than coal-fired plants because oil and natural gas cost more than coal.
   (d) Nuclear, hydroelectric, solar, and wind power plants emit relatively small amounts of greenhouse gases in comparison to fossil fuel–powered plants.
   (e) Renewable energy sources such as hydroelectric and wind are always more expensive per unit of energy generated than plants using fossil fuels.

5. Coal-fired power plants do *not* have
   (a) pulverized coal combustion.
   (b) fluidized bed combustion.
   (c) an integrated gasification combined cycle (IGCC) system.
   (d) fuel cells.
   (e) cooling towers.

6. Which of the following statements about the efficiencies of power plants is *true*?
   (a) Efficiencies of coal-fired plants have failed to improve significantly in the past few decades.
   (b) On average, in U.S. fossil fuel-based or nuclear-based power plants, nearly all of the energy in the fuel is converted into electricity.
   (c) Efficiency of a turbine electrical generator increases with the differences in temperature and pressure of the gases entering and leaving a turbine.
   (d) More efficient power plants can take advantage of less expensive materials for their boilers and turbines.
   (e) Cooling towers that are prominent features of many power plants have little influence on their efficiency.

7. Which of the following statements about carbon capture and storage is *false*?
   (a) Capturing greenhouse gases within a power plant is less expensive than capturing them after they are released to the atmosphere.
   (b) Carbon dioxide could be stored in depleted oil wells, in coal mines, or at depths in the oceans.
   (c) To remove CO_2 from a gas stream involves cryogenic, membrane, or chemical separation methods.
   (d) Two major modes of CO_2 transport are pipelines and tanker ships.
   (e) Costs of carbon capture and storage are negligible in comparison to other costs at power plants.

8. Which statement about nuclear power is *false*?
   (a) There have been no major accidents in nuclear power plants during the past two decades.
   (b) Nuclear waste disposal probably will have negligible costs.
   (c) Very few new nuclear power plants have been built in the United States and Western Europe during the past two decades.
   (d) The United States generates more electricity from nuclear power than any other country.
   (e) The price of uranium on world markets has been highly volatile during the past 40 years.

9. Which statement about hydroelectric power is *false*?
   (a) Hydroelectric projects not only provide power, but they also provide irrigation water and flood control.
   (b) Large hydroelectric projects may prove disruptive to local communities.
   (c) Hydroelectric projects incur relatively expensive construction costs but have low operating costs.
   (d) There are many remaining prime sites for dams in developed countries.
   (e) Hydroelectric dams usually interfere with normal flow of sediment in rivers.

10. Which statement about wind power is *true*?
    (a) Coupling wind to hydroelectric power grids is beneficial because of the intermittent nature of hydroelectric power.
    (b) The blades of wind turbines are situated high off the ground to minimize disruption of the land around them.
    (c) Modern wind turbines normally have two blades.
    (d) Wind turbines are responsible for a major fraction of bird and bat fatalities.
    (e) Wind turbines may obstruct views of natural beauty.

11. Photovoltaic solar electricity cells
    (a) cost less per power unit generated than most other power sources.
    (b) convert light directly into electricity.
    (c) cannot achieve more than 5% efficiency in terms of incoming solar energy versus outgoing electricity.
    (d) must operate for at least 4 years before they generate sufficient energy to recoup the energy expended in their manufacture.
    (e) are likely to power more and more airplanes.

12. High-temperature solar power plants depend on (select all that apply)
    (a) arrays of mirrors that track the sun.
    (b) filters that block unwanted wavelengths.
    (c) generators to convert mechanical energy into electricity.
    (d) low concentrations of greenhouse gases that do not absorb much sunlight.
    (e) hot weather.

13. Geothermal power plants
    (a) can be built almost anywhere.
    (b) operate only during periods of high seismic activity.
    (c) pump water 20 km down into the outer core of Earth.
    (d) use turbines and generators to convert energy from hot fluids into electricity.
    (e) are not currently cost effective.

14. Tidal and wave power plants
    (a) can be built almost anywhere.
    (b) have never been used before.
    (c) depend on fuel cells.
    (d) provide sanctuaries for fish and marine mammals.
    (e) are not yet cost effective.

Suggested Readings

Energy Information Administration. 2009. *International Energy Outlook, 2009*, U.S. Department of Energy, Washington, D.C., *http://www.eia.doe.gov/oiaf/ieo/pdf/0484(2009).pdf*.

This is an annual compilation of data about energy generation worldwide.

International Energy Agency. 2004a. *Prospects for CO_2 Capture and Storage*, Organization for Economic Cooperation and Development, Paris, *http://www.iea.org/textbase/nppdf/free/2004/prospects.pdf*.

A 252-page brochure describing the technology and economics of carbon capture and storage.

National Renewable Energy Laboratory. 2009. *Learning about Renewable Energy.* U.S. Department of Energy, *http://www.nrel.gov/learning/student_resources.html*, accessed June 10, 2009.

A web site with many short essays about renewable energy power plants.

9

MITIGATION STRATEGIES: OTHER SECTORS

The agricultural, forestry, industrial, commercial, and residential sectors produce the majority of greenhouse gases. Since 1970, their emissions have held relatively steady, however, while emissions from the transportation and electric power sectors have tripled (see Figure 7.10). Chapters 7 and 8 examined mitigation of greenhouse gas emissions from the transportation and electric power sectors, respectively. This chapter provides an analogous treatment of these other sectors. The final section of this chapter considers geoengineering, very large-scale modifications to Earth's environment that, if all else fails, might cool the planet or sequester huge amounts of atmospheric CO_2.

FIGURE 9.1 Global greenhouse gas emissions in 10^{12} g carbon equivalents per year from various types of human activities "Waste" includes municipal solid waste and human sewage. "Manufacturing" includes metal production and electricity transmission and distribution. "Agriculture" includes soil and manure management, rice cultivation, enteric fermentation, and other nonindustrial sources, such as land use changes and biomass burning. "Energy" includes coal mining, natural gas transmission and distribution, petroleum production, and electric power generation. "CFC/HCFC" denotes chlorofluorocarbons and hydrochlorofluorocarbons. (After U.S. Environmental Protection Agency 2006b.)

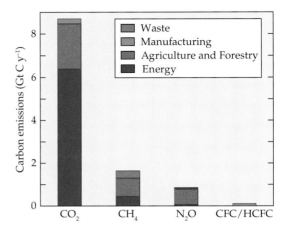

Agriculture and Forestry

Agriculture and forestry are responsible for between 10% and 32% of the world's total greenhouse gas emissions from human activities, depending on how much one attributes emissions from land use changes and biomass burning to these sectors (U.S. Environmental Protection Agency 2006b; Rogner et al. 2007; Smith et al. 2007; Backlund et al. 2008). Depending on this accounting, agriculture and forestry contribute as little as 1% and as much as 24% of carbon dioxide (CO_2) emissions, 47% to 52% of methane (CH_4) emissions, and 58% to 84% of nitrous oxide (N_2O) emissions (**Figure 9.1**). The largest source of CO_2 emissions is decay of above- and below-ground organic carbon after deforestation. Agricultural CH_4 emissions derive from the cultivation of wetland rice, burning of crop residues, and rearing of ruminants such as cows, sheep, goats, buffalos, and camels (see Figure 4.19). Soil and plant transformations of nitrogen fertilizers account for the majority of the agricultural N_2O emissions (see Figure 4.23).

Agriculture

Several external factors influence greenhouse gas emissions from the agricultural sector.

- Agricultural lands, although they have expanded 5% per decade in developing countries and 2.5% per decade in the world as a whole (**Figure 9.2**), have not kept pace with population growth (**Figure 9.3**), and so agricultural land per person has declined (see Figure 6.37A). To meet rising demands for food and fiber, farmers more intensively manage the lands that are already in production. Intensive management often results in a loss of soil car-

bon, particularly from lands recently converted from forests or wetlands to agriculture and primarily through plowing, which accelerates microbial decomposition. Such losses manifest as a net release of CO_2 to the atmosphere.

- People are now wealthier on average and can afford to consume more meat and dairy products (see Figure 9.3). This dietary shift has altered the balance from producing food for direct human consumption to producing feed for domestic animals. The net result is that crop residues, which previously remained in the field and amended soil carbon or slowly decomposed to CO_2, have become animal feed,

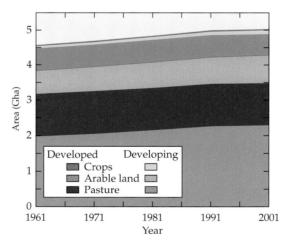

FIGURE 9.2 Agricultural land use (10^9 hectares) in developed and developing countries "Crops" and "Pasture" indicate areas that are in persistent use, whereas "Arable land" indicates areas that receive intermittent use for crops. (After FAOSTAT 2009.)

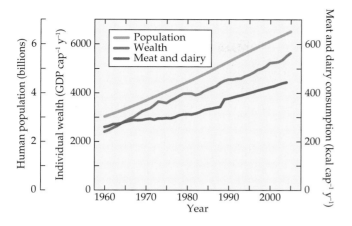

FIGURE 9.3 Worldwide trends Human population (10⁹ people), individual wealth (gross domestic product in constant, year 2000 $U.S. per person per year), and meat and dairy consumption (kilocalories per person per year). "Meat and dairy" includes meat from various animals, seafood, milk, and eggs. (After Development Data Group, The World Bank 2007; FAOSTAT 2007; United Nations Population Division 2008.)

- Policies, intended to nurture the emerging bioethanol industry in the United States and Europe, have shifted some agricultural production from food to fuel (see Chapter 7). Prices for staple crops such as maize have doubled in 2 years. Whether demand for biofuels will continue to influence the market for agricultural commodities is uncertain. Also uncertain is the potential for biofuels to mitigate greenhouse gas emissions.

The remainder of this section on agriculture ignores these external factors to focus on agricultural management practices that minimize greenhouse gas emissions.

NITROGEN MANAGEMENT Nitrogen fertilization—if one includes CO_2 released during the manufacture, distribution, and application of the fertilizer and N_2O released during microbial transformation of fertilizer in soils (denitrification)—is responsible for the majority of greenhouse gas emissions from agriculture (see Chapter 7). Some of the nitrate (NO_3^-) fertilizer applied to crops leaches from fields and contaminates groundwater (**Figure 9.4**); groundwater NO_3^- concentrations above 4 milligrams of nitrate nitrogen per liter (4 mg NO_3^--N L^{-1}) are associated with increased risk of cancer and other human health problems (Nolan and Hitt 2006). Moreover, U.S. maize farmers spend more than

and the animals release a portion of the organic carbon in the feed to the atmosphere as methane (CH_4), a greenhouse gas with 23 times the warming potential of CO_2.

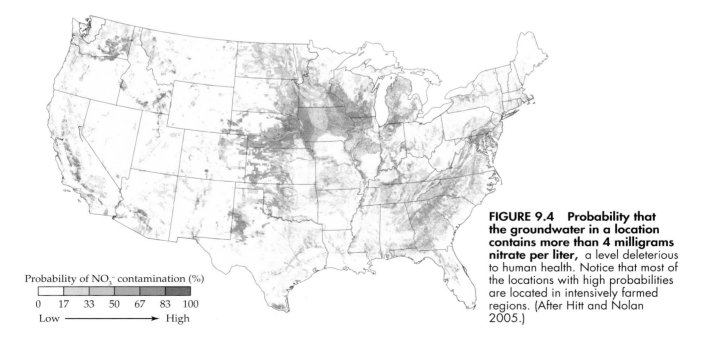

Probability of NO_3^- contamination (%)

0 17 33 50 67 83 100

Low ⟶ High

FIGURE 9.4 Probability that the groundwater in a location contains more than 4 milligrams nitrate per liter, a level deleterious to human health. Notice that most of the locations with high probabilities are located in intensively farmed regions. (After Hitt and Nolan 2005.)

$150 per hectare on nitrogen fertilizer each year. Therefore, higher nitrogen-use efficiencies—higher yields per nitrogen fertilizer applied—would not only decrease greenhouse gas emissions but would also decrease crop production costs and water pollution.

Breeding crops for better yields has inadvertently selected for cultivars with nearly double the nitrogen-use efficiency of their predecessors (see Figure 7.45). Newer cultivars attain higher yields per nitrogen absorbed from the soil (nitrogen-utilization efficiency) as well as acquire more of the nitrogen applied to the soil (nitrogen-uptake efficiency). Consequently, grain yields in the United States have increased about 3% per year without requiring additional nitrogen fertilization or emitting additional greenhouse gas emissions (see Figure 7.1).

Crop breeding programs are beginning to target high nitrogen-use efficiency as a primary objective. Optimizing nitrogen levels in various plant parts and enhancing nitrogen translocation within plants should enhance nitrogen utilization efficiency. Higher nitrogen-uptake efficiency may entail breeding crops that allocate a larger share of their resources to roots, a characteristic that may sacrifice some yield but would also benefit soil carbon sequestration and water acquisition from soils.

Complementing genetic improvement of crops is more sophisticated fertilizer management. Previously, farmers took a "one size fits all" approach, whereby most fields received a similar amount of fertilizer in a single application just before planting. To compensate for possible variation among locations or over years, farmers would err on the "safe" side and apply an excessive amount of nitrogen. Now, farmers optimize fertilizer rates among fields and sometimes within fields, and they add small amounts of nitrogen to the soil several times through the growing season. Fertigation, the practice in which farmers dissolve fertilizer in the irrigation water that they apply regularly through drip irrigation systems, has become commonplace in California. This practice both conserves water and minimizes fertilizer losses from **denitrification** or leaching.

The majority of agricultural N_2O emissions derive from the process of denitrification (see Figure 4.23). This process occurs when patches in a soil become anaerobic (oxygen deprived) during flooding or compaction. Certain soil microorganisms can sustain themselves in such anaerobic patches on the energy released during the stripping of oxygen from nitrogen compounds:

$$NO_3^- \text{ (nitrate)} \rightarrow NO_2^- \text{ (nitrite)} \rightarrow$$
$$NO \text{ (nitric oxide)} \rightarrow N_2O \text{ (nitrous oxide)} \rightarrow$$
$$N_2 \text{ (dinitrogen)}$$

Transfers between the steps of this pathway are not perfectly efficient, and some of the immediate compounds escape to the external environment. In particular, leaks before the last reaction step release substantial amounts of N_2O to the atmosphere. Indeed, the standard agricultural practice of heavy fertilization and irrigation at the beginning of the growing season to establish a crop creates temporary anaerobic conditions that generate a burst of N_2O (Matson et al. 1998). Management practices that diminish N_2O emissions from agricultural soils include distributing fertilizer and water more evenly across a field and throughout the growing season and improving field drainage through accumulation of soil carbon and avoidance of soil compaction.

Elevated CO_2 concentrations in the atmosphere inhibit the assimilation of NO_3^- in the shoots of most plants (see Chapter 5). To compensate for this phenomenon without exacerbating greenhouse gas emissions, farmers will need to manage soil nitrogen more carefully. In particular, multiple small applications of ammonium-based fertilizers will avoid the deleterious effects of excessive ammonium and promote slow but steady **nitrification** (microbial conversion of soil ammonium to nitrate) and root NO_3^- assimilation.

MINIMUM TILLAGE Combustion of fossil fuels by farm machinery is the second largest source of greenhouse gases from the agricultural sector (see Figure 7.44). Intensive agriculture typically involves the following field operations:

1. Clearing residues from the previous crop

2. Killing weeds

3. Breaking up soil into fine particles to prepare a uniform seed bed and encourage microbial release of soil nutrients

4. Depositing seeds at the proper soil depth and distance between adjacent plants

5. Placing a band of fertilizer where it nurtures germinating seeds without exposing them to excessive concentrations

6. Irrigating

7. Fertilizing after crop establishment

8. Removing more weeds

9. Harvesting the crop

In developed countries, farm machinery performs most of these operations. The operations require numerous passes of a tractor and implements over a field, and each pass consumes between 1 liter and 34 liters of diesel fuel per hectare of crop (**Table 9.1**). This translates to a fuel efficiency of 0.15 to 5.0 km L^{-1} (0.4 to 11.8 mpg).

TABLE 9.1 Typical diesel fuel requirements for various farming operations[a]

Operation	Rate
Shred stalks	6.7
Plow 8 inches deep	15.7
Heavy offset disk	8.9
Chisel plow	10.3
Tandem disk	
Stalks	4.2
Chiseled	5.1
Plowed	6.1
Field cultivate	5.6
Spring-tooth harrow	3.7
Spike-tooth harrow	2.8
Mulch treader	2.8
Rod weeder	2.8
Sweep plow	5.6
Cultivate row crops	4.2
Rolling cultivator	3.3
Rotary hoe	2.3
Anhydrous applicator	6.1
Planting row crops	4.7
No-till planter	3.3
Till plant (with sweep)	3.7
Grain drill	3.3
Combine	
Small grains	9.4
Beans	10.3
Corn and sorghum	15.0
Corn picker	10.8
Mower	
Cutterbar	3.3
Conditioner	5.6
Swather	5.1
Rake	
Single	2.3
Tandem	1.4
Baler	4.2
Stack wagon	4.7
Sprayer	0.9
Rotary mower	7.5
Haul small grains	5.6
Grain drying	56.1
Forage harvester	
Green forage	8.9
Haylage	11.7
Corn silage	33.7
Forage blower	
Green forage	3.3
Haylage	2.3
Corn silage	13.1
Moist ear corn	4.2

Source: Downs and Hansen 2007.
[a]L ha^{-1}

Minimum tillage refers to agricultural management practices that leave large pieces of plant residues in a field and reduce turnover of top soil layers. With minimum tillage, farmers eliminate one or two field operations and combine other operations into a single pass of a tractor. This diminishes soil compaction and soil erosion and conserves fuel. For example, converting wheat production from standard multiple-pass practices to minimum tillage (from a moldboard plow to stubble mulch system) saves 8.2 liters of diesel per hectare, or 20% of total fuel use (Downs and Hansen 2007). Minimum tillage, despite expectations to the contrary, has no significant effect on soil carbon sequestration (Veenstra et al. 2007).

Several recent technical developments facilitate minimum tillage. One is the availability of herbicide-resistant crops that permit herbicide applications to kill weeds even after the crop has emerged. Another is precision agriculture, whereby global positioning sensors (GPS) steer tractors and align implements to within 2 cm of the same spot, permitting the combination of several operations into a single pass or into multiple passes onto a specific location. For example, a farmer, in one pass, may till a narrow strip of soil to establish a uniform seedbed for the crop, plant seed in this strip, and inject fertilizer; on a subsequent pass, the farmer may clear out weeds precisely around this strip or apply additional fertilizer within the strip.

RICE PRODUCTION Humans consume more rice (5.4×10^{14} grams in 2005) than any other cereal (**Figure 9.5A**), and rice is grown on more than 150 million hectares worldwide (**Figure 9.5B**). Most of this rice is wetland rice (**Figure 9.6**). Cultivation of wetland rice involves planting into shallow standing water and raising the water level as the crop grows. This eliminates competition from most weeds as well as insures an adequate amount of water for this water-demanding crop. Rice fields, or paddies, are allowed to dry as the crop matures.

Soil microorganisms under anaerobic, flooded conditions harvest the energy released during the breakdown of organic carbon compounds into methane (CH_4), a process known as **methanogenesis**.

$$CH_3COOH \rightarrow CO_2 + CH_4 + \text{chemical energy} \qquad (9.1)$$

$$CO_2 + 4\,H_2 \rightarrow CH_4 + 2\,H_2O + \text{chemical energy} \qquad (9.2)$$

This process emits as much as 700 kg of **carbon equivalents** per hectare (Pathak and Wassmann 2007), or 1.5% of the world's greenhouse gas emissions from human activities (Neue 1997; U.S. Environmental Protection Agency 2006a).

Mitigation options for rice production are few. Water management practices, which drain paddies during the middle of the growing season instead of keeping them

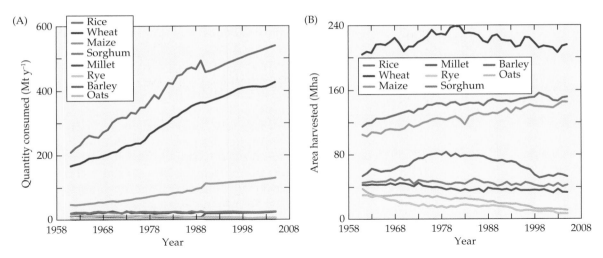

(A)

Legend:
Rice
Wheat
Maize
Sorghum
Millet
Rye
Barley
Oats

Quantity consumed (Mt y^{-1}) — y-axis: 600, 400, 200, 0
Year — x-axis: 1958, 1968, 1978, 1988, 1998, 2008

(B)

Legend:
Rice Millet Barley
Wheat Rye Oats
Maize Sorghum

Area harvested (Mha) — y-axis: 240, 160, 80, 0
Year — x-axis: 1958, 1968, 1978, 1988, 1998, 2008

FIGURE 9.5 Major cereal crops (A) Human consumption of cereal grains (10^6 metric tons per year). Shifts between 1989 and 1990 reflect a change in accounting procedures. (B) Area (10^6 hectares) devoted worldwide to cultivation of major cereal crops. (After FAOSTAT 2007.)

underwater continuously, diminish CH$_4$ emissions by 15% to 59% (Corton et al. 2000; Wang et al. 2000; Pathak and Wassmann 2007). These practices depend on the availability of additional water to reflood the paddies and add from 5% to 9% to production costs (Smith et al. 2007; Wassmann and Pathak 2007). Rice cultivars may differ in their root exudations of organic carbon and thus in their support for methanogenesis, but evidence that these differences influence CH$_4$ emissions under field conditions is inconclusive (Wassmann et al. 2002).

BIOMASS BURNING Agricultural burning serves several purposes. First, it quickly clears land of plant residues, a task that otherwise would require a large investment in labor, equipment, fuel, and time. Second, it returns some of the mineral nutrients in the biomass back to the soil. Finally, burning produces high temperatures that sterilize the top soil layers to reduce pest problems for subsequent crops.

The downside of agricultural burning is that most of the organic nutrients in the biomass literally go up in smoke rather than enrich the soil. This burning dis-

FIGURE 9.6 Terraced rice fields in Yunnan Province, China

FIGURE 9.7   Biogas generation from a mix of agricultural products in Germany The covered tank is a fermenter that excludes light and oxygen to promote anaerobic respiration and methane production.

sipates all of the nitrogen, some of it as N_2O (nitrous oxide), and all of the organic carbon, either as CO_2 (carbon dioxide), CO (carbon monoxide), or CH_4 (methane) depending on oxygen availability. Together, these greenhouse gas emissions amount to roughly 12% of the world's total from human activities (U.S. Environmental Protection Agency 2006a). Lost into thin air are 25% of the phosphorus, 20% of the potassium, and 5% to 60% of the sulfur in the biomass (Fairhurst et al. 2007). Soot and smoke particles from burning adversely affect air quality throughout agricultural regions. Lastly, burning reduces the albedo of the land surface (i.e., amount of sunlight reflected; see Chapter 3) for several weeks (Beringer et al. 2003) and thereby enhances the amount of solar radiation absorbed by about 5%.

Alternatives to agricultural burning include physically removing plant residues and using them for feed, bedding, or biofuels; chopping residues and mechanically incorporating them into soils; controlling pests through crop rotations, time of planting, and chemicals; and leaving fields fallow and letting natural decomposition occur. Usually, farmers employ a mix of these alternatives. Many local governments restrict burning to times when the fire danger is low and weather conditions readily disperse the soot and smoke that burning releases.

LIVESTOCK MANAGEMENT   Domestic animals kept in pens produce prodigious piles of poop (i.e., make massive mounds of manure) that become anaerobic and generate CH_4 and N_2O. This source amounts to 1.0% of the world's greenhouse gas emissions from human activities (U.S. Environmental Protection Agency 2006a). Composting, which regularly turns over manure and other organic material to promote aeration and the

faster decomposition of aerobic respiration, diminishes the emissions of CH_4 and other putrid products, which are generated during anaerobic respiration. Alternatively, farmers cover the manure and other organic material to induce anaerobic conditions and capture the CH_4 thus generated for use as a fuel (**Figure 9.7**). In Europe during 2006, such operations produced "biogas" equivalent to 1.3 million metric tons of oil (EurObserv'ER 2007), or about 0.5% of oil production from Western Europe (BP 2007).

Some domestic animals—specifically, cows, sheep, goats, buffalos, and camels—have a rumen, a compartment in their digestive system that harbors microorganisms that help break down organic carbon compounds such as cellulose, which are difficult to degrade. About 3% of these microorganisms conduct methanogenesis (see Equations 9.1 and 9.2). Ruminants exhale the CH_4 generated through this enteric fermentation, and it accounts for 4.4% of the world's greenhouse gas emissions (U.S. Environmental Protection Agency 2006a). Thus, four times more greenhouse gases emanate from the mouths of ruminants than from the other end.

A variety of mitigation options for livestock management are under investigation (Steinfeld et al. 2006; Smith et al. 2007). The CH_4 from enteric fermentation represents a loss of carbon from the feed and, therefore, a loss of productivity. Methane emissions per animal product are higher when the diet is poor. Less fiber and more oils or organic acids decrease CH_4 emissions per milk produced or weight gain, but such diets may be prohibitively expensive. Hormone treatments such as bovine somatotropin (bST) and growth hormone do not directly influence enteric fermentation, but they boost animal performance and thus reduce CH_4 emissions per milk produced or weight gain. Another possibil-

CHAPTER 9

ity is to administer antibiotics or vaccines that repress methanogenic microorganisms. Yet, broader use of hormones or antibiotics in domestic animals invokes health and environmental concerns.

Forestry

Forests cover about 30% of Earth's land area. Net loss of forested land between 2000 and 2005 averaged 7.3 million hectares, or 0.0002% per year. Certain areas of the world, however, such as equatorial South America, Africa, and Oceania, have suffered losses in excess of 0.5% per year, whereas others, such as the Iberian Peninsula and southeast Asia, have seen gains of more than 0.5% per year (Figure 9.8).

Deforestation (clearing land for agricultural, industrial, commercial, or residential purposes) not only results in the immediate loss of carbon in the lumber removed, but it also leads to decomposition of organic carbon from the leaves, twigs, and roots left behind (from 100 metric tons to 250 metric tons of carbon emitted per hectare) (Nabuurs et al. 2007). Several practices, such as slowing deforestation, promoting afforestation (conversion of lands into forests), and managing forests for high carbon intensity could sequester an additional 1.6×10^9 metric tons of carbon per year (Nabuurs et al. 2007), or mitigate about 14% of total human CO_2 emissions (see Figure 9.1). International agreements includ-ing the Kyoto Protocol provide economic incentives to promote such practices (see Chapter 11), and many of the organizations that supply carbon offsets (credits for projects that sequester greenhouse gases to balance activities that emit them) do so through reforestation.

Industrial Sector

Industrial emissions of greenhouse gases, if one includes emissions from electricity and steam generated off-site, amount to about 3.3 Gt (3.3×10^{12} g) carbon equivalents per year (Bernstein et al. 2007). Industries with the highest emissions are metal smelting, especially steel, iron, and aluminum; extraction and refining of petroleum; cement manufacture; pulp and paper processing; industrial nitrogen fixation to generate ammonia; and synthesis of ethylene, a precursor of many plastics (Figure 9.9). As the human population and per capita income have grown (see Figure 9.3), so has production of industrial goods. Since 1980, global annual production of steel has increased by 84%; petroleum by 40%; cement by 271%; paper by 180%; ammonia by 200%; aluminum by 223%; and ethylene by 194% (Price et al. 2006). The industrial sector, however, has not expanded as quickly as some other sectors (e.g., transportation), and its share of global greenhouse gas emissions actually declined, from 40% in 1971 to 37% in 2004 (Bernstein et al. 2007).

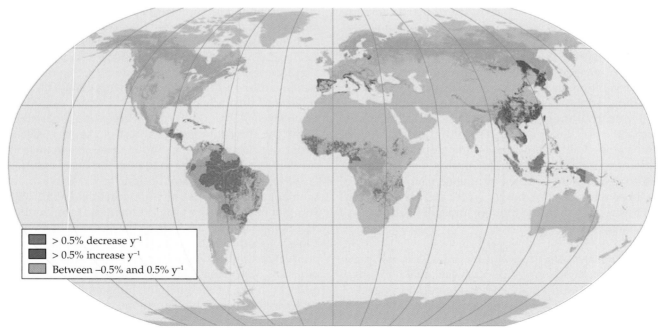

> 0.5% decrease y^{-1}
> 0.5% increase y^{-1}
Between −0.5% and 0.5% y^{-1}

FIGURE 9.8 Net change per year between 2000 and 2006 in the quantity of forested areas around the world

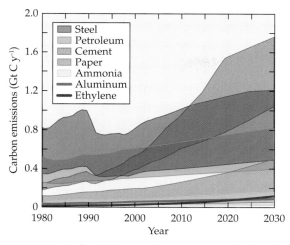

FIGURE 9.9   Industrial greenhouse gas emissions in carbon equivalents The upper line for an industry assumes high emissions per unit product and the A1 scenario for future emissions, whereas the lower line assumes low emissions per unit product and the B2 scenario (see Chapter 4 for the various emission scenarios). Estimates of emissions from aluminum and ethylene production do not range widely, and thus each is depicted by a single line. "Steel" includes iron, "Petroleum" includes extraction and refining, "Paper" includes pulp production, and "Ammonia" includes nitrogen fertilizer and explosives. The drop in emissions around 1990 reflects the economic decline of eastern Europe after the dissolution of the Soviet Union. (After Price et al. 2006.)

Many heavy industries have relocated from developed countries to developing ones such as China and India (see Chapter 8). Indeed, industrial greenhouse gas emissions from China and India rose from 12% of the world's total in 1971 to 42% in 2004 (Price et al. 2006). Because expected lifetimes of manufacturing equipment, power lines and pipelines, and power stations range from 10 years to 50 years (see Figure 8.1), industrial facilities now under construction in China and India will influence emissions for decades to come.

Over 80% of the greenhouse gas emissions from industry are energy related (Bernstein et al. 2007), yet energy efficiencies of most industrial processes are less than half of what is theoretically achievable (International Energy Agency 2006a). Increasing the energy efficiencies of the industrial sector thus presents an excellent opportunity to decrease greenhouse gas emissions as well as to save money. Strategies for increasing energy efficiencies include upgrading equipment, optimizing operating procedures, improving equipment maintenance, and enhancing the degree to which equipment runs at near capacity.

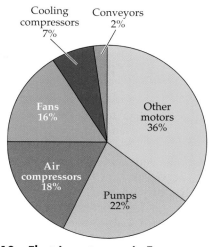

FIGURE 9.10   Electric motor use in European industries (After de Almeida et al. 2003.)

Motors

Most industrial equipment, including pumps, compressors, fans, and conveyors, contain electric motors (**Figure 9.10**). These motors consume more than 60% of the energy in the industrial sector (International Energy Agency 2006a). Therefore, the motors themselves, their controls, and their transmission systems are obvious targets for energy conservation in the industrial sector.

Premium motors have stronger magnets, higher conductivity brushes, and better bearings than their standard counterparts and are 3% to 30% more energy efficient (**Figure 9.11**) but about 20% more expensive

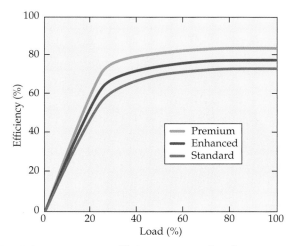

FIGURE 9.11   Energy efficiency versus load factor for premium, enhanced, and standard 1.1 kW (1.5 horsepower) electric motors. (After Neelis et al. 2007.)

(Copper Development Association 2008). In industrial applications that run motors for extended periods, the lower energy requirements of premium motors compensate for their higher purchase prices in less than 3 years (i.e., payback time < 3 y). For example, a 75-kilowatt (100 horsepower) AC standard induction motor that costs around $5,000 may use as much as $35,000 of electricity in a year; a premium motor of the same power costs $1,000 more, a 20% difference, but provides a 3% improvement in energy efficiency that saves $1,050 each year (Public Service of New Hampshire 2008).

Motors operate most efficiently when loaded at between 75% and 100% of their capacity (see Figure 9.11). Electronic adjustable speed drives vary the frequency and voltage of the electricity supplied to a motor to control its speed. Adjustable speed drives are expensive, costing roughly as much as the motors they control, yet under variable loads, they improve energy efficiencies by 10% to 20%, and the energy savings quickly compensate for the higher capital costs (Nadel et al. 2002; WSU CE Energy Program 2003; Bernstein et al. 2007). The 75-kilowatt motor of the previous example, if equipped with a $5,000 adjustable speed drive, might save $3,500 to $7,000 annually in electricity.

Transmission systems connect motors to their loads via belts, chains, or gears. V-belts (Figure 9.12A), the most common type, may reach efficiencies of 95% if properly installed and maintained, but they typically have efficiencies of less than 90%. Replacing V-belts with cogged belts (Figure 9.12B) or gears increases energy efficiency by an average of 4% (Nadel et al. 2002). In some circumstances, an adjustable speed drive permits a direct connection between the motor and the load, eliminating the need for a transmission system.

Steam heat

Steam is the preferred method of delivering heat to many industrial processes (Industrial Technologies Program 2004). Steam releases large quantities of energy when it condenses (539 calories per gram; see Chapter 3). It is low cost, nontoxic, and easy to transfer. Generating and delivering steam accounts for about 15% of the energy use in the industrial sector (International Energy Agency 2006a).

Modern electric power plants (see Chapter 8) cogenerate steam with efficiencies that approach 85% (Bernstein et al. 2007). A typical steam boiler exhibits a much lower efficiency, however. For instance, steam boilers in China average only 65% efficiency because they are mostly small units that cannot completely combust the poor-quality coal available as fuel (International Energy Agency 2006a).

Various procedures can improve the efficiency of existing steam systems (Table 9.2). The most cost-effective procedures include repairing steam traps, cleaning heat transfer surfaces, and insulating pipes, valves, and fittings. Others procedures may involve expensive retrofits (furnishing with new or modified parts), but even these may prove worthwhile as fuel costs climb and as more stringent emission regulations are imposed. Additional energy conservation can be

(A)

(B)

FIGURE 9.12 Transmission systems (A) V-belt drive between an electric motor and a vacuum pump. (B) Cogged belt and pulleys in an automobile engine.

TABLE 9.2 Procedures for improving the efficiency of existing steam systems

Procedure[a]	Typical gains (%)	Typical costs ($ GJ⁻¹)[b]	Use in OECD[c] (%)	Use in non-OECD (%)
Repair steam traps	5	1	50	25
Insulate pipelines	5	1	75	25
Install feedwater economizers	5	10	75	50
Reduce excess air in boiler	2	5	100	50
Clean heat transfer surfaces	1–2	–	75	50
Return condensate to boiler	10	10	75	50
Improve boiler blowdown	2–5	20	25	10
Recompress vapor	0–20	30	10	0
Flash condensate	0–10	10	50	25
Use vent condensers	1–5	40	25	10
Minimize boiler short cycling	0–5	20	75	50
Insulate valves and fittings	1–3	5	50	25

Source: International Energy Agency 2006a.

[a]The Industrial Technologies Program (2004) describes these procedures.

[b]In $U.S. per unit of steam energy

[c]Organisation for Economic Co-operation and Development. OECD has 30 economically developed countries as members. The Industrial Technologies Program (2004) describes each of these procedures.

achieved through a decrease in steam demand (International Energy Agency 2006a). The chemical industry, for instance, has updated certain processes with new catalysts that permit chemical reactions to occur at lower temperatures and pressures. As a result, these processes expend less steam.

New and well-maintained equipment helps. For example, multiple-stage dryers for wood, paper, or starch operate with less steam than single-stage dryers (Industrial Technologies Program 2004). Repair or replacement of steam traps—automatic valves that remove condensed water from a steam line and return it to the boiler, while preventing the loss of live steam (i.e., steam at full pressure)—yields a 7% to 29% energy savings (U.S. Department of Energy 2008a).

Industry-specific technologies

Electric motors and steam systems are common to many industries, whereas other opportunities for greenhouse gas mitigation are more industry specific. The following sections examine technologies in iron/steel and cement manufacturing processes. International Energy Agency (2006a) or Bernstein et al (2007) provide information on other industries.

IRON AND STEEL Iron and steel manufacturing is the largest industrial source of greenhouse gases (see Figure 9.9), contributing about 7% of the world's emissions from human activities. This stems from the vast amounts of energy required for iron and steel **smelting** (purifying iron or steel from iron ore) and from several chemical reactions that directly generate CO_2. A brief introduction to heavy metal follows (drum roll and pounding bass line, please).

Earth's crust contains deposits of iron ore, which is composed of iron combined with oxygen (e.g., iron hematite, Fe_2O_3) or sulfur (e.g., iron pyrrhotite, FeS, and iron pyrite "fool's gold," FeS_2). Heating iron ore in a low-oxygen environment and in the presence of carbon from charcoal, coal, or coke expels the oxygen or sulfur and produces a carbon–iron alloy: wrought iron, carbon steel, or pig iron, depending on the proportions of iron and carbon (Table 9.3). Higher carbon content yields a harder but more brittle metal. Stainless steel is a carbon–steel alloy that contains substantial amounts of chromium, nickel, and molybdenum to improve its corrosion resistance.

Box 9.1 describes a bloomery, the earliest method of smelting iron. Today, iron and steel production follow any of three pathways. The principal pathway is the blast furnace, which produces about 60% of the world's iron and steel (Figure 9.13). A blast furnace, in contrast to a bloomery, produces a steady stream of pig iron, which requires further processing to reduce its carbon content. Several reactions in a blast furnace convert iron ore into iron:

$$C + O_2 \rightarrow CO_2 + heat \qquad (9.3)$$

$$CO_2 + C \rightarrow 2\,CO \qquad (9.4)$$

$$CaCO_3 + heat \rightarrow CaO + CO_2 \qquad (9.5)$$

$$FeS + CaO \rightarrow FeO + CaS \qquad (9.6)$$

$$Fe_2O_3 + 3\,CO \rightarrow 2\,Fe + 3\,CO_2 \qquad (9.7)$$

$$FeO + CO \rightarrow Fe + CO_2 \qquad (9.8)$$

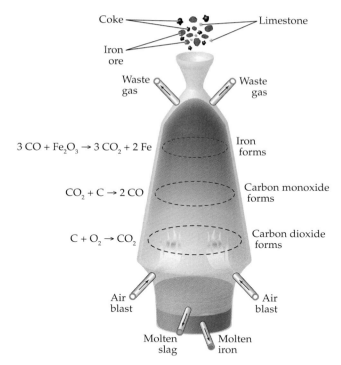

$$3\,CO + Fe_2O_3 \rightarrow 3\,CO_2 + 2\,Fe$$

$$CO_2 + C \rightarrow 2\,CO$$

$$C + O_2 \rightarrow CO_2$$

FIGURE 9.13 Blast furnace for iron and steel smelting Iron ore (primarily iron oxide), coke (purified coal), and limestone (primarily $CaCO_3$) is fed into the top of the furnace, where elemental iron forms from the reaction of the ore with carbon monoxide (CO). In the middle of the furnace, a reaction of carbon dioxide (CO_2) with the carbon in the coke produces CO. Oxidation of the coke near the air inlets heats the oven and generates CO_2. Limestone removes sulfur and other impurities from the ore and coke but releases CO_2. Molten pig iron flow out from the bottom of the furnace. Slag (a mixture of iron, oxygen, sulfur, silicon, and other impurities from the ore) is less dense and drains out from a tap hole at a higher point.

FIGURE 9.14 Coke prepared from coal

Several of these reactions generate CO_2. Blast furnaces, on average, emit 534 kg of carbon equivalents of greenhouse gases per metric ton of steel, whereas more energy efficient designs emit as little as 400 kg of carbon equivalents per metric ton (Anderson et al. 2008).

Most blast furnaces use coke as their primary carbon and energy source. To make coke, an oven heats coal in the absence of oxygen to high temperatures (up to 2000°C), driving off water and volatile hydrocarbons (coal tar and coal oil). What remains is coke, a porous material (**Figure 9.14**) that is 90% to 95% carbon, along with a few percent of various inorganic minerals such as sulfur. Preparing coke accounts for between 4% and 24% of the greenhouse gas emissions from the iron and steel industry (International Energy Agency 2006a). Newer blast furnaces that burn either coal directly, a mix of coal and coke, or a mix of coke and waste plastic decrease these emissions.

TABLE 9.3 Chemical composition of various iron products[a]

Element	Wrought iron	Carbon steel	Pig iron	Stainless steel
Iron	99–99.8	98.1–99.5	91–94	66–80
Carbon	0.05–0.25	0.07–1.3	3.5–4.5	0.1–1.0
Manganese	0.01–0.1	0.3–1.0	0.5–2.5	1.0–2.0
Sulfur	0.02–0.1	0.02–0.06	0.018–0.1	0.03
Phosphorus	0.05–0.2	0.002–0.1	0.03–0.1	0.04–0.05
Silicon	0.02–0.2	0.005–0.5	0.25–3.5	1.0
Chromium	—	—	—	10.5–20.0
Nickel	—	—	—	8.0–10.5
Molybdenum	—	—	—	0.2–7

Source: Camp and Francis 1920; Stainless Steel Information Center 2008.
[a]Units are in percent weight

BOX 9.1 Early iron smelting

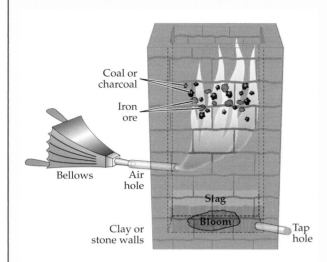

Figure A Bloomery, a furnace for iron production
A mixture of iron ore and either charcoal or coal is packed into the top. Air enters the furnace through clay pipes near the bottom by natural draft or forced with a bellows. As the carbon burns, drops of hot iron fall to the bottom and form a spongy mass called a bloom. The bottom of the bloomery also collects slag, a mixture of iron, oxygen, sulfur, silicon, and other impurities from the ore. A tap hole near the bottom of the furnace allows slag to drain. Extraction of the bloom requires the removal of bricks at the bottom of the bloomery or tipping over the bloomery.

The Iron Age commenced with the development of the bloomery (Figure A) a furnace that converts iron ore into a spongy mass of nearly pure iron called a bloom (Figure B). A bloomery heats iron to below its melting temperature, resulting in a low carbon content that makes the metal more malleable. A blacksmith then hammers or works a bloom into the desired shape, hence, the name "wrought iron" (wrought is the past and past participle of work). Advanced techniques of iron smelting and forging for strength and hardness were the secrets of the samurai sword (NOVA 2007).

Figure B A 40-pound iron bloom from a bloomery in Rockbridge, Virginia.

The second pathway, one that produces 35% of the iron and steel globally, is recycling of scrap metal. Recycling is relatively energy efficient: Electric arc furnaces that melt the scrap consume only 30% to 45% of the energy of blast furnaces per unit of iron or steel processed. Also, electric arc furnaces do not directly emit CO_2; rather, their greenhouse gas emissions depend on the source of electricity.

Metal recycling, however, requires a supply of scrap. China, which currently supplies 36% of the world's iron and steel, has little scrap and so must import most of what it uses. Consequently, prices of iron and steel scrap as well as U.S. exports of scrap have tripled during the last decade (**Figure 9.15**). The United States ships 30% of its iron and steel scrap to China, and among U.S. exports to China, this commodity ranks second in dollar value only to electronic components (U.S. Geological Survey 2008b).

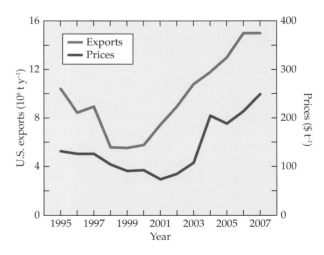

FIGURE 9.15 U.S. exports of iron and steel scrap in amounts (millions of metric tons per year) and average prices received ($U.S. per metric ton). (After U.S. Geological Survey 2008b.)

The third pathway, which produces about 5% of iron and steel, is the direct reduction of iron (DRI). This process first converts the methane (CH_4) in natural gas to carbon monoxide (CO) and hydrogen gas (H_2) (see Table 7.8):

$$CH_4 + CO_2 \rightarrow 2\,CO + 2\,H_2 \qquad (9.9)$$

Exposing iron ore to CO and H_2 at high temperatures (850°C to 1050°C) reduces the iron (i.e., strips oxygen from the ore):

$$FeO + CO \rightarrow Fe + CO_2 \qquad (9.10)$$

$$FeO + H_2 \rightarrow Fe + H_2O \qquad (9.11)$$

The iron thus generated is typically high in carbon (around 2.5%) and requires additional processing in an electric arc furnace to produce steel. Direct reduction of iron consumes more energy overall than iron smelting in blast furnaces but generates about half the CO_2 emissions because it uses natural gas instead of coke (Anderson et al. 2008).

CEMENT PRODUCTION Globally, greenhouse gases from cement production have risen to about 5% of human emissions and soon will surpass iron and steel manufacturing as the largest industrial source (see Figure 9.9). Half of cement emissions derive from energy use during processing. The other half is CO_2 released during the chemical reactions.

The chemistry of cement involves heating limestone (mostly $CaCO_3$) and small amounts of clay (mostly SiO_2) to 1450°C to make **clinker** (mostly crystals of calcium oxide silicates, $3\,CaO\cdot SiO_2$):

$$3\,CaCO_3 + SiO_2 + heat \rightarrow 3\,CaO\cdot SiO_2 + 3\,CO_2 \qquad (9.12)$$

Grinding the clinker to a fine powder and adding about 5% gypsum ($CaSO_4$) yields Portland cement, the most common type of cement. Portland cement plus gravel and sand make concrete.

Greenhouse gas abatement during cement production entails either decreasing the amount of limestone or conserving energy. A practice common in Western Europe, but not in the United States, is to replace some of the clinker in cement with oxides of silicon (SiO_2), aluminum (Al_2O_3), or iron (Fe_2O_3) from blast furnace slag (impurities in iron ore that are separated from the metal during smelting) or ash produced at coal-fired power plants. This changes some of the properties of the product and may require different curing procedures. Conserving energy may require purchasing new clinker kilns or grinding materials less finely. In addition, the United States uses mostly coal power for cement production, whereas some other countries substitute alternative sources such as nuclear power for a

TABLE 9.4 Cement production, CO_2 emissions, and use of alternative energy sources

Country	Production (10^6 tons y^{-1})	Emissions (kg CO_2 kg^{-1} cement)	Non-fossil fuels (%)
China	1300	0.90	—
India	160	0.93	—
United States	96	0.99	8
Japan	70	0.73	10
Russia	59	0.81	—
Korea, Rep.	55	0.90	
Spain	50	0.84	1.3
Turkey	48	0.85	—
Italy	44	0.84	2.1
Mexico	41	0.91	—
Brazil	40	0.82	—
Thailand	40	0.92	—
Germany	34	0.84	42
Iran	34	0.85	—
Vietnam	32	0.92	—
Egypt	29	0.85	—
Saudi Arabia	28	0.85	—
France	21	0.84	34.1

Sources: Cement Sustainability Initiative 2005; U.S. Geological Survey 2008a; Battelle Memorial Institute 2002.

substantial portion of the energy and thereby emit less CO_2 per amount of cement produced (Table 9.4).

Commercial and Residential Buildings

Buildings, if one includes the greenhouse gases produced from off-site generation of electricity and heat, are the source of 2.9 Gt C y^{-1} (2.9×10^{12} g carbon equivalents per year), or nearly one-quarter of the worldwide total from human activities (Levine et al. 2007). For comparison, commercial transportation emits less than 0.3 Gt C y^{-1}. Space heating and cooling, water heating, and lighting consume most of the energy in commercial and residential buildings (Figure 9.16).

Space heating and cooling

Although Chapter 8 contains a brief discussion of solar heating, the following section presents a broader treatment. Key to energy conservation in buildings is minimizing unwanted exchanges of thermal energy between the interior and the environment, either by conduction

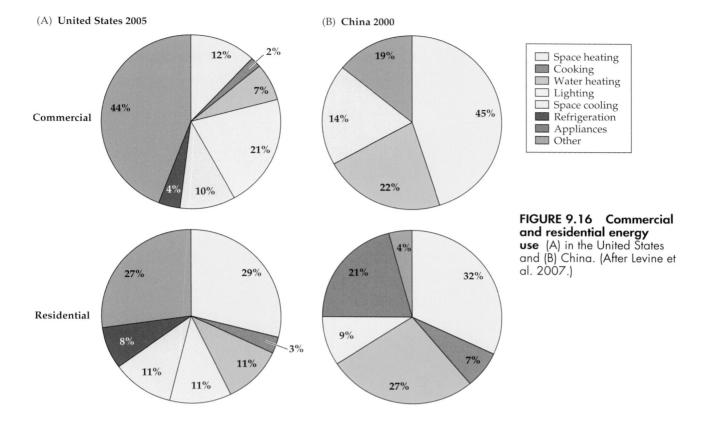

(A) **United States 2005**

(B) **China 2000**

Commercial

Residential

Space heating
Cooking
Water heating
Lighting
Space cooling
Refrigeration
Appliances
Other

FIGURE 9.16 Commercial and residential energy use (A) in the United States and (B) China. (After Levine et al. 2007.)

(transfers of kinetic energy when one molecule bumps into another), by convection (transfers of kinetic energy when a molecule moves from one place to another), or by radiation (transfers of electromagnetic energy having wavelengths that excite the motion of molecules; see Chapter 3). The extent of these energy transfers depends on air and moisture leaks, insulation levels, thermal properties of windows and doors, and HVAC (*h*eating, *v*entilating, and *a*ir *c*onditioning) equipment.

THERMAL LEAKAGE The outer shell of a building serves as a barrier to impede thermal exchange between the building and its surroundings. Air leaks driven by winds, by pressure differentials from mechanical ventilation systems, and by temperature differentials continually challenge this thermal barrier. Moisture leaks degrade the performance of insulation (water is somewhat thermally conducting) as well as influence humidity in a building and affect comfort levels (see Figure 6.36). Infrared images, which visualize temperature gradients, are helpful in pinpointing thermal leaks from buildings and guide the installation of supplemental insulation, moisture barriers, and weather stripping (**Figure 9.17**).

Temperature differential (°F)

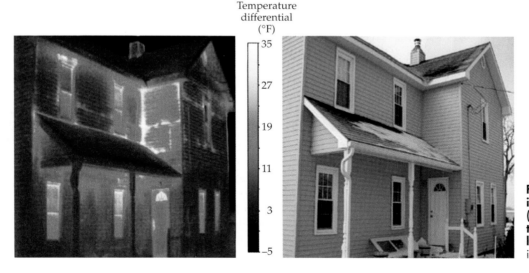

FIGURE 9.17 Infrared image (left) of a house (right) that indicates the locations of thermal leaks Image taken with an infrared camera.

TABLE 9.5 Properties of various insulating materials: thermal conductivity and embodied energy

Material	Thermal conductivity		
	Minimum RSI[a]	Maximum RSI	Embodied energy[b]
Air with no external wind	0.07 (circulating)	0.35 (still)	—
Wood chips and other loose-fill wood products	0.07	—	—
Snow	0.07	—	—
Straw bale	0.10	0.17–0.21	
Wood panels, such as sheathing	0.17	—	—
Vermiculite loose-fill	0.15	0.17	—
Perlite loose-fill	0.19	—	—
Rock and slag wool loose-fill	0.14	0.23	—
Rock and slag wool batts	0.21	0.27	16–88
Fiberglass loose-fill	0.15	0.26	—
Fiberglass rigid panel	0.17	—	—
Fiberglass batts	0.14	0.27	17–31
High-density fiberglass batts	0.25	0.35	—
Cementitious foam	0.14	0.27	—
Cellulose loose-fill	0.21	0.26	2–3
Cellulose wet-spray	0.21	0.26	—
Icynene spray	0.25	—	—
Cotton batts (blue jean insulation)	0.26	—	—
Icynene loose-fill (pour fill)	0.28	—	—
Urea-formaldehyde foam	0.28	0.32	—
Urea-formaldehyde panels	0.35	0.42	—
Polyethylene foam	0.21	—	—
Phenolic spray foam	0.33	0.49	—
Phenolic rigid panel	0.28	0.35	—
Molded expanded polystyrene (EPS)	0.26	0.28	—
Extruded expanded polystyrene (XPS)	0.25–0.33	0.35–0.37	61–113
Open-cell polyurethane spray foam	0.25	—	—
Closed-cell polyurethane spray foam	0.38	0.47	96–144
Polyurethane rigid panel (pentane expanded)	0.38 aged[c]	0.47 initial	—
Polyurethane rigid panel (CFC/HCFC expanded)	0.43 aged	0.55 initial	—
Polyisocyanurate spray foam	0.30	0.58	—
Foiled polyisocyanurate rigid panel (pentane expanded)	0.38 aged	0.47 initial	—
Silica aerogel	0.69	—	—
Vacuum insulated panel	—	2.08	—
Cardboard	0.21	0.28	113–127

Sources: Wikipedia 2008a, b; Harvey 2007.
[a]RSI is in units of K m^2 W^{-1} per centimeter of material, although conductivity often is not linear with thickness. To convert to American R units (ft$^2 \cdot °$F·h Btu^{-1} in^{-1}), multiply by 14.42.
[b]Energy required during manufacture. In units of MJ m^{-2} RSI^{-1} for a 1 m^2 isolation panel with an RSI of 1.
[c]Aged refers to values of foam 5 to 10 years after expansion.

INSULATION Adequate levels of insulation not only conserve energy, but also permit buildings to downsize their heating and cooling equipment. Building codes in most localities have extensive sections on insulation and, with every update, tend to dictate more insulation. To comply with building codes, contractors must tally the thermal properties of all building materials.

Building insulation comes in many forms (Table 9.5). The primary requirements of an insulating material are that it has low thermal conductance and fills space to minimize convective transfers via air movements. Some insulation is covered with shiny foil to decrease radiative transfers, and others contain moisture barriers to minimize leaks. Certain types of insulation (e.g., polyurethane) have desirable thermal properties, yet their manufacture may require an inordinate amount of energy for materials and processing.

Installing insulation during the initial construction of a building is less expensive than retrofitting an existing building. The mean lifetime of buildings, however, is in the range of 40 years to 120 years (see Figure 8.1), and adding insulation to older buildings may reduce their energy requirements by 12% to 40% (International Energy Agency 2006a; Levine et al. 2007). In some cases, a building has access spaces that will accommodate additional insulation, or an insulating layer may be applied internally or externally. In other cases, foam may be injected into air cavities. Unfortunately, many of these foams release greenhouse gases such as CO_2 and halocarbons (CFCs or HCFCs). The payback time—time required for heating-related emission savings to offset emissions during manufacturing and installation—ranges from 10 years for non-halocarbon foams to over 100 years for halocarbon foams (Harvey 2007).

WINDOWS AND DOORS Buildings usually exchange a disproportionate amount of thermal energy, 10% to 40%, through windows and doors (International Energy Agency 2006a). Insulation values are assessed in units of the temperature differential that can be maintained over a surface area per unit of power (energy per time); the metric units are degrees $K \cdot m^2 \ W^{-1}$ and the American customary units are R-values in $ft^2 \cdot °F \cdot h$ Btu^{-1}: Higher values mean greater insulation. A window of single-pane glass has a value of 0.16 K m^2 W^{-1} (R-0.9), and a simple panel door has a value of 0.33 K m^2 W^{-1} (R-1.9) (Table 9.6). In contrast, walls, ceilings, and floors in U.S. buildings typically have at least 2.0 K m^2 W^{-1} (R-11).

Window upgrades include:

1. Replacing solid aluminum window frames with either aluminum frames having thermal barriers or frames made of wood, vinyl, wood-plastic composites, or fiberglass;

2. Increasing the number of panes of glass (e.g., replacing single-pane glass with double-pane glass);

3. Increasing the air space between panes of glass;

4. Using glass that has reflective coatings or films between the panes that block the entry of ultraviolet light and the exit of infrared light; or

5. Filling the space between panes with argon or krypton, gases that have less thermal conductivity than air.

TABLE 9.6 Insulation value of walls, windows, and doors[a]

Material	RSI
Walls, wood frame	
90 mm fiberglass, 12 mm foam	2.04
152 mm fiberglass, 12 mm foam	2.63
Windows with aluminum frames	
Single-pane	0.14
Double-pane, 1.3 mm air space	0.21
Double-pane, 1.3 mm, coating	0.26
Windows with vinyl or wood frames	
Single-pane	0.16
Single-pane and storm window	0.35
Double-pane	
6 mm air space	0.29
1.3 mm air space	0.36
1.9 mm air space	0.42
1.3 mm, coating	0.55
1.3 mm, coating, argon	0.61
1.3 mm, films	0.68
1.3 mm, film, coating	0.71
Triple-pane	
6 mm air spaces	0.45
1.3 mm air spaces	0.57
Doors	
Storm, metal and glass	0.17
Wood	
panel	0.33
hollow core	0.38
solid core, 44 mm	0.53
solid core, 57 mm	0.65
Metal, insulated with 50 mm foam	2.64

Sources: Warner 1995; Oak Ridge National Laboratory 2004; ColoradoENERGY 2008.
[a]In RSI units (K m^2 W^{-1}). To convert values to American R units ($ft^2 \cdot °F \cdot h$ Btu^{-1}), multiply by 5.675.

Windows that use a combination of these approaches can achieve insulation values as high as 1.23 K m^2 W^{-1} (R-7), an eight-fold improvement over single-pane glass.

Most buildings, with the exception of motels, have far fewer exterior doors than windows. Nevertheless, doors are responsible for a large portion of the thermal transfers between buildings and their environment. Insulation values for doors differ greatly (see Table 9.6). Designing entryways with a revolving door or a series of doors can save more than 14% of the energy use in a building (Sustainability@MIT 2007).

HVAC Heating, ventilation, and air conditioning, once a rarity in developing countries, is now widespread and will more than double the building energy usage in these countries during the next few decades (Levine et al. 2007). Fortunately, the energy efficiency of HVAC

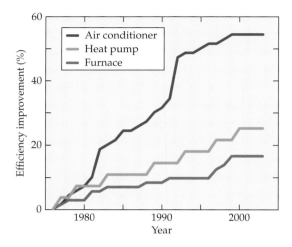

FIGURE 9.18 Improvement (%) in the energy efficiency of air conditioners, heat pumps, and furnaces in the United States since 1975 The standard measure for air conditioners and heat pumps is COP, coefficient of performance, the useful heat movement (joules) per energy input (joules). For furnaces, the measure is AFUE, annual fuel utilization efficiency, the percentage of energy in a fuel that is extracted as heat. (After HVACWebsite.com 2008.)

equipment has improved steadily over the past 30 years (**Figure 9.18**), and this equipment has a faster replacement rate (15 years to 20 years) than the buildings themselves (40 years to 120 years) (see Figure 8.1).

- The average furnace for residential heating in the United States has increased in efficiency (thermal energy divided by chemical energy in fuel) from 73% in 1975 to 85% in 2006 (HVACWebsite.com 2008). Larger units that remove some of the water vapor from the exhaust typically exceed 88% efficiency (Levine et al. 2007).

- Mechanical ventilation systems that exploit premium electric motors and electronic adjustable speed drives (see the discussion of motors earlier in this chapter) reduce building energy requirements by 10% to 15% (International Energy Agency 2006a).

- New buildings are again taking advantage of natural ventilation through courtyards, atria, wind towers, solar chimneys, and windows that actually open; for example, appropriate ventilation of houses in California could cut their energy use by 16% to 53% (Loisos and Springer 2000).

- Air conditioners in the United States have improved in efficiency (cooling power divided by

fan and compressor power) from 1.8 to 2.8 over the past 30 years (HVACWebsite.com 2008). More efficient mini-split systems available in Japan have efficiency values as high as 6.8, and large commercial units can achieve values up to 7.9 (Levine et al. 2007).

- Heat pumps expend electrical energy to transfer thermal energy between a building and its surroundings and thus can heat or cool a building. Soil temperatures below about 10 m in depth approach the average temperature of a location.

  If the annual average temperature of a location is 20°C, then thermal exchange with deeper soils can heat a building in winter and cool it in winter. This reversibility makes heat pumps less efficient than dedicated air conditioners or heaters, however. Still, the efficiency of heat pumps in the United States improved by 25% between 1975 and 2006 (see Figure 9.18).

WEB TOPIC 9.1 **Temperature changes with depth in the soil** As one descends to a depth of a few meters, the temperature approaches the annual average for that location.

- "Thermostat wars" rage daily in houses and office buildings all over the world. Peace in our time may be achievable through flexible control systems that monitor and independently adjust temperatures in multiple locations. Manufacturers of such control systems claim energy savings of 35% to 40%.

Hot water

Fossil fuel or electricity heats most of the water in commercial and residential buildings, an activity that comprises between 7% and 27% of the energy that buildings consume (see Figure 9.16). Several technological advances can decrease this energy use:

1. Solar thermal water heaters are becoming more common (see Figure 8.49) and can meet from 50% to 90% of the hot water needs in a building, depending on climate.

2. Newer appliances, such as more water-efficient washing machines and dishwashers, use 50% less hot water than older models.

3. More-efficient and better-insulated tank water heaters can attain energy savings of 10% to 20%.

4. Tankless water heaters that are located close to the points of use (e.g., near a shower or a kitchen sink) eliminate heat losses during storage or distribution and require 30% less energy. These, together with recovery of heat from waste water and exhaust air, can save 90% of the energy now expended for hot water (Levine et al. 2007).

Lighting

"How many Americans does it take to screw in a lightbulb?" The answer is, "Nearly every American will need to screw in a lightbulb." The U.S. government passed a law (the Energy Independence and Security Act) on December 19, 2007 banning the sale of incandescent lightbulbs by the year 2014, with a few exceptions (decorative lights, appliance bulbs, and three-way bulbs). Many other countries are enacting similar legislation. So why should incandescent lightbulbs be replaced?

INCANDESCENTS Incandescent lightbulbs emit as visible light only 5% to 10% of the electricity that they consume; the remainder of their electromagnetic radiation is longer wavelengths, which we perceive as heat (see Chapter 3). Incandescent bulbs expend almost 5% of global energy and contribute 0.4 Gt (4×10^{11} g) carbon equivalents of greenhouse gases annually, or about 4% of the total (International Energy Agency 2006b). Other lighting sources are many times more efficient in terms of the light provided per energy input (**Figure 9.19**).

Incandescent lightbulbs (**Figure 9.20A**) currently hold 80% of the market for residential lighting in the United States, 80% in the United Kingdom, 50% in Germany, and 20% in Japan (Harden 2007; EPA Newsroom 2008). These bulbs have a much cheaper purchase price than any of the alternatives, fit all common lighting fixtures, and deliver the "warm" light enriched in the longer wavelengths to which we are accustomed. Consequently, consumers have resisted changing to other types of bulbs (Harden 2007).

Halogen lamps (**Figure 9.20B**) are incandescent bulbs in which the tungsten filament is sealed in a fused quartz or high melting-point glass envelope that is filled to high pressures (7 atmospheres to 8 atmospheres) with an inert gas such as argon plus a small amount of a halogen gas such as iodine or bromine. When the bulb temperature exceeds 400°C, tungsten evaporates from the filament and reacts with the halogen. The tungsten halide circulates back to the hot filament (above 2500°C), where it disassociates, depositing the tungsten back on the filament and extending

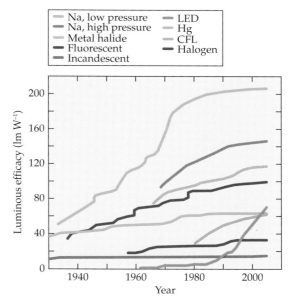

FIGURE 9.19 Luminous efficacy (lumens per Watt) of lamps This is the fraction of electromagnetic power that is useful for lighting and considers the sensitivity of the human eye to various wavelengths. "Fluorescent" denotes long fluorescent tubes; "LED," white light–emitting diode; "Hg," high-pressure mercury; "CFL," compact fluorescent lamp; and "Halogen," tungsten halogen. (After International Energy Agency 2006a; Steele 2007; Taguchi 2008.)

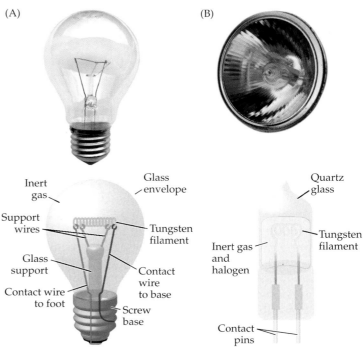

FIGURE 9.20 Filament lamps (A) Components of an incandescent lightbulb. (B) Components of a halogen lamp.

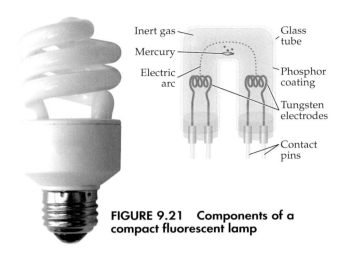

FIGURE 9.21 Components of a compact fluorescent lamp

FIGURE 9.22 Color appearance of lamps (left to right): GE 13-Watt "sunlight," 6500 K compact fluorescent lamp (CFL); Sylvania 60-Watt "extra soft white" incandescent; Bright Effects 15-Watt, 2644 K CFL; and Sylvania 14-Watt, 3000 K CFL.

the lifespan of the bulb. The higher temperatures of halogen lamps provides light that is richer in the visible wavelengths than regular incandescent bulbs (see Equation 3.3), and thus halogens attain higher efficiencies; that is, they provide more useable light per energy input (see Figure 9.19). On the other hand, the higher temperatures of halogens pose a greater risk of burns and fire. Many institutions, including college dormitories, ban the use of halogen lighting. Currently, halogen lamps comprise only 4.6% of the international lighting market (International Energy Agency 2006b).

FLUORESCENTS Fluorescent lamps (**Figure 9.21**) produce 64% of the electric light in the world (International Energy Agency 2006b). A fluorescent light fixture consists of a bulb and a **ballast**. The ballast is an electronic circuit that provides high voltage to start the bulb and then a steady current to keep the bulb illuminated. The initial high voltage ionizes the inert gas (usually argon) in the bulb. As current begins to flow from one end of the bulb to the other, mercury in the bulb vaporizes (i.e., changes phase from a liquid to a gas), fills the bulb, conducts current, and emits ultraviolet radiation. This ultraviolet radiation excites the phosphors coating the inside of the bulb, and they emit visible light. Although the phosphors that were available in the 1940s through 1970s emitted wavelengths rich in blue (cool-white), now a range of phosphors provide a choice of light spectra from those that nearly match incandescents to those that nearly match sunlight (**Figure 9.22**).

Sales of compact fluorescent lamps (CFLs) have increased exponentially over the past decade (**Figure 9.23**) and by 2003 represented 6% of the global lighting market (Lefèvre et al. 2006) and 19% of the deliv-ered light (International Energy Agency 2006b). This growth in sales is likely to continue as new regulations about the use of incandescent bulbs take effect. Although CFLs are less efficient than traditional, long fluorescent tubes (see Figure 9.19) because the bulbs and ballasts in CFLs are contorted into shapes that fit into standard fixtures, they are four- to five-fold more efficient than the incandescent bulbs that they replace. Moreover, CFLs typically last 5 to 20 times longer than incandescent bulbs (Lefèvre et al. 2006).

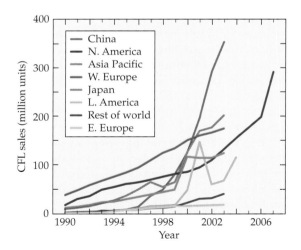

FIGURE 9.23 Sales of compact fluorescent lamps (millions of units) in various countries (After Lefèvre et al. 2006; EPA Newsroom 2008.)

Nonetheless, CFLs present several major problems:

- CFLs cost three to ten times more than their incandescent counterparts. The saving in energy and replacement costs for CFLs, however, will payback the additional purchase price in less than 6 months, assuming typical electricity rates and lamp usage (Energy Star 2008).

- Most CFLs cannot be dimmed with standard circuits, and those that are dimmable adjust from only 20% to 80% of full output. By contrast, incandescent bulbs can be dimmed over their full range, from 0% to 100%.

- CFLs, like all fluorescent bulbs and other high intensity discharge lamps, contain small amounts of mercury and, if broken, may release vapors that can exceed U.S. federal guidelines for chronic exposure (AB1109 2007). Newer bulbs contain less mercury and, with proper disposal, pose minimal health risks.

- Standard CFLs fail prematurely if exposed to the higher temperatures experienced in enclosed or recessed light fixtures. Special CFLs are available for such purposes.

- CFLs may emit only low light levels when first ignited and may take a minute to reach full brightness.

- CFLs grow dimmer with use and, by the middle to end of their lifespan, produce 70% to 80% of their original light output.

CFLs seem to be a viable short-term solution, however, and their widespread adoption would eliminate 7% of household greenhouse gas emissions (Lefèvre et al. 2006).

LEDS Light-emitting diodes are semiconductor devices that work almost like photovoltaic solar cells (see Figure 8.36) in reverse. A semiconducting material (e.g., gallium indium nitride, gallium arsenide, or gallium phosphide) is infused with impurities to create a p-n junction (see Chapter 8). Setting the p-side (anode side) to a higher voltage (i.e., higher electrical potential) than the n-side (cathode side) causes charge carriers to flow into the p-n junction. Charge carriers can be either electrons or holes, places in the crystal lattice where electrons are absent. If an electron meets a hole, it falls to a lower chemical energy level and releases electromagnetic energy in the form of a photon (a particle of light). The wavelength of the photon emitted depends on the semiconductor material.

Today, LEDs that emit a wavelength ranging from the infrared to the ultraviolet are available (**Figure**

FIGURE 9.24 **Light-emitting diodes (LEDs) of various colors**

9.24). Most suitable for interior lighting are "white" LEDs (**Figure 9.25**). These are modified blue LEDs in which the semiconductor chip is coated with a phosphor that absorbs blue wavelengths of around 465 nanometers (nm; 10^{-9} m) and emits into a broad spectrum centered around yellow, from about 500 nm to 650 nm (**Figure 9.26**). The human eye perceives the combination of blue and yellow as white.

LEDs currently account for a negligible fraction of lighting because of performance and cost issues (International Energy Agency 2006b), but they are improving rapidly: Light output from a single LED has doubled every 2 years, cost per light output has halved every 3 years, and luminous efficacy has doubled every 6 years

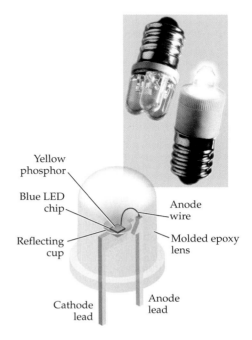

Yellow phosphor

Blue LED chip

Reflecting cup

Anode wire

Molded epoxy lens

Cathode lead

Anode lead

FIGURE 9.25 **Components of a white light-emitting diode**

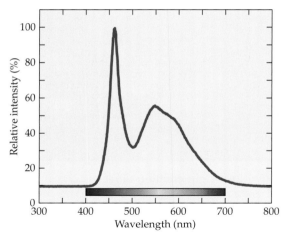

FIGURE 9.26 Light spectrum of a white LED This consists of a blue LED (hence the peak at 465 nm) with phosphors that emit yellow light. The rainbow inset shows the apparent colors of the various wavelengths.

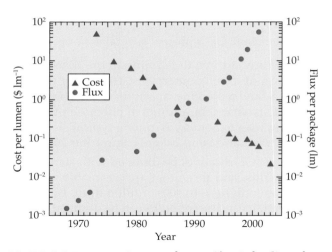

FIGURE 9.27 LEDs: Cost per lumen ($U.S. lm⁻¹) and light flux (lumens) per each package containing one or several semiconductor chips Note that the cost and flux axes are on a logarithmic scale. (After Steele 2007.)

(**Figure 9.27**; see also Figure 9.19). Sales of high-output LEDs expanded by 42% between 1995 and 2005 (Steele 2007). They are now common in traffic lights, vehicle lights, flashlights, retail display cases, and backlights for screens of cell phones and notebook computers. In comparison to fluorescent bulbs, LEDs are longer lasting, more robust, faster responding, easier to dim, smaller, and mercury free. LEDs are destined to become a larger segment of the lighting market as their technology advances.

HIDS High-intensity discharge lamps include mercury, metal halide (**Figure 9.28A**), and sodium lamps (**Figure 9.28B**). They produce light from an electric **arc** (i.e., current flowing through a normally non-conductive medium) between tungsten electrodes housed inside a fused quartz or high melting-point glass tube. This tube contains both an inert gas (usually argon) and metal salts. An electronic or magnetic ballast initially generates a high voltage that ionizes the inert gas. This starts the electric arc, heats the inner tube, and vaporizes the metal salts to produce a light-emitting plasma (a gas in which a portion of the atoms have electrical charges). Once the plasma forms, the ballast sustains the arc at a low, steady current.

Mercury lamps emit light that is rich in ultraviolet and blue wavelengths (i.e., having peaks at 253.7 nm, 365.4 nm, 404.7 nm, 435.8 nm, 546.1 nm, and 578.2 nm). These cast a ghostly pall over the illuminated area (**Figure 9.29**) and require UV filters in the outer glass envelope to

minimize the discharge of dangerous radiation. Metal halide lamps add to the mercury other chemicals, such as sodium iodide and scandium iodide, that serve to broaden the spectral distribution of light emitted, giving it the appearance of natural daylight; these lamps

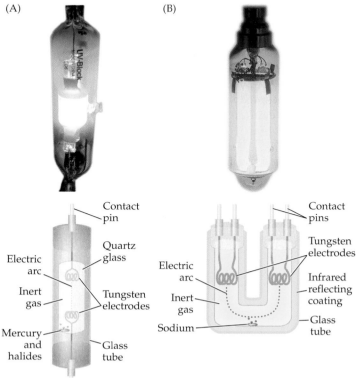

FIGURE 9.28 Arc lamps (A) Metal halide lamp. (B) Sodium lamp.

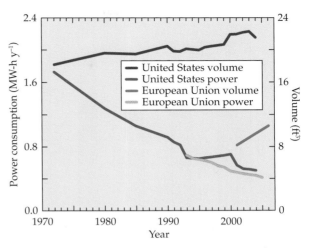

FIGURE 9.29 Mercury vapor lighting (left) and high-pressure sodium lighting (right) on a highway ramp near Everett, Washington

FIGURE 9.30 Power consumption (10^6 Watt-hours per year) and volume (cubic feet) of refrigerators in the United States and European Union. (After Appliance Magazine.com 2006; Bertoldi and Atanasiu 2007; Ellis 2007; U.S. Department of Energy 2007a.)

provide illumination for nighttime sporting events. Low-pressure sodium lamps emit nearly pure yellow light (having peaks at 589.0 nm and 589.6 nm). High-pressure sodium lamps also contain mercury, and the blue light from mercury in addition to the accentuation of minor wavelengths from sodium under pressure gives these bulbs a whiter appearance.

HID lamps are relatively expensive, about $150 to $250 for a unit that includes a fixture, ballast, and bulb. Prices do not vary much with size from 75 Watts (W) to 400 W, and so most installations have a few large units rather than many small units. HIDs account for only 1.2% of global lamp sales, but they deliver 29% of the electric light because they have higher efficacy (see Figure 9.19), greater average power, and longer lifetimes than most other lighting sources (International Energy Agency 2006b). Their high purchase price, large size, and hot bulbs make it unlikely, however, that HIDs will further penetrate the residential lighting market (except in closet greenhouses).

Appliances

Appliances consume 40% of the electric power in residential and commercial buildings if one includes household appliances, office equipment, and consumer electronics (Levine et al. 2007). The major electricity-sappers are electronics, refrigerators, computers, ovens/stoves, and washers/dryers (Table 9.7). Technological advances have improved the energy efficiencies of many of these.

Refrigerators, for example, are several times more energy efficient today than they were in the 1970s, even though they are larger (**Figure 9.30**). The higher efficiencies are the consequence, in part, of tougher efficiency standards that governments have imposed (Nadel 2002; Ellis 2007). These standards dictate the use of better insulation, tighter seals, premium electric motors, less frequent heating of the walls to eliminate frost, and more efficient exchange coils. Because refrigerators in the United States have an average lifespan of 14 years (U.S. Department of Energy 2007a), turnover to high-efficiency models is slow, and refrigeration still consumes 4% to 8% of the electricity in U.S. buildings (see Figure 9.16). The number of refrigerators is increasing more than 5% per year in developing countries, and so adoption of efficiency standards in these countries

TABLE 9.7 Percentage of energy consumed by different appliances in U.S. residential and commercial buildings

Appliance	Residential	Commercial
Electronics	29.1	40.1
Refrigerators	29.7	26.4
Computers	4.4	20.8
Ovens and stoves	17.8	12.6
Washers and dryers	19.0	—

Source: U.S. Department of Energy 2007a.

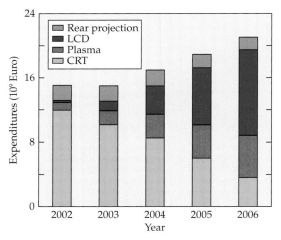

FIGURE 9.31 Purchases (billions of euros) of different types of televisions in western Europe "CRT" designates cathode ray tube and "LCD," liquid crystal display. (After Bertoldi and Atanasiu 2007.)

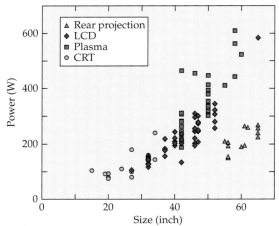

FIGURE 9.32 Power consumption (Watts) versus size (inch diagonal) of television "CRT" designates cathode ray tube and "LCD," liquid crystal display. (After CNET 2008.)

could be a highly effective way of conserving energy (McNeil and Letschert 2005; Yusuf et al. 2007).

A television is usually the first purchase after a household obtains electricity (McNeil and Letschert 2005). In only 5 years, the market has shifted from televisions based on cathode ray tubes (CRTs) to those based on plasma and liquid crystal displays (LCDs) (**Figure 9.31**). LCDs have a higher energy efficiency at a given display size than CRTs (**Figure 9.32**), but LCDs are typically larger than CRTs, negating any benefits in terms of energy consumption per set. LCDs generally use fluorescent tubes to serve as backlights but are beginning to incorporate LEDs for this purpose, and this should improve their energy efficiency. Rear-projection televisions that use HIDs or LEDs as projection lamps enjoy the highest energy efficiencies.

Inner space

Size matters and may supersede efficiency. For instance, vehicles are 70% more fuel efficient per unit weight than they were 30 years ago (see Figure 7.25), yet engine sizes have doubled, negating any change in vehicle fuel efficiency (see Figure 7.21). Similarly, improved energy efficiencies of U.S. residences have led to about a 10% decline in **energy intensity** (i.e., energy expended per unit of floor area) over the past 2 decades (**Figure 9.33**), yet floor area of a typical residence and number of residences have both increased by over 30% during this period. Hence, residential buildings use 30% more energy than they did 20 years ago.

Commercial buildings in the United States use 30% more energy per floor area (i.e., they have a higher

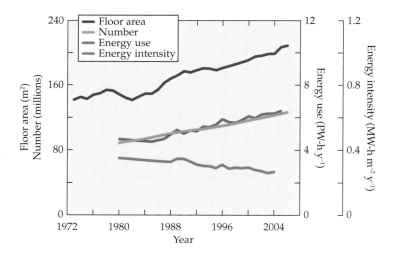

FIGURE 9.33 Median floor-area of one-family houses (m²), number of households (millions), total residential energy consumption (10^{15} Watt-hours per year), and energy intensity (10^6 Watt-hours per m² per year) in the United States. (After U.S. Department of Energy 2007a; U.S. Census Bureau 2008; U.S. Department of Energy 2008c.)

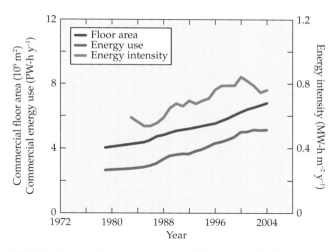

FIGURE 9.34   Total commercial floor area (billion m², total commercial energy consumption (10¹⁵ Watt-hours per year), and energy intensity (10⁶ Watt-hours per m² per year) in the United States (After U.S. Department of Energy 2007a; Energy Information Administration 2008c; U.S. Department of Energy 2008b.)

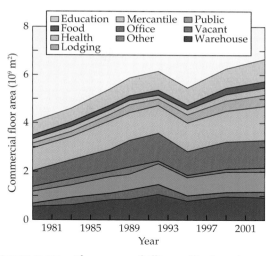

FIGURE 9.35   Floor area (billion m²) of various commercial activities in the United States (After Energy Information Administration 2008c.)

energy intensity) than they did 2 decades ago (**Figure 9.34**), a consequence of the shift from low energy uses, such as vacant space and warehouses, to high energy uses, such as offices and shops (**Figure 9.35**). This higher energy intensity, together with the addition of new floor space, doubled energy consumption by commercial buildings from 1984 to 2004.

Geoengineering

Some fantastic proposals to mitigate global climate changes fall into the realm of geoengineering. **Geoengineering** is the study of large-scale modifications in Earth's environment for human benefit. The following section examines two geoengineering approaches—changing Earth's albedo and fertilizing Earth's oceans—and considers whether they might enhance the future habitability of our planet.

Global dimming

Albedo is the proportion of sunlight reflected from Earth (see Chapter 3). All else being equal, a higher albedo means a cooler planet. Geoengineering methods to increase Earth's albedo, or dim the planet's surface, include placing sunshades into orbit, releasing sulfur dioxide into the upper atmosphere, and seeding clouds.

SUNSHADES IN SPACE   Placing 16 trillion (1.6 × 10¹³) refractive screens (screens that bend the light pass-

ing through them), each 0.6 meters in diameter, into a solar orbit 1.5 million km above Earth would cast a diffuse shadow (penumbra) that would shade the entire Earth (**Figure 9.36**) and diminish the solar radiation that impinges on Earth by 1.8%, an amount that would approximately compensate for the climate warming from a doubling in atmospheric CO_2 concentrations (Angel 2006). To place the screens in orbit, a vertical tube 2 km long would use electromagnetic pulses to launch a rocket filled with screens every few minutes. Once released into space, the refractive screens would orient themselves via an on-board computer, camera, and solar sails and form an elliptical array measuring 6200 km by 7200 km. Use of refraction rather than reflection would diminish sunlight-induced pressure that would otherwise shift the screens' orbit. Development and deployment of this system would require about 25 years and cost several trillion dollars (Angel 2006).

SULFUR DIOXIDE IN THE UPPER ATMOSPHERE   When Mount Pinatubo in the Philippines erupted in June 1991, it spewed so much sulfur dioxide into the upper atmosphere that Earth's albedo increased by 0.7% (see Chapter 3). This cooled global temperatures by 0.5°C for over a year but did not seriously disrupt climate systems. Humans could mimic this phenomenon and launch balloons or shoot artillery shells to release sulfur (S₂) or hydrogen sulfide (H₂S) into the upper atmosphere (Crutzen 2006; Wigley 2006). This would form

FIGURE 9.36 Proposal to send screens that deflect sunlight into orbit (A) Array of screens in solar orbit 1.5×10^9 m above Earth that would cast a shadow to cool the planet (not to scale). (B) Some of the 16 trillion sunlight-refracting screens proposed for deployment 1.5 Gm from Earth. Each screen would be about 0.6 m in diameter.

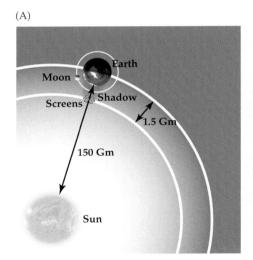

(A)

Earth
Moon
Screens   Shadow
1.5 Gm
150 Gm
Sun

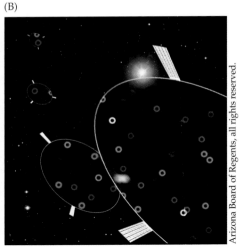

(B)

aerosols (nanometer-sized particles) of sulfate (SO_4^{2-}) that reflect sunlight and cool the planet. Combustion of fossil fuel also emits sulfur dioxide (SO_2), but this generates relatively large sulfate particles in the lower atmosphere (**Figure 9.37**) that stay airborne for only about a week. Sulfate aerosols in the upper atmosphere, by contrast, have residence times of several years, and so small amounts—about 7% of current sulfur dioxide (SO_2) emissions from the combustion of fossil fuels—would balance the global warming from a doubling in atmospheric CO_2 concentrations (Crutzen 2006; Wigley 2006).

CLOUD SEEDING As cloud cover increases, so does Earth's albedo, and the planet cools (see Figure 3.40). One approach to enhance cloud cover is cloud seeding, whereby very small particles (typically 0.2 μm in diameter) sprayed into clouds act as nucleation centers that promote additional condensation of water vapor into a liquid. This does not make new clouds; it just makes existing clouds denser. For example, exhaust from ship engines releases very small particles that stimulate the formation of dense clouds in still ocean air (**Figure 9.38**).

Unmanned cloud-seeding sailing ships (**Figure 9.39**) might ply the seas to intensify low-level marine clouds (Salter et al. 2008). These ships would run downwind and drag turbines through the water to generate electricity for turning the rotor sails and for spraying seawater from the tops of the rotors. Fine droplets of saltwater serve as ideal nucleation centers. A working fleet of 2000 ships, each with a displacement of 300 metric tons, might increase cloud density sufficiently to compensate for a doubling in atmospheric CO_2 concentrations (Salter et al. 2008).

Fertilizing oceans

Resources such as sunlight, water, and nutrients are vital to aquatic and terrestrial plants alike (see Figure 5.4). Oceans, although they receive the same amount of sunlight as land areas and have greater access to

FIGURE 9.37 Brownish haze of sulfate, nitrate, and carbon aerosols that streams down from the southern edge of the Himalayas over western India, Bangladesh, and the Bay of Bengal. NASA's Terra satellite acquired this image on December 4, 2001.

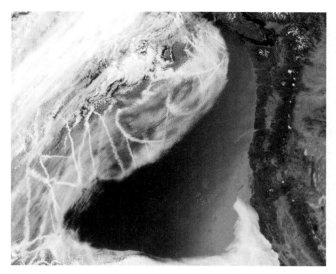

FIGURE 9.38 Ship tracks—long, dense white clouds against an uneven, lighter cloud background—that develop from the exhaust of ships on the west coast of the United States from Seattle to San Francisco.

FIGURE 9.39 Cloud-seeding ship with Flettner rotors that spin about a vertical axis and act as powerful computer-controlled sails. Seawater sprays from the tops of the rotors to seed clouds. (After Salter et al. 2008.)

water, have lower average photosynthetic productivity (**Figure 9.40**) because of nutrient limitations. Some places—specifically, the eastern equatorial Pacific, northeastern subarctic Pacific, and Southern Ocean—have relatively high concentrations of most nutrients, yet support only low densities of phytoplankton (free-

floating plants). Addition of iron to these waters generates massive phytoplankton blooms (**Figure 9.41**). These phytoplankton blooms convert additional atmospheric CO_2 into organic carbon, some of which sinks into the deep blue sea, where turnover back to the atmosphere may take millennia.

0   1   2   3
Net primary productivity
(kg C m^{-2} y^{-1})

FIGURE 9.40 Average net productivity (kg carbon sequestered per square meter of area per year) of vegetation on land and in the oceans during 2002. Dashed lines indicate the three high-nutrient, low-chlorophyll zones. Gray areas are those lacking data.

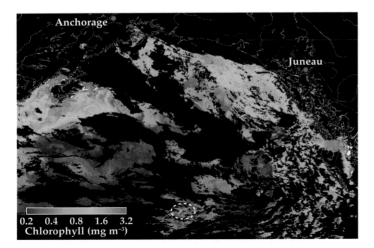

FIGURE 9.41 Iron fertilization experiment in the northeastern Pacific Research vessels added iron to an 8 km² area in the middle of the dotted white ellipse (bottom center) on July 9 and 16, 2002. The OrgView-2 satellite measured surface chlorophyll levels on July 29, 2002 and found ten-fold increases in the fertilized area. Black areas represent those without data because of cloud cover.

The potential of this approach for greenhouse gas mitigation depends on how much carbon is sequestered per unit of iron applied. Large-scale experiments in the Southern Ocean found that this ratio (kg C kg⁻¹ Fe) was only about 1% of that predicted from laboratory studies. Apparently, phytoplankton cannot absorb most of the iron added to the ocean, physical reactions in the ocean remove a substantial portion of the iron, and only 5% to 25% of the organic carbon from phytoplankton photosynthesis exits surface waters because other organisms near the surface efficiently recycle it. Iron fertilization of oceans is therefore unlikely to influence significantly atmospheric CO_2 concentrations (Buesseler and Boyd 2003; Boyd 2004).

Summary of geoengineering

The scale, complexity, and cost of geoengineering worry nearly everyone. "How can you engineer a system whose behavior you don't understand?" asks Ronald Prinn, a climate scientist at the Massachusetts Institute of Technology (Morton 2007). All of the geoengineers cited in this section consider their proposals to be last resorts, actions to be taken only if abatement of greenhouse gas emissions fails to avert catastrophic climate change. Fortunately, as detailed in Chapters 7 through 9, many alternatives involving energy conservation are available.

Summary

Greenhouse gas emissions from the agricultural, forestry, industrial, commercial, and residential sectors produce the majority of greenhouse gases. Agriculture and forestry contribute most of the methane and nitrous oxide: Methane derives from the flooding of soils and the belching of domestic animals; nitrous oxide derives from fertilization and flooding of soils. Opportunities to mitigate greenhouse gas emissions from the agricultural sector include more judicious use of farm machinery and better management of irrigation water and fertilization, such as through fertigation. Heavy industries such as metal smelting and cement manufacture cause carbon dioxide emissions from the electricity needed to fuel the processes as well as from the production procedures themselves. The use of premium motors and motor controls, newer and cleaner steam generators, and more efficient furnaces is diminishing greenhouse gas emissions from many industries. The incorporation of better insulation in buildings, the installation of solar hot water heaters, the shift away from incandescent lighting, and the replacement of old appliances such as refrigerators hold the promise to reduce building energy consumption and, thereby, greenhouse gas emissions.

Review Questions

1. Agriculture and forestry are responsible for most of the emissions from human activities of (select all that apply)
   (a) carbon dioxide.
   (b) methane.
   (c) nitrous oxide.
   (d) chlorofluorocarbons.
   (e) hydrochlorofluorocarbons.

2. Agriculture lands have *not* been
   (a) expanding in developing countries.
   (b) expanding as fast as human population growth.
   (c) intensely managed for high productivity.
   (d) dedicated to producing more feed for domestic animals.
   (e) losing soil carbon.

3. Nitrogen fertilization is (select all that apply)
   (a) increasing rapidly in the United States.
   (b) the major source of greenhouse gas emissions from agriculture.
   (c) releasing physiologically significant amounts of laughing gas.
   (d) a major source of groundwater contamination.
   (e) a major expense.

4. Minimum tillage does *not*
   (a) require that agriculture resume primitive methods.
   (b) diminish carbon dioxide emissions from farm operations.
   (c) reduce use of fossil fuels.
   (d) decrease soil compaction.
   (e) depend on technological advances such as the development of herbicides or the use of global position systems.

5. Methanogenesis
   (a) is the conversion of nitrous oxide to methane.
   (b) is more prevalent under aerobic conditions.
   (c) occurs in rumens and flooded soils.
   (d) is responsible for subterranean fires.
   (e) occurs in wheat fields.

6. Most of the methane emissions from domestic animals
   (a) comes out their anus.
   (b) comes out their mouth.
   (c) perspires from their skins.
   (d) metabolizes internally.
   (e) is captured during biogas production.

7. Forested land is generally (select all that apply)
   (a) decreasing in area worldwide.
   (b) expanding in area worldwide.
   (c) a potential source of nitrous oxide emissions.
   (d) a potential sink for carbon dioxide emissions.
   (e) a potential sink for methane emissions.

8. Industrial emissions of greenhouse gases are
   (a) decreasing worldwide in absolute amount.
   (b) decreasing worldwide as a percentage of total emissions.
   (c) decreasing from China and India.
   (d) independent of energy efficiencies.
   (e) independent of the expected lifetimes of equipment.

9. Which one of the following statements about electric motors and steam generators is *false*?
   (a) They are common to many industrial processes.
   (b) They present several opportunities for energy conservation.
   (c) They have not improved significantly during the past 30 years.
   (d) They require regular maintenance.
   (e) They account for about 75% of the energy use in the industrial sector.

10. Iron and steel industries generate significant amounts of greenhouse gases in all but one of the following processes. Which one does not?
    (a) Preparing coke
    (b) Melting iron ore in a blast furnace
    (c) Melting iron ore in a bloomery
    (d) Transporting scrap metal to mills for reprocessing
    (e) Converting methane to carbon monoxide and hydrogen gas for direct reduction of iron

11. Cement production generates the greatest amount of greenhouse gases during the process of
    (a) making clinker from limestone and clay.
    (b) grinding clinker.
    (c) adding gypsum to clinker to produce Portland cement.
    (d) rotating Portland cement and water in a cement mixer.
    (e) adding gravel and sand to Portland cement to make concrete.

12. Energy conservation in the heating and cooling of residential and commercial buildings benefits from (select all that apply)
    (a) installation of additional insulation.
    (b) window or door upgrades.
    (c) replacement of older furnaces or air conditioners.
    (d) use of solar water heaters.
    (e) appropriate ventilation.

13. Which one of the following statements is *false*?
    (a) Incandescent lamps consume 1% of the world's electricity.
    (b) Halogen lamps are a type of incandescent bulb that operates at a higher temperature than a standard incandescent.
    (c) Fluorescent tubes and CFLs contain small amounts of mercury.
    (d) Fluorescent, mercury, sodium, and metal lamps need ballasts that regulate the flow of current through them.
    (e) LED lamps are rapidly improving in cost and efficacy.

14. Energy usage in buildings over the past 2 decades have increased because
    (a) the energy intensity of residences has declined.
    (b) the energy intensity of commercial buildings has risen.
    (c) the efficiency of appliances such as refrigerators has not changed significantly.
    (d) there are more residences, and each one is larger on average.
    (e) more floor area of commercial buildings is vacant or being used for warehousing.

15. Proposals for geoengineering include
    (a) measures that should be taken immediately.
    (b) spreading sulfur dioxide into the upper atmosphere to decrease Earth's albedo.
    (c) seeding clouds to enhance infrared emissions from the atmosphere to the surface.
    (d) distributing sunshades between Earth and the sun.
    (e) fertilizing the oceans with nitrogen.

Suggested Readings

Bernstein, L., J. and 9 others. 2007. Industry. In: *Climate Change 2007: Mitigation. Contribution of Working Group III to the Fourth Assessment Report of the Intergovernmental Panel on Climate Change*, B. Metz, O. R. Davidson, P. R. Bosch, R. Dave, and L. A. Meyer, eds. Cambridge University Press, Cambridge. pp. 447–496.

Levine, M., D. and 11 others. 2007. Residential and commercial buildings. In: *Climate Change 2007: Mitigation. Contribution of Working Group III to the Fourth Assessment Report of the Intergovernmental Panel on Climate Change*, B. Metz, O. R. Davidson, P. R. Bosch, R. Dave, and L. A. Meyer, eds. Cambridge University Press, Cambridge. pp. 387–446.

Smith, P. and 11 others. 2007. Agriculture. In: *Climate Change 2007: Mitigation. Contribution of Working Group III to the Fourth Assessment Report of the Intergovernmental Panel on Climate Change*, B. Metz, O. R. Davidson, P. R. Bosch, R. Dave, and L. A. Meyer, eds. Cambridge University Press, Cambridge. pp. 497–540.

These three chapters of the IPCC 2007 report treat mitigation of greenhouse gas emissions from industrial sources, buildings, and agriculture.

International Energy Agency. 2008. *Energy Technology Perspective 2008*. Organisation for Economic Co-operation and Development, Paris.

This 485-page report describes many technologies that will diminish greenhouse gas emissions from human activities.

10

ECONOMICS OF GLOBAL CLIMATE CHANGE

The final chapters of this book focus on the social sciences and global climate change, beginning with economics. Economics is the study of how humans distribute resources (Mansfield and Yohe 2004). Economic terms such as *costs, market forces*, and *gross domestic product* have appeared sporadically in previous chapters, but only in the context of earth science, biology, or engineering. Some readers have undoubtedly been asking "Where's the bottom line?" They may demand "Show me the money!" These phrases reflect that economic considerations often take precedence in determining the course of human events. "Indeed the world is ruled by little else," claimed John Maynard Keynes, one of the most influential economists of the twentieth century (Keynes 1936).

This chapter first introduces some fundamental economic concepts. It then applies these concepts to regulatory policy for allocating natural resources such as Earth's atmosphere that influence global climate change. Next, it discusses cost–benefit analysis as a method for selecting mitigation projects and the difficulties in evaluating intangibles such as "aesthetics," "human welfare," "peace of mind," or "reputation" in an uncertain future. It ends with a few economic models of global climate change.

Many of the issues raised here are long-standing. Thomas Robert Malthus, an eighteenth-century economist, predicted that the human population, if it faithfully obeyed the biblical dictum to "be fruitful and multiply" (Genesis 1:28), would soon deplete the world's resources (Malthus 1798). Although such concerns persist to this day, little consensus has developed about which economic policies are most effective in conserving resources. A witticism attributed to the twentieth-century Irish playwright George Bernard Shaw is that if all the economists in the world were laid end to end, they would not reach a conclusion. Part of the problem is that subtle differences in economic assumptions lead to highly divergent conclusions. This becomes evident near the end of the chapter, in the sections that contrast the analyses of the economists Nordhaus and Stern.

Market Forces

Market forces refer to the influence of supply and demand on the distribution of goods or services. **Supply** is the willingness and ability to provide a good or service, and **demand** is the desire for a good or service, as indicated by the willingness and ability to pay. Belief in market forces has at times reached a religious fervor, yet most economists agree that market forces are poorly suited to the allocation of natural resources such as Earth's atmosphere. To appreciate this dichotomy and its role in the economics of global climate change requires familiarity with a few terms and concepts.

Definitions

A **market**, in economic parlance, is a common arena where firms or individuals voluntarily exchange goods and services. Markets vary greatly in size and procedures. At one end of the range is a "flea market" where hundreds of vendors may sell or trade inexpensive goods to thousands of passersby (**Figure 10.1A**). At the other end is a Christie's auction in which only a few selected museums or collectors participate (**Figure 10.1B**). Traders buy and sell raw goods such as coal and electricity in commodities markets and buy and sell investments and debts in financial markets such as stock exchanges and bond markets. Some markets, such as eBay, an online auction and shopping website, and NASDAQ, an electronic stock exchange, conduct business solely over the Internet. Each market establishes its own rules of engagement such as who may participate and how goods or services are exchanged.

A **free market** is part of a market economy in which the participants arrange transactions entirely by mutual consent. Governments do not intervene directly in a free market (there are no taxes, subsidies, minimum

(A)

(B)

FIGURE 10.1 Different types of markets (A) Berlin flea market in 2002. (B) James Bruce Gardyne conducts an Old Masters auction at Christie's Auction house at King Street, London, April 2007.

BOX 10.1 Supply and demand for a good or service

Demand for a good or service is a function of its asking price whereby the quantity demanded decreases with increasing price (D1 and D2 in Figure). In contrast, supply of a good or service is a function of its selling price whereby the quantity available for sale (S1 and S3) increases with price. With a demand curve of D1 and supply curve of S1, the equilibrium quantity consumed is Q1, and the equilibrium price is P1. (Such diagrams without numbers are common in microeconomics.)

Green and orange dots and dashed lines indicate deviations from the equilibrium price. If asking price temporarily rises to the green dot at left, demand decreases; if only a few sell at this higher price, supply increases to the green dot at right, and an excess supply (pink area) develops until price settles back to the equilibrium level P1.

Conversely, if asking price temporarily lowers to the orange dot at right, demand increases; if many sell at this new price, supply decreases to the orange dot at left, and a shortage (blue area) develops until price climbs back to the equilibrium level P1.

Increasing overall demand for a good or service would result in a shift from D1 to D2. The quantity sold shifts from Q1 to Q2, and the equilibrium price shifts from P1

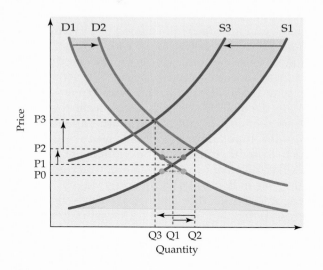

to P2. Decreasing overall supply or imposing a tax of (P3 minus P0) would cause a shift from S1 to S3, a decrease in the quantity sold from Q2 to Q3 (assuming demand curve D2), and an increase in selling price from P2 to P3.

---

wages, or price ceilings) but do participate in defining and enforcing basic rules. In a free market, the price of most goods or services reflects their value in comparison to other goods and services. Competition among vendors promotes reasonable prices and quality. A subset of a free market is a **perfect market**, one that has both perfect competition (i.e., no one party unduly influences the price of a good or service) and perfect information (i.e., every participant has access to complete, up-to-date knowledge about the actions of other participants and about the goods or services available). Of course, seldom is anything absolutely free or perfect, but free or perfect markets remain useful theoretical constructs.

The opposite of a free market operating in a market economy is a **controlled market** operating in a planned economy. In a controlled market, the government manages major sectors of the economy, dictating the goods and services produced and the distribution of income. A **mixed economy** combines elements of both controlled and free markets. Most economies in the world today are mixed.

Market demand refers to the desirability of a good or service in the marketplace. Demand for a good or

service depends primarily on its price. As its price increases, consumers will purchase less of the good or service, and the quantity consumed will decrease (curve D1 in **Box 10.1**). The entire demand curve will shift to the right (from D1 to D2 in Box 10.1) if consumers become wealthier (e.g., greater wealth in China increases the demand for electricity and thereby the demand for coal), if alternative goods or services become more expensive (e.g., higher oil prices increase the demand for coal), or if advertising creates greater consumer demand (e.g., ads for inexpensive, clean-burning, automatic coal stoves attract new customers).

Market supply refers to how much of a good or service is available for sale in the marketplace. Higher prices generally increase the supply (curve S1 in Box 10.1). The entire supply curve will shift to the left (from S1 to S3 in Box 10.1) if the price of inputs (labor, capital, and land), which production of the good or service requires, increases. Conversely, the supply curve will shift to the right (from S3 to S1 in Box 10.1) if technological advances decrease input costs. For example, a new labor contract that provides higher salaries for coal miners shifts the supply curve to the left, whereas the invention of sophisticated coal-mining equipment that decreases labor requirements shifts the supply curve to the right.

Equilibrium price for a good or service is the price at which the quantity desired of a good or service equals the quantity available (Q1 and P1 for curves D1 and S1 in Box 10.1). Market forces of supply and demand tend to sustain the equilibrium price. Were the price any higher than the equilibrium price (green dots in Box 10.1), the quantity desired would decline, and the quantity available would increase until an excess in supply developed. Eventually, this glut would lead sellers to lower their prices to the equilibrium level. In contrast, if the price were below the equilibrium price (orange dots in Box 10.1), the quantity desired would increase, and the quantity available would decline until an excess in demand developed. This shortage would stimulate a price increase to the equilibrium level.

Market forces are economic factors that influence prices in a free market through their effects on supply and demand. A shift in the demand curve from increased wealth or more expensive alternatives (from D1 to D2 in Box 10.1) moves the equilibrium price and quantity upward (from P1 and Q1 to P2 and Q2 in Box 10.1). A shift in the supply curve from increased input costs (from S1 to S3 in Box 10.1) also moves the equilibrium price upward (from P2 to P3 in Box 10.1) but the quantity downward (from Q2 to Q3 in Box 10.1).

Governmental policies

Governments often intervene in the marketplace to achieve societal objectives. For example, to promote energy conservation and greenhouse gas mitigation, governments have tried a variety of policies involving taxes, price floors, permits, and regulations. These policies, because they directly and indirectly influence supply and demand, have broad consequences.

A tax on a good or service shifts the supply curve to the left (from S1 to S3 in **Figure 10.2**). The extent to which the buyer or the seller pays the tax depends on the demand curve and the **elasticity of demand**, the availability of substitutes for a good or service. If demand for a good or service is entirely independent of price because no substitutes are available (demand is totally inelastic, D1 in Figure 10.2), the buyer pays the added cost of the tax (P2 minus P1 in Figure 10.2). A sales tax (value-added tax) approximates this situation. At the other extreme, if the demand for a product varies strongly with price because many substitutes are available (demand is completely elastic, D2 in Figure 10.2), the seller bears the added cost of the tax (P1 minus P0 in Figure 10.2). An excise tax (one levied

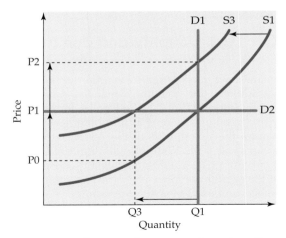

FIGURE 10.2 Influence of taxes on the price of a good or service Levying a tax of either (P2 minus P1) or (P1 minus P0) would cause a shift in the supply curve from S1 to S3. If demand is mostly independent of price (D1), a condition called "inelastic demand," the price increases from P1 to P2, and the buyer pays for all of the tax, P2 minus P1. If demand depends strongly on price (D2), a condition called "elastic demand," the quantity consumed decreases from Q1 to Q3, and the seller pays for all of the tax, P1 minus P0.

on the producer of goods such as tobacco, alcohol, or gasoline) approximates this situation. More commonly, the demand for a good or service has an intermediate response to price (see D2 in Box 10.1), and so a buyer pays part (P3 minus P2 in Box 10.1) of a tax, and the seller pays the remainder (P2 minus P0 in Box 10.1).

According to neoclassical economic analysis of supply and demand, reducing taxes has several indirect effects. For example, several U.S. presidential candidates in 2008 proposed a "tax holiday" to lift U.S. excise taxes on gasoline during the summer in order to provide some relief for the electorate from the burden of high gasoline prices. This would shift the supply curve to the right (from S3 to S1 in Box 10.1) and would save the electorate some money (P3 minus P2 in Box 10.1) but would also stimulate gasoline consumption (from Q3 to Q2 in Box 10.1), contribute to higher revenues for oil producers (P2 minus P0 in Box 10.1), and incur greater government deficits. In essence, a tax holiday would transfer funds from the U.S. government to oil producers—not the intended purpose.

Governments sometimes impose a price floor to prop up the price of a particular good or a price ceiling to insure that its price does not exceed a certain level; either policy results in a deviation from equilibrium prices. For example, the U.S. government sets minimum prices for certain agricultural commodities,

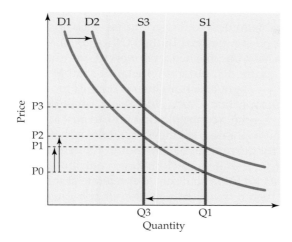

FIGURE 10.3   Influence of a permit system on the price of a good or service Setting the quantity of permits available at Q1 often is accompanied by expectation of shortages, shifts the demand curve to the right (from D1 to D2), and causes a price increase (from P0 to P1). Decreasing the number of permits lowers the quantity available (from Q1 to Q3) and increases the price (from P0 to P2). A shift in the demand curve from D1 to D2 increases the price to P3.

a policy that almost ensures a surplus supply (green dots in Box 10.1) and forces the government to store the agricultural surplus or to limit agricultural production. In the 1970s, the U.S. government set maximum prices for some fossil fuels, a policy that triggered shortages (orange dots in Box 10.1), forced fuel rationing, and fostered an illicit black market.

Similarly, if governments issue a limited number of permits for a good or service, its supply becomes independent of price (S1 and S3 in **Figure 10.3**). Demand

for the good or service increases based on expectations of future shortages in supply, shifting its demand curve to the right (from D1 to D2 in Figure 10.3) and increasing its price (from P0 to P1 in Figure 10.3). Issuing fewer permits (from Q1 to Q3 in Figure 10.3) also increases the price of the item (from P0 to P2 in Figure 10.3); if its demand curve shifts, its price rises still further (from P2 to P3 in Figure 10.3).

In addition to taxes, price floors, and permits, governments enact and enforce regulations that set a minimum standard of quality to protect public health, safety, and welfare. The World Bank measures the performance of the sovereign states of the world according to their ability "to formulate and implement sound policies and regulations that permit and promote private sector development" (Kaufman et al. 2008). The Bank bases this measurement of regulatory quality on surveys of commercial business information providers, private firms, non-governmental organizations, and public-sector organizations who evaluate a number of factors, including fairness of government practices, ease of starting a business, trade policy, access to capital, and treatment of non-residents. This measurement differs widely among countries (**Figure 10.4**). Presumably, countries with a higher value of regulatory quality would be more effective in implementing environmental regulations to address global climate change.

Failure of Market Forces

The economics of global climate change do not fit comfortably into the classical mold of a market economy because of several factors:

- Divergent national interests distort global markets for goods or services.

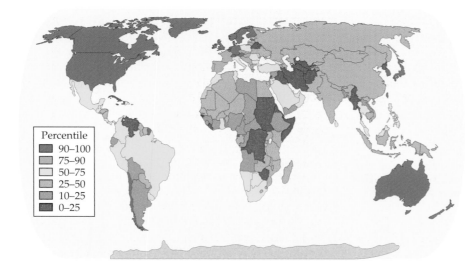

Percentile
90–100
75–90
50–75
25–50
10–25
0–25

FIGURE 10.4   A measure of regulatory quality for the sovereign states of the world in 2007 Colors designate the percentile range in which the value of regulatory quality for a state lies. For example, regulatory quality in Canada lies in the 90th to 100th percentile, or in the top 10% of sovereign states. (After Kaufman et al. 2008.)

- Monopolies or cartels control the supply of some goods or services.
- Some relevant goods or services are free and invite indiscriminate use.
- Crucial information about some goods or services is sparse, uncertain, or not publicly available.
- Important processes span several generations of human beings.

The following sections discuss each of these factors and their role in economic policies on global climate change.

National interests

Addressing global environmental issues such as climate change requires the cooperation of many sovereign states. Developed countries such as the United States, Japan, and members of the European Union have been responsible for most of the greenhouse emissions to date (see Chapter 4). These countries are reluctant to proceed unilaterally with projects to reduce emissions, however, when much less expensive alternative projects that reduce an equivalent amount of emissions are

available in developing countries such as China and India (Congressional Budget Office 2003). China and India have become major emitters of greenhouse gases (see Figure 8.3) but are reluctant to implement policies that might constrain their economic growth. From one perspective, China, India, and other developing countries have contributed so little to past additions of greenhouse gases that they should not be held accountable for current global warming. From another perspective, if emissions from large contributors such as China and India continue to accelerate at their current rate, then it will hardly matter what mitigation measures other countries implement. This issue and possible remedies are a major topic in Chapter 11, which describes international law.

Monopolies or cartels

Utilities that supply electricity; natural gas; drinking water; sewage disposal; telephone service; cable television; and general mail delivery are natural monopolies because one provider who benefits from economies of scale can serve a local market more efficiently than multiple providers can. Economies of scale are significant for utilities because they have high **capital costs**

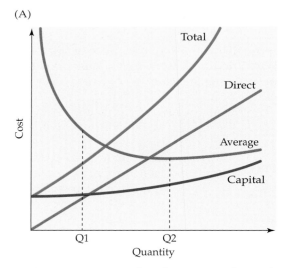

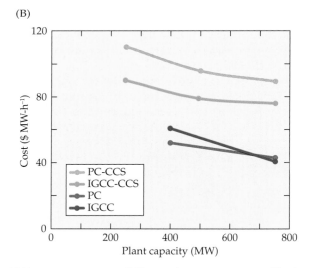

FIGURE 10.5 Economies of scale (A) Cost of providing a utility service as a function of quantity provided. (A) Capital costs include acquiring land, constructing facilities, and purchasing equipment. Direct costs include inputs, such as raw materials, labor, and maintenance, to produce the service. Total costs equal the sum of capital, direct, and indirect costs. (Indirect costs include payments for expenses such as advertising and security, which are relatively minor for utilities.) Average cost per item is total costs divided by the quantity provided. The economies of scale, in which the lowest average cost per item is achieved at a higher quantity

(Q2 in comparison to Q1), are the consequence of high capital costs for utilities. (B) Influence of electric plant size on electricity generating costs ($U.S. per 10^6 Watt-hours) for a pulverized coal plant (PC), an integrated gasification combined cycle plant (IGCC), and these types of plants equipped with carbon capture and storage (PC-CCS and IGCC-CCS). CO_2 storage is in geological formations. Larger plants have lower energy costs because of lower capital costs per unit energy: A 200-MW generator does not cost twice as much as a 100-MW one. (B after Rubin et al. 2007.)

but have **direct costs** that increase relatively slowly with size (**Figure** 10.5). Capital costs include those of acquiring land, constructing facilities, and purchasing equipment; direct costs include those of materials, labor, and maintenance to produce a good or service. Moreover, utilities provide services that are largely homogeneous: A watt of electricity from one company should be indistinguishable from a watt of electricity from another company. Economies of scale together with homogeneity of services act as barriers to any company attempting to enter a market in a new locality. The newcomer has difficulty not only in competing on price, but also in distinguishing its product from that already available.

Natural monopolies tend to expand and form cartels that surreptitiously collude and manipulate markets. For example, by the end of the 1920s, three electric power utility companies controlled half of the industry in the United States and were guilty of stock watering (artificially inflating the value of assets; named after the practice of forcing a cow to bloat itself with water before weighing it for sale), misuse of subsidies, and dubious accounting (Energy Information Administration 2000). When consumers no longer tolerate such improper behavior, governments intervene. They may divide a company into smaller entities (e.g., the breakup of AT&T, the U.S. telephone monopoly), appoint or elect commissions to regulate private companies (e.g., Public Utility Commission of Ohio), take over segments of an industry (e.g., Los Angeles Department of Water & Power), or negotiate trade agreements with international entities.

The following section describes the electric power industry in some detail because of its large contribution to greenhouse gas emissions (see Figure 7.10). This industry has vacillated between regulation and deregulation and between public ownership and privatization during its century of existence. Inconsistent economic policies toward the electric power industry have slowed the adoption of technologies that diminish greenhouse gas emissions from power plants.

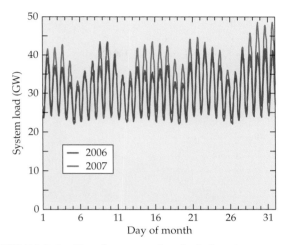

FIGURE 10.6   Hourly system load of electric power (10⁹ Watts) in California during August of 2006 and 2007 (After California Independent System Operator Corporation 2008.)

ELECTRIC POWER UTILITIES   Electric power utilities in the Unites States are a mix of publicly owned entities, privately owned companies, and rural cooperatives (**Table** 10.1). Publicly owned utilities are not-for-profit agencies that provide service mostly around major cities such as Los Angeles, San Antonio, Seattle, and Orlando. Privately owned utilities are located usually in urban and suburban areas. Rural cooperatives are private, independent electric utilities owned by the members that they serve, but they are able to borrow money from the federal government at below-market rates for building new transmission and distribution lines in rural areas.

Electricity is an unusual commodity. Supply of electricity and, especially, demand for it fluctuate daily, seasonally, and annually (**Figure** 10.6). Electricity is difficult to store. At present, storage methods include batteries, which have limited capacity (see Table 7.10); electrolysis of water to generate hydrogen, which just transfers the problem to storage of a substance with a

TABLE 10.1   Electric power industry in the United States

| Type of provider | Number of providers | Percentage | | | Revenue per mile ($) |
		Customers	Power sales	Power lines	
Publicly owned	2010	15	16	7	86,302
Privately owned	220	73	74	50	62,665
Rural cooperatives	930	12	10	43	10,565

Source: American Public Power Association 2007; National Rural Electric Cooperative Association 2008.

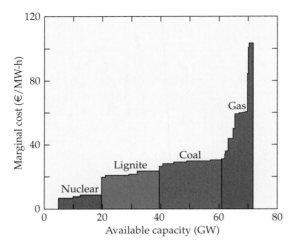

FIGURE 10.7 Marginal cost (euros per 10⁶ Watt-hours) of producing electricity in Germany, 2006 Almost 20 GW (10⁹ Watts) of nuclear power is available in Germany. To generate 21 GW, the German electrical utilities bring online lignite (brown coal) power plants, and this last 1 MW-h of electricity costs double that from nuclear power plants. "Gas" denotes natural gas. (After Zoettl 2008.)

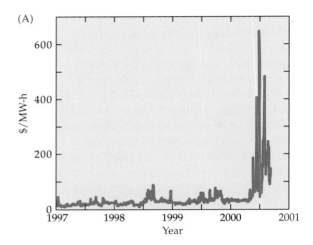

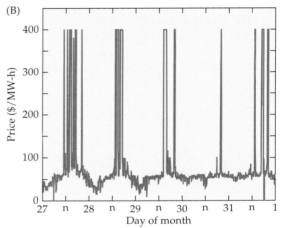

FIGURE 10.8 Price for electric power in California ($U.S. per 10⁶ Watt-hours) (A) Daily average spot prices from 1997 to 2001 at the California Oregon Border, a proxy for sales into northern California. The tick marks indicate January 1 and June 1 of the year. (B) Real-time prices during a heat wave at the end of August 2007. Figure 10.6 shows total load for this period. Each "n" designates noontime. Price spikes in the afternoon reflect when electricity was purchased on spot markets to meet shortfalls. Low prices in the middle of the night reflect surplus generating capacity. (A after California Independent System Operator Staff Report 2003; B, after FERC staff report 2000.)

low heat content per volume (see Table 7.7); and pumping water behind dams, which require massive infrastructure (see Figure 8.27). Without storage, if supply falls short of demand, brownouts or blackouts ensue, affecting nearly everyone harshly and somewhat indiscriminately. Electric power utilities go to great expense to meet peak demand and avoid such occurrences.

Utilities have access to electric power from a variety of generating facilities (**Figure 10.7**). When demand for electricity is light, they use renewable energy or nuclear power, which has low operating costs and, therefore, low marginal costs. **Marginal cost** is the added cost of producing one more of an item; specifically, the marginal cost of electricity is the added cost of generating one more watt for one more hour.

When demand for power surpasses the capacity of local renewable or nuclear power plants, utilities increase the electricity generated by coal-fired power plants. Coal-fired power plants have higher marginal costs because of fuel expenditures (see Figure 7.5). Utilities also purchase electricity from distant renewable or nuclear power plants that have lower operating costs but incur higher transmission costs. For example, dams on the Colorado River in Nevada and Arizona and on the Columbia River in Washington State, Oregon, and British Columbia provide, respectively, 15% and 7% of the electricity in California (California Energy Commission 2008). Once renewable, nuclear, and coal-fired power plants are running near full capacity, utilities increase the electricity generated by natural gas power

plants. These have even higher marginal costs (see Figure 10.5B) because of the higher cost of natural gas, but they can rapidly vary output with demand and emit fewer air pollutants than coal-fired power plants, an important property because periods of high electricity demand often coincide with periods of bad air quality (high air pollution).

As a last resort, utilities purchase electricity on the **spot market**. A spot market sells commodities for cash and immediate delivery, sometimes at exorbitant prices (**Figure 10.8A**). For example, California utilities paid more than $1400 per MW-h (in year 2007 $U.S. ; 1 MW-h equals 10⁶ Watt-hours) in June 2000 when companies such as Enron illegally manipulated the electricity spot market in the western United States; by contrast, the going rate was $60 per MW-h in June 1999 (McNamara 2002). In 2007, government regulations limited the top rate to $400 per MW-h (**Figure 10.8B**).

To moderate peak demand for electricity, utilities may institute rationing, priority service, or real-time pricing (Crew et al.1995).

- Rationing of electricity when demand exceeds supply always hurts someone. The blackouts during the summer of 2000 in California forced many small businesses to close temporarily. Utilities attempt to minimize public effects of outages during shortages. For example, utilities put hospitals, police and fire stations, and other public safety services on separate power grids that are not subject to brownouts or blackouts.

- Priority service for electricity, like that for mail or airline tickets, differentiates customers by their willingness to pay extra for a higher level of service. In one type of priority service (capacity subscription), customers pay a fee for a certain level of capacity (determined by fuse size) as well as a fee for actual consumption. Capacity subscription offers customers more certainty about their future power bill than many other alternatives (Doorman 2005).

- Real-time pricing indexes the rate that a customer pays for electric power to the marginal cost of generating it at that moment: Electricity at peak demand times is more expensive than at slack times. Often a utility will predict a day or two in advance what the hourly rates will be, so customers can plan their activities accordingly and schedule discretionary energy use when rates are low. Real-time pricing has high **transaction costs** (i.e., costs incurred in making an economic exchange) because of the requirement to meter a customer's power usage continually. Such transaction costs amount to a smaller fraction of the total for large industrial and commercial customers than for small businesses and residences. California utilities recently invested $4.28 billion to equip 11.3 million customers with advanced electric meters that provide real-time monitoring of power use, but the utilities have yet to determine a pricing policy (Woo et al. 2008).

Does priority service or real-time pricing diminish total power consumption or merely shift consumption from peak to off-peak hours? The limited data that are available indicate a significant decrease in overall consumption and only a slight shift in usage patterns (**Figure 10.9**). Therefore, these economic policies might mitigate global climate change in two ways. First, greater uniformity in daily and seasonal electricity loads will promote the construction of larger, but less flexible power plants that produce low greenhouse

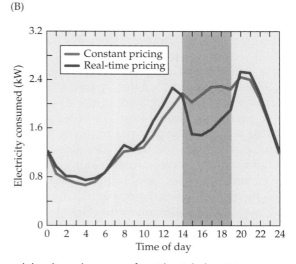

(A)

(B)

FIGURE 10.9 Pricing strategy and electricity consumption (A) Pricing strategy and the diurnal pattern of industrial electricity consumption (10⁶ Watts). Real-time pricing increased costs of electricity by 30% between 16:00 and 18:00 (gray banded area) during the summertime in the Carolinas, Ohio, Indiana, and Kentucky. (B) Pricing strategy and the diurnal pattern of residential electricity consumption (10³ Watts per household). Real-time pricing doubled the cost of electricity between 14:00 and 19:00 (gray banded area) whenever maximum temperatures exceeded 32°C near San Diego, California. (A after Taylor et al. 2005; B after Herter et al. 2007.)

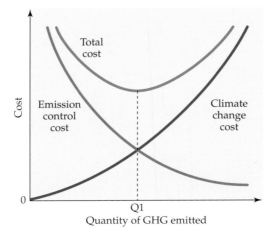

FIGURE 10.10 **Cost of electricity as a function of greenhouse gas emissions** Cost of controlling emissions increases exponentially as quantity of emissions approaches zero. Conversely, cost of damages from climate change increase exponentially as quantity of emissions becomes large. Total costs of electricity, the sum of the two other costs, reaches a minimum Q1 at some intermediate quantity of emissions.

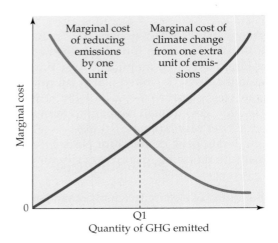

FIGURE 10.11 **Marginal costs of emission control and climate change as a function of greenhouse gas emissions** Marginal cost is the effect of one more unit of greenhouse gas. Total costs are minimal at Q1, where the curves intersect. At this point, the cost of damages resulting from additional climate change equals the cost of additional mitigation of greenhouse gases.

gas emissions, such as those based on nuclear energy or coal-fired plants equipped with carbon dioxide capture and storage (CCS). Second, conservation of electric power usage will cut greenhouse gas emissions.

Economic models predict how electric power costs will respond to various methods for controlling greenhouse gas emissions. Modest controls that slightly diminish emissions are less expensive than ones that are more aggressive, but the higher emissions under modest controls result in more extreme climate changes that incur greater costs (**Figure 10.10**). For example, pulverized coal power plants without CCS capability are less expensive to build and operate than those with CCS (see Figure 10.7), but they release additional carbon dioxide that may lead to warmer temperatures. A warmer environment may lead to consumption of more electricity for cooling or to fiercer storms that will down more power lines. Total costs, the sum of control costs and climate change costs, reach a minimum at the quantity of emissions where the cost of climate change produced by one unit of emissions equals the cost of reducing emissions by this one unit (**Figure 10.11**); in economic parlance, the optimum occurs when the marginal cost of climate change equals the marginal cost of emission control. This economic theory seems straightforward, yet assessing actual costs and their responses to changing circumstances—putting real numbers on the axes of Figures 10.10 and 10.11—is difficult.

DEREGULATION OF ELECTRIC UTILITIES **Vertically integrated monopolies** dominate the electric power industry, which means that in most localities, a single company or government-owned enterprise generates most of the electricity, transmits this electricity over high voltage lines from power plants to substations, and distributes and meters all of this electricity to end users (Griffin and Puller 2005). The monopolistic nature of electric power derives from several factors.

- Electric utilities enjoy economies of scale in which capital costs per unit energy decrease with plant size (see Figure 10.5A) and transmission lines are more efficient with thicker wire and higher voltages.

$$P_{loss} = \frac{RP^2}{V^2} = \frac{4L\rho P^2}{\pi d^2 V^2} \qquad (10.1)$$

where

P_{loss} = power loss from a transmission line

R = resistance of the transmission wire

P = power transmitted

V = voltage transmitted

L = length of wire

ρ = resistivity of the wire material

d = diameter of wire

- Established power companies already have power plants in close proximity to major customers and hold right-of-ways for high voltage power lines from power plant to substations. These shorter distances lower transmission costs (smaller L in Equation 10.1). An upstart power company would need to set up power plants and high-voltage transmission lines near customers, both formidable tasks.

- Electric utilities provide homogenous services. As mentioned previously, 1 watt of electric power generated by one company is the same as 1 watt generated by another.

- A single set of power lines distributes electricity to each residence, business, and factory. Running multiple distribution lines from competing utilities down a street would be wasteful. Instead, an upstart company would need to lease existing distribution lines from established utilities.

Despite these barriers, some economists (most notably, the twentieth-century economist Milton Friedman) espoused that free market competition in the electric power industry would spur innovation, improve efficiencies, and lower prices. Consequently, several governments privatized and deregulated parts of their utilities in the 1980s and 1990s. Often, they separated the companies that generated electricity from those that were responsible for its transmission and distribution.

While the successes of privatization and deregulation have been subtle, the failures have been spectacular. Chile privatized its electric utilities in the mid-1980s, dividing the two main government power companies into six generating companies and eleven distribution companies. The military junta in power at the time sold the public resources (power plants, transmission lines, and substations) to partisans at bargain prices. Nonetheless, some economists argue that, except for some major power shortages in 1998 and 1999, private ownership and operation of the electricity industry in Chile has been "very successful" (Pollitt 2005).

California was the first state in the United States to deregulate its energy market, beginning in the mid-1990s. It required Investor Owned Utilities, or IOUs (primarily Pacific Gas and Electric, Southern California Edison, and San Diego Gas and Electric), to sell off a significant portion of their power generation to private, unregulated companies such as AES, Dynegy, Enron, Mirant, Reliant, and Williams. The buyers of those power plants then became the wholesalers from which the IOUs purchased electricity. During 2000 and 2001, electric power brokers manipulated the supply of electricity in the so-called free market for electric

power to create shortages that forced several large-scale blackouts in California, unprecedented spot prices for electricity (see Figure 10.8A), and the insolvency of the IOUs.

The electric power brokers used several schemes to manipulate electricity markets. Because regulations allowed brokers to charge higher prices for electricity produced out-of-state, the brokers would make it appear that the electricity, which they sold to the IOUs, was being generated somewhere other than California. In another scheme, wholesalers scheduled "routine maintenance" at peak times to shut down power plants in California, create shortages, and force the IOUs to purchase out-of-state electricity from the wholesalers on the spot market at inflated prices. In yet another scheme, wholesalers scheduled power deliveries through the limited number of transmission lines serving California to create the appearance of congestion and then were able to charge the IOUs an extra fee for the use of those lines. The energy wholesalers were politically well connected, and so federal regulators, who were supposed to prevent such illegal manipulations, failed to intervene in a timely fashion.

Advocates of privatization contend that public regulators still held too much control over the California market and thwarted corrective free market processes, whereas opponents of privatization assert that the fully regulated system had worked well for a half century and that deregulation permitted unscrupulous energy speculators such as Enron to operate without government oversight and wreck a viable system. One point of agreement, however, is that privatization and deregulation have failed to stimulate large-scale investment in technological upgrades for electric power utilities to meet the future needs of their customers or to mitigate greenhouse gas emissions.

Free goods and services

Adam Smith, an eighteenth-century economist, forwarded the idea that the "invisible hand" of a free market promotes the overall good of a society because individuals who pursue their own self-interest and produce what may be of greatest value also maximize the total revenues of their community. Unfortunately, free markets break down—the invisible hand gets caught in the cookie jar—in situations with large externalities. An **externality** occurs when a buyer and seller in an economic transaction do not necessarily bear all of the costs or reap all of the benefits of the transaction.

Pollution and technology innovation are classic examples of processes with large externalities (Jaffe et al. 2005). A polluter reaps benefits indirectly from polluting while imposing the pollution costs on others.

FIGURE 10.12 Cartoon account of the tragedy of the commons (Copyright Ken Avidor.)

The polluter therefore lacks incentive to reduce those costs. The problem reverses for a firm that adopts new technology. The firm typically creates benefits for others while incurring most of the costs and therefore loses incentive to invest in technology. Pollution creates a negative externality, and so the invisible hand allows too much of it, whereas technology innovation creates positive externalities, and so the invisible hand produces too little of it.

With respect to global climate change, Earth's atmosphere has two properties that interfere with market forces. It is to some extent nonrival; one person using the atmosphere does not perceptibly diminish the quality or quantity of atmosphere available to others. It is nonexcludable; one person or entity cannot keep another person or entity from using the atmosphere. Who owns the sky? If nobody owns a natural resource, nobody will take care of it; if it is free for the taking, people will tend to overuse it (Congressional Budget Office 2003). This is known as "the tragedy of the commons."

Garrett J. Hardin, an ecologist concerned with human overpopulation, described the typical scenario for the tragedy of the commons (Hardin 1968). Imagine a pastoral setting in which shepherds are grazing their flocks on a common green (Figure 10.12). This commons can sustain a certain number of sheep in perpetuity. If one shepherd sneaks in a few sheep above the number that is sustainable, he or she benefits significantly, whereas the damage of the few extra sheep is distributed among all the shepherds and may be imperceptible, at least initially. In other terms, the common green initially seems to be a non-rival resource. Once the other shepherds discover that one of their own is profiting at their expense, they will add sheep to remain competitive. With all the additional sheep, the commons suffers severe overgrazing, and everyone—both sheep and shepherds—dies. "Freedom in a commons brings ruin to all," wrote Hardin. The main point is that market forces are poorly suited for conserving common property resources, and governments must intervene.

Government intervention may take the form of collecting taxes on each sheep that grazes on the commons or issuing permits that limit the number of sheep grazing on the commons at any one time. Tax rates may

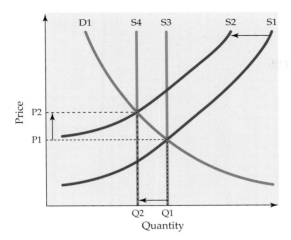

FIGURE 10.13 Influence of taxes or permits on the price or quantity of a natural resource Levying an excise tax of P2 minus P1 increases the price to P2, causes a shift in the supply curve from S1 to S2, and decreases the quantity consumed from Q1 to Q2. Decreasing the number of permits lowers the quantity consumed from Q1 to Q2 and increases the price from P1 to P2. Therefore, both approaches produce similar increases in prices and similar reductions in consumption of the resource.

differ; for example, a shepherd with hundreds of sheep may pay a steeper rate per sheep than a shepherd with only a few sheep or vice-versa. Permit numbers may differ; for example, a shepherd with a large family may be permitted to graze more sheep than a shepherd who does not have a spouse and children. A variation of a tax system is **tax-and-dividend**, where a government redistributes some of the taxes it collects to encourage certain behaviors; for example, taxes collected from buyers of "gas-guzzling" vehicles pay for rebates to buyers of gas-electric hybrid vehicles. A variation of the permit system is **cap-and-trade**, in which Little Bo Peep who has lost her sheep may sell her unused permits to another shepherd who wants to graze more sheep on the commons.

Both taxes and permits will regulate the total number of sheep grazing on the commons. For any level of taxes, there is a corresponding level of permits that will achieve the same purpose (Congressional Budget Office 2003). In principle, both approaches should lead to similar levels and prices of grazing (**Figure 10.13**). Taxes offer the advantage that the shepherds can anticipate their costs more accurately (situation D2 in Figure 10.2), whereas permits offer the advantage that governments can anticipate the quantity of sheep on the commons more accurately (situation D1 in Figure 10.3). Which system is preferable depends on which type of certainty—certainty of costs versus certainty of benefits—is more important and how rapidly costs rise as

the government tightens restrictions on grazing. Excise taxes also have lower transaction costs than permits because the government can collect from the relatively few sheep suppliers rather than from every shepherd (this makes more sense when applied to collecting from petroleum suppliers versus petroleum consumers).

Sheep grazing and greenhouse gas emissions are analogous, except that counting sheep is relatively easy. By contrast, greenhouse gases are invisible, spread freely and capriciously over political borders, and influence the environment globally, not just locally. Most economists propose a mix of taxes and permits to regulate greenhouse gases (Aldy et al. 2003), a topic that receives more attention later in the chapter.

Lack of information

In a free market, providers and consumers make rational decisions based on information about the cost, quantity, and quality of the available goods and services. Information about global climate change, however, suffers from being sparse, uncertain, or not publicly available. For instance, estimates of polar bear populations are sparse, estimates of greenhouse gas emissions from various sources are uncertain, and estimates of petroleum reserves in various locations are not publicly available. Without reliable information, economic decisions are prone to error.

About 25 satellites that currently orbit Earth provide crucial data about the planet's climate trends (National Research Council 2005). More than half of these are near or past their expected service lifetimes and will no longer be operational by 2010 (**Figure 10.14**). Failure to launch replacement satellites will severely limit the information available about the state of the world. Nonetheless, during the early 2000s U.S. government officials diverted much of the funding from Earth-monitoring satellites to human exploration of other worlds such as Mars.

Multigenerational time scales

Average global temperatures have warmed discernibly only after two centuries of large-scale emissions of greenhouse gases (see Figure 2.40). Even if these emissions abate today, temperatures will continue to rise through the next century (see Figure 4.33). Such long delays between cause and effect, action and reaction, increase the difficulty of developing efficient economic policies. For example, total costs of greenhouse gas emissions reach a minimum when marginal costs of emission control equal marginal costs of climate change (see Figure 10.11), but evaluating the costs of climate change may take decades or longer, during

FIGURE 10.14   Estimated number of civil optical land imaging satellites orbiting Earth at the end of the designated year This group of satellites is not identical to climate satellites: Some optical satellites do not provide climate data, whereas some climate satellites are not optical. "DMC" is an acronym for Disaster Monitoring Constellation, a group of satellites managed by DMC Interational Imaging for Algeria, Nigeria, Turkey, the United Kingdom, China, and Spain. (After Stoney 2008.)

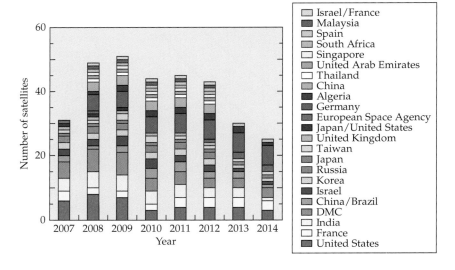

which time costs of emission control may decrease with technological advances or increase with labor costs.

The long time scales of climate change also influence the cost of borrowing money to construct the large infrastructure projects that address climate change. Loans, even at a low interest rate (more officially known as the "discount rate"), reach phenomenal sums when compounded continuously over many decades. Encumbering future generations with huge debts presents an ethical dilemma. These issues receive fuller treatment in the following sections.

Free market forces and climate change: a summary

Most economists believe that market forces alone cannot achieve significant reductions in greenhouse gas emissions because of national interests, monopolies, free goods or services, lack of information, and multigenerational time scales. They conclude that governments therefore have the right and responsibility to take corrective economic measures. There is far less agreement about what types of measures—taxes, permits, tax-and-dividend, or cap-and-trade—would be most effective. Proponents of each type usually support their arguments with cost–benefit analyses.

Costs versus Benefits

Mitigation of greenhouse gas emissions may require large (priced at greater than $100 million), government-sponsored projects (see Chapter 7 through Chapter 9). Governments, in theory, choose the course of action that yields the highest benefits at the lowest costs. In practice, cost–benefit analyses suffer from several prob-

lems. Analyses of large projects must account for the complexity and longevity of such endeavors as well as their dependence on leading edge—and often "bleeding edge"—technologies. In addition, project planning is an imperfect process (Lovallo and Kahneman 2003; Flyvbjerg 2007). Planners might:

- Base estimates on past projects that seem similar to the current one but prove to be distinct;
- Struggle with assessing the value of intangibles such as "aesthetics," "human welfare," "peace of mind," or "reputation";
- Be pressured by various interest groups to include or exclude various costs or benefits;
- Have a personal stake in whether a project secures funding;
- Suffer few penalties for mistakes; and
- Tend to be overly optimistic, allocating insufficient funds and time to remedy mishaps.

Consequently, large projects tend to underestimate costs and overestimate benefits (**Box 10.2**). For example, the actual costs for 90% of transportation projects were significantly above those estimated (**Figure 10.15**). The actual benefits in 90% of railroad projects—particularly, estimates of train ridership—fell significantly below those anticipated (**Figure 10.16**). Unfortunately, time has not improved the accuracy of cost–benefit analyses for transportation projects.

Improving accuracy

Inaccurate cost–benefit analyses contribute to both wasteful distributions of resources and delays in completion. These, in turn, are responsible for additional

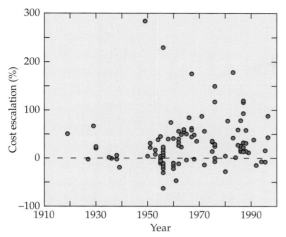

FIGURE 10.15 Underestimation of costs (fixed prices) for 111 transportation projects versus year of commencement. A cost escalation of 100% means that costs were double the amount estimated, whereas an escalation of –50% means that costs were half the amount estimated. (After Flyvbjerg et al. 2002.)

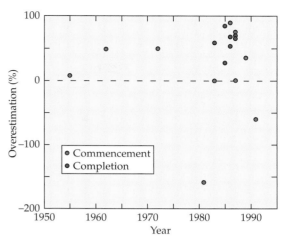

FIGURE 10.16 Overestimation of ridership for 17 railroad projects versus year of commencement or completion when commencement was unknown. (After Flyvbjerg et al. 2005.)

costs and fewer benefits. Wasteful spending, delayed completion, additional costs, and fewer benefits than advocates of large-scale government projects originally advertised discourage the citizenry from funding such projects, including those for climate-change mitigation. To improve the planning process for large-scale projects such as those for climate change mitigation will

---

BOX 10.2 A difficult cost–benefit analysis

An example of how aesthetic values may interfere with the accuracy of cost–benefit analyses is the Sydney Opera House, home to the Opera Australia and the Australian Ballet. When construction began in 1959, the facility had an estimated cost of $7 million ($U.S.) and a completion date of 1963. After some initial cost overruns, the state government decided that further funding supplements would require their approval. Subsequent overruns—and there were many—triggered increasingly acrimonious debates about the project that delayed construction, compromised the facility, and eventually drove the architect off the project and ruined his career (Flyvbjerg 2007).

The Opera House finally opened in 1973 at a cost of $107 million ($U.S.), 1500% more than the original estimate. It is poorly suited for opera and ballet (Flyvbjerg 2007). One-fourth of the seats have obscured views. The orchestra pit, which is half-buried under the stage, subjects musicians to sound levels that are deafening even when the brass section sits in a separate enclosure. The wings of the stage are so small that stagehands must catch dancers as they leap from stage into the wings to prevent them from crashing into walls (Morgan 2006). Nonetheless, this edifice is widely recognized as one of the human-wrought wonders of the world.

Sydney Opera House The original cost–benefit analysis of this large project bore little relation to actual costs and benefits.

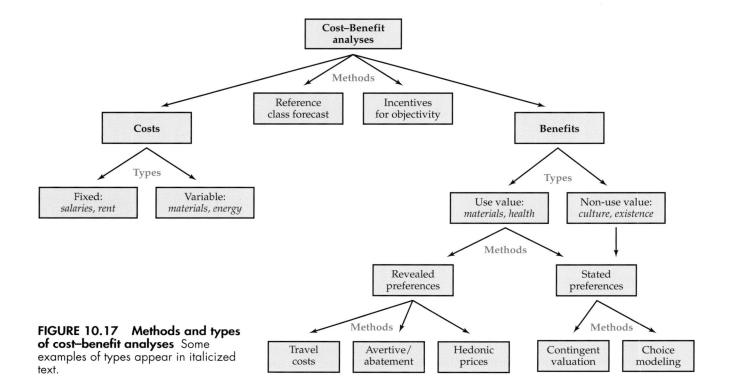

FIGURE 10.17 Methods and types of cost–benefit analyses Some examples of types appear in italicized text.

require more accurate forecasting methods and better incentives for objectivity (Flyvbjerg 2007).

REFERENCE CLASS FORECASTING One method that may increase the accuracy of cost–benefit analyses is reference class forecasting (**Figure 10.17**). It has five steps (Lovallo and Kahneman 2003):

1. *Select a reference class of comparable completed projects.* The class should be broad enough to provide a large sample size for robust statistical testing but narrow enough to share many attributes with the project under evaluation. Important attributes may include type of product, location, market forces, and maturity of the technologies.

2. *Assess the distribution of outcomes for these completed projects.* Quantitative data on projects in the reference class should fall into a distribution that can provide the appropriate statistical measures. For example, projects on coal-fired power plants should provide distributions of construction costs, electricity generated per day, prices received for this electricity, and operating expenses.

3. *Predict the position of the current project in each distribution.* Similarities between the current project and others in the reference class should indicate where it falls along a distribution of outcomes. For example, a proposed coal-fired power plant should fall on the distribution of construction costs according to the positions of similar plants. This positioning is subject to bias, however, and the following two forecasting steps make adjustments to achieve greater accuracy.

4. *Evaluate the reliability of the predictions.* Past performances on a variety of forecasts can establish a correlation between prior estimates and actual outcomes. For example, comparisons between predicted and actual construction costs of coal-fired power plants or of mass-transit systems indicate a propensity for bias.

5. *Correct the intuitive estimate.* Bias in step 3 shifts the estimate for the current project further away from the average outcome of the reference class than appropriate. The less reliable prior predictions have been, the closer the current estimate should be to the average outcome. For example, if construction costs of coal-fired power plants in

the reference class average $100 million, but the estimate for a proposed plant is $60 million, past performance as evaluated in step 4 should determine how much the estimate should be adjusted upward from $60 million toward $100 million.

Despite the potential advantages of this approach and its official endorsement by the American Planning Association, the British government, and the European Commission (Flyvbjerg and COWI 2004; American Planning Association 2005; HEATCO 2007), few major projects use reference class forecasting (Flyvbjerg 2007).

BETTER INCENTIVES FOR OBJECTIVITY Planners of large projects play two roles that often conflict (Wachs 1989). First, they are "scientists" who collect and analyze data to determine the best course of action. Second, they are "advocates" who prepare information to support their clients' preferred course of action. Clients, understandably, shop for an expert who will tell them what they want to hear and who thereby increases the probability that their project will compete successfully for funding.

Several procedures could improve this situation (Flyvbjerg 2007).

- Cost–benefit analyses for large public project should be subject to independent peer review conducted by governmental offices such as the General Accounting Office in the United States or the National Audit Office in the United Kingdom.

- Governments should make all information about a project publicly available and hold open hearings.

- Planners who consistently produce deceptive analyses should receive professional sanctions and, occasionally, criminal penalties.

- Planners or their organizations must share financial responsibility for cost overruns or benefit shortfalls.

Intangible benefits

Market forces work effectively to set the economic value of many goods or services. For example, the prices of toasters or car washes tend to fall within a narrow range in any location. What is an appropriate price for more unique goods or services? This section examines methods to assess aesthetic and cultural values, human welfare, and risk avoidance.

AESTHETICS AND CULTURE "Beauty [may be] in the eye of the beholder," yet we routinely appraise goods or services based on aesthetic and cultural values. Every

FIGURE 10.18 Guennol Lioness, a 5000-year-old, 8.3 cm high, Mesopotamian limestone figurine, which sold in 2007 at Sotheby's New York for more than $57 million, the highest price ever paid for a sculpture.

purchase of high-heeled shoes is a victory for aesthetics and culture over function. Hotels often charge more for rooms with an ocean view than those that overlook a highway; this price differential may disappear if the ocean view contains hundreds of wind turbines (see Figure 8.38). Sotheby's and other auction houses regularly arrange the sale of fine artwork (**Figure 10.18**). Unique goods or services once declared to be priceless may just turn out to be pricey.

Still, cost–benefit analyses for large-scale projects to mitigate climate change may require assessment of goods or services for which market data is unavailable. How does one put a price tag on the Sydney Opera House or the temple at Abu Simbel (see Figure 8.22), structures that are national treasures? On what basis did someone select a few archeological sites for relocation out of the thousands flooded by the Aswan and Three Gorges Dams (see Chapter 8)? What circumstances would justify the construction of new dams on the Colorado River in the Grand Canyon (**Table 10.2**) or the Merced River in Yosemite (**Figure 10.19**)? Ana-

(A)

(B)

FIGURE 10.19 Potential dam sites (A) The Colorado River flowing through Marble Canyon, Arizona, a section of the Grand Canyon. The U.S. government conducted cost–benefit analyses in the mid-1960s that justified the construction of a dam here and at Bridge Canyon. These dams, together with the Glen Canyon and Hoover dams, would have turned much of the Grand Canyon into a series of lakes. Independent analyses concluded that the proposed dams were economically unsound (see Table 10.2) and contributed to the eventual defeat of the proposals. (B) The Merced River in Yosemite National Park, California. At one time, the federal government considered constructing a dam on this river and flooding Yosemite Valley.

lysts and planners estimate the value of non-market goods or services based on either revealed preference or stated preference techniques (Pearce and Özdemiroglu 2002; HM Treasury 2003; HEATCO 2007; Powe 2007).

Revealed preference techniques Revealed preference techniques associate a non-market good or service with the actual markets for a complementary (i.e., surrogate or proxy) good or service. Because complementary markets depend on the use of a good or service, indirect revealed preference techniques can evaluate some of the value associated with historical significance, reputation, diversity, or uniqueness. Methods to gauge revealed preferences include analyses of travel costs, avertive/abatement costs, and hedonic prices (see Figure 10.17).

- Travel cost analysis estimates the monetary value of recreational or educational sites (Clawson and Knetsch 1966). Surveys of visitors to a site obtain information about travel distance, travel costs per distance, travel time, value of time spent traveling, and demographic factors such as age, income, gender, and education levels. Analysts or planners calculate the number of visits to the site from people living in various locations; the average travel cost per visit from various locations; and the number of visits per capita from various locations as a function of average travel costs and other demographic factors. Expressed mathematically, the relationship is:

$$V = f(T, X) \qquad (10.2)$$

where

V = number of visits per capita

f = function

T = travel costs

X = demographic factors

TABLE 10.2 Two estimates of cost-to-benefit ratios for dams proposed in the Grand Canyon[a]

Dam	Bureau of Reclamation[b]	Carlin-Hoehn
Bridge Canyon	0.50	1.64
Marble Canyon	0.59	1.32

Source: Carlin 1968.

[a]A ratio below 1.0 favors a project, whereas a ratio above 1.0 does not.

[b]Bureau of Reclamation and Carlin-Hoehn assumed interest rates of 3% and 3⅛%, respectively.

(A)

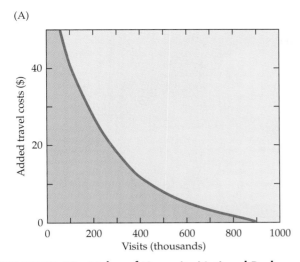

(B)

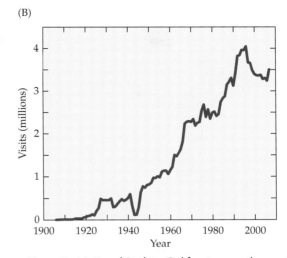

FIGURE 10.20 Value of Yosemite National Park
(A) Demand curve (red) of added costs ($U.S.) versus thousands of visits to Yosemite National Park in California during 1953. Area under the curve (light red) estimates the annual recreational value of Yosemite, about $15 million in 1953 in terms of 1953 $U.S. This amount of money equals $120 million in year 2008 $U.S. (B) Millions of visits to Yosemite National Park in California over the past century. The number of visits increased from 0.94 million to 3.50 million from 1953 to 2008. If the demand curve only needs adjustment for inflation (unlikely), the annual recreational value of Yosemite would be about $450 million in 2008 ($120 million x 3.50 million/0.94 million). (A after Clawson and Knetsch 1966; B after National Park Service 2008.)

- From this relationship, analysts or planners predict how the number of visits to the site of interest from a given location will vary with rising costs such as higher entrance fees or higher gasoline prices. The sum of visitors from all locations as a function of rising costs produces a demand curve. Demand curves usually are downward sloping where inexpensive costs encourage a high number of visits, and vice-versa (**Figure 10.20A**). Area under the demand curve (consumer surplus) provides an estimate for the recreational or educational value of a site at a given time. For example, the annual recreational value of Yosemite National Park is around $450 million in year 2008 $U.S. (**Figure 10.20B**), and the value of the Great Barrier Reef in Australia (see Figure 6.24) is between $700 million and $1.6 billion in year 2002 $U.S. (Carr and Mendelsohn 2003).

- Avertive/abatement cost analysis examines the expenditures of consumers on preventive measures to reduce damages from undesirable factors such as pollution or accidents (HEATCO 2007). It assumes that these measures are optimal at the point where their marginal cost (expenditure on the last unit purchased) equals their marginal value of reduced impact (reduced damages resulting from purchasing one more unit) (see Figure 10.10). Goods that individuals buy to avert pollution include air filters, water purifiers, and noise insulation; those that they buy to avert accidental injury include seat belts and fire detectors.

- Hedonic price analysis divides a good or service into its constituent parts and estimates the market values of each part. Because environmental factors play a major role in the price of a property and in wages earned for a job, differences in property values or wages provide an estimate of the market value of environmental factors. For example, relative house prices in two neighborhoods may reflect differences in average ambient noise levels. Wages are higher for jobs in unpleasant environments than for those that require similar skills but are conducted under better working conditions: Divers who clean industrial waste traps (**Figure 10.21A**) receive higher wages than divers who clean the hulls of yachts in tropical ports; highway toll collectors (**Figure 10.21B**) earn more than movie ticket collectors.

Stated preference techniques Stated preference techniques depend on constructed markets developed from interviews or surveys. These interviews or surveys ask people about the economic values they attach to environmental goods and services (Pearce and Özdemiroglu 2002). In-person or telephone interviews, in contrast to mail or web surveys, allow researchers

(A)

(B)

FIGURE 10.21 Hazard pay and hedonic price analysis (A) Diver descending into the wastewater lagoon of a paper mill (cellulose processing; see Chapter 7) to clean out plugged nozzles. (B) Toll collection on Alligator Alley, the highway that crosses the state of Florida through the Everglades.

to clarify questions and probe interviewees to obtain a fuller understanding of their responses. Focus groups (group interviews) tend to address general problems and foster sharing of information among participants. Individual interviews are better suited to details and confidential issues such as potential financial payments. Telephone interviews are relatively inexpensive but allow no visual aids, and many people will not provide sensitive information over the telephone; thus, telephone interviews are becoming less common. Based on the responses, analysts and planners estimate people's willingness to pay for a particular benefit or their willingness to accept payment for bearing a particular loss. Stated preference techniques can estimate non-use values, values that people derive from something independent of whether they ever use it. For example, many people value protecting polar bears from extinction although they may never see a polar bear in its native habitat.

Designs of stated preference surveys follow two styles: contingent valuation and choice modeling (see Figure 10.17).

- Contingent valuation broadly examines the non-market good or service. Analysts and planners limit the choices of the interviewee to primarily one option, asking questions such as "What are you willing to pay to insure that the Grand Canyon is left in its current state for another year?" or "Are you willing to pay $5 to insure that the Grand Canyon is left in its current state for another year?" "If yes, how about $10?" "If no, how about $1?" Other questions address whether the interviewee has visited the Grand Canyon recently or intends to visit it in the near future and whether these visits or other reasons influence the respondent's willingness to pay.

- Choice modeling surveys people's rankings and ratings of alternative characteristics or attributes of a non-market good or service. This technique assumes that the entire worth of the good or service equals the sum of its parts, and it is less reliable when the rankings of different characteristics are strongly correlated. Various forms of choice modeling are choice experiments, contingent ranking, contingent rating, and paired comparisons (Pearce and Özdemiroglu 2002). Choice experiments present a baseline scenario corresponding to the status quo and several alternative options and ask a respondent to choose among them. ("Choose A, B, or neither, where the choice of 'neither' maintains the status quo."). Contingent ranking proceeds in the same way but asks a respondent to rank alternatives in terms of desirability. Contingent rating asks a respondent to rate each alternative on a scale, for example, from 1 to 10. Pairwise comparisons ask a respondent to indicate a strength of preference for one alternative over another.

Surveys consistently show that indirect factors can influence people's preferences about environmental issues (Gowdy and Erickson 2005). There are endowment effects (people value things they already possess more than those they will acquire), the scope issue (people value the parts more than the whole), loss aversion (people are more averse to suffering a loss than they favor receiving an equal gain), and hyperbolic discounting (people favor short-term payoffs over long-term ones and expect a higher interest rate for short-term loans than for long-term loans). These tendencies may distort evaluations of environmental goods or services (Getzner et al. 2005; Gowdy and Erickson 2005).

HUMAN WELFARE   Global climate change will influence human health and well-being (Gamble 2008). Heat-related deaths will mount as heat waves become more severe. Rising sea level will threaten coastal cities and their fresh water supplies. Greater severity of storms and forest fires will put more people at risk. Certain insect vectors that carry human diseases will extend their ranges to higher latitudes and altitudes. Warmer temperatures and altered precipitation patterns will shift food production. Policies to abate greenhouse gases will likely force people to spend more time at routine tasks such as commuting to work, heating water, and discarding wastes.

Economic assessments of human welfare depend on complex value judgments. What is the value of a human life? How important is quality of life? "Time is money," the saying goes—but how much money is involved?

Value of statistical life   A standard measure for the monetary value of a human life, the so-called **value of statistical life**, estimates what society is willing to pay for reducing each member's risk of fatality by a small amount (Dockins et al. 2004). Planners use hedonic price or contingent valuation analyses to estimate values of statistical life. Hedonic price analyses examine what firms pay employees as a function of the risk of fatalities in different industries or jobs (**Figure 10.22**). Contingent valuation analyses survey the willingness of people to pay for decreasing the risk of fatalities. For example, respondents indicate whether they would pay $10 for the reduction in risk from two deaths per 10,000 people to one death per 10,000 people. If 100,000 people would be willing to pay $10, the value of statistical life would be $1 million ($10 multiplied by 100,000) (Kochi et al. 2006).

Estimates of the value of statistical life from such surveys are highly variable (**Table 10.3**). Various gov-

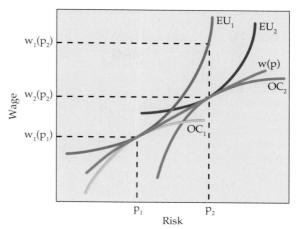

FIGURE 10.22   Hedonic price estimate of value of statistical life Labor costs increase with the level of safety, and thus, for any given level of profits, firms pay workers less as safety levels rise or risks decline. Employer 1's willingness to pay wages as a function of risk (p_1 is a probability of death) follows OC_1 (offer costs). A worker at employer 1 is willing to accept wages as a function of risk according to EU_1 (expected utility). These functions meet at the optimal wage $w_1(p_1)$. A similar exercise for employer 2 yields $w_2(p_2)$. Wages as a function of risk, $w(p)$, has a slope equal to the value of statistical life. (After Viscusi and Aldy 2003.)

ernmental agencies rely on different meta-analyses of surveys, and so their estimates of the value of statistical life in year 2008 $U.S. range from $1.3 million for the U.S. Federal Aviation Administration and $3.7 million for the British Department of Transport to $8.0 million for most branches of the U.S. Environmental Protection Agency (Viscusi and Aldy 2003; U.K. Department for Transport 2007). The air pollution branch of the U.S. Environmental Protection Agency recently lowered their value of statistical life from $8.0 million to $6.9 million (Borenstein 2008). This has a large effect on policy. For example, this branch in 2005 estimated that the U.S. Clean Air Interstate Rule provided $130 billion (in year 2008 $U.S.) in annual health benefits (U.S. Environmental Protection Agency 2005b). With their new value of statistical life, annual health benefits of cleaner air decline to $112 billion.

Quality of life   One common approach for assessing the cost effectiveness of medical treatments is the **quality-adjusted life year** (Phillips and Thompson 2001; HM Treasury 2003; Mortimer and Segal 2008). This measure takes into account changes in both human life expectancy and quality of life. EuroQol, a network of European researchers, has developed a survey that ranks quality of life on five issues: mobility, pain/discomfort, self-care, anxiety/depression, and general activi-

TABLE 10.3 Value of statistical life[a]

| Country | Number of studies | Value | | Standard deviation |
		Average	Median	
United States	22	8.7	7.8	6.3
Canada	7	11.6	5.1	13.2
United Kingdom	4	33.2	28.6	26.7
Australia	2	14.2	14.2	12.2
Austria	1	10.5	10.5	—
South Korea	1	2.0	2.0	—
India	1	20.4	20.4	—
Japan	1	16.3	16.3	—
Taiwan	1	1.5	1.5	—
Total	**40**	**12.1**	**8.4**	**12.9**

Source: Bellavance et al. 2007.
[a]In millions of year 2008 $U.S.

ties such as those involved in work, study, housework, and play (EoroQol 2008). For each issue, respondents choose one of three levels: "no problems," "some problems," or "major problems." Responses to the survey yield composite scores that range from 0.0, which indicates that a person is as good as dead, to 1.0, which indicates a person has no health and happiness problems. For example, if a clinical treatment extends life expectancy by 2 years and if the quality of life during that period is 0.8, the quality-adjusted life year would be 2.0 year × 0.8 quality score = 1.6 quality-adjusted life year. To compare cost effectiveness of alternative treatments, one divides their costs by the quality-adjusted life years.

A quality-adjusted life year, if one estimates the monetary value for a life year, can provide a measure of human welfare for cost–benefit analyses of environmental policies. This measure differs from the value of statistical life in many respects, however (Hammitt 2002; Freeman 2006). For example, quality-adjusted life year explicitly evaluates quality of life but assumes that it is independent of longevity, whereas value of statistical life empirically determines the relationship between quality and longevity. Nonetheless, both measures tend to reach similar conclusions about policies that impact mortality (Hubbell 2006; Kenkel 2006).

Time = money Time spent at work has a well-defined market value equal to wages plus benefits. Time spent at other activities may have direct market equivalents. For example, some people pay others to do their laundry or to change the oil in their vehicle, and so one can estimate the value of the time saved from the cost of these services.

Other activities may be more difficult to evaluate. Time spent commuting to and from work or traveling for personal reasons has no direct market value, so planners use revealed preference or stated preference techniques to survey the real or hypothetical choices of travelers between faster, pricier modes and slower, cheaper modes. For example, if a traveler selects a train over a bus when the train journey takes 2 hours at a cost of $30 and the bus trip takes 3 hours at a cost of $20, the traveler values their travel time at $10 per hour because they are willing to spend an additional $10 to save an hour of travel time.

Estimates of the value of travel time range from 52% of wages in Oceania (mostly in Australia) to 101% of wages in southern Europe (Table 10.4). Time spent commuting between home and work is assessed a value 10% greater than travel for other purposes (Mackie et al. 2003). People walking to, waiting for, or taking public transport demand greater compensation for their time—as much as 250% more—than when they spend

TABLE 10.4 Value of travel time[a]

Region	Number of studies	Mean	SD[b]
N. Europe	40	83	64
S. Europe	25	101	85
N. America	18	68	47
Oceania	7	52	45
Total	**90**	**83**	**67**

Source: Zamparini and Reggiani 2007.
[a]Percentage of wages.
[b]Standard deviation.

the same amount of time riding in a private vehicle. Unreliability, the deviation from expected travel time, also elicits an additional penalty (HM Treasury 2003). All of these factors influence cost–benefit analyses of environmental policies that involve transportation.

RISK AVOIDANCE "Risk" and "uncertainty," terms that are used nearly interchangeably in casual conversation, have different meanings in economics (Bjornstad 2004). Risk is the probability that one particular outcome will occur out of all possible future outcomes. An astute blackjack cardplayer knows the probability of being dealt a face card but does not know what the next card dealt will actually be. Similarly, an analyst may know the probability distribution of a variable but not the actual value of the variable that will occur. Uncertainty, on the other hand, is the lack of knowledge about the probability of future outcomes. Even an astute blackjack cardplayer has difficulty in a casino that varies the number of cards in the deck or places an unknown number of face cards in the deck.

Risk that an accident will significantly threaten the success of a project becomes more likely as the project grows in size and complexity. To minimize such possibilities, planners attempt to manage risk through identifying and characterizing as many risks as possible early in the project; conducting pilot studies on parts of the project for which the risks are uncertain; allowing for alternative designs at most junctures of the project; and transferring risks contractually to other parties such as insurance companies (HM Treasury 2003). Risk management for large projects must take into account the **precautionary principle**, whereby people support strong measures to avoid catastrophic outcomes even if their chance of occurrence seems remote (Harris and Roach 2007). Consequently, taxpayers often bear most of the risk for large public works such as electric power plants and levees for flood control.

For example, nuclear accidents do happen. Virtually all property and liability insurance policies issued in the United States exclude damages resulting from nuclear accidents. Under the Price-Anderson Act of 1957, utilities pay a premium each year for $300 million in private insurance for offsite liability coverage for each nuclear reactor unit that they operate (U.S. Nuclear Regulatory Commission 2008). So far, the nuclear insurance pool has paid out a total of $151 million to cover claims, including $70 million for the accident at Three Mile Island (see Chapter 8) and about $65 million for the nuclear operations of the U.S. Department of Energy. If a nuclear accident causes damages in excess of $300 million, each licensed nuclear reactor would be assessed a prorated share of the excess up to

$95.8 million. With 104 plants in operation, this secondary pool contains about $8.6 billion.

An accident on the scale of Chernobyl, however, would incur roughly $300 billion in damages (Eeckhoudt et al. 2000; Rothwell 2002) and would rapidly empty this secondary pool. In such cases, the president of the United States is required to petition Congress for additional funds for disaster relief in which the federal government would pay 75% of the needs of the victims and state and local governments would pay the remainder. Given the probability of a Chernobyl-sized accident, liability protection from the U.S. government amounts to an annual subsidy of about $240 million to the nuclear power industry (Rothwell 2002).

Countries differ in what they will spend to avoid risk. Most levees in the United States are currently being built to a 100-year flood standard, which means that these levees are expected to protect against floodwaters rising to levels that occur only once in 100 years on average. In contrast, the minimum standard for levees in the Netherlands is 1,250 years, and many of their levees are rated to withstand a 10,000-year storm. This indicates that people in the Netherlands are willing to pay more to diminish risks than those in the United States, a topic that will be examined at length in Chapter 12.

Uncertainty

In life, as in blackjack, people fear uncertainty more than risk (Bjornstad 2004). Cardplayers can determine their best bet based on the odds. Similarly, planners can manage risks according to the likelihood of alternative events, and insurance companies can set their premiums on the probabilities of negative outcomes that would leave them liable. Cardplayers, planners, or insurance companies cannot assign odds, likelihoods, or probabilities to unforeseen—and perhaps unforeseeable—events.

Virtually all economic decisions about global climate change are rife with uncertainty. To make matters worse, climate changes in the past have led to mass extinctions (see Chapter 2). How does one place a value on avoiding the end of humankind?

On a smaller scale, cost–benefit analyses of environmental projects depend on factors such as the value of statistical life (see Table 10.2) and travel time (see Table 10.3) that vary greatly (i.e., the spread among estimated values may exceed their average value). Moreover, analyses of such projects are highly dependent on the rate of technological change, discount rate, and degree of intergenerational equity. Small differences in these rates, when compounded over the long time periods during which climates change, can lead to huge differences in realized costs or benefits.

TECHNOLOGICAL CHANGE "Change is hard" is a common expression, and technological change is no exception. Technological change includes creation and adoption of new products as well as improvement in the quality and efficiency of existing products. It is subject to a large externality, whereby an innovator who is responsible for technological change assumes most of the research and development costs but might not reap most of its benefits. Once an innovator reveals a piece of technological knowledge, it becomes available to all.

WEB TOPIC 10.1

Magic trick Revealing the technological basis of a magic trick in a patent application diminished the value of the trick.

Moreover, technological knowledge, unlike other goods or services, does not wear out with repeated or simultaneous use. Because consumers in a free market are not willing to pay for something that may become freely available and inexhaustible (behavior known as **free riding**), investment in research and development dwindles, and technology stagnates (Mansfield and Yohe 2004). An example, mentioned earlier in the chapter, is that deregulation of the electric power industry has made it more responsive to market forces yet has not stimulated sufficient investment in research and development.

Patents and copyrights are methods for compensating innovators in order to encourage research and development. In a patent or copyright, a government grants an inventor or author exclusive rights for a fixed period in exchange for public disclosure of the invention or work. Unfortunately, patent or copyright holders usually set the price for licensing their invention or work so high that it discourages people who would otherwise make productive use of it. From the perspective of benefits to society, patent or copyright holders should grant permission at a low price to anyone who could use their invention or work, because the marginal cost of this use is practically nil. Then again, such a compensation scheme would provide little incentive for innovation.

Technological change to address global climate change suffers from a second externality: the tragedy of the commons. Free markets place little value on Earth's atmosphere and thereby place little value on efforts to decrease greenhouse gas emissions. Governmental policies that would raise the price of greenhouse gas

TABLE 10.5 Distribution of U.S. Federal funds among various climate-related technologies in 2004

Category of funding	Percentage
Research and development ($1.3 billion annually)	
Energy conservation	41
Renewable energy supply	33
Fossil-fuel greenhouse gas reductions	14
Carbon capture and sequestration	8
Nuclear	2
EPA science and technology	1
Forest, range, agriculture	1
Energy Information Administration	1
Technology adoption ($1.0 billion annually)	
State energy efficiency grants	34
Renewables production tax incentives	27
Hybrid/fuel cell car tax credits	16
Cogeneration tax incentives	10
EPA information/voluntary programs	9
Landfill gas production tax credit	3
Solar homes tax credits	1

Source: Jaffe et al. 2005.

emissions, such as an emissions tax or a cap-and-trade system would stimulate technological change in this area. The United States and most of the world, however, have delayed implementing such policies (see Chapter 11). Thus, policy-induced technology change has been slower than is socially desirable (Jaffe et al. 2005; Newell et al. 2006).

Fortunately, governments—even those that balk at implementing climate change policies—have found it politically expedient to fund climate-related research and development and to push adoption of energy-efficient technologies (Table 10.5). This includes research grants or contracts to public institutions and general tax credits or matching funds to firms who undertake research in this area or who adopt these technologies. The U.S. National Science Foundation, for example, currently allocates over $1 billion to study global climate change.

The U.S. National Academy of Sciences conducted a cost–benefit analysis of the research funded by the U.S. Department of Energy in the areas of fossil energy and energy efficiency (National Research Council 2001). This research played an important role in many important technological innovations from 1978 to 2000 (Table

TABLE 10.6 Influence of U.S. Department of Energy funding on the most important technological innovations for fossil energy and energy efficiency[a]

Technology now in the marketplace	Influence[b]
Fossil energy	
Efficient gas turbine in stationary systems	A/M
3-D seismic imaging	A/M
Deep water drilling and production	A/M
Oil and gas reservoir modeling	A/M
Oil and gas drilling: horizontal and deviated	A/M
Diamond drill bits	D
Coal-bed methane	I
Flue gas cleanup	I
Atmospheric fluid-bed combustion	I
Fracture technology for tight gas	I
Oil refinery optimization	A/M
Longwall coal mining	A/M
Coal cleaning	A/M
Energy efficiency	
More efficient electric motors	A/M
Higher mileage automobiles	A/M
More efficient electronic ballasts	D
More efficient household refrigerators	D
More effective insulation	I
Synthetic lubricants	A/M
More efficient gas furnaces	A/M
More energy-efficient windows	I
More efficient industrial processes	A/M
More efficient buildings	I

Source: National Research Council 2001.
[a]From 1978 to 2000.
[b]Influence levels: A/M, absent or minimal; I, influential; D, dominant.

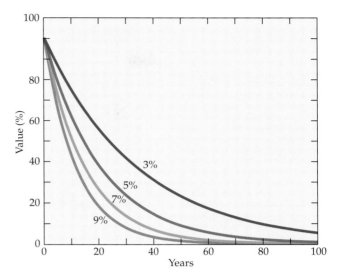

FIGURE 10.23 Value of a good or service with a discount rate of 3%, 5%, 7%, or 9% as a function of time (years) With a 5% discount rate, the present day value of a $100 benefit realized after 10 years is $61.

The pace of technological innovation can be breathtaking. For example, light output from a single light-emitting diode (LED) lamp doubles every 2 years, cost per light output halves every 3 years, and luminous efficacy doubles every 6 years. Energy efficiency of refrigerators doubles every 15 years. Fuel efficiencies of vehicles per unit weight and coal-fired power plants improve steadily. Nonetheless, slow turnover of equipment or facilities tempers the overall rate of technological change.

"If a man can…make a better mousetrap…the world will make a beaten path to his door" (Emerson 1889). Who will build a better mousetrap? When will it be ready? How much better will it be? How quickly will the world adopt it? These questions highlight some of the uncertainties about the rate of technological change that confound cost–benefit analyses of environmental projects.

DISCOUNT RATE Banks lend money at interest rates that incorporate the risk of nonpayment, the rate of inflation, and a decent rate of return on investment. In contrast, government borrowing carries a low risk of nonpayment, and cost–benefit analyses for government projects calculate everything in terms of a constant currency already adjusted for inflation (e.g., year 2008 $U.S.). Consequently, to convert values of future costs or benefits to present ones (**Figure 10.23**), governments use a different rate, most commonly called the **discount rate**, which depends primarily on the rate of return on investment.

10.6). In the area of fossil energy, innovations such as advances in atmospheric fluidized bed combustion and nitrogen oxide controls to decrease air pollution required research expenditures totaling $10.4 billion, but realized economic benefits of $10.7 billion and environmental benefits of $60 billion (in year 1999 $U.S.). In the energy efficiency area, innovations such as advances in compressors for refrigerators and freezers, energy-efficient electronic ballasts for fluorescent lighting, and better-insulating window glass spent $7 billion for research, but realized $30 billion in economic benefits.

TABLE 10.7 Annual growth of stock markets, interest rates[a], and equity premium for selected countries

| Country | Period | Percentage | | |
		Market index	Riskless security	Equity premium
United States	1802–2004	8.4	3.0	5.4
	1871–2005	8.3	2.7	5.6
	1889–2005	7.7	1.3	6.4
	1926–2004	9.3	0.7	8.6
United Kingdom	1900–2005	7.4	1.3	6.1
Japan	1900–2005	9.3	–0.5	9.8
Germany	1900–2005	8.2	–0.9	9.1
France	1900–2005	6.1	–3.2	9.3
Sweden	1900–2005	10.1	2.1	8.0
Australia	1900–2005	9.2	0.7	8.5
India	1991–2004	12.6	1.3	11.3

Source: Mehra 2006.
[a]For relatively risk-free securities.

What constitutes a decent rate of return on investment and thus an appropriate discount rate? Several processes come into play (Lind 1982; Markandya and Pearce 1991; Cowen 2007; Heal 2007). First, there is the **impatience principle**, whereby people expect some compensation if they must wait for something. Say, for example, you order a hybrid vehicle today, but the car dealer tells you it is back-ordered and cannot be delivered until 3 months from now. You might cancel the order or request a discount. The size of the discount that you are willing to accept for waiting 3 months equals the annual discount rate compounded continuously over one-fourth of a year.

A second process is the **marginal productivity of capital**, the extent to which an extra dollar invested in resources today produces more than a dollar's worth of additional goods or services in the future. For example, if the government spends another dollar to increase the height of a hydroelectric dam, the dam should generate more than a dollar's worth of additional electricity during its operational life.

Third, there is the **opportunity cost of capital**, the rate of return from potential private savings accounts or investments that might otherwise use the capital spent on a government project. Part of the difficulty in evaluating opportunity cost of capital is the **equity premium**, the differences between the rate of return earned by a broad market index (such as from investment in publicly traded stocks) and the rate earned by a relatively risk-free security (such as from investment in a government bond). An equity premium derives from several factors including compensation for the larger risks of stocks versus bonds; constraints on borrowing such as high transaction costs or large down payments, which smooth temporal fluctuations in the consumption of goods or services; government rules and tax policies that favor one type of investment over another; and divergence between borrowing and lending rates (Mehra 2006). Equity premiums ranged from 5.4% to 11.3%, depending on country and data set (Table 10.7). A survey of 226 economists predicted an equity premium of 7% over 10-year and 30-year horizons (Welch 2000).

Theoretically, the purpose of governmental projects is to increase public well-being. Borrowing for projects should therefore not significantly divert funds from private savings by consumers or lower the effective national rate of return on private business investments. Consumers tend to decide how much to save based on the rate of return for bonds, whereas business investors decide how much to invest based on the rate of return for stocks.

Raising money for a large public project may displace capital from private savings, private investments, or some of both. If the public project displaces mostly private savings, then the appropriate discount rate is the bond rate of return, which from 2007 to 2009 fluctuated between 0.9% and 2.5% in the United States. (U.S. Office of Management and Budget 2009). At this discount rate, a consumer is equally likely to buy a U.S. Savings Bond or to put money into a savings account at a private bank. If a large public project must raise capital at a higher discount rate, then it would be at a disadvantage in comparison to raising capital through the sale of government bonds or deposits in savings

TABLE 10.8   Annual discount rates as a function of length of program

Duration (years)	Discount rate (%)		
	United States cost–benefit[a]	United States cost effectiveness/lease[b]	United Kingdom cost–benefit
3	3.0 or 7.0	2.1	3.5
5	3.0 or 7.0	2.3	3.5
7	3.0 or 7.0	2.4	3.5
10	3.0 or 7.0	2.6	3.5
20	3.0 or 7.0	2.8	3.5
31	3.0 or 7.0	2.8	3.0
76	3.0 or 7.0	2.8	2.5
126	3.0 or 7.0	2.8	2.0
201	3.0 or 7.0	2.8	1.5
301	3.0 or 7.0	2.8	1.0

Source: U.S. Office of Management and Budget 2003, 2008; HM Treasury 2003.
[a]Cost–benefit analyses of government projects.
[b]Government analyses of cost-effectiveness and lease purchases.

accounts. If the project displaces mostly private investment, then the discount rate should be the rate of return for stocks, which has averaged 4.3% in the United States over the past 4 decades (Moneychimp 2009); again, raising capital for a public project at a higher discount rate would put it at a disadvantage in comparison to private projects that raise capital through selling stocks. One might calculate a discount rate from a weighted average of the rates of return for bonds and stocks, where the weights reflect the responses of private savings and private investments to public capital expenditures. In practice, such responses are difficult to assess, and discount rates tend to vary more with attitudes about both spending on public works and government regulations (Bjornstad 2004).

The British government sets the discount rate between 3.5% and 1.0% for government projects, depending on their duration (Table 10.8). The U.S. Office of Management and Budget asks planners to prepare two cost–benefit analyses for proposed government regulations, one based on a discount rate of 3% and the other on a rate of 7% (U.S. Office of Management and Budget 2003). Yields on U.S. Treasury bonds and notes averaged 2.2% between April 2004 and April 2009 (U.S. Office of Management and Budget 2009), and average yields on U.S. stocks were 4.3% (Moneychimp 2009). Discount rates of 3% and 7% thus exceed the rates of return for bonds and stocks, respectively, and discourage the initiation of new government projects. The Office of Management and Budget does admit, however, that smaller discount rates might be appropriate for environmental projects with long time horizons (Graham 2007).

Most of the costs in environmental projects incur near the beginning and will be subject to the discount rate for only a short period, whereas most of the benefits accrue near the end and will be subject to the discount rate for a long period. Consider a 30-year project that has a projected cost of $1 million and benefits of $3 million. "What a deal!" might be your first reaction. Yet if the costs are mostly startup expenses and the benefits are realized only near the end of the project, the costs remain $1 million, but the $3 million of benefits at the end translate to $1.24 million at the onset with a discount rate of 3% and only $0.39 million with discount rate of 7% [Future value = Present value $(1 + \text{Discount rate})^{\text{Years}}$; $3.00 = $1.24 (1 + 0.03)^{30}$ and $3.00 = $0.39 (1 + 0.07)^{30}$] (see Figure 10.23). A high discount rate thus makes environmental projects appear economically unfeasible.

Because the discount rate alone can doom environmental projects and regulations, it has stimulated volumes of debate (e.g., Lind 1982; Portney and Weyant 1999; Weisbach and Sunstein 2007). Some economists argue that the global economy has grown steadily despite dire warnings (Figure 10.24) and that technological changes are likely to enable quicker, more thorough, and less expensive mitigation of global climate change in the future; therefore, a high discount rate is appropriate. Others argue that growth of the global economy has been highly dependent on the availability of cheap energy and that as we deplete readily available fossil fuels, the opportunity cost of capital will

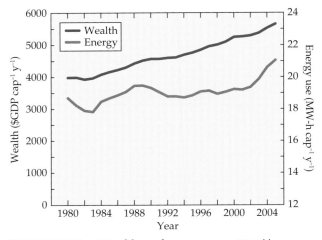

FIGURE 10.24 Wealth and energy use Wealth is in terms of annual gross domestic product in year 2000 $U.S. per capita and annual energy consumption in megaWatt-hours per capita. The correlation is between the two is 80%. (After Development Data Group; The World Bank 2007; Energy Information Administration 2007c.)

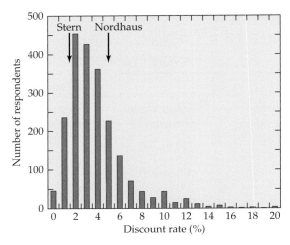

FIGURE 10.25 Distribution of the opinions of 2160 economists on an appropriate discount rate (%) for projects that mitigate global climate change The mean (average) is 3.96%, the median (where there is an equal number of responses higher and lower) is 3%, the mode (the most common response) is 2%, and the standard deviation (a measure of the extent to which values are spread around the mean) is ± 2.94%. Arrows indicate the long-term rates assumed in the models of Stern and Nordhaus. (After Weitzman 2001.)

decline; therefore, the discount rate should reflect only the impatience principle and deserves a low value.

INTERGENERATIONAL EQUITY Most societies believe that each generation bears some responsibility for the quality of life that it leaves to its descendants and that every generation deserves an equal opportunity of success (see Chapter 12). **Intergenerational equity** in a financial context means that an institution, if it is supported by fixed endowments, should not spend more than the after-inflation rate of return on its investments to sustain its assets at similar levels for present and future beneficiaries (Tobin 1974). In a broader context, Earth's natural resources are an endowment, and intergenerational equity means that societies should strive to sustain natural resources for present and future inhabitants of the planet.

Some analysts address intergenerational equity by adding a **bequest value** to the benefits of an environmental project; a bequest value places a price on the satisfaction derived from preserving a natural resource for future generations. More commonly, analysts substitute a low "social discount rate" for the real discount rate (Lind 1982; Portney and Weyant 1999; Weisbach and Sunstein 2007). A social discount rate reflects the willingness of a society to trade present consumption of resources for the consumption of resources in the future.

Nonetheless, future generations are likely to be wealthier than the present one if the world's economy continues to grow even at modest rates. Thus, policies that sacrifice present consumption in favor of future

consumption essentially transfer wealth from the poorer present generation to richer future generations, an intergenerational inequity. A positive social discount rate compensates for such inequity. Conversely, a shrinking economy such as occurred in 2008 and 2009 might justify a zero or even negative social discount rate.

Harvard economist Martin Weitzman surveyed economists about their opinions on an appropriate discount rate (social or real) for projects that mitigate the possible effects of global climate change (Weitzman 2001). He received responses from 2160 economists in 48 different countries. The most common response (i.e., the mode) of these economists was a discount rate of 2%, although the average was 4% because some economists called for very high discount rates (**Figure 10.25**). About 98% of the respondents believed that the economy would continue to grow and recommended a positive social discount rate. Many of the respondents also favored a hyperbolic discount rate, one that declines significantly with time when a project extends over many decades (Weitzman 2001).

Distributional issues

Intergenerational equity treats the distribution of wealth over time. More commonly, economists consider the distribution of wealth over space, that is, the allocation of resources among all living people. Ideally, everyone would profit equally from a change in economic policies. To profit equally might mean that, for example, the government either issues tax rebates (e.g., everyone receives a $600 check) or cuts the tax rate (e.g., everyone receives back the same percentage of what they contributed). Each course of action is unfair: A tax rebate favors the poor over the rich, whereas a tax-rate cut favors the rich over the poor. The choice between tax rebates and tax rate cuts might be straightforward for Robin Hood, who took from the rich and gave to the poor, but it is less obvious to most governments.

Neoclassical economics places great faith in market forces. Evaluation of market forces involves assessments of **marginal utility**, changes in people's well-being before and after some small change in resource allocations. Marginal utility assumes that all other conditions remain relatively constant. Global climate change is likely to wreak far-reaching and long-lasting shifts in resource allocations. Evaluating marginal utility becomes elusive under such circumstances.

One measure of success in resource allocation is a Pareto improvement (Vilfredo Pareto, 1848–1923, was an Italian economist) that betters the conditions of at least one individual without harming anyone else. Economic performance reaches **Pareto optimality** when no additional Pareto improvements are possible; that is, no other changes would make someone better off without sacrificing the well-being of someone else. Theoretically, resource allocation in a free market should converge to Pareto optimality.

Pareto optimality does not address redistribution of wealth, however. For example, both tax rebates and tax rate cuts would be Pareto improvements. A society might attain Pareto optimality even if it does not provide basic amenities to some of its citizens because this would require a small sacrifice from others. Moreover, resource allocations converge to Pareto optimality only if markets exist for all possible goods and services, all markets are perfectly competitive and in full equilibrium, and transaction costs and externalities are negligible. None of these assumptions hold true for projects or regulations concerned with global climate change. Global climate change involves nonmarket goods or services from monopolies or cartels. Supply and demand for these goods and serves fluctuate wildly (far from equilibrium) and suffer from high transaction costs and large externalities. Therefore, resource allocations to remedy global climate change require a criterion other than Pareto optimality.

Kaldor-Hicks optimality (Nicholas Kaldor, 1908–1986, and John Hicks, 1904–1989, were British economists) expands Pareto optimality to include projects or regulations that will sacrifice the well-being of some people if they receive compensation for the harm they suffer. Suppose one shepherd initially grazes 10 sheep on the common green, while another grazes 100 sheep. If proposed changes allow the first shepherd to graze 20 sheep but limit the second shepherd to 90 sheep, the first shepherd might offer the second shepherd some compensation, such as a side of beef, to support the changes. The second shepherd might accept the offer if the compensation seems adequate, if beefsteaks were an attractive alternative to lamb chops. Revealed preference and stated preference techniques, as described earlier in the chapter, attempt to determine what compensation would suffice to seal the deal.

Compensation schemes for addressing global climate change are hard to envision. Under most scenarios, short-term costs for climate change mitigation will fall disproportionately on the larger emitters of greenhouse gases (see Figure 4.14), but short-term benefits will accrue disproportionately to other parties, such as those most vulnerable to sea level rises. Long-term costs and benefits are even more difficult to assess. Whether the beneficiaries of global climate change could financially compensate those bearing most of the costs is doubtful, particularly when many of the beneficiaries are among the poorest of the world, whereas those bearing most of the costs are among the richest.

Game theory examines situations in which certain people gain at the expense of others and where people choose their actions based on the actions of others. **Nash equilibrium** (John Forbes Nash, born 1928, is an American mathematician whose life was fictionalized in the movie *A Beautiful Mind*) is the condition where no person benefits from a change in strategy if the other people keep theirs unchanged. People, when faced repeatedly with the same circumstances, become less cooperative over time and converge towards Nash equilibrium unless governmental bodies can encourage cooperation through penalties to free riders (Fehr and Gachter 2000a,b). A free rider in an economic context is someone who consumes more than their fair share of a resource or fails to pay for their fair share of the costs of its production. For example, a country may ignore international standards for greenhouse gas emissions and therefore incur no added expenses yet enjoy the benefits of diminished climate change if other countries abide by the standards. Indeed, the United States has spent relatively little to abate greenhouse gas emissions but still has benefited from the efforts of European countries in this regard. Collective actions to discourage such behavior and to avoid further environmental degradation are the topic of Chapter 11.

Economic Models

Predicting the influence of global climate change on the world's economy requires computer modeling (e.g., Mendelsohn et al. 2000; Alberth and Hope 2007; Nordhaus 2008a). These models incorporate at least a rudimentary GCM (global climate model or general circulation model; see Chapter 4). GCMs vary in methodologies (e.g., finite differences versus spectral transforms) and assume different scenarios about future human greenhouse gas emissions (e.g., A1FI versus B1). This variation in climate futures, added to the complexities of market economics and governmental fiscal policies, results in a broad range of possibilities.

Nordhaus versus Stern

In 1999, William Nordhaus (William D. Nordhaus, born 1941, is the Sterling Professor of Economics at Yale University and coauthor with Paul Samuelson of a popular introductory economics textbook) published an economic analysis of global climate change that he prophetically entitled "Requiem for Kyoto." He reached the following conclusions.

1. Costs of implementing strong mitigation measures such as the Kyoto Protocol (an international agreement reached in 1997 in Kyoto, Japan to address the problems of climate change) would exceed their benefits by between $1.1 trillion and $2.2 trillion (in year 2007 $U.S.).

2. Strong mitigation policies, in comparison to business-as-usual policies, would decrease damages from climate change by only 6%.

3. Strong mitigation policies would have dire economic consequences because they would reduce global GDP (gross domestic product) by 6%.

4. The United States would bear almost two-thirds of the costs of the Kyoto Protocol but enjoy only a small portion of the benefits (Nordhaus and Boyer 1999).

People who oppose greater U.S. participation in international efforts to mitigate greenhouse gas emissions commonly cite this and subsequent studies by Nordhaus (Nordhaus 2008b).

Six years later, the British government commissioned Nicholas Stern (Lord Nicholas H. Stern, born 1946, is the Patel Chair at the London School of Economics and Political Science and the former chief economist of the World Bank) to review the same issue. His report, the *Stern Review*, made the following points.

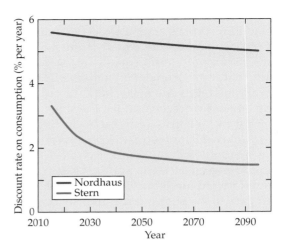

FIGURE 10.26 Global average discount rate on consumption (percentage per year) as a function of year in the analyses of Nordhaus and Stern This is equivalent to the rate of return on capital assuming no inflation, risk, or taxes. (After Nordhaus 2008b.)

1. The benefits of implementing strong mitigation policies would exceed their costs by about $2.5 trillion (in year 2007 $U.S.).

2. A business-as-usual policy would increase the risk of accelerated economic damages from climate change and would diminish global GDP by at least 4%.

3. The expenses for stabilizing greenhouse gas concentrations in the atmosphere would amount to only about 1% of global GDP and would not inhibit future economic growth (Stern 2007).

Advocates for greater international efforts to mitigate greenhouse gas emissions regularly cite this study.

It is somewhat disconcerting that two eminent economists who use similar tools and who live on the same planet should reach such divergent conclusions. At times, the debate between the protagonists has turned nasty. Nordhaus accuses Stern of taking "the lofty vantage point of the world social planner, perhaps stoking the dying embers of the British Empire, in determining the way in which the world should combat the dangers of global warming" (Nordhaus 2008b). Stern responds that Nordhaus dictates that "the utility of a person born in 1995 would be worth roughly half that of a person born in 1960. This discrimination seems very hard to justify as an ethical proposition" (Stern 2008).

Despite their divergent conclusions, Nordhaus's and Stern's analyses differ in only a few aspects. One

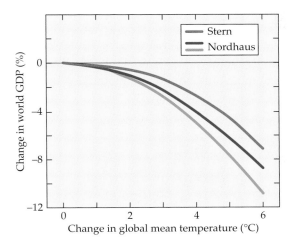

FIGURE 10.27 Economic damages from potential temperatures anticipated during this century in the analyses of Stern and Nordhaus in 1999 (blue and green lines in the Nordhaus analyses represent two methods of weighting impacts). (After Stern 2007.)

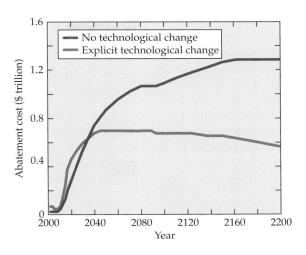

FIGURE 10.28 Total abatement costs (year 2000 $U.S.) to maintain atmospheric CO$_2$ levels at 450 ppm with and without explicit technological change, derived from the computer model in the *Stern Review*. (After Alberth and Hope 2007.)

of these—as you may have already guessed—is their choice of discount rate. Nordhaus assumes that low-income countries have a discount rate of nearly 6% and that high-income countries have an initial rate of 3%, which declines linearly to 2.3% by 2100 and to 1.4% by 2300 (Nordhaus and Boyer 1999). In contrast, Stern assumes a discount rate that starts just above 3% but lowers exponentially to 1.4% by 2100 (**Figure 10.26**). Economists most commonly name 2% as an appropriate long-term discount rate for projects to mitigate potential global climate change; therefore, Nordhaus chooses a higher discount rate and Stern a lower discount rate than most economists (see Figure 10.25).

The sensitivity of the world's GDP (gross domestic product) to average global temperature is similar in the Nordhaus and Stern analyses (**Figure 10.27**). Nordhaus, however, selects a single temperature value for the year 2100, 2.8°C warmer, whereas Stern uses a range of values, 1.5° to 4.5°C warmer for the year 2100. Consequently, global warming decreases global GDP by about 2% in the Nordhaus analysis and by as much as 4% in the Stern analysis.

Another difference between Nordhaus and Stern is their treatment of technilogical change. Research and development incur costs today but yield benefits in the future; therefore, Stern's lower discount rate encourages research and development and thereby accelerates technological change. Moreover, Nordhaus does not explicitly account for the influence of market forces on the rate of technological change, whereas Stern does. Total abatement costs for maintaining atmospheric CO$_2$

concentrations at a certain level are initially higher in an economic model that incorporates technological change versus one that does not, but technological change significantly decreases these costs in the longer term (**Figure 10.28**).

Appropriate policies

What are reasonable economic policies to address global climate change, given the diversity of expert opinions? The U.S. Government Accountability Office recently surveyed 18 U.S. economists, including Nordhaus and other major critics of the *Stern Review*, about their perspectives (U.S. Government Accountability Office 2008). All of these economists agreed that the U.S. government should use a market-based mechanism to establish a price on greenhouse gas emissions, but they differed on the type of mechanism and its stringency.

With respect to type of mechanism, eight of the economists preferred a cap-and-trade system with a safety valve, which means the government implements cost controls or issues additional permits if the price of permits exceeds a certain level; three preferred a cap-and-trade system without a safety valve; and seven preferred a tax on emissions (U.S. Government Accountability Office 2008). A cap-and-trade program would be more effective than a tax in achieving a desired level of greenhouse gas emissions because, unlike a tax, it would provide certainty that emissions would not exceed a certain level. Taxes, however, might be economically more efficient because the price of emission taxes would be less susceptible to market fluctuations than the price of permits would be. In addition, a tax system might be administratively less burdensome than a cap-and-trade system.

With respect to the stringency, seven of these economists said the price per metric ton of greenhouse gas emissions in carbon dioxide equivalents should range from less than $1 to $10, seven said it should range from $11 to $20, and three said it should exceed $20 (in year 2007 $U.S.) (U.S. Government Accountability Office 2008). In addition, the majority believed that prices of emissions should gradually increase over time. Furthermore, all of these economists suggested that the U.S. government should regulate emission by 2015, regardless of the actions of other nations, and that emission prices should apply to all sectors of the economy. Finally, 14 of the 18 economists were at least moderately certain that the benefits of their preferred portfolio of actions would outweigh the costs, and the other four did not respond to this question (U.S. Government Accountability Office 2008).

Overall, the vast majority of economists now advocate that governmental economic policies to address global climate change are not only feasible, but also desirable.

Summary

Market forces that balance supply and demand cannot adequately address the issue of global climate change because divergent national interests distort global markets, monopolies control the supply of goods such as electricity, relevant services such as disposal of wastes in the atmosphere are virtually free, crucial information is not publically available, and important processes span several generations. Therefore, governments have the right and responsibility to impose regulations such as taxes or permits for greenhouse gas emissions and to finance large infrastructure projects such as railroad lines or hydroelectric dams that mitigate emissions. Evaluating the costs versus benefits of such regulations and projects is difficult because of their complexity, longevity, and dependence on intangibles. To evaluate intangibles involves revealed preference and stated preference techniques. All economic assessments of climate change mitigation suffer from uncertainty about the rate of technological change and about the appropriate value for a discount rate on large projects. Most economists believe that the discount rate for projects that mitigate global climate change should be around 2%. A higher rate discourages the undertaking of any

such projects, whereas a lower one might justify projects that are economically unsound. The vast majority of economists now advocate that governmental economic policies to address global climate change are not only economically feasible, but also desirable.

Review Questions

1. Select all of the following correct definitions.
   (a) A market is a common arena where firms or individuals voluntarily exchange goods or services.
   (b) Market forces are the influences that supply and demand have on the distribution of goods and services.
   (c) Supply is the willingness and ability to provide a good or service.
   (d) Demand is the desire for a good or service as indicated by the willingness or ability to pay.
   (e) A free market is one in which participants arrange transactions entirely by mutual consent, without government intervention.

2. In a controlled market operating in a planned economy,
   (a) governments may be seen but not heard.
   (b) governments insure that there is perfect competition and perfect information.
   (c) governments may manage the distribution of income.
   (d) goods and services will reach their equilibrium price.
   (e) demand determines supply of goods and services.

3. The demand curve will shift to the left (i.e., a good or service will cost less at a given supply) as
   (a) consumers become wealthier.
   (b) alternative goods or services become less expensive.
   (c) advertising creates greater consumer demand.
   (d) human population growth accelerates.
   (e) research and development of new technologies slow down.

4. The supply curve will shift to the right (i.e., the amount of a good or service available for sale will increase at a given price) as
   (a) laborers receive higher wages.
   (b) technology advances.
   (c) real estate prices rise.
   (d) interest rates increase.
   (e) governments fail to intervene.

5. The equilibrium price
   (a) increases as the demand curve shifts to the left.
   (b) increases as the supply curve shifts to the right.
   (c) never changes.
   (d) increases as a sales tax decreases.
   (e) increases as the number of permits decreases.

6. Market forces do not precisely define the economics of global climate change because (select all that apply)
   (a) divergent national interests distort global markets for goods or services.
   (b) monopolies or cartels control the supply of some goods or services.
   (c) some relevant goods or services are free and invite indiscriminate use.
   (d) crucial information about some goods or services is sparse, uncertain, or not publicly available.
   (e) important processes span several generations of human beings.

7. Developing countries such as China and India argue that they
   (a) currently emit only a small fraction of greenhouse gases.
   (b) are not responsible for most of the greenhouse gases in the atmosphere today.
   (c) have no interest in maintaining Earth's environment.
   (d) will not experience any problems from the anticipated changes in Earth's climate.
   (e) have no control over the amount of greenhouse gases emitted by industries in their countries.

8. Some utilities are natural monopolies because they
   (a) do not benefit from economies of scale.
   (b) have low capital costs but variable direct costs.
   (c) provide services that are generally homogeneous and indistinguishable as to their source.
   (d) provide services that contain nothing artificial.
   (e) can easily enter a market in a new locality.

9. Electricity is an unusual commodity because
   (a) its supply and demand are relatively constant over time.
   (b) it is difficult to store in significant quantities.
   (c) it is relatively easy to replace with other goods and services.
   (d) it has been sold in the same form for several centuries.
   (e) it does not benefit from technological change.

10. Power utilities first use the electricity
    (a) generated nearby by renewable or nuclear power plants.
    (b) generated by distant hydroelectric dams.
    (c) generated by coal-fired power plants.
    (d) generated by natural gas–fired power plants.
    (e) purchased on the spot market.

11. To moderate peak demand for electricity, power utilities may (select all that apply)
    (a) institute brownouts or blackouts.
    (b) differentiate customers by their willingness to pay extra for a higher level of service.
    (c) charge more for electricity during peak hours.
    (d) schedule routine maintenance during peak hours.
    (e) encourage congestion on major transmission lines.

12. Which one of the following is *not* a method that governments use to regulate allocation of natural resources?
    (a) Taxes
    (b) Tax-and-dividend
    (c) Permits
    (d) Cap-and-trade
    (e) Search-and-seizure

13. Planners of large projects tend to
    (a) lack aesthetic values.
    (b) allocate insufficient funds and time for mishaps.
    (c) be insulated from outside pressures.
    (d) overestimate costs.
    (e) underestimate benefits.

14. Planners of large projects generally do *not*
    (a) compare the current project with previous ones.
    (b) collect and analyze data to determine the best course of action.
    (c) prepare information to support their clients' preferred course of action.
    (d) carry the entire financial burden of cost overruns or benefit shortfalls.
    (e) take pride in their work.

15. Assessments of aesthetic and cultural values, human welfare, and risk avoidance may use (select all that apply)
    (a) revealed preference techniques such as travel cost and hedonic price analyses.
    (b) stated preference techniques such as contingent valuation and choice modeling.
    (c) values of statistical life based on willingness to pay for slightly reducing the risk of fatality.
    (d) changes in quality of life in terms of mobility, discomfort, self-care, and anxiety.
    (e) pilot studies on parts of a project for which the risks are uncertain.

16. Free riding
    (a) takes advantage of something that is freely available and inexhaustible.
    (b) refers to a day pass for a municipal transit system.
    (c) does not occur in a free market.
    (d) involves sawing a person in half.
    (e) stimulates investment in research and development.

17. The tragedy of the commons
    (a) is one of Shakespeare's plays that describes shepherds tending their flocks.
    (b) assumes that excluding additional users of public goods is trivial.
    (c) assumes that each additional user of public goods puts a significant burden on resources
    (d) explains, in part, the failure of market forces with regard to natural resources.
    (e) promotes capitalism.

18. The discount rate
    (*a*) is usually higher for large-scale public works than for private business ventures.
    (*b*) is usually equal to the prime interest rate set by the Federal Reserve Bank.
    (*c*) determines to a major extent the economic feasibility of large-scale public works.
    (*d*) is similar in the economic analyses of global climate change by Nordhaus and Stern.
    (*e*) expects people to wait for something without compensation.

19. Intergenerational equity
    (*a*) is a justification for Medicare.
    (*b*) considers the wealth of future generations and their technical capacities.
    (*c*) assumes that gender bias will disappear by the end of the century.
    (*d*) is the right to inherit the same diversity in natural and cultural resources enjoyed by previous generations.
    (*e*) expects people to wait for something without compensation.

20. Which of the following statements about taxes and permits for greenhouse gas emissions is *false*?
    (*a*) Effects of taxes on the total amount of greenhouse gas emissions are relatively more difficult to predict.
    (*b*) Effects of permits on the total amount greenhouse gas emissions are relatively less difficult to predict.
    (*c*) Effects of additional taxes on business expenses are relatively less difficult to predict.
    (*d*) Effects of permits on total on business expenses are relatively more difficult to predict.
    (*e*) A permit system would have negligible transaction costs (low administrative costs).

21. Google Corporation's philosophy "You can make money without doing any evil" is consistent with
    (*a*) Pareto optimality.
    (*b*) Kaldor-Hicks optimality.
    (*c*) Nash equilibrium.
    (*d*) most compensation schemes for addressing global climate change.
    (*e*) progressive taxation.

22. Most economists believe that governmental economic policies to address global climate change
    (*a*) should use a discount rate above 5%.
    (*b*) should rely on taxes on greenhouse gas emissions.
    (*c*) are feasible and desirable.
    (*d*) must involve all sovereign states.
    (*e*) will never happen.

Suggested Readings

Congressional Budget Office. 2003. *The Economics of Climate Change: A Primer*. U.S. Congressional Budget Office, Washington, D.C., *http://www.cbo.gov/ftpdocs/41xx/doc4171/04-25-ClimateChange.pdf*.

This introduction to the economics of climate change was written by a government office when the Republican party held a majority in the U.S. government.

Nordhaus, W. D. 2008. *A Question of Balance: Weighing the Options on Global Warming Policies*. Yale University Press, New Haven, CT.

This book presents the economic arguments of Nordhaus.

Stern, N. 2008. The economics of climate change. *American Economic Review* 98: 1–37 doi:10.1157/aer.98.2.1.

This review article presents the economic arguments of Stern.

U.S. Government Accountability Office. 2008. *Climate Change: Expert Opinion on the Economics of Policy Options to Address Climate Change*. GAO-08-605, Washington, D.C., *http://www.gao.gov/new.items/d08605.pdf*.

This report surveys the opinions of prominent U.S. economists on the economics of policies to mitigate climate change.

11

GLOBAL CLIMATE CHANGE AND THE LAW

Climate change, like Earth's atmosphere, does not respect territorial borders. Greenhouse gas emissions from one sovereign state, although they affect the whole planet, may not be proportional to the severity of problems that the state itself experiences from climate change. For example, the world's poorest states emit negligible amounts of greenhouse gases but are often the most vulnerable to changes in sea level, precipitation, major storms, heat waves, and spread of infectious diseases (see Chapter 4). Moreover, multinational corporations tend to locate operations that emit large amounts of greenhouse gases in sovereign states like China and India where the infrastructure is sufficient to support such operations, but where the environmental regulations are less restrictive than in developed states.

The previous chapter (Chapter 10), on economics, argues that private enterprise in a free market cannot adequately address the problem of global climate change, and thus governments have the right and responsibility to do so. The extent to which governments assume this responsibility, however, varies with their national interests. Some sovereign states behave as free riders, whereby they make little effort to abate greenhouse gas emissions yet reap benefits from the abatement efforts of other states; free riders avoid the cost of greenhouse gas abatement and thereby can produce products for less and attract additional business. The solution to the free-rider problem is collective action that penalizes such behavior.

How do the people of the world arise and take collective action that encourages sovereign states and multinational corporations to behave in a globally responsible manner? This brings us into the realm of public international law, which sets both rules of behavior between sovereign states and standards for multinational corporations. The following introduces some of the framework of international law and then examines its application to environmental issues in general and to global climate change in specific.

Public International Law

Public international law differs from other branches such as criminal, civil, and administrative law in enactment, interpretation, compliance, and enforcement. The next sections describe some of these differences.

Enactment and interpretation

A **sovereign state** is a political association with independent authority over a geographical area and the human population living there. The world contains 193 sovereign states that enjoy general international recognition and about 10 others whose legitimacy is disputed. These practice many forms of government (**Figure 11.1**), which vary not only in the environmental regulations that they enact but also in the diligence with which they enforce such regulations (**Figure 11.2**). They also differ in their propensity to participate in

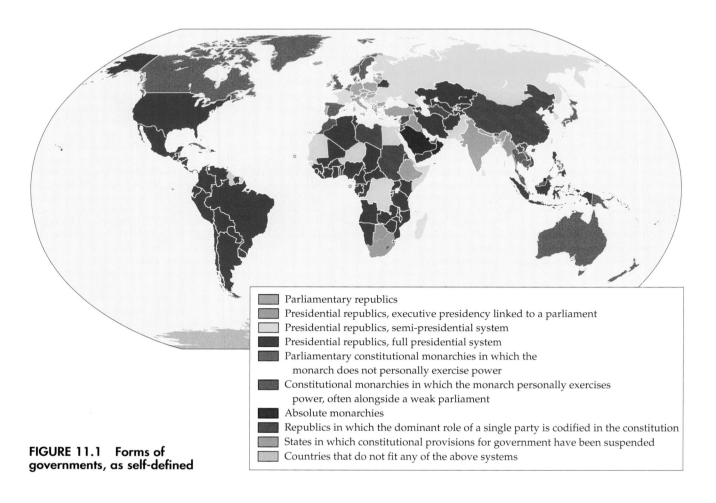

FIGURE 11.1 Forms of governments, as self-defined

Parliamentary republics
Presidential republics, executive presidency linked to a parliament
Presidential republics, semi-presidential system
Presidential republics, full presidential system
Parliamentary constitutional monarchies in which the monarch does not personally exercise power
Constitutional monarchies in which the monarch personally exercises power, often alongside a weak parliament
Absolute monarchies
Republics in which the dominant role of a single party is codified in the constitution
States in which constitutional provisions for government have been suspended
Countries that do not fit any of the above systems

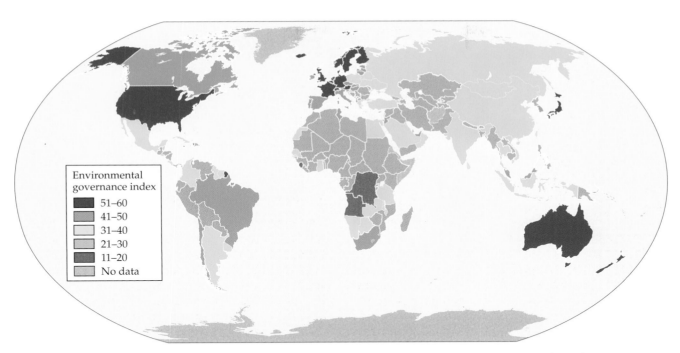

FIGURE 11.2 Index of environmental governance evaluated by the World Economic Forum A higher number indicates better environmental governance. This index evaluates regulations on air pollution, chemical wastes, toxic wastes, and water pollution; clarity, stability, flexibility, stringency, and innovation of regulations; consistency of regulation enforcement; and leadership in environmental policy. The average score for this index was 37.7, median was 35.8, maximum was 59.7, and minimum was 15.3. (After Esty et al. 2005.)

international agreements that address environmental issues (**Figure** 11.3).

The United States, like most sovereign states, delegates international issues primarily to its executive branch of government. The president of the United States negotiates and makes treaties and agreements with foreign governments (provided two-thirds of the Senate concurs) and holds the position of commander in chief of the armed forces. The secretary of state, a presidential appointee, serves as the foreign minister who conducts state-to-state diplomacy. The U.S. Constitution endows the legislative branch of government with sole authority to approve treaties, declare war, approve presidential appointments of the secretary of state and ambassadors, and regulate commerce with foreign states. The continual shifting balance of power between the U.S. executive and legislative branches sometimes leads to inconsistent policies on international issues such as global climate change.

When national interests clash, as they invariably do, sovereign states have several options for resolving their differences, including the use of military force. One might scoff at the possibility that disagreements about greenhouse gases might provoke a war, but the anticipated climate changes are likely to displace large numbers of people and to alter distributions of natural resources (see Chapter 4). Such disturbances have a history of precipitating armed conflict. Diminishing the likelihood of armed conflict provides yet another reason for mitigating global climate change.

At the close of World War II, the major world powers established the United Nations (U.N.) and the International Court of Justice "to save succeeding generations from the scourge of war" (United Nations 1945). The International Court of Justice, which is located in The Hague, Netherlands, relies on four sources of law to resolve differences between states peacefully (International Court of Justice 1945):

1. *International conventions, whether general or particular, establishing rules expressly recognized by the contesting states.* These are treaties, agreements, and other legal instruments to which the contesting states are signatories.

2. *International custom, as evidence of a general practice accepted as law.* Customary international laws are widely established norms that sovereign states follow, not out of habit or expediency, but because the states consider them binding legal obligations.

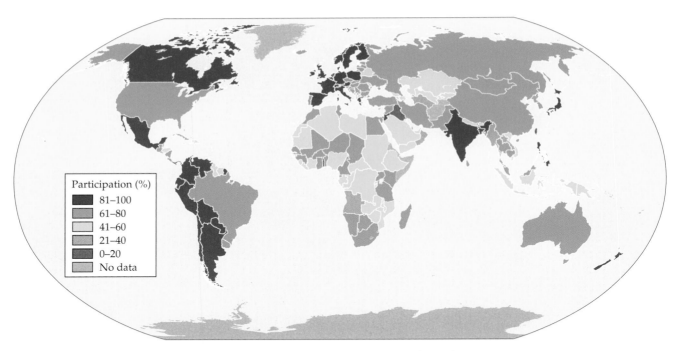

FIGURE 11.3 Participation in international environmental treaties Sovereign states were allocated one point for signature, accession, and ratification without signature of a treaty and another point for its ratification with signature, acceptance, approval, or succession. Treaties that were scored were the UNCCD, Vienna Convention, Montreal Protocol and its amendments, CITES, UNFCCC, Kyoto Protocol, Basel Convention, UNCBD, Ramsar Convention, and Cartagena Protocol for a total of 30 points. States not listed under the list of parties to a treaty received 0 points for the respective treaty. The mean value was 52%, median was 57%, maximum was 100%, and minimum was 0%. (After Esty et al. 2005.)

3. *General principles of law recognized by civilized nations.* These are beliefs common to mature legal systems.

4. *Judicial decisions and the teachings of the most highly qualified publicists of the various nations, as subsidiary means for the determination of rules of law.* These are determinations of law by the courts of member nations, such as decisions of the Supreme Court of the United States or writings of **publicists**, experts in public international law.

The following examines each of these sources with regard to environmental issues.

TREATIES, AGREEMENTS, AND OTHER INTERNATIONAL LEGAL INSTRUMENTS An international legal instrument is a document that binds the participating sovereign states to behave in a specified manner. The International Court of Justice and most states do not distinguish between the terms "treaty" and "agreement." The U.S. Constitution acknowledges that a treaty differs from an agreement or a compact but does not specify what these differences are (U.S. Supreme Court Center 2008). The Constitution stipulates that ratification of a treaty requires a two-thirds majority of the U.S. Senate (article 2, section 2, clause 2), whereas decisions of the U.S. Supreme Court and acts of Congress deem that ratification of a congressional and executive agreement needs only a simple majority of both houses of Congress, and an executive agreement needs no congressional approval whatsoever. During the first half century of its history, the U.S. government entered into twice as many treaties as executive agreements; during the last half century, it has entered into ten times more executive agreements than treaties. Nonetheless, the remainder of the chapter will refer to all such legal instruments as "treaties."

The term "**convention**" signifies a formal multilateral treaty negotiated under the auspices of an international organization and with a large number of participants (United Nations Treaty Collection 2008). A convention usually outlines broad objectives that enjoy widespread support without providing details that may be more contentious. A "**protocol**" is the next step: a treaty that specifies substantive obligations to implement the broad objectives of a previous convention (United Nations Treaty Collection 2008). For example, the U.N. Framework Convention on Climate Change (UNFCCC) established the broad objective to limit greenhouse gas emissions, and the subsequent Kyoto Protocol contained the specific provisions and regulations about greenhouse gas emissions discussed in the second half of this chapter. The two-step process of convention and protocol seems to simplify and accelerate treaty making.

The Vienna Convention on the Law of Treaties (United Nations 1969) codified pre-existing customary international law (i.e., general practices accepted as binding legal obligations). The first part of this convention specifies who can negotiate treaties, how parties can address their reservations about certain provisions in a treaty, and when do treaties enter into force (i.e., become binding). It distinguishes between signing and ratifying a treaty: When a designated representative of a sovereign state signs a treaty, it is largely ceremonial, although the state is obliged to refrain from acts that would defeat the object and purpose of the treaty; in contrast, ratification is the act through which a sovereign state establishes in international law its consent to be bound by a treaty. The next part of the Vienna Convention deals with observance, application, and interpretation of treaties. Subsequent parts present the procedures for amending and modifying existing treaties, invalidating and terminating treaties, and handling instances of state succession and aggression. The Vienna Convention entered into force on January 27, 1980, and currently, 108 states have ratified it.

The United States—despite playing a leading role in the formulation of the Vienna Convention on the Law on Treaties—still has not ratified it. The United States accepts much of this convention as customary international law, however, and U.S. federal and state courts routinely cite it. Nevertheless, the Vienna Convention and U.S. common law differ in three aspects (Criddle 2004).

1. The Vienna Convention allows courts less freedom than U.S. courts to interpret a provision in a treaty outside the ordinary meanings of words; that is, international courts take a more literal interpretation of the provisions in a treaty than U.S. courts sometimes do.

2. The convention discourages courts from deferring to the uncorroborated interpretation of a treaty by a single sovereign state, whereas U.S. courts regularly give substantial deference to interpretations of the U.S. executive branch, even when other treaty partners disagree with these interpretations.

3. In treaty cases, the convention envisions municipal courts as quasi-international tribunals committed to traditional rule-of-law values. U.S. courts in such cases try to reconcile U.S. and international law whenever possible but favor U.S. law and national strategic interests if such reconciliation proves impossible.

Some of the differences between the Vienna Convention and U.S. common law derive from American exceptionalism (Criddle 2004). Alexis de Tocqueville coined the term "American exceptionalism" in his 1835 book *Democracy in America* to describe the belief that the United States is unique among sovereign states in its commitment to liberty, egalitarianism, and individualism (de Tocqueville 1835). Counterbalancing American exceptionalism is the globalization of economic markets that benefits from harmonization of domestic and international laws.

CUSTOMARY INTERNATIONAL LAW   Standard practices that governments follow consistently from a sense of legal obligation may become customary international laws. Customary international laws were predominant at one time, but because they often lack precision about what constitutes sufficient evidence, treaties have become increasingly important (United Nations Forum on Forests 2004). For instance, customary international law obligates a government to warn other states promptly about emergencies of an environmental nature and about environmental damages to which other states might be exposed. A clear violation of this law occurred in August 1986 when a reactor at the Chernobyl Nuclear Power Plant in Ukraine exploded (see Chapter 8) and sent a plume of highly radioactive material over large portions of Europe (**Figure 11.4**). At first, the Soviet Union released a brief public statement (less than 250 words) stating that an accident of manageable proportions had occurred (Sands 1988). Three days later, only after radiation sensors in Western Europe and satellites images revealed the true magnitude of this catastrophe to the world, the Soviet Union shifted its stance and began to provide accurate information. The Vienna Convention on Early Notification of a Nuclear Accident entered into force 6 months later to codify appropriate procedures for handling such incidents (United Nations 1986).

Governmental declarations, orders, or memos, in the absence of other legal documents, can become binding to the same extent as a treaty. For example, the provisions of the 1948 Universal Declaration on Human Rights, although not specifically intended to be a legally binding instrument, is now generally accepted as customary international law. Even if governments

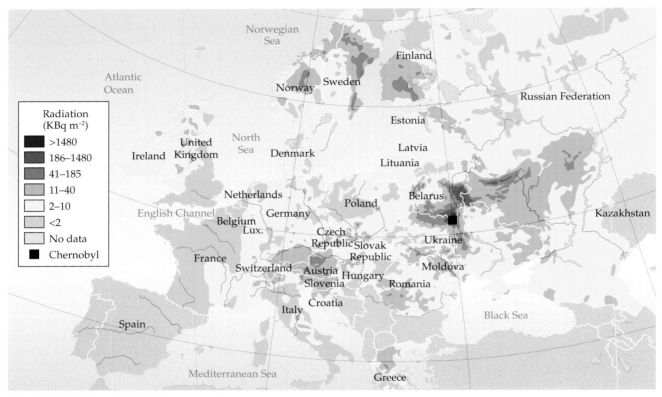

FIGURE 11.4 Spread of radiation from the Chernobyl nuclear accident in August 1986 Radiation is in units of kiloBecquerels per square meter.

do not sign a multinational treaty, they may still be bound to it as customary international law unless they actively and persistently repudiate the treaty (Burney 2008).

GENERAL PRINCIPLES OF LAW International law holds a few truths to be self-evident and uses them to fill gaps between treaties, customary law, and judicial decisions:

- *Promises are binding.* States have an obligation to act in good faith: They cannot take a position inconsistent with a previous one, especially when other states have relied on the earlier representation. This is the legal concept of **estoppel.**

- *One should not benefit from one's own wrongdoing.* A state seeking redress for the actions of another must come with "clean hands"; that is, a state guilty of illegal conduct has no right to complain about corresponding illegalities on the part of other states, especially if these were to counter its own illegality. A shepherd who has grazed too many sheep on the village commons cannot seek compensation if other shepherds begin to graze too many sheep (see Chapter 10).

- *Composition and application of rules should be just and fair.* This is the legal concept of **equity.** Negotiations about mitigation of global climate change usually forward as a general principle that any permit system for greenhouse gas emissions should be equitable and take into account that states are at different levels of economic development and that their contributions to current atmospheric levels are highly disparate (Sands and Jennings 2003).

- *Governments should not gratuitously endanger the lives of people.* The first case brought before the International Court of Justice involved this principle. In 1946, someone littered the Strait of Corfu (**Figure 11.5**) off the coast of Albania with mines. British ships struck the mines, damaging the ships and killing naval personnel (**Figure 11.6**). The Court declared that Albania, in failing to warn other shipping nations about the mines, had violated "elementary considerations of humanity" and ordered Albania to pay the United Kingdom £843,947

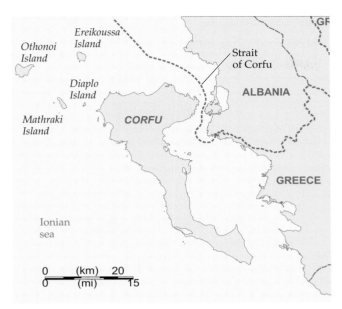

FIGURE 11.5 **Strait of Corfu** The dashed line designates the Albanian border.

in compensation. The unauthorized release of environmental pollutants would be an analogous offense.

- *Widespread and systematic acts of persecution or atrocities against a group of people is a "crime against humanity," the highest level of criminal offense.* Violations of treaties that address global climate change are unlikely to fall into this category.

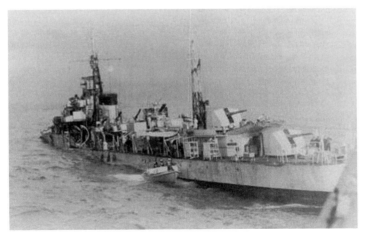

FIGURE 11.6 **HMS Saumarez after it hit a mine in the Strait of Corfu in 1946** The ship lost 43 of her crew in the incident and was so badly damaged that she had to be scrapped.

JUDICIAL DECISIONS The International Court of Justice states "The decision of the Court has no binding force except between the parties and in respect of that particular case" (International Court of Justice 1945). Nonetheless, the court itself and other judicial bodies often regard prior judicial decisions and writings of publicists as explaining the state of international law (Burney 2008). In addition, arbitration tribunals and "municipal courts" such as the U.S. Supreme Court also serve as sources of international law, even though their decisions are not binding on other countries. Finally, declarations or resolutions of the United Nations General Assembly, even those that are nonbinding, may provide another source of international law.

Compliance and enforcement

Compliance is the degree to which a state conforms to a treaty (Zaelke et al. 2005). Enforcement is the set of actions that states take to ensure compliance (Office of Enforcement 1992). The American diplomat Chester Bowles (1901–1986) observed, "20 percent of the regulated population will automatically comply with any regulation, 5 percent will attempt to evade it, and the remaining 75 percent will comply as long as they think that the 5 percent will be caught and punished" (Bowles 1971). Catching and punishing this 5% requires monitoring for compliance and enforcing penalties for noncompliance.

Perhaps the most important factor in achieving compliance and enforcement is the formulation of a treaty itself. Compliance with or enforcement of a poorly crafted or unpopular treaty is difficult. An effective treaty depends on support from the general populace as well as support from local, regional, and national law enforcement. To garner such support often requires education, promotion, technical assistance, and financial incentives (**Tables 11.1 and 11.2**). Consequently, most modern multinational treaties not only enumerate obligations for the state participants but also create an administrative structure to assist participants in monitoring compliance and enforcing its provisions (United Nations Forum on Forests 2004).

MONITORING FOR COMPLIANCE Four sources of information about compliance with environmental regulations are (1) inspections conducted by program officers; (2) self-monitoring by the regulated community; (3) citizen complaints; and (4) monitoring areas near facilities (Office of Enforcement 1992). Inspections provide the most relevant and reliable information, but are the most costly. Self-monitoring transfers the burden of monitoring to the regulated community, but must rely on the integrity and capability of this community to provide

TABLE 11.1 Suggested actions to educate, promote, and offer technical assistance for groups subject to environmental regulations

Information source	Process
Publications	Create brochures and guidance manuals for distribution to regulated groups.
Training programs	Design programs to train regulated groups about requirements and compliance.
Conferences	Bring together officials from enforcement agencies, regulated groups, and other interested parties.
"Hotlines"	Dedicate telephone numbers or web sites that regulated groups can consult to receive information and assistance.
Technical assistance	Schedule visits by trained personnel or inspectors who assist regulated groups in making changes necessary for compliance.
Cooperative arrangements	Facilitate cooperative arrangements among small businesses that may want to comply but do not have the necessary resources to do so. For example, establish a central processing center for recycling.
Media announcements	Distribute information through newspaper, television, radio, and the Web about requirements and enforcement activities against violators.
Universities	Provide universities with materials and guest lecturers to incorporate information about regulations into their curricula.
Trade and professional associations	Participate in publications and meetings of associations. Encourage interactions between regulated groups and enforcement personnel.

Source: Office of Enforcement 1992.

accurate information. Citizens, including NGOs (non-governmental organizations), serve as independent observers, but the amount, frequency, and quality of information are variable. Monitoring areas near facilities helps determine if whether regulations are achieving an adequate level of protection, but cannot reliably trace environmental deviations to a specific source.

Monitoring emissions of greenhouse gases presents several additional challenges. Both natural and human sources emit the three major greenhouse gases—carbon dioxide, methane, and nitrous oxide. Many of these are non-point sources; that is, they release small amounts of gas over large geographical areas. These gases are virtually colorless, odorless, and tasteless. The concentrations of carbon dioxide, methane, and nitrous oxide in the atmosphere are tiny (0.04%, 0.0002%, and 0.00003%, respectively), and monitoring emissions requires sophisticated equipment.

ENFORCEMENT Government enforcement programs seek to correct violations as well as encourage compliance through prosecution of violators (Office of

TABLE 11.2 Financial incentives for compliance with environmental regulations

Incentives	Implementation
Offsets	Require those who are building a new facility to pay for renovations that reduce the environmental impact of an existing facility.
Loans	Require that the recipient of a loan devote a certain portion of it to restore or protect environmental quality.
Environmental bonds	Issue bonds that finance renovation of facilities to meet environmental requirements.
Fees and taxes	Levy fees or taxes on a facility based on its environmental impact.

Source: Office of Enforcement 1992.

Enforcement 1992). Laws in some countries empower private citizens and groups to bring enforcement actions against violators. Insurance companies and financial institutions may require compliance with all regulations before a utility or industry is eligible for insurance or a loan. Finally, a society may impose strong social sanction for noncompliance; for example, the public may boycott products from a certain company if it harms the environment.

Successful enforcement mechanisms should achieve one or more of the following outcomes: return a violator to compliance, impose a sanction (i.e., penalty), nullify any economic benefits of noncompliance, require that a violator provide accurate information, remediate past environmental damages, and correct the internal management problems of a violator (Office of Enforcement 1992). Every sovereign state has its own mechanisms for enforcement, and international treaties usually consider each state participant as a sovereign entity and leave the details about local, regional, and national enforcement to the individual state.

DISAGREEMENTS OR DISPUTES Most treaties set procedures for resolving disagreements or disputes. A disagreement is a difference in a point of view and is less contentious than a dispute, which is a difference on a point of law or a point of fact. The U.N. Charter affirms that "All Members shall settle their international disputes by peaceful means in such a manner that international peace and security, and justice, are not endangered" (United Nations 1945). Peaceful means of dispute resolution include negotiation, inquiry, conciliation, arbitration, and judicial settlement.

- *Negotiation.* International treaties often specify negotiations as the preferred means of dispute resolution. An obligation to negotiate means that each party must make an honest effort to negotiate. Failure to negotiate in good faith can be held against a party in subsequent actions such as arbitration or judicial settlement.
- *Inquiry.* An inquiry establishes an official version of the facts so that dispute resolution can proceed.
- *Conciliation.* In conciliation, a respected third party helps two parties in conflict reach a mutually agreeable resolution of the dispute. This resolution is just a recommendation, not binding unless the two sides wish to make it so.
- *Arbitration.* An arbitrator is an expert who is court appointed to decide which of the solutions proposed by the parties in conflict is best. This decision is usually binding.

- *Judicial settlement.* Bringing a dispute to court such as the International Court of Justice for a judicial settlement usually is more expensive and more time consuming than other peaceful means of conflict resolution. To begin such a process, all parties must have standing; that is, the court must recognize that each party has a valid stake in the dispute. The case must not be considered moot because it duplicates a case that has already been resolved. Each party argues its version of the facts, and the court determines an official version. Each party argues what the law should be, and the court provides its interpretation. Finally, the court applies its interpretation of the law to the official version of the facts and decides what the outcome should be (Burney 2008).

Treaties

The following section describes in chronological order three treaties that influence international policies about global climate change: the United Nations Law of the Sea, the Montreal Protocol on Substances that Deplete the Ozone Layer, and the Kyoto Protocol to the U.N. Framework Convention on Climate Change. It then contrasts these treaties and examines possible reasons for differences in their effectiveness.

The United Nations Law of the Sea

In past centuries, territorial waters of a maritime state extended 3 nautical miles (5.56 km) from shore, the range of eighteenth-century land-based cannons. Within this 3-mile limit, the state had exclusive rights to regulate trade and extract natural resources. Beyond the 3-mile limit lay international waters that were open to all nations but belonged to none.

By the twentieth century, not only did the range of cannons greatly exceed the 3-mile limit, but also maritime states were eager to expand their territorial waters and assert control over additional offshore resources. The United States in 1945 laid claim to all of its continental shelf, and other nations quickly followed suit. Between 1946 and 1950, Argentina, Chile, Peru, and Ecuador extended their territorial waters as far offshore as 200 nautical miles to include their Humboldt Current fishing grounds. Other nations extended their territorial seas to 12 nautical miles. Today, Jordan is the only state that still uses the 3-mile limit.

The United Nations convened the first U.N. Conference on the Law of the Sea (UNCLOS I) in Geneva, Switzerland from 1956 through 1958 and drafted four treaties: the Convention on the Territorial Sea and Contiguous Zone (entered into force in 1964); the Convention on the Continental Shelf (entered into force in 1964); the Convention on the High Seas (entered into force in 1962); and the Convention on Fishing and Conservation of Living Resources of the High Seas (entered into force in 1966). Although UNCLOS I was successful in addressing many issues, it did not resolve the extent of territorial waters.

A second U.N. Conference on the Law of the Sea (UNCLOS II) met for 6 weeks in Geneva in 1960 but made little progress. A third U.N. Conference on the Law of the Sea (UNCLOS III) began in New York with 160 participant states in 1973 and continued until 1982, when it culminated in the Law of the Sea Convention Treaty (Wang 1992). This treaty came into force in 1994, after the 60th state ratified it. Yes, you read it right: The treaty was 9 years in the making and required another 12 years before it took effect. International negotiations demand great patience.

The Law of the Sea Convention introduced a number of provisions, the most significant of which established definitions of coastal boundary zones, navigation rights on the high seas, rights of archipelagic states such as the Philippines and Indonesia, transit passage through straits such as the Strait of Corfu, exclusive economic zones, continental shelf jurisdiction, regulations for deep seabed mining, regimes for exploitation of the seabed in international waters, protections for the marine environment, pro-

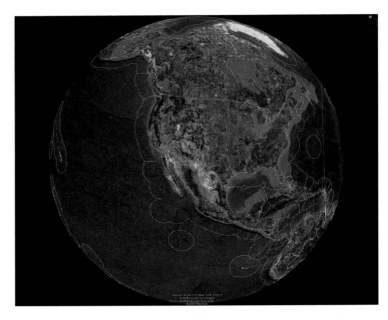

FIGURE 11.7 Exclusive economic zones along the coastlines of North and Central America Yellow lines demarcate recognized boundaries.

tocols for scientific research, rights of tax-free access to and from the sea for landlocked states, and procedures for settlement of disputes (Division of Ocean Affairs and the Law of the Sea 2008).

The treaty defines various coastal boundary zones (**Figures 11.7 and 11.8**), measured from a carefully defined territorial sea baseline. Normally, this baseline follows the coastline at low tide, but a state may draw

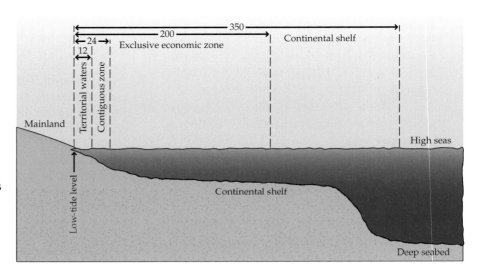

FIGURE 11.8 Definition of boundary zones along coastlines of maritime states The numbers indicate distances in nautical miles from low-tide level. The continental shelf may extend as far as 350 nautical miles (648 km) from shore.

a straight baseline if a coastline is deeply indented, has fringing islands, or is highly unstable. The treaty provides specifications for six types of coastal boundary zones (Division of Ocean Affairs and the Law of the Sea 2008).

- *Internal waters* are all waters and waterways on the landward side of the territorial sea baseline. A coastal state is free to set laws, regulate use, and use any resource in its internal waters. Foreign vessels have no right of passage through internal waters.

- *Territorial waters* extend out to 12 nautical miles from the sea baseline. A coastal state is free to set laws, regulate use, and use any resource in its territorial waters. Vessels have the right of "innocent passage" through territorial waters, provided they pass in an expeditious and continuous manner that does not threaten the peace, good order, or security of the coastal state. Submarines must navigate on the surface and show their flag. In coastlines with strategic straits, the treaty allows military vessels "transit passage," in which they can maintain postures such as keeping the crew at battle stations that would be otherwise be illegal in territorial waters.

- *Archipelagic waters* are all waters inside a baseline drawn from the outermost points of the outermost islands. An archipelagic state maintains full sovereignty over these waters, as it does for its internal waters, but foreign vessels have the right of innocent passage through them, as they do for the state's territorial waters.

- *Contiguous zones* extend from 12 nautical miles to 24 nautical miles from the sea baseline. A coastal state can enforce laws regarding smuggling or illegal immigration within this zone.

- *Exclusive economic zones* extend 200 nautical miles from the sea baseline. A coastal state has sole exploitation rights over all natural resources, such as fish or oil, within this zone. Foreign states have freedom of navigation and overflight, subject to the regulation of the coastal states, and may lay submarine pipes and cables.

- *The continental shelf* refers to the coastal plain associated with a continent. The continental shelf may never exceed 350 nautical miles from

the baseline nor exceed 100 nautical miles beyond the distance from shore where the waters reach a depth of 2500 meters. A coastal state has exclusive rights to extract nonliving materials in the subsoil of its continental shelf, such as oil or precious minerals, and to harvest living resources attached to the continental shelf, but it does not have exclusive rights to sea life in the water column.

The Law of the Sea Convention, in addition to defining coastal boundary zones, affirms that "States have an obligation to protect and preserve the marine environment" (United Nations 1982). They must take all measures possible to prevent, reduce, and control pollution that endangers rare or fragile marine ecosystems and habitats of depleted or threatened marine species. This includes control of pollution that enters waters from the shore, pipelines, cables, vessels, or atmosphere.

The Law of the Sea Convention establishes an International Seabed Authority that authorizes seabed exploration and mining and that collects and distributes royalties from seabed mining. The United States strongly objected to some of these provisions and refused to ratify the treaty. Nevertheless, the United States considers most of the treaty to be customary international law. Further modification of the treaty in 1994 to address the objections of the United States convinced President William J. Clinton and President George W. Bush to urge the U.S. Senate to approve the treaty, and on October 31, 2007, the Senate Foreign Relations Committee voted 17 to 4 to send the treaty to the full U.S. Senate for a vote. This has not happened as of June 2009 because the treaty still faces some opposition from senators who consider participation in such a treaty to be detrimental to U.S. national interests.

The Law of the Sea Convention and policies regarding global climate change intersect in several ways. The Law of the Sea Convention defines pollutants as substances introduced by humans into the seas that harm marine life and declares, "States shall adopt laws and regulations to prevent, reduce and control pollution of the marine environment from or through the atmosphere" (United Nations 1982). Atmospheric CO_2, as it dissolves in the oceans, dramatically increases their acidity (see Figure 4.39) and endangers sea life (see Chapter 6). Therefore, the Law of the Sea Convention provides another argument for the United States to reduce greenhouse gas emissions.

The Law of the Sea Convention and global climate change also intersect at proposals to capture CO_2 gen-

FIGURE 11.9 Probability of at least one undiscovered oil and/or gas field with recoverable resources greater than 50 million barrels of oil equivalent. Black lines show longitude and latitude and red lines show current coastlines.

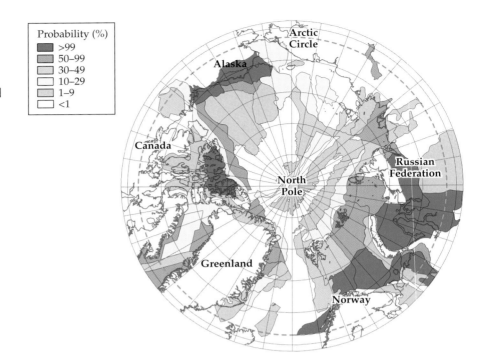

erated from combustion of fossil fuels at electric power plants and to store it in deep oceans (see Figure 8.14). Unfortunately, storing even a small percentage of the potential CO_2 emissions in the oceans would significantly alter water oxygen availability and pH (see Figure 8.16). This would threaten fragile deep-sea ecosystems and thereby violate the Law of the Sea Convention. Any major implementation of carbon capture and offshore storage would require amendment of this treaty (Flory 2005).

Finally, global climate change has shifted the boundaries of maritime states delineated in the Law of the Sea Convention. With rising sea levels and melting polar sea ice, additional waters have become open to navigation and to extraction of natural resources. The U.S. Geological Survey estimates that about 11% of the world's undiscovered oil, 25% of its undiscovered natural gas, and 17% of its undiscovered natural gas liquids lie beneath the Arctic Ocean (**Figure 11.9**).

To stake claim to some of this territory in a dramatic fashion, Russia had a miniature submarine in 2007 plant a flag in the seabed 2 miles below the North Pole. In response, Norway, Denmark, Canada, and the United States, the other states that ring the Arctic Ocean, filed counterclaims (**Figure 11.10**). Remember that sovereign states have mineral rights (i.e., rights to minerals, oil, and natural gas) to the seabed within 200 nautical miles from their shores but can extend these rights to as far as 350 nautical miles along a continental shelf. All of these states are conducting detailed surveys of the Arctic continental shelf and have proposed various division methods to maximize their territorial claims. The United States, because it has yet to ratify the Law of the Sea Convention, is permitted only limited participation in some discussions. This explains why the U.S. senators from Alaska, a state that potentially has a lucrative stake in Arctic mineral rights, have pushed for U.S. ratification.

Substances that deplete the ozone layer

Chlorofluorocarbons (CFCs) and hydrochlorofluorocarbons (HCFCs) are chemicals that at one time were the main ingredients of refrigerants, propellants for aerosol sprays, cleaning solvents for dry cleaning and electronic circuits, and bubbles injected into foams. Through the 1970s and 1980s, scientific evidence mounted that release of these chemicals was depleting the ozone layer in Earth's upper atmosphere (National Research Council 1982). This layer absorbs ultraviolet solar radiation (see Figure 3.34) that otherwise could harm life on land. For example, the incidence of human skin cancer increases with exposure to ultraviolet radiation. More specific to the topic of this book, however, is that CFCs and HCFCs are also potent greenhouse gases (see Table 3.1).

(A) Current

United States
claims
this area

(B) Proposed division by median line

United States
could claim
this area

(C) Proposed division by sector

United States
could claim
this area

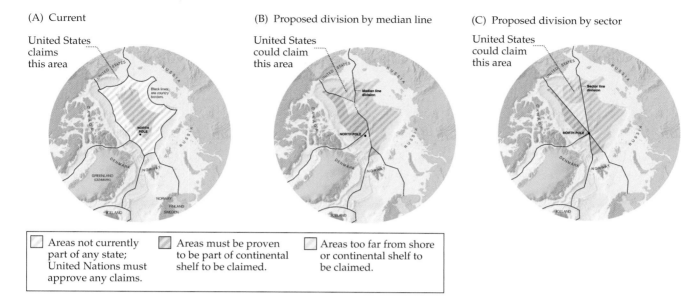

Areas not currently part of any state; United Nations must approve any claims.

Areas must be proven to be part of continental shelf to be claimed.

Areas too far from shore or continental shelf to be claimed.

FIGURE 11.10 Proposals to divide mineral rights in the Arctic Ocean Mineral rights include the right to extract oil and natural gas. (A) Currently, the Law of the Sea Convention allocates exclusive rights in economic zones that extend 200 nautical miles (370 km) from the sea baseline. Rights to resources in the seabed and sub-seabed can extend as far as 350 nautical miles (648 km) along a continental shelf, but underwater ridges that may be used to demarcate the continental shelf must be shown to be extensions of the shelf itself rather than separate features of the seafloor; therefore, states bordering the Arctic Ocean are carefully mapping underwater ridges. (B) Canada, Denmark, and the United Nations have proposed a median line method that divides the Arctic Ocean proportional to each state's coastline. (C) Russia and Norway advocate a sector method that divides along lines of longitude, with the North Pole as the center. (©The New York Times.)

In 1985, 28 states, including most of the major producers of CFCs and HCFCs, negotiated the Vienna Convention on Protection of the Ozone Layer to provide a general framework for international regulations on ozone-depleting substances. This was followed, in 1987, by the Montreal Protocol on Substances that Deplete the Ozone Layer, a treaty with a timetable to phase out and eventually eliminate the production and consumption of these chemicals. The Montreal Protocol allowed developing states a 10-year delay before they needed to meet the same limits to production and consumption as developed states. The Montreal Protocol entered into force in 1989 after ratification by over 11 states representing more than two-thirds of the world's consumption of ozone-depleting substances. It has undergone seven revisions (1990 London; 1991 Nairobi; 1992 Copenhagen; 1993 Bangkok; 1995 Vienna; 1997 Montreal; and 1999 Beijing). By 2008, 193 sovereign states—all but Andorra, San Marino, and Timor-Leste—had ratified the Montreal Protocol.

The London revision in 1990 established a multilateral fund to finance in developing states the conversion of existing manufacturing processes from ozone-deplet-ing substances, the training of personnel, the payment of royalties and patent rights on new technologies, and the establishment of national ozone offices (Multilateral Fund 2008). Industrialized states make donations to replenish the fund every 3 years. As of July 2008, some 49 industrialized states have contributed over 2.4 billion $U.S.

The Montreal Protocol adopted nonconfrontational enforcement procedures (Sarma 2005) that some consider weak (Victor 1998). Parties to the Montreal Protocol are required to report annually on their production and consumption of ozone-depleting substances. If a state is out of compliance in its commitments, it may receive assistance for the collection and reporting of data, technical assistance, and financial assistance (Ozone Secretariat 2006). If a state remains noncompliant, an implementation committee may suspend its rights and privileges under the treaty. This may include cutting off financial and technical assistance or prohibiting trade, both import and export, of ozone-depleting substances with other parties to the Montreal Protocol. The Implementation Committee has yet to impose such sanctions on any party, despite some flagrant abuses of the treaty.

The 1997 Montreal Amendment to the Montreal Protocol introduced a system in which parties to the Montreal Protocol license each manufacturer within their jurisdiction for every ozone-depleting substance it produces or uses (Curlin et al. 2005). Many parties have implemented this licensing system and are able to identify illegal manufacturers by monitoring the balance of production and use. Unfortunately, there are few supporting enforcement mechanisms for the licensing system, and this has allowed black-market production of ozone-depleting substances to flourish around the world. The black market for chlorofluorocarbons may be the second most lucrative criminal smuggling operation after illegal drugs (Kesselaar et al. 1996).

Despite all the difficulties in enforcing the Montreal Protocol, the world's major producers and consumers of ozone-depleting substances have been compliant with it. As a result, emissions of CFCs dropped precipitously after the Montreal Protocol entered into force in 1989 (see Figure 4.26A), and their concentrations in the atmosphere has followed suit (see Figure 4.26B). The treaty encourages certain users to substitute HCFCs for CFCs because HCFCs have shorter life spans in the atmosphere (see Table 3.1); as a result, HCFC emissions continued to rise after 1989 but now appear to have leveled off. The Montreal Protocol has set 2015 as the year to begin limiting the production and consumption of HCFCs.

In 2003, Kofi Annan, then secretary-general of the United Nations, declared, "Perhaps the single most successful international agreement to date has been the Montreal Protocol" (United Nations Environmental Programme 2007). Indeed, the loss of stratospheric ozone, which became significant in the 1980s, has started to turn around (**Figure 11.11**), and global ozone concentrations are anticipated to return to pre-1980 levels sometime after 2060 (World Meteorological Organization 2007). Without the Montreal Protocol, the ozone levels in the upper atmosphere would be 20% to 40% lower than they are today (**Figure 11.12**), and average temperatures at sea level would be up to 1°C warmer in the Northern Hemisphere (**Figure 11.13**). An economic analysis of the Montreal Protocol indicates that its benefits vastly exceed its costs (**Table 11.3**).

Global climate change

In 1988, two United Nations organizations, the World Meteorological Organization and the U.N. Environment Programme, established the Intergovernmental Panel on Climate Change (IPCC), an international

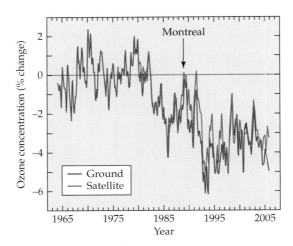

FIGURE 11.11 Global ozone concentrations as percentage change from the 1964–1980 average Plotted are values from ground-based and satellite data. The arrow indicates when the Montreal Protocol came into force. (After Fioletov et al. 2002; World Meteorological Organization 2007.)

group of scientists and other experts to assess the causes of climate change, its potential environmental and socioeconomic consequences, and adaptation and mitigation options to respond to it (IPCC 2009). The

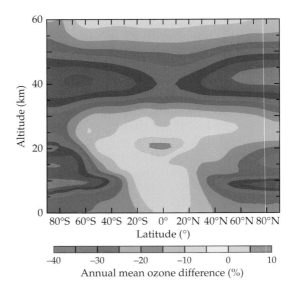

FIGURE 11.12 Predictions from a computer model of the changes in ozone concentrations (percentage) at various latitudes and altitudes that would have occurred if the Montreal Protocol had not been enacted. For example, sea level concentrations at the poles would be 10% to 20% lower. (After Morgenstern et al. 2008.)

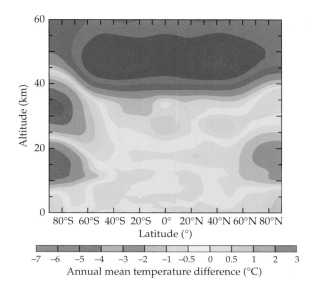

FIGURE 11.13 Predictions from a computer model of the changes in temperatures (°C) at various latitudes and altitudes that would have occurred if the Montreal Protocol had not been enacted. For example, sea level temperatures at the poles would be 0.5°C to 1.0°C warmer. (After Morgenstern et al. 2008.)

IPCC issued its first assessment report in 1990, concluding that "emissions from human activities are substantially increasing the atmospheric concentrations of greenhouse gases and that this will enhance the greenhouse effect and result in an additional warming of the Earth's surface" (IPCC 2004).

These conclusions prompted the drafting of the United Nations Framework Convention on Climate Change (UNFCCC) in 1992 at the Earth Summit in Rio de Janeiro. This treaty begins with a sentence that should sound familiar: "The Parties to this Convention… [are] concerned that human activities have been substantially increasing the atmospheric concentrations of greenhouse gases, that these increases enhance the natural greenhouse effect, and that this will result on average in an additional warming of the Earth's surface and atmosphere and may adversely affect natural ecosystems and humankind." Only four states in the world have failed to ratify this treaty: Andorra, the Holy See (Vatican), Iraq, and Somalia.

The UNFCCC, like other international conventions, primarily outlines procedural matters (United Nations 1992). It encourages signatories to take measures that anticipate, prevent, or minimize the causes of climate change and that mitigate its adverse effects, but all such measures are voluntary. It obligates signatories to develop, periodically update, and publish a national inventory of human sources and sinks of greenhouse gases. It designates sovereign states as either developing, Annex I industrialized states, or Annex II developed states (**Figure 11.14**).

- Developing states have no immediate limits to their greenhouse gas emissions but may receive funds and technologies from Annex II states to help mitigate their emissions.
- Annex I states have the objective of limiting their greenhouse gas emissions to 1990 levels.

TABLE 11.3 Projected global benefits versus costs of the Montreal Protocol 1987–2060

Category	Number or value
Human health benefits	
Number of deaths from skin cancer averted	333,500
Number of skin cancer cases averted	20,600,000
Number of cataract cases averted	129,100,000
Overall financial benefits[a]	
Value of skin cancer deaths averted	$333
Value of nonfatal skin cancers and cataracts averted	$339
Value of averting damages to fisheries, agriculture, and materials	$459
Overall financial cost[a]	$235
Net benefits	**$896**

Source: Sunstein 2007.
[a]Billions $U.S.

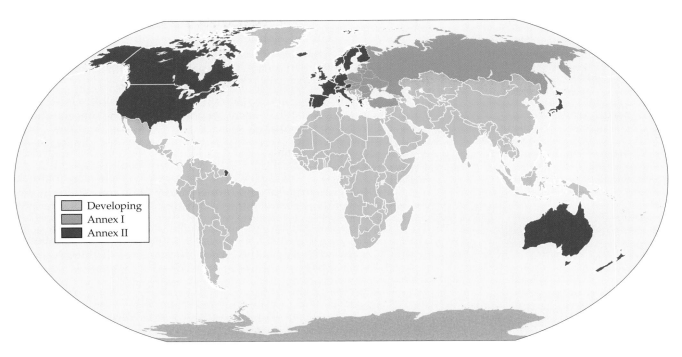

FIGURE 11.14   Designation of states as Developing, Annex I, or Annex II in the United Nations Framework Convention on Climate Change (After United Nations 1992.)

- Annex II states are Annex I states with the added responsibility of providing funds and technologies to developing states for conducting greenhouse gas inventories and adapting to adverse effects of climate change.

Concerned states convened in Kyoto, Japan in 1997 and drafted the Kyoto Protocol to set legally binding commitments for the United Nations Framework Convention on Climate Change (United Nations 1998). The Kyoto Protocol entered into force in 2005, after ratification by over 55 states, including enough Annex I states to account for more than 55% of the world's 1990 CO_2 emissions. As of 2009, 184 states have ratified it. A notable exception is the United States, which signed the treaty but has not ratified it.

The Kyoto Protocol specifies a number of requirements for its contracting parties (i.e., states that ratify the treaty) (United Nations 1998):

- All parties should implement programs that mitigate and adapt to climate change.
- Annex I parties must provide detailed inventories of their current greenhouse emissions as well as their emissions in 1990 that serves as the baseline for comparison.
- Annex I parties between 2008 and 2012 must limit their greenhouse gas emissions to a rate that varies from 10% higher (e.g., Iceland) to 8% lower (e.g., European Union) than their rate in 1990, depending on the state (**Figure 11.15**).
- Annex I parties may trade emission credits, whereby Annex I parties that emit less than their assigned amount may sell any surplus to Annex I parties that emit more than their assigned amount, a mechanism called joint implementation.
- Annex I parties may also gain emission credits through a clean development mechanism in which they sponsor projects in a non–Annex I party that mitigate greenhouse gas emissions.
- The first session of an annual Conference of the Parties shall approve mechanisms to determine and address cases of noncompliance with the provisions of the Kyoto Protocol, including a list of consequences.

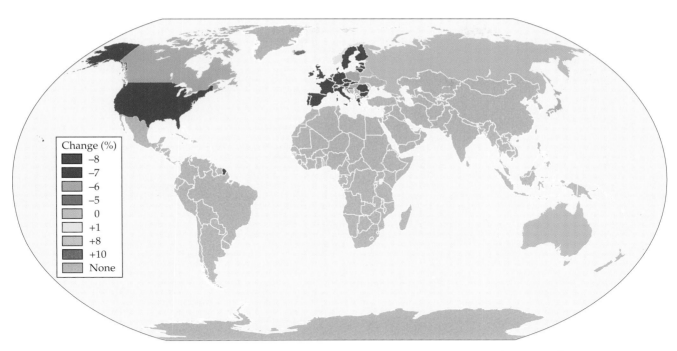

FIGURE 11.15   Commitments to regulate greenhouse gas emissions as specified in the Kyoto Protocol Colors designate changes in percentage from 1990 emission levels. States shown in gray have no legally binding commitments. (After United Nations 1998.)

- In subsequent years, the Conference of the Parties shall review the commitments of states in light of new scientific findings and experience gained in implementing climate-change policies.

The Kyoto Protocol aggregates all greenhouse gas emissions into one composite value equal to the sum of the amount of individual gases times their warming potentials (see Table 3.1).

The Conference of the Parties to the Kyoto Protocol established a Compliance Committee with a facilitative branch and an enforcement branch (UNFCCC 2008). The facilitative branch provides advice and assistance to parties in order to promote compliance. The enforcement branch determines whether an Annex I party complies with its greenhouse gas emissions during a particular commitment period and the consequences of being out of compliance. If the emissions of a party exceed its assigned amount, the enforcement branch must declare that that party is in noncompliance and require the party to make up the difference between its emissions and its assigned amount during the second commitment period, plus an additional penalty deduction of 30%. Also, the enforcement branch shall require

the party to submit a compliance action plan and suspend the eligibility of the party to make transfers under emissions trading (UNFCCC 2008).

Although the Kyoto Protocol did not enter into force until 2005, many of the Annex I parties met their commitments to reduce greenhouse gas emissions in the interval from 1990 to 2005 (**Figure 11.16**). During this period, however, total emissions grew about 24% worldwide. Current implementation of the Kyoto Protocol does not limit emissions from developing states, and from 1990 to 2005, emissions from Turkey increased by 76%, from China by 65%, and from India by 54% (van Vuuren et al. 2003; Rogner et al. 2007; United Nations 2007; World Resources Institute 2007). Several Annex I parties are not close to meeting their commitments: Turkey was allowed to convert from Annex I to non–Annex I; emissions from Spain increased by 60% from 1990 to 2005, from Canada by 54%, from Portugal by 40%, from Greece and Ireland by 25% each, and from New Zealand by 23% (United Nations 2007). The Kyoto Protocol has not yet imposed major sanctions for noncompliance.

Carbon emissions trading expanded under the Kyoto Protocol to a value of $64 billion U.S. (Capoor

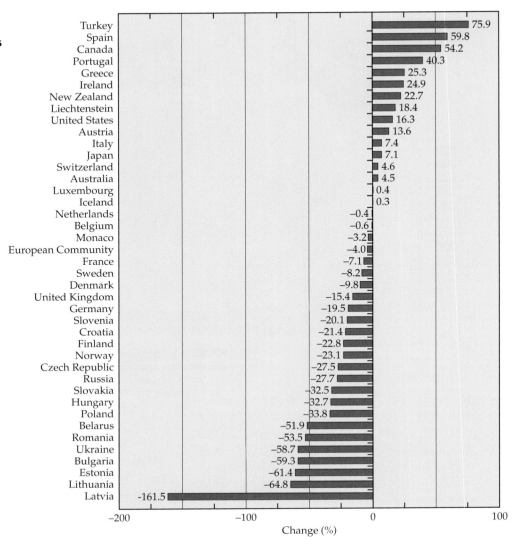

FIGURE 11.16 Change from 1990 to 2005 (%) in greenhouse gas emissions from Annex I states (After United Nations 2007.)

and Ambrosi 2008) and to a volume of 170 Mt carbon equivalents (**Figure 11.17**), or about 2% of the world's total emissions. The United Kingdom purchases over half of the world's carbon emission credits (**Figure 11.18A**). The vast majority of emission credits support projects in China that diminish greenhouse gas emissions such as the construction of higher efficiency coal-fired electric power plants (**Figure 11.18B**).

Effectiveness of the Law of the Sea Convention versus Montreal versus Kyoto

The Kyoto Protocol has achieved only limited success in slowing the rise in atmospheric concentrations of greenhouse gases (Victor et al. 2005; Prins and Rayner 2007; Barrett 2008). These concentrations continue to increase significantly (see Figure 4.18) along with the emissions from many states (see Figure 4.14), most of whom are parties to the Kyoto Protocol. The Law of the Sea Convention, although it established the maritime rights of states, has been less effective in protecting the marine environment. For example, over 75% of the world's fish stocks are either fully exploited or overexploited, and the sustainable use of marine living resources will require additional measures (Report of the Secretary-General 2008). In contrast, the Montreal Protocol has successfully curtailed production and consumption of ozone-depleting substances and has begun to reverse the loss of atmospheric ozone (see Figure 11.11).

One difference among the treaties is the participation of the United States. The United States has adopted most of the Law of the Sea as customary international law. The U.S. Senate in 1988 ratified the Montreal Protocol by a unanimous vote. On the contrary, not a single U.S. senator in 1999 publically supported the Kyoto Protocol (Gerrard 2007). Even Al Gore, who was vice president at

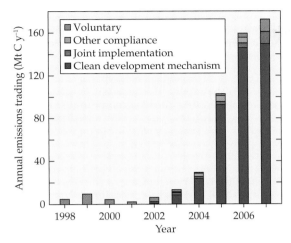

FIGURE 11.17 Global carbon emissions trading (10^6 metric tons per year) Trading was "Voluntary" before treaties came into force. "Other compliance" includes trading to meet local and regional regulations. "Joint implementation" shows how much Annex I states trade credits. "Clean development mechanism" shows how much Annex I states support projects in developing states. (After Capoor and Ambrosi 2008.)

the time, indicated in 1999 that the United States would not ratify the Kyoto Protocol without modification to several provisions (Sunstein 2007). This same person subsequently won an Academy Award and the Nobel Peace Prize for his efforts to address the issue of global climate change. The provisions of the Kyoto Protocol that the United States found objectionable are:

• *Kyoto's lack of regulations for emissions from developing states.* China has become the world's largest emitter of greenhouse gases (see Figure

4.14), despite receiving most of the funding from Annex I states to support projects that diminish emissions (see Figure 11.18). In defense of China, one-quarter of the country's greenhouse gas emissions derive from manufacturing goods for Annex I states (Wang and Watson 2007); its emissions per person are still much lower than the global average (see Figure 7.3A); and its historical contribution to atmospheric concentrations of greenhouse gases is one-third that of the United States or the European Union (Baumert et al. 2005). Nonetheless, emissions from developing states such as China, India, South Korea, and Brazil are increasing so fast that they will soon surpass the total emissions from Annex I states (Energy Information Administration 2008e). Undoubtedly, mitigating global climate change will require limits on greenhouse gas emissions from these developing states.

• *Kyoto's cap-and-trade system for emissions.* Many economists believe that the cap-and-trade system prescribed in the Kyoto Protocol would seriously impair the U.S. economy (see Chapter 10) and that alternative measures for mitigating greenhouse gas emissions would achieve higher benefits at lower costs (Aldy et al. 2003).

• *Kyoto's choice of 1990 as the baseline year.* Greenhouse gas emissions from Eurasia (Eastern Europe) reached a maximum between 1986 and 1990 (**Figure 11.19**). Emissions from Germany and Russia dropped dramatically after East Germany merged with West Germany at the end of 1990 and the Soviet Union dissolved in 1991. Consequently, during the drafting of the

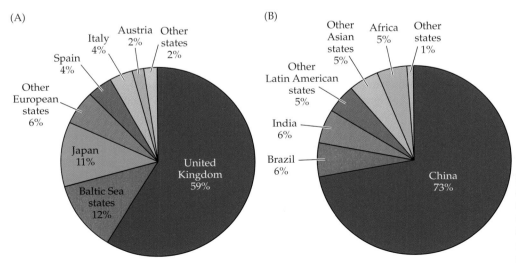

FIGURE 11.18 Buyers and suppliers of carbon emission credits in 2007 as a percentage of credits sold (A) Buyers. (B) Suppliers. (After Capoor and Ambrosi 2008.)

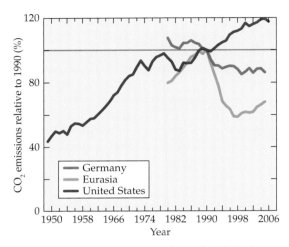

FIGURE 11.19 CO$_2$ emissions from fossil fuel burning with respect to 1990 levels for the United States, Germany (sum of East Germany and West Germany before 1990 and merged afterward), and Eurasia (U.S.S.R. before 1991 and sum of separate states afterward). (After Energy Information Administration 2006b.)

Kyoto Protocol in 1997, Germany and Russia realized that it would be relatively easy to improve upon 1990 emission levels and to generate emissions credits for trade. By contrast, emissions from the United States remained relatively stable from 1972 until 1991, before rising sharply. During the drafting of the Kyoto Protocol in 1997, the United States lobbied for setting 1995 as the baseline year but was outvoted (Pring 2001).

Key provisions of the Kyoto Protocol—in particular, the limits to greenhouse gas emissions from Annex I states—are set to expire in 2012. Negotiations on a post-Kyoto treaty are in progress. Many analysts believe that such negotiations benefit from comparisons between the present Kyoto Protocol and the Law of the Sea Convention or the Montreal Protocol (Barrett 2003; Victor and Coben 2005; Highum 2006; Benedick 2007; Sunstein 2007). The following sections contrast these treaties in terms of their history, scope and complexity, national self-interests, fairness, and enforcement.

HISTORY The Law of the Sea Convention addressed long-standing problems for which an extensive body of customary international law already existed, but still, negotiators labored for more than two decades on the Law of the Sea Convention before it entered into force. By contrast, stratospheric ozone depletion and global climate change are issues of recent origin. Scientific

evidence about the role of chlorofluorocarbons (CFCs) in ozone depletion was first obtained in the 1970s (National Research Council 1982), negotiations about ozone depletion began among 28 states in 1985, and the Montreal Protocol entered into force in 1989. Scientists reached a consensus about global climate change and the role of human activities in the 1980s. Many states convened in 1997 to negotiate the Kyoto Protocol, and it entered into force in 2005.

SCOPE AND COMPLEXITY The Law of the Sea Convention delineated territorial waters and extraction of natural resources from the seas and seabed—matters broad in scope but conceptually straightforward. The Montreal Protocol focused on human production and consumption of the few manufactured substances that deplete ozone; ozone depletion is associated with higher incidence of human skin cancer, a harm that is both significant and easily envisioned (Sunstein 2007). The Kyoto Protocol addressed the interactions between greenhouse gases and Earth's climate, a much more difficult task because many natural and human activities generate greenhouse gases, and greenhouse gases are only one of many factors responsible for changes in climate (see Chapter 3).

NATIONAL INTERESTS States must serve the interests of their leaders and citizens. Because sovereign states have no obligation to ratify a treaty, the benefits of doing so must outweigh those of not ratifying it (Barrett 2003).

The Law of the Sea Convention appealed to the self-interest of most states because it clarified territorial claims to natural resources and rights of passage. Matters of self-interest with respect to the protection of marine life in international waters are less obvious, and achieving this objective has proved more difficult.

The Montreal Protocol was consistent with the interests of several states, particularly, the United States (Benedick 2007). By 1977, American consumers had become so worried about the loss of ozone in the upper atmosphere that they had reduced the U.S. market for ozone-depleting substances in spray cans by two-thirds, even without governmental regulation. The following year, U.S. legislation banned CFCs as propellants for nonessential uses. By 1987, the year in which the Montreal Protocol was drafted, U.S. production of ozone-depleting substances had plummeted from one-half of the world's total to one-fourth. Moreover, President Ronald Reagan, who suffered from skin cancer and had several operations to remove cancerous lesions from his face, became a strong supporter of the

Protocol (Benedick 2007). U.S. chemical companies such as DuPont soon expanded their manufacture of HCFCs and HFCs, moderate-cost alternatives to CFCs.

In contrast, the Kyoto Protocol served the self-interests of few, if any, major world powers. On March 13, 2001, less than two months after he was sworn into office, President George W. Bush declared that he opposed the Kyoto Protocol because it "would cause serious harm to the U.S. economy" (Bush 2001).

FAIRNESS Sovereign states must perceive a treaty as being fair to all parties. The Law of the Sea Convention established uniform rules for territorial claims. The Montreal Protocol set standard limits to production and consumption of ozone-depleting substances, although it allotted developing states more time to meet these limits. The Kyoto Protocol, however, placed the burdens of compliance entirely on developed states, while it offered developing states financial benefits with few obligations.

Differences between the Montreal Protocol and the Kyoto Protocol in their treatment of developing states derived in part from the manner of the initial negotiations. Many of the negotiations for the Montreal Protocol were restricted to the major producers of ozone-depleting substances (Benedick 2007). Negotiations for the Kyoto Protocol, on the contrary, included representatives from most states, a situation where the interests of the more numerous developing states easily overwhelm those of developed states such as the United States.

ENFORCEMENT Incidents of noncompliance with the Law of the Sea Convention, because they usually involve territorial rights, can quickly escalate. The treaty, therefore, provided a mechanism for states to avoid armed conflict. Enforcement in the Montreal Protocol and the Kyoto Protocol emphasized the "carrot over the stick," providing incentives for compliance rather than sanctions for noncompliance. Incentives in the Montreal Protocol were adequate to compensate for abandoning ozone-depleting substances, particularly when alternative substances became available at modest costs, whereas incentives in the Kyoto Protocol did not at all compensate for most of the costs in mitigating greenhouse gas emissions.

In both the Montreal Protocol and the Kyoto Protocol, sanctions for noncompliance are relatively "toothless" in that they do not compel states to honor their commitments. For example, the Russian Federation, a party to the Montreal Protocol, has been cited repeated for its production of ozone-depleting substances (Ozone Sec-

retariat 2006). Similarly, Canada openly admits that it is unlikely to meet its commitments under the Kyoto Protocol for greenhouse gas emissions (Environment Canada 2007).

Regional Governments and Nongovernmental Organizations

The United States

Although the U.S. government has neither regulated greenhouse gas emissions nor ratified the Kyoto Protocol, a number of states and cities in the United States have enacted legislation to bring their jurisdictions into compliance with the spirit of the Kyoto Protocol (**Table 11.4**). These are largely symbolic gestures because state and city governments do not have broad jurisdiction over greenhouse gas emissions and cannot become parties to the Kyoto Protocol, a treaty among sovereign states. Nonetheless, such legislation does influence emissions from the operations of these entities themselves. For example, some state and local governments require that their departments purchase more fuel-efficient vehicles.

In 2005, ten Northeast and Mid-Atlantic states of the United States formed a Regional Greenhouse Gas Initiative (Regional Greenhouse Gas Initiative 2008), whereby they pledged to stabilize greenhouse gas emissions at around 2003 levels by 2015. The participants would seek a 10% reduction in emissions between 2015 and 2020. The proposal would also allow participants to purchase emission credits to meet 50% of their emission reductions.

California emits an equivalent of 134 million metric tons of carbon per year (California Energy Commission 2006), or 1.8% of the world total. If California were considered a separate sovereign state, it would rank somewhere between ninth and sixteenth highest worldwide in greenhouse gas emissions. In 2006, the California government passed the Global Warming Solutions Act (Assembly Bill 32), a law that commits the state to reduce its emissions by 25% by the year 2020. The California Air Resources Board is charged with releasing in 2009 a scoping plan to achieve the 2020 statewide emissions limit and with implementing the plan the following year. The plan will probably include a cap-and-trade system for emissions.

California also passed legislation (Assembly Bill 1493) in 2002 that imposes mileage standards on personal vehicles sold in California. This would reduce greenhouse gas emissions by about 22% by 2012 and

TABLE 11.4   North American agreements that address global climate change

Focus	Law	Parties	Start date
Vehicle fuel efficiencies	Assembly Bill 1493	California Arizona, Colorado, Connecticut, District of Columbia, Florida, Maine, Maryland, Massachusetts, New Mexico, New Jersey, New York, Oregon, Pennsylvania, Rhode Island, Utah, Vermont, and Washington intend to follow suit	2006
GHG[a] emissions	Assembly Bill 32	California	2010
GHG emissions	Regional Greenhouse Gas Initiative	Connecticut, Delaware, Maine, Maryland, Massachusetts, New Hampshire, New Jersey, New York, Rhode Island, and Vermont	2009
GHG emissions	Mayors Climate Protection Center	902 U.S. cities in all 50 states, District of Columbia, and Puerto Rico, representing a total population of over 81,603,449 citizens, adopted the provisions of the Kyoto Protocol	2008
GHG emissions	Western Climate Initiative Participants	Arizona, California, Montana, New Mexico, Oregon, Utah, and Washington and the Canadian provinces of British Columbia, Manitoba, Ontario, and Quebec Observers: Alaska, Colorado, Idaho, Kansas, Nevada, Wyoming, Saskatchewan, Baja California, Chihuahua, Coahuila, Nuevo Leon, Sonora, and Tamaulipas	2012
Clean and diversified energy	Western Governor's Association Resolution 07-16	Alaska, American Samoa, Arizona, California, Colorado, Guam, Hawaii, Idaho, Kansas, Montana, Nebraska, Northern Mariana Islands, Oklahoma, Oregon, South Dakota, Texas, Utah, Washington, and Wyoming	2015

[a]Denotes greenhouse gas.

30% by 2016. The automobile manufacturers and the U.S. Environmental Protection Agency under President George W. Bush's administration contested this legislation. In 2007, the U.S. Supreme Court held in the case of Massachusetts vs. Environmental Protection Agency that carbon dioxide (CO_2) and other greenhouse gases are pollutants in the specific context of the U.S. Clean Air Act (Supreme Court of the United States 2007). Therefore, the EPA has jurisdiction over greenhouse gas emissions and could grant California permission to regulate vehicle emissions. Many other states have pledged that they will enact similar legislation once California has cleared these legal hurdles (Table 11.4).

Other countries

Canada, as discussed in the previous section, is unlikely to meet its greenhouse gas emission targets for the Kyoto Protocol. Four Canadian provinces, however, have joined seven U.S. states to form the Western Cli-

mate Initiative. Six other U.S. states, six Mexican states, and one other Canadian province serve as observers. In 2008, the group agreed to phase in a cap-and-trade system for greenhouse gas emissions, beginning in 2012, with the goal of reducing emissions by 15% from 2005 levels by 2020 (Western Climate Initiative 2008).

ICLEI (founded in 1990 as the International Council for Local Environmental Initiatives and now officially called ICLEI - Local Governments for Sustainability) is an international organization with a membership of over 1080 cities, towns, and counties (ICLEI 2009). ICLEI provides technical consulting, training, and information services to achieve local, national, and global sustainability objectives. One of these objectives is reducing local greenhouse gas emissions.

Nongovernmental organizations

Nongovernmental organizations (NGOs) are legally constituted, nonprofit associations that operate without direct representation of any government. There are roughly 40,000 such organizations working internationally, and about 3% of them focus on environmental issues (Anheier et al. 2001). For example, the Climate Action Network (CAN) is a worldwide network of over 430 NGOs that promotes government and individual action for limiting human-induced climate change to ecologically sustainable levels (Climate Action Network 2008). Most NGOs depend on corporate sponsorships to fund their operations, but all NGOs in one survey believe that corporations are guilty of **greenwashing**, placing an inordinate emphasis on enhancing their reputations for environmental sustainability rather than taking substantial actions (National Public Relations 2008).

This uneasy relationship between NGOs and their corporate sponsors is likely to exacerbate as NGOs play an ever-increasing role in advocating expansion of environmental regulations as well as serving as independent watchdogs to insure compliance with existing regulations. Social policy on global climate change presents many challenges: The worst effects of global climate change are often not experienced by those most responsible for the problem; the people who are most vulnerable to the effects may be almost powerless to influence the problem; the causes and effects of climate change are complex and contested; and governmental mechanisms to address the problems are often weak, under-resourced, and fall under the purview of many, poorly coordinated bureaucracies (Newell 2008).

Under such circumstances, NGOs may have distinct advantages over governments because of their independence and ability to operate across borders, the narrow focus of their support base, and their moral convictions. For example, negotiations on a treaty to succeed the Kyoto Protocol have engaged many NGOs in the process, particularly environmental and public interest groups. The next and final chapter of this book (Chapter 12) will examine how NGOs and other public and private institutions have attempted to sway public opinion about global climate change.

Summary

Global climate change is, by the very nature of its causes and effects, an issue that requires the cooperation of many sovereign states. Enactment and interpretation of international treaties to address climate change rely on conventions, customs, general principles of law, and judicial decisions. Compliance with and enforcement of such international treaties depends on monitoring of greenhouse gas emissions and a clear procedure for bringing violators back into compliance. Also important are procedures for peaceful resolution of disagreements or disputes among sovereign states.

Three international treaties concerned with global climate change are the United Nations Law of the Sea, the Montreal Protocol on Substances that Deplete the Ozone Layer, and the Kyoto Protocol to the U.N. Framework Convention on Climate Change.

The United Nations Law of the Sea intersects with climate change in several ways: Higher carbon dioxide concentrations in the atmosphere increase the acidity of oceans and endanger sea life, carbon capture and storage in the oceans threaten fragile deep-sea ecosystems, and rising sea levels and melting sea ice open additional waters to navigation and extraction of natural resources. The Montreal Protocol presents a timetable to phase out and eventually eliminate the production and consumption of chemicals that deplete Earth's ozone layer; some of these chemicals (CFCs and HCFCs) also are potent greenhouse gases. The Kyoto Protocol establishes limits for greenhouse gas emissions from some sovereign states and methods for these states to trade emission credits. There is general agreement that the Montreal Protocol has been the most successful of the three treaties because of its narrow scope and the easily envisioned consequences of a failure to take effective action.

Review Questions

1. Global change abatement cannot depend solely on the laws of one sovereign state because
   (a) disruptions from climate change that a state experiences is proportional to its greenhouse gas emissions.
   (b) multinational corporations have a headquarters in one state.
   (c) free riders are those that travel freely from state to state.
   (d) greenhouse gases emitted from any state influence the whole planet.
   (e) poor states are likely to emit more greenhouse gases than rich ones.

2. Which one of the following does the International Court of Justice in The Hague *not* rely on as a source of law?
   (a) Conventions
   (b) Customs
   (c) General principles
   (d) Judicial decisions and teachings of publicists
   (e) Medical instruments

3. Mark all of the following that apply to the concept of American exceptionalism.
   (a) Term was coined by Alexis de Tocqueville in 1835
   (b) Counterbalanced by the economic benefits of harmonizing domestic and international laws
   (c) Justifies the behavior of multinational corporations
   (d) Maintains that the United States is unique in its commitment to liberty and individualism
   (e) Explains the attitude of U.S. courts to international law

4. Which of the following are examples of general principles of law? (Mark all that apply.)
    (a) Promises are binding (estoppel).
    (b) One should not benefit from wrongdoing (clean hands).
    (c) Rules should be just and fair (equity).
    (d) No one should gratuitously endanger lives.
    (e) No one should commit atrocities against a group of people (crimes against humanity).

5. Monitoring for compliance usually does *not* involve
    (a) warrants.
    (b) inspections by officials.
    (c) self-monitoring by the regulated community.
    (d) citizens' complaints.
    (e) monitoring neighboring areas.

6. Successful enforcement should (mark all that apply)
    (a) return a violator to compliance.
    (b) impose a penalty.
    (c) nullify any benefit of noncompliance.
    (d) remediate any past damages.
    (e) correct any management problems of the violator.

7. Peaceful means of settling disputes may include (mark all that apply)
    (a) negotiation.
    (b) inquiry.
    (c) conciliation.
    (d) arbitration.
    (e) judicial settlement.

8. Select two of the following international treaties that had some success in addressing issues arising from global climate change.
    (a) The First through Fourth Geneva Conventions, beginning in 1864
    (b) The First and Second Hague Peace Conventions, in 1899 and 1907
    (c) The United Nation Conferences on the Law of the Sea in 1956, 1960, and 1973
    (d) The Montreal Protocol on Substances that Deplete the Ozone Layer in 1987
    (e) The United Nations Convention on the Rights of the Child, in 1989

9. UNCLOS defines all but one of the following coastal boundary zones. Which one is *not* defined?
    (a) Internal waters
    (b) Territorial waters
    (c) Conservation zone
    (d) Exclusive economic zone
    (e) Continental shelf

10. Mark all of the following that apply to the current situation with CFCs.
    (a) CFCs have become the drug of choice among college students.
    (b) Emissions are decreasing.
    (c) Atmospheric concentrations are decreasing.
    (d) CFCs are sold illegally on a large, international black market.
    (e) CFCs have no moderate-cost alternatives.

11. Mark all of the following that had to occur before the Kyoto Protocol came into force.
    (a) It needed ratification by 55 states.
    (b) It needed ratification by developed (Annex I) states that accounted for 55% or more of total CO_2 emissions in 1990.
    (c) It needed ratification by the United States.
    (d) It needed ratification by all the states that originally signed the treaty.
    (e) It needed ratification by the General Assembly of the United Nations.

12. Which of the following goals for greenhouse gas emissions was *not* part of the original Kyoto Protocol?
    (a) The United States would reduce emissions to 7% below its year 1990 emissions.
    (b) Iceland could increase its emissions to 10% above its year 1990 emissions.
    (c) Russian could emit the same amount of greenhouse gases as in 1990.
    (d) China could increase its emissions to 20% above its year 1990 emissions.
    (e) Germany would reduce emissions to 8% below the year 1990 combined emissions of West Germany plus East Germany.

13. Greenhouse gas emissions are
    (a) decreasing worldwide.
    (b) decreasing from all Annex I party states.
    (c) not related to atmospheric concentrations of the gases.
    (d) being traded among Annex I parties to the Kyoto Protocol.
    (e) to be regulated for all parties to the Kyoto Protocol after 2015.

14. In the United States, local and state governments can
    (a) become parties to the Kyoto Protocol.
    (b) regulate greenhouse gas emissions from electric power plants.
    (c) specify the fuel efficiency of personal vehicles.
    (d) specify the fuel efficiency of their government vehicles.
    (e) negotiate with Canada or Mexico.

Suggested Readings

Gerrard, M. 2007. *Global Climate Change and U.S. Law*, American Bar Association Section of Environment Energy and Resources, Chicago, Ill., *http://www.abanet.org/abapubs/globalclimate*.

This book covers the body of U.S. law related to global climate change.

Highum, E. G. 2006. *Success vs. Failure in International Environmental Treaty Negotiations: A Comparative Analysis of the Montreal Protocol and Global Climate Change Negotiations*, International Studies Association, San Diego, *http://www.allacademic.com/one/isa/isa06/index.php?cmd=Download+Document&key=unpublished_manuscript&file_index=2&pop_up=true&no_click_key=true&attachment_style=attachment&PHPSESSID=c123141638d43d6ceec7b220c27e33d7*.

This paper provide a comparative analysis of the factors that led to success versus failure in international environmental treaty negotiations.

Office of Enforcement. 1992. *Principles of Environmental Enforcement*, U.S. Environmental Protection Agency, Washington, D.C., *http://nepis.epa.gov/Exe/ZyPURL.cgi?Dockey=500003C8.txt*.

This is a general report on approaches to enforce environmental regulations.

Sunstein, C. R. 2007. Of Montreal and Kyoto: A tale of two protocols. *Harvard Environmental Law Review* 31: 1–65.

This review compares the relative effectiveness of the Montreal and Kyoto Protocols.

12

GLOBAL CLIMATE CHANGE: THE HUMAN RESPONSE

Global climate change involves many complex scientific issues. C. P. Snow, a physicist and novelist, brought attention to the schism that developed during the twentieth century between scientists and the rest of society and to the discomfort most people feel about complex scientific issues (Snow 1959). Readers who have perused the first 11 chapters of this book are now familiar with the science of global climate change and can discuss many of the issues with some authority. Others who have not had this transformative experience may find the science of climate change intimidating or boring and must therefore rely on so-called experts to interpret the issues for them. Organizations ranging from petroleum companies to environmental groups have engaged the services of experts who expound points of view consistent with their own interests. The resulting diversity of "authoritative" opinions about global climate change confuses the public.

FIGURE 12.1 Beliefs about human activity as a contributor to global temperature change Percentage of respondents in two polls who believe, do not believe, or are not sure that human activity is contributing to an increase in global temperatures. The BBC poll interviewed about 1000 adults from each country via telephone or face-to-face in July 2007. The Harris poll surveyed, via the Web, 2563 U.S. adults in April 2007 and about 1000 adults from each of several European countries in November 2006. (After BBC World Service Poll et al. 2007; Harris Interactive 2007.)

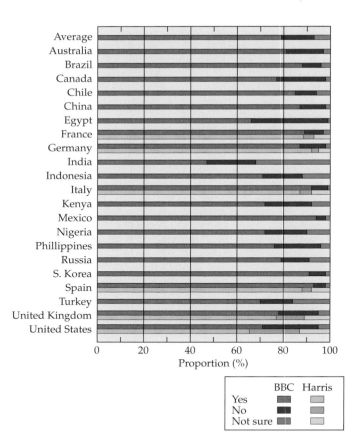

This chapter first surveys public opinions about global climate change. It then examines how these opinions vary with societal factors such as national origin, education, gender, race, income, age, and religion. It describes the battle among various interests groups for the "hearts and minds" of the public. Next it looks to the lessons of history as they pertain to environmental problems such as climate change. It concludes with descriptions of some trends that affirm human resilience and adaptability in the face of potential danger.

Public Awareness of Climate Change

Most people, when surveyed about current issues, express at least a fair amount of concern about global climate change (**Table 12.1**). They also believe that human activities are largely responsible for changing the climate and the problems that result from such changes (**Figures 12.1 and 12.2**). Nonetheless, they rank global climate change as less important than the economy, jobs, terrorism, education, health care, poverty, crime, and other issues (**Table 12.2**).

Surveys—even those asking a similar number of similar people similar questions at similar times—may obtain different results. For example, 71% of U.S. adults responded that "human activity, including industry and transportation, is a significant cause of climate change" (BBC World Service Poll et al. 2007); 65% of U.S. adults responded that "the activities of human beings are contributing to an increase in global temperatures" (Harris Interactive 2007); and 47% of U.S. adults responded that "global warming is happening due to human activity" (Pew Research Center 2008). Notice that the phrasing of the question in each survey differs slightly.

Surveys (**Figure 12.3**) and media coverage (**Figure 12.4**) indicate that public concern about global climate change has been waning since about 2006. A string of new cataclysmic climatic events of the magnitude of the European heat wave in 2003 (see Table 6.5) or Hurricane Katrina in 2005 would place global climate change higher on everyone's priority list (Krosnick et al. 2006), but we hope that such stimuli will not prove necessary. The recent decline in concern, however, highlights the difficulties in keeping the public eye focused on the problem of global climate change. Several factors, including the nature of the human psyche and the complexity of the topic, contribute to this lack of focus.

The human psyche is not well suited for addressing the problem of global climate change (Leiserowitz 2006; Weber 2006). Our mental skill set is highly attuned to solve urgent, well-defined problems with clear solutions, such as procuring our next meal or finding a safe place to sleep. Global climate change is the exact opposite, a problem that extends over several generations, seems vague and abstract, and has no surefire remedies. Indeed, potential remedies to this problem only diminish the odds of undesirable outcomes and may seem too meager a reward for all the hard work and sacrifice that they entail.

(A)

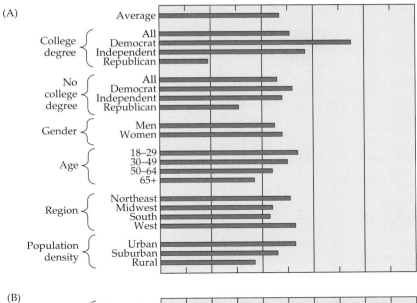

(B)

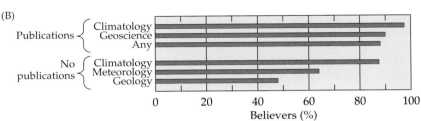

FIGURE 12.2 Percentage of respondents in two polls who believe that global warming results from human activities (A) One poll surveyed 1502 U.S. adults in April 2008 and categorized the respondents by whether they had earned a college degree, their political party affiliation (Republican, Democrat, or Independent), gender, age, region, and population density. (B) A second poll surveyed professional geoscientists in 2008 via the Web: 90% were from the United States, 6% from Canada, and 4% from other countries; 90% had completed a Ph.D., and 7% held a master's degree. Under Publications: "Climatology" designates climatologists who are publishing on global warming, "Geoscience" designates geoscientists publishing on global warming, "Any" designates geoscientists publishing on any topic. Under No publications: "Climatology" designates climatologists who are not publishing, "Meteorology" designates meteorologists who are not publishing, and "Geology" designates economic geologists who are not publishing. (A after Pew Research Center 2008; B after Doran and Zimmerman 2009.)

TABLE 12.1 Percentage of respondents in various countries who have heard about global climate change and degree of concern about the issue by those who have heard

		Level of concern (%)[a]			
Country	Have heard	Great deal	Fair amount	Little/ none	Do not know
China	78	20	41	37	2
Egypt	47	24	51	23	1
France	97	46	41	14	0
Germany	95	30	34	36	1
India	57	65	20	13	2
Indonesia	35	28	48	23	1
Japan	99	66	27	7	0
Jordan	48	26	40	34	–
Nigeria	42	45	33	20	2
Pakistan	12	31	25	39	5
Russia	80	34	31	34	–
Spain	93	51	34	14	2
Turkey	75	41	29	23	8
United Kingdom	100	26	41	32	1
United States	91	19	34	47	1

Source: Pew Research Center 2006a.

[a]Numbers are rounded.

TABLE 12.2 Top policy priorities for the U.S. president and congress, by political party affiliation, in four recent surveys[a,b]

Republicans					Democrats					Independents				
Issue	Jan '09	Jan '08	Jan '07	Jul '06	Issue	Jan '09	Jan '08	Jan '07	Jul '06	Issue	Jan '09	Jan '08	Jan '07	Jul '06
Economy	83	76	65	80	Job situation	89	76	67	78	Economy	85	74	60	78
Terrorism	79	86	93	84	Economy	88	76	77	80	Terrorism	81	65	77	72
Job situation	72	43	39	52	Health care	71	81	77	89	Job situation	80	61	60	63
Military	64	62	56		Terrorism	71	74	74	69	Social Security	62	60	60	71
Social Security	60	60	62	73	Education	71	72	74	86	Energy	61	61	58	67
Energy	51	53	45	56	Uninsured	66	65	70		Medicare	61	60	61	
Budget deficit	51	52	42	47	Energy	66	59	64	66	Education	59	66	66	83
Morality	50	51	54	43	Medicare	65	66	70		Budget deficit	57	57	53	55
Medicare	48	47	53		Poverty	62	62	67		Health care	52	68	66	79
Immigration	46	64	63	64	Social Security	60	70	72	79	Uninsured	50	57	52	
Education	46	54	65	75	Environment	54	67	67	64	Poverty	48	51	46	
Health care	45	53	58	69	Budget deficit	52	64	57	62	Crime	47	50	57	
Crime	41	49	56		Taxes	48	50		66	Taxes	45	44		63
Gov't ethics	37	42	28		Tax cuts	48	37	54		Tax cuts	45	31	41	
Poverty	34	34	48		Crime	47	62	69		Immigration	42	51	49	57
Taxes	31	46		74	Morality	46	44	45	31	Gov't ethics	42	42	35	
Tax cuts	31	44	49		Climate change	45	47	48	56	Environment	41	56	59	58
Global trade	28	37	33		Military	38	37	42		Military	41	34	45	
Uninsured	28	27	44		Immigration	34	43	48	52	Morality	40	39	42	28
Environment	20	39	41	30	Global trade	33	37	35		Global trade	32	38	31	
Climate change	16	12	23	23	Gov't ethics	30	36	44		Climate change	25	38	40	49

Source: Pew Research Center 2006b, 2007, 2008, 2009b.
[a]Issues are listed in descending importance according to the January 2009 survey.
[b]Percentage of U.S. respondents who rated the issue as very important.

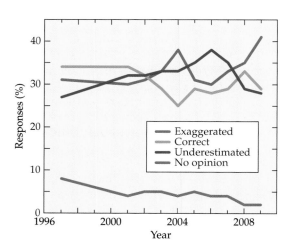

FIGURE 12.3 Public view of the seriousness of global warming Percentage of respondents in the United States who believe that reports about the seriousness of global warming are exaggerated, correct, or underestimated or who have no opinion. (After Saad 2009.)

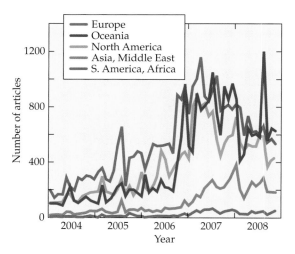

FIGURE 12.4 Newspaper coverage of global climate change Number of articles in 50 newspapers in 20 countries that mentioned global climate change during a given month. Oceania is composed of the islands in the Pacific including Australia, New Zealand, and New Guinea. (After Revkin 2008.)

For an analogy, consider that both poor diet and spoiled food can be life threatening. Public education programs that promote a healthy diet are less successful than those that promote proper food safety because a poor diet poses long-term and indefinite risks, whereas the threat of food poisoning evokes an immediate, visceral reaction in nearly everyone. Imagine what we would eat if the main effects of a poor diet were to manifest only after 50 years, as is anticipated for many effects of global climate change.

In fact, many people do not perceive global climate change to be an imminent danger that warrants direct attention (Lorenzoni et al. 2005) (see Figure 12.3). Global warming of a few degrees may seem relatively benign to people who prefer warmer winters and earlier springs or who escape to higher latitudes or higher elevations in the summer. Similarly, a sea level rise of a few feet or more violent hurricanes may not pose a threat to those who occupy higher ground. Moreover, a significant segment of the population still does not accept that human activities are the major cause of climate change and do not perceive any direct benefits of mitigating climate change, although they directly experience the costs of such actions at the gasoline pump or in utility bills.

In comparison, the Montreal Protocol on Substances that Deplete the Ozone Layer has received continuous, strong public support (see Chapter 11). A key difference between ozone depletion and global climate change is that ozone depletion increases the incidence of skin cancer, and virtually everyone recognizes the importance of averting skin cancer. In addition, mitigation of ozone depletion includes relatively simple measures such as using alternative refrigerants in compressors or converting from aerosol spray cans to squeeze bottles.

Societal Factors and Global Climate Change

We are all, to some degree, products of our environment. Consequently, national origin, education, gender, race, income, age, and religion influence our opinions about global climate change. The following sections examine each of these factors.

National origin

People from various countries have different opinions about the risks of global climate change. At the extremes, people in Japan express the most concern about the problem, whereas people in the United States express the least (see Table 12.1). Many indicators suggest that people in the United States, relative to citizens

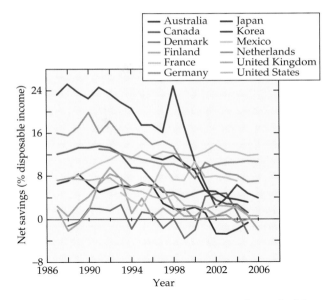

FIGURE 12.5 Net monetary savings per household as a percentage of household disposal income in various countries. (After OECD 2008a.)

of other developed countries, are risk takers and thus are less worried about long-term risks such as global climate change.

- Savings accounts provide a financial buffer against fluctuations in income or expenses as well as funds for retirement. Net savings per U.S. household have declined to nearly zero, whereas households in France and Germany deposit over 10% of their disposable income into saving accounts (**Figure 12.5**).

- Obesity—defined as when someone's weight in kilograms is greater than 30 times the square of their height in meters—is associated with increased risk of hypertension, diabetes, cardiovascular diseases, respiratory problems, musculoskeletal diseases, or some forms of cancer (OECD 2008b). Therefore, human life expectancy decreases with obesity (Fontaine et al. 2003; Olshansky et al. 2005). The rate of obesity is rising in many countries, but it is highest in the United States, where the condition afflicts one-third of adults (**Figure 12.6**).

- The United States is the only wealthy, industrialized nation that does not provide nearly universal health care for its citizens (**Figure 12.7**). An increasing number of people in the U.S. population lack health insurance (**Figure 12.8**) and face large financial liabilities in case of a major illness. About 27% of personal bankrupt-

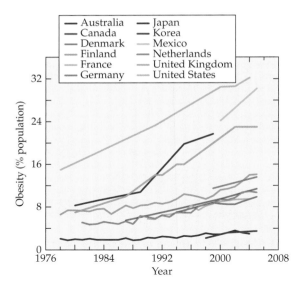

FIGURE 12.6 Incidence of obesity in people aged 15 or older (% of population) in various countries (After OECD 2008b.)

cies in the United States cite medical expenses as a major cause (Himmelstein et al. 2005).

• Unemployment benefits are lower in the United States than in other wealthy, industrialized nations (**Figure 12.9**).

These indicators reflect a positive, "can-do" attitude in which one takes advantage of current opportunities rather than dwells on negative potentialities. In other words, hard work and ingenuity will prevail over misfortunes that may arise. For example, most Americans believe that they could lose weight if the need arises (Saad 2008). This can-do attitude derives from the rapid upward and downward mobility of U.S. society as well as its geographical mobility (de Tocqueville 1835). It includes the American frontier spirit, the tension between the civilization of a settlement and the savagery of a wilderness, which fosters strength and self-reliance (Turner 1921). Unfortunately, neither a "can-do" attitude nor frontier spirit cultivate a long-term and a global perspective.

Countries such as France, Germany, and Japan—whose frontiers closed centuries ago, who have since weathered severe resource limitations, and who fought major wars on their soil in the twentieth century—have become more risk averse than the United States. In these countries, mitigation of global climate change is a central issue of several political parties. Greenhouse gas emissions from these countries have diminished (see Figure 11.16).

FIGURE 12.7 Types of health insurance coverage in different countries Percentage of the population who have public or private health insurance coverage. (After Institute of Medicine 2004; OECD 2007.)

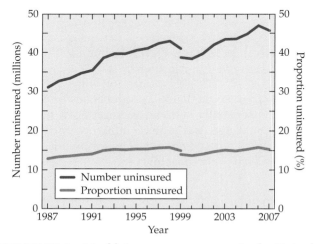

FIGURE 12.8 Health insurance coverage in the United States Millions of people without health insurance and percentage of the total population who do not have health insurance. The shift in 1999 derives from a change in the accounting method. (After U.S. Census Bureau 2007.)

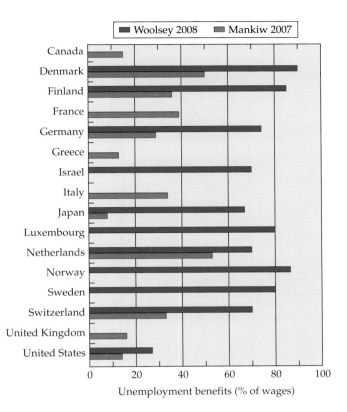

FIGURE 12.9   Unemployment benefits in different countries as a percentage of full employment wages according to two recent analyses. (After Mankiw 2007; Woolsey 2008.)

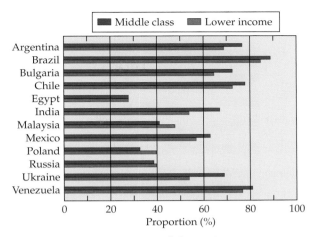

FIGURE 12.10   Perception of the seriousness of climate change by people with different incomes Percentage of people with middle or lower income in various countries who believe that global warming is a "very serious" problem. (After Pew Research Center 2009a.)

Education, gender, race, income, and age

The belief that human activities are responsible for global warming generally increases with education (see Figure 12.2). In the United States, 51% of respondents with college degrees hold this belief, whereas 45% of those without degrees do. Similarly, professional earth scientists who focus on climate issues are more likely to hold this belief than are other earth scientists. Indeed, 97.4% of professional climatologists who publish peer-reviewed research about global climate change have concluded that human activity is a significant contributing factor in changing mean global temperatures, whereas only 46.6% of economic geologists, who have not published in peer-reviewed journals, have come to this conclusion (see Figure 12.2). An exception to this trend are people in the United States who identify themselves as Republicans: Only 19% of Republicans with college degrees believe that global warming is happening due to human activity, whereas 31% of those without college degrees do (Pew Research Center 2008).

Most polls indicate that females in the United States are slightly more likely than males to believe that human activities are responsible for global warming (see Figure 12.2) and to support greater expenditures on mitigating climate change (ecoAmerica 2008). Race does not appear to have a major effect on opinions about global climate in the United States, but there are differences among regions and communities. In particular, respondents from urban communities in the northeastern and western United States were more likely to believe that human activities are responsible for global warming. Belief that global warming is a very serious problem generally increases with income (**Figure 12.10**); this results in part from the relationship between income and education.

Public opinion about the causes of global warming also varies with age of the respondent (see Figure 12.2). Young adults in the United States accept responsibility for global warming more readily than do older people: 54% for 18-to-29-year-olds versus 37% for people over 65 (Pew Research Center 2008). Moreover, only 31% of 18-to-29-year-olds in the United States believe that reports about global warming are exaggerated, whereas 47% of those over 65 years old believe so (Saad 2009).

Religion

Religion may provide guidance in situations that extend outside of our immediate sphere of influence, last longer than our lifetime, or pertain to our stewardship of Earth. Indeed, the moral obligation that we feel to the world at large, to future generations, and to

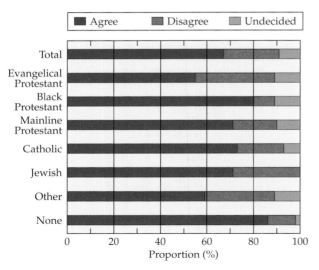

FIGURE 12.11 Perception of the seriousness of climate change based on religious affiliation Percentage of U.S. adults of different faiths who believe that global climate change will have disastrous effects if we do not act dramatically, based on the Baylor Religion Survey of 1700 people in fall 2007. (After Grossman 2008.)

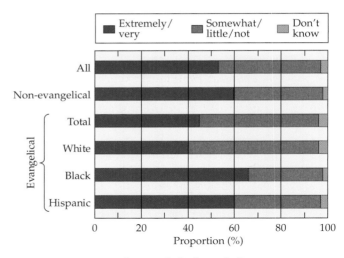

FIGURE 12.12 Religious beliefs and climate change Percentage of U.S. respondents of different faiths who believe that combating global warming and other environmental threats is extremely to very important versus somewhat, little, or not important. Based on a survey of 1610 people: 612 white, 217 black, and 238 Hispanic evangelicals and 543 non-evangelicals. (After Greenberg and Berktold 2004.)

Mother Nature interweaves with our spiritual values. Therefore, opinions about climate change—a global, multigenerational, natural resource issue—often depend on religion.

Religious tenets are open to many interpretations, and so are their applications to global climate change. Evangelical Protestants in the United States provide an example of the range of beliefs on global climate change. As a whole, evangelicals are less likely to support measures for mitigating greenhouse gases than are members of other religious denominations (**Figure 12.11**), but most black and Hispanic evangelicals believe that global warming is extremely or very important; most white evangelicals believe that it is only somewhat, little, or not important; and the belief of the general populace lies somewhere in between (**Figure 12.12**).

In 2006, 86 senior U.S. evangelical Protestant leaders endorsed a statement called the Evangelical Call to Action on Climate Change that declares

The consequences of climate change will be significant, and will hit the poor the hardest…Christian moral convictions demand our response to the climate change problem…The need to act now is urgent. Governments, businesses, churches, and individuals all have a role to play in addressing climate change—starting now. The basic task for all of the world's inhabitants is to find ways now

to begin to reduce the carbon dioxide emissions from the burning of fossil fuels that are the primary cause of human-induced climate change (The Evangelical Climate Initiative 2009).

Soon afterward, the Interfaith Stewardship Alliance, a coalition of more than 110 evangelical Protestants, issued a rebuttal that states

Global warming will have moderate and mixed (not only harmful but also helpful), not catastrophic, consequences for humanity…Human emissions of carbon dioxide and other greenhouse gases are probably a minor and possibly an insignificant contributor to its causes…Government-mandated carbon dioxide emissions reductions not only would not significantly curtail global warming or reduce its harmful effects but also would cause greater harm than good to humanity—especially the poor…Therefore we pledge to oppose quixotic attempts to reduce global warming (Interfaith Stewardship Alliance 2006).

As of April 2009, over 260 senior evangelical leaders have endorsed the Evangelical Call to Action on Climate Change. The second group, now called the Cornwall Alliance for the Stewardship of Creation, claims endorsements from 150 leaders of the religious and scientific community (Cornwall Alliance for the Stewardship of Creation 2009).

A diversity of opinions on climate change is not unique to U.S. evangelical Protestants. In the Catholic Church, Cardinal Pell of Sydney declares "Global warming has ceased" and cites experts who believe "Futile attempts to prevent global climate change would be a tragic misallocation of resources" (Pell 2008). The Holy See (the Vatican; the main body of the Catholic Church), however, supports international efforts to address climate change, commending "the United Nations Framework Convention on Climate Change (UNFCCC) for providing a global framework for concerted international action to mitigate climate change and to adapt to its impacts" (Migliore 2008).

Public Relations

Policies about global climate change may have enormous economic repercussions and may affect the survival of the human species. With the stakes so high, the battle for the hearts and minds of the general public becomes fierce. The diametrically opposed statements quoted in the previous section highlight some major points of contention:

- To what degree is Earth's climate changing?
- How responsible are human activities for these changes?
- What consequences do these changes hold for humanity?
- How worthwhile is mitigation of greenhouse gas emissions?

The preponderance of scientific evidence now supports that Earth's climate is changing significantly, that these changes are largely the result of greenhouse gas emissions from human activities, that these changes could produce catastrophic effects for humans, and that modest mitigation measures could prevent the worst of these changes. These conclusions are supported by 97.4% of professional climatologists who are active in global climate change research (Doran and Zimmerman 2009). The counterevidence, which skeptics of climate change have forwarded, either fails to withstand closer examination (see Chapter 4) or is not currently testable.

So why does so much controversy about global climate change persist? One explanation is that the changes in climate to date have been subtle; for example, the warming that Earth has experienced over the past century averages about 0.6°C (see Figure 2.40), but it varies from place to place and from year to year. Moreover, predictions about future changes depend on complex computer models and offer a range of possible outcomes; this virtual reality and lack of certainty is not convincing to the public. Finally, the public finds controversy more interesting than consensus (Lynch and McGoldrick 2005), and news media "strive to reflect a wide range of opinion and explore a range and conflict of views so that no significant strand of thought is knowingly unreflected or under represented" (BBC 2009). Consequently, a few skeptics of global climate change have had many opportunities to expound their views. The public thus obtains the false impression that there is very little agreement about global climate change.

Civilizations and the Environment

Traditionally, historians have emphasized the role of political, military, sociological, economic, and religious forces in the fate of human civilizations. The classical model is *The History of the Decline and Fall of the Roman Empire*, a six-volume treatment by Eduard Gibbon (Gibbon 1776). Recent studies of civilizations expand upon such treatment by including the influence of environmental factors.

In particular, Jared Diamond (professor of geography and physiology at the University of California, Los Angeles) describes in his book *Collapse* a number of civilizations that have committed "unintentional ecological suicide," or "ecocide" (Diamond 2006). He proposes that five major factors contributed to this fate: (1) environmental damage; (2) climate change; (3) hostile neighbors; (4) loss of friendly trade partners; and (5) poor treatment of environmental problems. He documents the interplay of these factors in the downfall of the statue builders of Easter Island, the Anasazi of the North American deserts, the Mayans of Central America, and the Vikings of Greenland.

An overall scenario for these civilizations is that agricultural advances such as irrigation (**Figure 12.13**), terracing (see Figure 9.7), or introduction of exotic crops such as maize (**Figure 12.14**) permitted unprecedented human population growth during periods of benign climate. For example, the population of the Anasazi expanded rapid from A.D. 700 to 1130 through advances in irrigation, terracing, and maize cultivation.

Larger human populations forced the spread of food production onto marginal lands such as newly cleared forests that were not well suited to this intensive use, and resulted in overhunting, overfishing, and overgrazing. This overexploitation of marginal lands led to water and soil problems including salinization and erosion of fertile topsoil. Land degradation, periods of unfavorable climate, attacks by hostile neighbors, loss of trading partners, and combinations thereof drove these civilizations to collapse. Even the rich and famous of these civilizations, despite their access

FIGURE 12.13 Anasazi irrigation canal in central Arizona dated to about A.D. 800.

FIGURE 12.14 Teosinte (the wild ancestor of maize) on the left, maize ear on the right, and ear of their F1 hybrid in the center.

to better land, food, and weapons and their religious authority, were unable to withstand a sudden surge of starving masses.

Dr. Diamond contrasts these failed civilizations against those of the New Guinea highlands and Japan that have persisted for millennia through better population control and better treatment of environmental problems. He compares modern Haiti with the Dominican Republic, sovereign states who share the island of Hispaniola but who have managed their natural resources differently. Today, only 1% of Haiti remains forested, whereas 28% of the Dominican Republic remains forested (**Figure 12.15**). This deforestation has had dire consequences for Haiti: lack of building materials, loss of soil fertility, high sediment loads in rivers, loss of watersheds and hydroelectric power, and decreased rainfall.

Has humankind learned from the lessons of the past? Dr. Diamond, although he describes himself as a "cautious optimist," expresses doubts about worldwide conservation and restoration efforts. The next and final section of this chapter addresses this question with respect to global climate change and reaches a different conclusion.

Sources of Optimism

Reports about global climate change often fall into the category of "more bad news." Nonetheless, major factors that influence the human response to this issue

show encouraging trends. The following sections present four of these.

Everybody talks about weather

The vast majority of people in developed countries are aware about the issue of global climate change (see Table 12.1). One reason for this is the release of several major motion pictures about climate change. A fictional disaster film released in 2004 entitled *The Day After*

FIGURE 12.15 Relative deforestation Satellite photo of the border between Haiti (at left) and the Dominican Republic (at right) showing differences in the extent of deforestation.

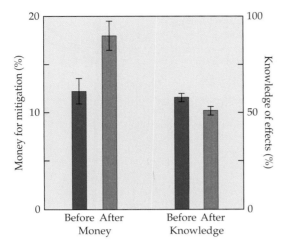

FIGURE 12.16 Influence of watching the movie *The Day After Tomorrow* on concern and knowledge about climate change Concern was assessed by comparison of how much respondents would donate, out of a hypothetical £1000, to climate mitigation efforts versus other charities. Knowledge was assessed by a test on the anticipated effects of global climate change. Shown are means and the variation of the respondents before and after they viewed the movie. (After Balmford et al. 2004.)

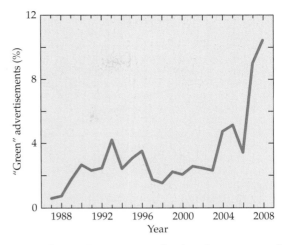

FIGURE 12.17 Percentage of advertisements in the magazines *Time, Fortune, National Geographic, Forbes, Sports Illustrated,* and *Vanity Fair* for companies that make environmental claims. (After Terrachoice Group Inc. 2009.)

Tomorrow increased moviegoers' concern about climate change but reduced their knowledge about it (**Figure 12.16**). The 2006 documentary *An Inconvenient Truth,* narrated by former U.S. Vice President Al Gore, won Academy Awards for the Best Documentary Feature and the Best Original Song in 2007. Subsequent efforts include a documentary *The 11th Hour* produced and narrated by Leonardo DiCaprio in 2007, a mockumentary *Sizzle* produced and directed by Randy Olson in 2008, and a science fiction drama *Age of Stupid* directed by Franny Armstrong in 2009.

Heightened awareness about global climate change is also evident in advertising. Many companies now tout their environmental practices or the environmental benefits of their products and services (**Figure 12.17**). For example, GE (General Electric), one of the world's largest companies, documents that in its operations, greenhouse gas emissions per revenue dollar have decreased by over 10% per year (**Figure 12.18**). Although some of this marketing (not GE's) may constitute greenwashing (i.e., exaggerated environmental claims), even greenwashing promotes environmentally responsible behavior.

Affirmation of public service for global climate change culminated in the awarding of the 2007 Nobel Peace Prize to the Intergovernmental Panel on Climate Change (IPCC) and former Vice President Al Gore. The Nobel Foundation praised "their efforts to build up and disseminate greater knowledge about man-made climate change, and to lay the foundations for the measures that are needed to counteract such change" (Nobelprize.org 2009).

Human population growth abates

Every human requires and deserves a share of Earth's natural resources, but such resource use is not sustainable if the human population continues to grow expo-

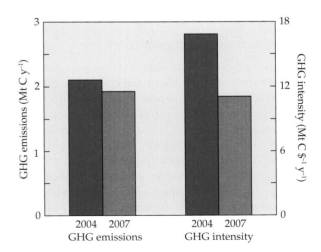

FIGURE 12.18 Total greenhouse gas (GHG) emissions and emissions per dollar generated (intensity) from the operations of the GE corporation in 2004 and 2007. Emissions are in 10^6 metric tons per year and intensity is in 10^6 metric tons per $U.S. per year. (After GE 2008.)

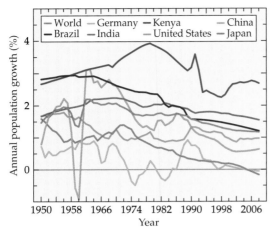

FIGURE 12.19 Human population growth (% per year) for the world and selected countries The trends for Kenya are representative of Africa, and those for Brazil are representative of South America. (After U.S. Census Bureau 2009.)

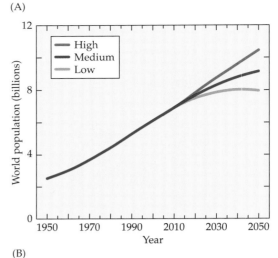

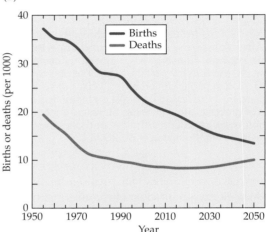

FIGURE 12.20 Estimated population growth (A) World human population in billions according to high, medium, and low estimates of human population growth. The human population is expected to level off around 2050 under the medium and low growth scenarios. (B) Births and deaths per 1000 people in the world assuming a medium estimate of population growth. (After Population Division 2009.)

nentially. Fortunately, human population growth has declined to about half of what it was in 1963 (**Figure 12.19**). The world's population should stabilize in the middle of this century because of a rapidly declining birth rate (**Figure 12.20**). Better yet, populations are beginning to shrink in developed countries such as Germany and Japan, which have relatively high resource use per person. In developing countries such as Kenya, Brazil, and India, population growth remains high but is decreasing. Once the world's population stabilizes, mitigation of greenhouse gas emissions will become easier.

Global education levels increase

Education levels of both men and women have risen over the past decade in all regions of the world (**Figure 12.21**). Generally, as education increases, so does the belief that human activities are responsible for global warming (see Figure 12.2). Education correlates with the wealth of nations (**Figure 12.22**), and a wealthier, better-educated populace can more readily adapt to global climate change. Better-educated women tend to have fewer children (**Figure 12.23**), and this will promote lower resource use. In summary, rising education levels should promote mitigation of and adaptation to global climate change.

Technology advances

Many existing technologies can significantly diminish greenhouse gas emissions from human activities (see

Chapters 7, 8, and 9). A number of these technologies are already cost effective, and other will become so as fossil fuel prices increase or governments impose tighter restrictions on greenhouse gas emissions. Governments and private industries are also developing new technologies. Clearly, we will require a mix of technological advances and conservation efforts to avoid the worst-case scenarios for global climate change.

Final Words

Astute readers will notice that this book, although it covers a broad range of topics related to global climate change, does not address several issues. It avoids most

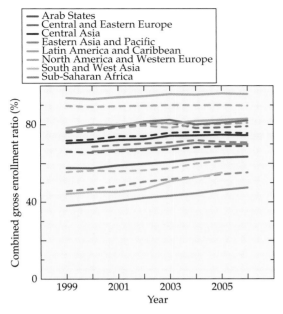

FIGURE 12.21 **Education worldwide** Proportion of males (represented by dashed lines) and females (represented by solid lines) receiving primary through high school education in different regions of the world. (After UNESCO Institute of Statistics 2009.)

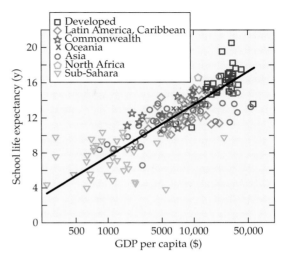

FIGURE 12.22 **Relationship between a nation's average education level and its per capita wealth** "School life expectancy" is the average number of years a child of school entrance age is expected to spend in primary, middle-school, or high-school education. "GDP per capita" is the total value of all goods and services produced in a country, divided by its population. Note the logarithmic scale for GDP. The black line shows the best fit of the data to a line. (After Huebler 2008.)

of the politics of global climate change such as policy differences among recent U.S. presidential administrations because this politics tends to polarize, rather than instruct. The book generally treats human adaptation to global climate change as ancillary to mitigation. It does not discuss the international negotiations currently in progress on a treaty to succeed the Kyoto Protocol. These and other issues will undoubtedly deserve greater attention in future texts.

Summary

Most people believe that global climate change is a significant problem but struggle with what, if anything, to do about it because the risks from climate change seem complex, very long-term, and indefinite. The diversity of information on climate change aired by various interest groups further serves to confuse the public. Many cultural factors, such as national origin, education, income, age, and religion, influence public opinion. Nevertheless, as public awareness of global climate change grows, human population growth diminishes, education levels increase, and technology advances, governments and private enterprises will take appropriate actions to solve this pressing problem.

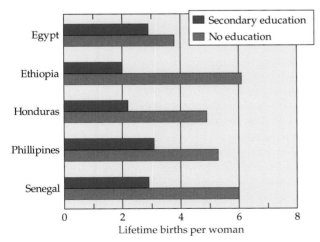

FIGURE 12.23 **Lifetime births per woman in selected countries by level of education** Women with a middle-school education or higher versus women with no formal education. (After Measure DHS 2009.)

Review Questions

1. Global climate change is difficult for many people to address because it (select all that apply)
   (a) extends over several generations.
   (b) offers little substantial evidence.
   (c) seems vague and complex.
   (d) disproportionately affects the world's wealthiest people.
   (e) has no surefire remedies.

2. Which one of the following, according to Jared Diamond, is *not* one of the five major factors that contribute to "ecocide"?
   (a) Environmental damage
   (b) Decreased fecundity
   (c) Climate change
   (d) Hostile neighbors
   (e) Loss of friendly trade partners

3. People in the United States, in comparison with those of Western Europe,
   (a) are more risk adverse.
   (b) save more of their disposable income.
   (c) are more likely to be obese.
   (d) enjoy universal health care.
   (e) receive greater unemployment benefits.

4. As people gain education, most
   (a) have more children.
   (b) earn less money.
   (c) read less.
   (d) become less healthy.
   (e) become more likely to believe that human activities are responsible for global warming.

5. Religious leaders
   (a) all support efforts to mitigate global climate change.
   (b) all oppose efforts to mitigate global climate change.
   (c) may differ markedly on the issue of global climate change.
   (d) typically find global climate change to be independent of their religious beliefs.
   (e) generally believe that the separation of church and state prevents them from voicing an opinion on global climate change.

6. Most professional climatologists who are active in global climate change research believe that (mark all that apply)
   (a) Earth's climate is changing significantly.
   (b) recent climate changes are largely the result of greenhouse gas emissions from human activities.
   (c) recent climate changes could produce catastrophic effects for humans.
   (d) modest mitigation measures could prevent the worst of the potential climate changes.
   (e) the separation of church and state prevents them from voicing an opinion on global climate change.

7. The world's human population is
   (a) shrinking.
   (b) growing at the fastest rate ever.
   (c) growing faster in Western Europe than in Africa.
   (d) likely to reach a maximum in the middle of this century.
   (e) not related to the issue of global climate change.

8. Education level is
   (a) increasing in most regions of the world.
   (b) higher in males all over the world.
   (c) not related to the wealth of nations.
   (d) not related to human population growth.
   (e) not related to the issue of global climate change.

Suggested Readings

Diamond, J. M. 2006. Collapse: *How Societies Choose to Fail or Succeed.* Penguin, New York.

> This book describes the importance of ecological factors in the failure of past human civilizations.

Leiserowitz, A. 2006. Climate change risk perception and policy preferences: The role of affect, imagery, and values. *Climatic Change* 77: 45–72.

> This article discusses the psychology of humans dealing with the risk of climate change.

Terrachoice Group, Inc. 2009. *The Seven Sins of Greenwashing: Environmental Claims in Consumer Markets*, TerraChoice Environmental Marketing, Philadelphia, *http://sinsofgreenwashing.org/?dl_id=4.*

> This 26-page report tracks the advertising claims of corporations that overstate their environmental commitments.

APPENDIX
Primer on Scientific Graphs

The Makings of a ~~Graphic Novel~~ Novel Graphic: What Are Those Squiggly Lines and Dots?

Say, for the sake of argument, that you are compelled to read a textbook that features illustrations with squiggly lines and dots that bear little resemblance to anything real. What should you do?

One option is to immediately drop the course and try to return the textbook. I do not recommend this course of action. Another option is to skip these figures entirely in the hope that your instructor will not require you to learn any of this information. After all, what you don't know won't hurt you. I do not recommend this course of action, either. Last, but not least, you might actually try to appreciate these critical pieces of evidence that support the conclusions in the textbook. For those readers who are willing to give this the old college try, the following is a primer on reading scientific graphs.

X-Y Graphs

The vast majority of data figures in this textbook are *x-y graphs* that show the relationship between two variables.

The value of one variable is expressed horizontally on the *x-axis*. This is usually the *independent variable*, one that changes without the con-

straint of other factors. In the example below (**Figure A**), the independent variable is the month. The value of the other variable is expressed by its vertical position on the *y-axis*. This is usually the *dependent variable*, one that changes with the value of the independent variable. In this example, precipitation is the dependent variable and the value for a particular month is designated by the distance of the data point above the *x-axis*. The left-most red dot is the typical amount of precipitation for January and the right-most is the typi-

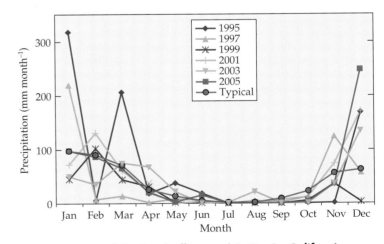

FIGURE A   Precipitation (millimeters) in Davis, California
Plotted are data for each month of every other year during the last decade and typical precipitation values (long-term averages).

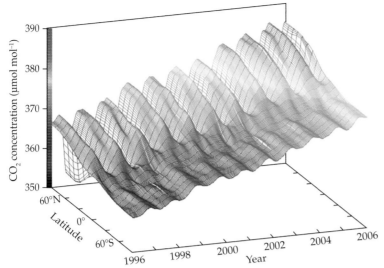

FIGURE B **Global distribution of atmospheric CO$_2$**

length and width of the horizontal plane and the height along the vertical plane. On the 3D graph (**Figure B**), the height along the vertical plane (and the color) represents the magnitude of the CO$_2$ concentration, whereas the length along the horizontal plane represents the year, and the width along the horizontal plane represents the latitude.

If you are having difficulty in projecting from the squiggly lines onto the various planes to determine the actual values on an axis, you are normal.

One last wrinkle is, when scientists conduct measurements, the data usually vary among samples in a given experiment and among experiments. To indicate the amount of variation in a particular data set, graphs may show bands that represent the variation around the mean (**Figure C**) or error bars (**Figure D**). This variation may be expressed in terms of the range of values or statistical measures such as the standard deviation, standard error of the mean, or 95% confidence limits.

cal amount of precipitation for December. The superscript "$^{-1}$" in the y-axis label designates that the label should be divided by this measure, and so the units are interpreted "millimeters/month" or "millimeters per month." The left-most red dot shows that the typical amount of precipitation in Davis, California is about 100 millimeters per month during January.

The *key* for the graph is shown in the white box. The key lists what each color or symbol represents. Dark blue diamonds represent the values during the year 1995.

In the graph of sea level (Figure C), the red line designates the average values, whereas the dark blue band around the line designates the standard deviation, a measure of variation around the average values. A wide band indicates that the data was highly variable. The smaller graph inset in the upper left-hand corner shows an expansion of the boxed portion in the larger graph.

Variations on a Theme

Sometimes a graph will present the values of two separate dependent variables at once. It may designate the values of the second dependent variable via the axis labels on the right-hand y-axis. On rare occasions, a third dependent variable might be represented via axis labels on a separate left-hand y-axis.

The only time it gets crazier than this is when graphs are in 3D. Put on your 3D goggles and try to imagine a vertical plane and a horizontal bottom plane that projects along and into the page. Take a piece of paper, fold it at a right angle, and place it on your desk: the surface of the paper facing upwards is the vertical plane, the surface of the paper on the desk is the horizontal plane; we are assessing the position along the

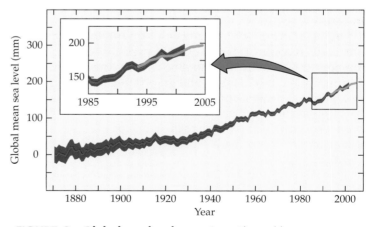

FIGURE C **Global sea level over time** The red line represents the yearly averages derived from tide gauges; the blue area indicates the standard deviation. The inset shows an expanded view of the past two decades.

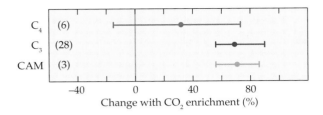

FIGURE D   Water-use efficiency in response to atmospheric CO_2 levels Water use efficiency is the amount of CO_2 assimilated per amount of water lost through transpiration. Shown are the changes in water use efficiency from leaves of plants grown at ambient CO_2 concentration ($\approx$365 ppm) to those grown at elevated CO_2 concentration ($\approx$550 ppm). A change of 100% would mean that growth at elevated CO_2 doubled water use efficiency. Circles and error bars designate means and confidence intervals, respectively.

In the graph of water use efficiencies (Figure D), the dependent variable, *Change with CO_2 enrichment*, is represented on the *x*-axis. For different types of plants (C_4, C_3, and CAM), the dots represent the average change that occurs when plants are exposed to elevated CO_2 atmospheres. The bars represent the variation in the results among the number of studies denoted in parentheses. For example, the water use efficiency of C_4 plants increased an average of 33% when grown at 550 ppm CO_2 in comparison to 365 ppm, but this response in 6 studies varied 95% of the time between a 16% decrease and a 73% increase.

This primer should get you started along the happy trail of graph reading. This textbook could have presented most of its information in tables instead of graphs, but graphs are so much more colorful and permit more intuitive detection of trends and relationships. For example, Figure C shows that global average sea levels have been rising steadily for over a century. In contrast, tables excel in presenting exact numbers immediately. This textbook strives to achieve the appropriate mix of tables and graphs that leaves no reader feeling data deprived.

ANSWERS
TO REVIEW QUESTIONS

CHAPTER 1
1. d
2. c, d
3. a, b, c
4. b
5. a, c

CHAPTER 2
1. e
2. a, b, c, d, e
3. c
4. a, c
5. a
6. a, d, e
7. d
8. b
9. b
10. a, c

CHAPTER 3
1. e
2. c, d, e
3. b
4. a
5. a, c, d
6. a
7. a, b, c, d, e
8. e
9. d
10. d
11. b, c
12. a, c, d, e
13. a, b, e
14. c

CHAPTER 4
1. e
2. e
3. e
4. b
5. c
6. b
7. b
8. c
9. d, e
10. c, d ,e
11. b, d, c, e, a
12. a
13. a, b, c, d, e

CHAPTER 5
1. a
2. e
3. c
4. a, b, c, d, e
5. c
6. a
7. e
8. e
9. c
10. c
11. b

CHAPTER 6
1. a, b, c, e
2. a, c, d, e
3. c, e
4. a
5. a, e
6. d
7. c
8. a
9. b
10. c
11. a
12. e
13. b
14. a, d
15. b, c, d
16. a, b, d, e
17. b, c, e
18. a, e

CHAPTER 7
1. a, c, d
2. c
3. d
4. a, b, c, d, e
5. e
6. a, b, d
7. d
8. a, b, c, d
9. a
10. a
11. e
12. b
13. a, b, e
14. a, b, d

CHAPTER 8
1. d
2. c
3. a, b
4. e
5. d
6. c
7. e
8. b
9. d
10. e
11. b
12. a, c
13. d
14. e

CHAPTER 9
1. b, c
2. b
3. b, d, e
4. e
5. c
6. b
7. a, d
8. e
9. c
10. e
11. a
12. a, b, c, d, e
13. a
14. d
15. d

CHAPTER 10

1. a, b, c, d, e
2. c
3. b
4. b
5. e
6. a, b, c, d, e
7. b
8. c
9. b
10. a
11. a, b, c
12. e
13. b
14. d
15. a, b, c, d, e
16. a
17. d
18. c
19. d
20. e
21. a
22. c

CHAPTER 11

1. d
2. e
3. a, b, d, e
4. a, b, c, d, e
5. a
6. a, b, c, d, e
7. a, b, c, d, e
8. c, d
9. b
10. b, c, d
11. a, b
12. d
13. d
14. d

CHAPTER 12

1. a, c, e
2. b
3. c
4. e
5. c
6. a, b, c, d
7. d
8. a

GLOSSARY

acid dissociation constant (pK_a) pH at which the proto-nated and deprotonated forms of a substance are present in equal amounts.

active site   Portion of an enzyme involved in the primary chemical reactions.

aerobic respiration   Breakdown of high-energy organic compounds that occurs in the presence of oxygen.

affinity   Degree to which an enzyme associates with a chemical substrate.

aggregates   Masses or clods of soil.

albedo   Percentage of incident sunlight reflected by an object.

ammonium (NH_4^+) Positively charged form of inorganic nitrogen that is available to organisms from the soil.

anaerobic respiration   Breakdown of high-energy organic compounds that occurs in the absence of oxygen.

angle of incidence   Angle between an incoming ray of sunlight and the line perpendicular to surface it strikes.

aphelion   Point at which an orbital body is farthest from the sun.

arc   Current flowing through a normally non-conductive medium.

ballast   Electronic circuit for arc lamps that provides high voltage to start the bulb and then a steady current to keep the bulb illuminated.

bequest value   Monetary worth of the satisfaction derived from preserving a natural resource for future generations.

biofuels   Substances derived from recently living organisms that release energy when combusted.

biomass   Biologically generated material.

biosphere   Portion of Earth that supports life.

blackbodies   Ideal emitters of electromagnetic radiation.

blooms   Population explosions of algae.

C_3 carbon fixation   Biochemical pathway that converts ATP and NADPH, which plants produce during the light reactions of photosynthesis, into sugars; also called the Calvin-Benson cycle and the light-independent reactions of photosynthesis.

C_4 carbon fixation   Biochemical pathway that converts CO_2 into an organic acid, which flows to a specialized part of the plant where the organic acid is broken down to release CO_2 at relatively high concentration for efficient conversion into sugars via C_3 carbon fixation.

cap-and-trade   Permit system in which those who have unused permits may sell them to other consumers.

capital costs   Costs of acquiring land, constructing facilities, and purchasing equipment.

carbon equivalents   Accounting for emissions of green-house gases through a conversion to the warming potential in comparison to CO_2 and putting it in mass units of the carbon in CO_2.

carbon sequestration   Accumulation of organic carbon compounds in soils and depletion of CO_2 from the atmosphere.

catalyst   Substance that promotes a chemical reaction without itself being consumed.

catalytic converter   Chamber in the exhaust system of an internal combustion engine that diminishes air pollution via the facilitation of three chemical reactions: (1) carbon monoxide to carbon dioxide; (2) unburnt fuel to carbon dioxide and water; (3) nitrous oxides to nitrogen and oxygen.

catalyzes   Facilitates a chemical reaction so that it proceeds faster than it would otherwise.

charismatic megafauna   Large animals with popular appeal.

chemical equation   Formula that list the names and relative amounts of reactants and products in a chemical reaction.

chloroplast Compartment within a plant cell in which photosynthesis is conducted.

climate Long-term trends in weather.

clinker Main component of cement; mostly crystals of calcium oxide silicates, $3\ CaO \cdot SiO_2$.

CO_2 acclimation The decline in the stimulation of net CO_2 assimilation and growth after longer exposures to CO_2 enrichment.

conservation of energy law Energy is neither created nor destroyed; it can only be transformed from one form into another.

controlled market Government manages major sectors of the economy, dictating the goods and services produced and the distribution of income.

convention Formal multilateral treaty negotiated under the auspices of an international organization and with a large number of participants.

Coriolis force Apparent force resulting from Earth's rotation.

correlation Relationship between two or more factors in which a change in one factor is associated with changes in the others.

cosine law Solar energy per unit area depends on the incoming solar energy times the cosine of the angle of incidence.

crassulacean acid metabolism Biochemical pathway that converts CO_2 into an organic acid, which is stored overnight in a specialized part of a plant cell; the next day while the stomata are closed to minimize water loss, this organic acid is broken down to release CO_2 for conversion into sugars via C_3 carbon fixation.

cryogenic liquid Chemical that evaporates at very low temperatures and thus must be stored at a very low temperature.

cyanobacteria Large group of bacteria that have the capacity to conduct photosynthesis.

decomposition Breakdown of complex organic matter to simpler compounds.

demand Desire for a good or service, as indicated by the willingness and ability to pay.

dendrochronology Study of tree rings to evaluate past environmental conditions.

denitrification Process through which microbes conduct anaerobic respiration and generate energy using nitrogen compounds such as nitrate and nitrite instead of oxygen, to accept electrons during the breakdown of organic compounds; this process may produce nitrous oxide.

derivative Measure of how a function instantaneously changes as its input changes.

diffraction Bending of electromagnetic radiation around small objects.

dilution through growth Concentration of a substance within an organism declines as the organism grows but fails to bring in additional amounts of substance from the environment.

direct costs Costs of materials, labor, and maintenance to produce a good or service.

discount rate Rate used to convert values of future costs or benefits to present ones.

eccentricity Extent to which Earth's orbit around the sun deviates from a perfect circle.

elasticity of demand Degree to which the demand for a good or services changes with price.

electrical potential Potential energy of a charge in an electrostatic field.

electromagnetic radiation Type of energy composed of particle waves that can propagate through a vacuum.

electrons Negatively charged sub-atomic particles.

elements Atoms having a unique number of protons (a sub-atomic particle) in the nucleus.

embrittlement Reaction of hydrogen with metals that weakens them.

emissity Efficiency with which a substance emits electromagnetic radiation, ranging from 0 (inefficient) to 1 (efficient).

endemic Common to a defined geographical area.

energy budget Tally of all the energy entering and leaving a system.

energy intensity Energy expended per unit of floor area in buildings.

engine displacement Volume swept by the pistons in an internal combustion engine; generally, the larger the displacement, the greater the power, but the lower the fuel efficiency.

enzymes Proteins that serve as catalysts for biochemical reactions.

epeirogeny Formation and distribution of continents on Earth.

equations of state Mathematical equations that relate density, pressure, temperature, and water balance.

equilibrium price Price at which the quantity desired of a good or service equals the quantity available.

equity premium Differences between the rate of return earned by a broad market index and that earned by a relatively risk-free security.

equity Composition and application of rules should be just and fair.

estoppel States cannot take a position inconsistent with a previous position, especially when other states have relied on the earlier representation.

eukaryotes Organisms composed of cells that have nuclei and other specialized compartments (organelles).

evapo-transpiration Process through which water changes phase from a liquid to a gas.

externality Situation in which a buyer and seller in an economic transaction do not necessarily bear all of the costs or reap all of the benefits of the transaction.

fats Class of biochemicals composed of three carbon chains of various lengths joined together at one end.

finite differences Method for solving differential equations by examining changes in a variable such as temperature over small increments of space or time.

flue gas Gas produced by combustion that exhausts to the atmosphere.

forcing factors Environmental parameters that influence climate.

fossil fuels   Substances (such as petroleum, coal, and natural gas) derived from organic matter deposited in the ground and subjected to heat and pressures for millennia that release energy when combusted.

free market   Part of a market economy in which the participants arrange transactions entirely by mutual consent; governments do not intervene directly in a free market but do participate in defining and enforcing basic rules.

free-riding   Situation in which someone does not pay for a good or service, yet benefits from others who do pay.

Gaia hypothesis   Theory that life on Earth has served as an active control system that stabilizes the physical environment and chemical composition of the planet.

gamma rays   Electromagnetic radiation with wavelengths shorter than 10 pm (1 pm = 10^{-12} m).

GCM   Acronym for either general circulation model or global climate model.

general circulation model   Mathematical model that simulates the atmospheric cycles and oceanic currents that influence climate.

geoengineering   Very large-scale modifications to Earth's environment.

geostrophic balance   Assumption that, above a friction layer, the Coriolis force is responsible for the horizontal differences in air pressure across a volume of air.

global climate model   Mathematical model that calculates climatic forcing factors over space and time to simulate climate in different locations.

Gondwana   Supercontinent that formed about 550 million years ago.

greenhouse gases   Gases in the atmosphere that transmit nearly visible light, but absorb strongly at certain wavelength in the infrared.

greenwashing   Placing an inordinate emphasis on enhancing one's reputation for environmental sustainability rather than taking substantial actions.

gymnosperms   Plants with seeds that cover the embryo in a desiccation-resistant cover.

half-life   Amount of time required before one-half of a radioactive isotope decays into another element.

herbivores   Organisms that eat plants.

host vectors   Organisms that transmit diseases.

hydrate   Chemical compound encased in a crystal lattice of water molecules.

hydrides   Chemical compound containing hydrogen covalently bound to a metal.

hydrocarbons   Chemical compounds containing only multiple hydrogen and carbon atoms connected by high-energy bonds.

hydrolysis   Breakdown of a large chemical compound through a reaction with water.

hydrostatic balance   Assumption that the force of gravity is responsible for the vertical differences in air pressure across this volume of air.

immobilization   Process through which soil microorganisms incorporate mineral nitrogen and thereby diminish its availability to other organisms.

impatience principle   People expect some compensation if they must wait for something.

infrared radiation   Electromagnetic radiation with a wavelength from about 0.7 μm to 1000 μm.

intergenerational equity   Principle that every generation deserves an equal opportunity of success and equal quality of life.

intermediate   Organic compounds that degrade over years.

inverse square law   Amount of solar energy striking an area decreases with the square of the distance from the sun.

invertebrates   Animals that lack a backbone.

isotopes   Forms of an element that have different numbers of neutrons in their nucleus.

Kaldor-Hicks optimality   Condition in which new projects or regulations may sacrifice the well-being of some people if these people receive compensation for the harm they suffer.

Kepler's second law   Gravitational bodies rotate through equal areas of space in equal times.

labile   Organic compounds that degrade relatively rapidly.

magnetometer   Sensitive scientific instrument that measures the strength and direction of the magnetic field in rocks.

marginal cost   Added cost of producing one more of an item.

marginal productivity of capital   Extent to which an extra dollar invested in resources today produces more than a dollar's worth of additional goods or services in the future.

marginal utility   Changes in people's well-being before and after some small change in resource allocations.

market   Common arena where firms or individuals voluntarily exchange goods and services.

market forces   Influence of supply and demand on the distribution of goods or services.

membrane permeability   Extent to which a biological membrane permits a substance to pass through it.

membrane   Barrier that separates biological compartments.

mesophyll   Leaf tissue active in photosynthesis.

meta-analysis   Statistical approach that combines results from many independent studies to increase experimental sample size and thus enhance the strength of statistical tests.

metabolism   Life-sustaining biochemical reactions.

methanogenesis   Soil microorganisms under anaerobic, flooded conditions harvest the energy released during the breakdown of organic carbon compounds into methane.

mineralization   Microbial breakdown of soil organic matter into inorganic nitrogen, notably, ammonium or nitrate.

minimum tillage   Agricultural management practices that leave large pieces of plant residues in a field and reduce turnover of top soil layers.

mitigation   Steps taken to make the anticipated changes less severe.

mixed economy Combines elements of both controlled and free markets.

Nash equilibrium Condition in which no person benefits from a change in strategy if the other people keep their strategy unchanged.

net CO_2 assimilation Difference between the CO_2 consumed during photosynthetic carbon fixation and that released during respiration.

net forcings Total effects of certain parameters on the climate.

neutrons Uncharged sub-atomic particles.

nitrate (NO_3^-) Negatively charged form of inorganic nitrogen that is available to organisms from the soil.

nitrification Microbial conversion of soil ammonium to nitrate.

noble gases Compounds that are gases at room temperature and are chemically inert.

nuclear fission Self-sustaining reaction in which radioisotopes disintegrate and release energy.

obliquity Axial tilt of Earth or the angle between its daily rotation and its orbital plane around the sun.

opportunity cost of capital Rate of return from potential private savings accounts or investments that might otherwise use the capital spent on a government project.

orbital plane Plane on which Earth orbits around the sun.

orogeny Process of mountain building on Earth.

oxidizing agent Substance that remove electrons from many other molecules.

paleomagnetism Study of residual magnetism in ancient rocks.

Pangaea Supercontinent formed about 300 million years ago.

Pareto optimality Condition in which no new projects or regulations would make someone better off without sacrificing the well-being of someone else.

partial pressure Pressure that one gas in a mixture of gases would exert if it were the only gas present.

perfect competition No one party unduly influences the price of a good or service.

perfect information Every participant has access to complete, up-to-date knowledge about the actions of other participants and about the goods or services available.

perfect market One that has both perfect competition and perfect information about the goods or services available..

perihelion Point at which an orbital body is closest to the sun.

pH Activity of protons in a solution and, thus, a measure of acidity or alkalinity.

photoelectric effect Emission of electrons from matter after it absorbs electromagnetic radiation.

photorespiration Biochemical pathway in plants in which a high-energy sugar is oxidized and CO_2 is initially released; the purpose of this pathway is controversial.

photosynthesis Biological process in plants and cyanobacteria that converts sunlight into chemical energy.

photovoltaic effect Process in which exposure of a certain material to electromagnetic radiation results in the buildup of electrical potential between electrodes in the material; solar cells use this effect to generate electricity from sunlight.

phyla Taxonomic classification that ranks below kingdom and above class; there are approximately 35 different animal phyla.

phytoplankton Microscopic algae that are suspended in the water column.

plankton Microscopic organism suspended in the water column.

precautionary principle Support of strong measures to avoid catastrophic outcomes even if their chance of occurrence seems remote.

precession Alignment of Earth's axis of diurnal rotation with its distance from the sun.

protocol Treaty that specifies substantive obligations to implement the broad objectives of a previous convention.

protons Positively charged sub-atomic particles.

proxy measures Indirect measurements such as the width of tree rings that provide an estimate of environmental parameters such as temperature.

publicists Experts in public international law.

quality-adjusted life year Estimate of the monetary value of a life based on changes in both human life expectancy and quality of life.

radio waves Electromagnetic radiation with wavelengths longer than 1 mm.

recalcitrant Organic compounds that degrade only over centuries.

refraction Bending of electromagnetic radiation as it passes through different media.

resolution Observable spatial or temporal detail.

respiration Biological process in all organisms that breaks down high-energy organic compounds to release energy for growth and maintenance.

Rodinia Supercontinent that formed near the equator about 1 billion years ago.

sequestration Process in which organic compounds become stored away from active chemical processes.

sinks Entities that remove materials such as greenhouse gases from the environment.

smelting Purifying iron or steel from iron ore.

soil solution Solution that fills the spaces between soil particles.

solar insolation Amount of solar energy reaching Earth per unit area.

solubility Extent to which a substance can dissolve in a solution.

sorbents Solid materials on which chemical substances attach.

sources Entities that release materials such as greenhouse gases to the environment.

sovereign state Political association with independent authority over a geographical area and the human population living there.

spectral ringing Difficulty in approximating functions via spectral transforms that may develop at the points where the variables change abruptly.

spectral transforms Method for solving differential equations by approximating a function as the sum of a sequence of sine and cosine functions.

spot market Market that sells commodities for cash and immediate delivery.

stomata Openings in the leaves of plants through which CO_2 enters and water escapes.

stratosphere Upper atmosphere between 10 km and 50 km above the Earth's surface.

sunspots Patches on the surface of the sun where temperatures are fluctuating.

supply Willingness and ability to provide a good or service.

tax-and-dividend Government redistributes some of the taxes it collects to encourage certain behaviors.

tectonic Related to the movement and deformation of Earth's crust.

thermals Process through which energy absorbed by Earth's surface heats the air that is adjacent to it, and this warmer air rises.

thermohaline circulation Large-scale ocean currents driven by density gradients in the water created by changes in the surface temperature and salinity.

torque Rotational force down a shaft; for example, the ability to turn the wheels of a vehicle.

transaction costs Costs incurred in making an economic exchange.

transpiration Loss of water vapor from plants.

troposphere Lower atmosphere between the Earth's surface and 10 km.

ultraviolet radiation Electromagnetic radiation with a wavelength from about 10 nm to 400 nm (1 nm = 10^{-9} m).

upwellings Upward currents that bring nutrients from the ocean depths up near the surface.

value of statistical life Estimate of the monetary value of a life based on what society is willing to pay for reducing each member's risk of fatality by a small amount.

wind turbines Mechanical devices that convert air movements into the rotation of electrical generators.

zooplankton Microscopic animals that are suspended in the water column.

REFERENCES

Aars, J., N. J. Lunn, and A. E. Derocher, eds. (2006) *Polar Bears: Proceedings of the 14th Working Meeting of the IUCN/SSC Polar Bear Specialist Group, 20–24 June 2005, Seattle, Washington, USA.*, International Union for Conservation of Nature and Natural Resources, Gland, Switzerland.

AB1109 (2007) *California Lighting Efficiency and Toxics Reduction Act*, California Assembly, Sacramento, CA, *http: //www.leginfo. ca.gov/pub/07-08/bill/asm/ab_1101-1150/ab_1109_bill_20071012_ chaptered.pdf.*

ACIA (2005) *Arctic Climate Impact Assessment*, Cambridge University Press, New York, *http: //www.acia.uaf.edu.*

ACIL Consulting (1999) *Study on Factors Impacting on Australia's National Average Fuel Consumption Levels to 2010*, Australian Greenhouse Office, *http: //www.greenhouse.gov.au/transport/ publications/pubs/impactingfactors.pdf.*

Adamec, L. (1997) Mineral nutrition of carnivorous plants: a review. *Botanical Reviews* 63: 273–295.

Adamson, K.-A. (2007) *2007 Automotive Infrastructure Survey*, Fuel Cell Today, *http: //www.fuelcelltoday.com/FuelCellToday/FCT-Files/FCTArticleFiles/Article_1187_Automotive%20Infrastructure. pdf.*

Addison, K. and M. Hiraga (2007) *Oil yields and characteristics.* Journey to Forever, *http: //journeytoforever.org/biodiesel_yield.html*, accessed Sept. 10, 2007.

Agrawala, S., ed. (2007) *Climate Change in the European Alps: Adapting Winter Tourism and Natural Hazards Management*, Organisation for Economic Co-operation and Development, Paris.

Agrell, J., E. P. McDonald, and R. L. Lindroth (2000) Effects of CO_2 and light on tree phytochemistry and insect performance. *Oikos* 88: 259–272.

Ainsworth, E. A., P. A. Davey, C. J. Bernacchi, O. C. Dermody, E. A. Heaton, D. J. Moore, P. B. Morgan, S. L. Naidu, H. S. Y. Ra, X. G. Zhu, P. S. Curtis, and S. P. Long (2002) A meta-analysis of elevated $[CO_2]$ effects on soybean (*Glycine max*) physiology, growth and yield. *Global Change Biology* 8: 695–709.

Ainsworth, E. A. and S. P. Long (2005) What have we learned from 15 years of free-air CO_2 enrichment (FACE)? A meta-analytic review of the responses of photosynthesis, canopy. *New Phytologist* 165: 351–371.

Ainsworth, E. A. and A. Rogers (2007) The response of photosynthesis and stomatal conductance to rising $[CO_2]$: mechanisms and environmental interactions. *Plant Cell and Environment* 30: 258–270.

Alberth, S. and C. Hope (2007) Climate modelling with endogenous technical change: Stochastic learning and optimal greenhouse gas abatement in the PAGE2002 model. *Energy Policy* 35: 1795–1807 DOI 10.1016/j.enpol.2006.05.015.

Aldy, J. E., S. Barrett, and R. N. Stavins (2003) Thirteen plus one: a comparison of global climate policy architectures. *Climate Policy* 3: 373–397 DOI10.1016/j.clipol.2003.09.004.

Alfe, D., M. J. Gillan, and G. D. Price (2007) Temperature and composition of the Earth's core. *Contemporary Physics* 48: 63–80 DOI 10.1080/00107510701529653.

Algaze, G. (2001) Initial social complexity in southwestern Asia—The Mesopotamian advantage. *Current Anthropology* 42: 199–233.

Ali, A. (1996) Vulnerability of Bangladesh to climate change and sea level rise through tropical cyclones and storm surges. *Water Air and Soil Pollution* 92: 171–179.

Allen, P. A. and J. L. Etienne (2008) Sedimentary challenge to Snowball Earth. *Nature Geoscience* 1: 817–825.

Alliance Technical Services (2007) *Hydrogen Workshop for Fleet Owners*, National Hydrogen Association, Washington, D.C., *http: // www.hydrogenassociation.org/general/fleet_Module5.pdf.*

Alsema, E. A., M. J. de Wild-Scholten, and V. M. Fthenakis. 2006. Environmental Impacts of PV Electricity Generation—a Critical Comparison of Energy Supply Options. Paper read at 21st European Photovoltaic Solar Energy Conference Sept. 4 – 8, 2006, at Dresden,Germany.

Alternative Fluorocarbons Environmental Acceptability Study (2006) *Production, Sales, and Emissions Data. http: //www.afeas. org/prodsales_download.html*, accessed January 18, 2007.

America Wind Energy Association (2008) *U.S. Wind Energy Projects. http: //www.awea.org/projects/*, accessed February 20, 2008.

American Planning Association (2005) *JAPA Article Calls on Planners to Help End Inaccuracies in Public Project Revenue Forecasting. http: //www.planning.org/newsreleases/2005/ftp040705.htm*, accessed July 1, 2008.

American Public Power Association (2007) *Public Power: Shining a Light on Public Service*, Washington, D.C., *http: //appanet.files. cms-plus.com/PDFs/PPFactSheet.pdf*.

American Public Transportation Association (2007) *Public Transportation Fact Book*, 58th Edition. Washington, D.C., *http: //www. apta.com/research/stats/factbook/documents/factbook07.pdf*.

American Public Transportation Association (2009) *Public Transportation Fact Book*, 60th Edition. Washington, D.C., *http://www. apta.com/research/stats/documents09/2009_apta_fact_book_with_ outer_covers.pdf*.

Amthor, J. S. (2001) Effects of atmospheric CO_2 concentration on wheat yield: Review of results from experiments using various approaches to control CO_2 concentration. *Field Crops Research* 73: 1–34.

An, F. and A. Sauer (2004) *Comparison of Passenger Vehicle Fuel Economy and Greenhouse Gas Emission Standards around the World*, Pew Center on Global Climate Change, Arlington, VA, *http: // www.pewclimate.org/docUploads/Fuel%20Economy%20and%20 GHG%20Standards_010605_110719.pdf*.

Anderson, D. L. (2007) *New Theory of the Earth*, Cambridge University Press, Cambridge, UK.

Anderson, S. H., G. E. Metius, and J. M. McClelland (2008) *Future Green Steelmaking Technologies*, Midrex Technologies Inc., Charlotte, NC, *http: //www.midrex.com/uploads/documents/ FutureGSTechnologies-Final2.pdf*.

Angel, R. (2006) Feasibility of cooling the Earth with a cloud of small spacecraft near the inner Lagrange point (L1). *Proceedings of the National Academy of Sciences of the United States of America* 103: 17184–17189.

Anheier, H., M. Glasius, and M. Kaldor (2001) Introducing Global Civil Society. In: *Global Civil Society 2001*, Anheier, H., M. Glasius, and M. Kaldor, eds. Sage Publications, London. pp. 3–22, *http://www.lse.ac.uk/Depts/global/Publications/ Yearbooks/2001/2001chapter1.pdf*.

Appliance Magazine.com (2006) *British Consumers Like Bigger Appliances*. Canon Communications, *http: //www.appliancemagazine.com/news.php?article=10033&zone=0&first=1*, accessed May 10, 2008.

Aquastat (2007) *FAO's Information System on Water and Agriculture*. Food and Agriculture Organization of the United Nations, *http: //www.fao.org/AG/AGL/aglw/aquastat/dbase/index.stm*, accessed May 29, 2007.

Archer, C. L. and M. Z. Jacobson (2005) Evaluation of global wind power. *Journal of Geophysical Research-Atmospheres* 110: D12110 1–20 DOI10.1029/2004jd005462.

Atkinson, A., V. Siegel, E. Pakhomov, and P. Rothery (2004) Long-term decline in krill stock and increase in salps within the Southern Ocean. *Nature* 432: 100–103.

Attiwill, P. M. and M. A. Adams (1993) Tansley Review No. 50: Nutrient cycling in forests. *New Phytologist* 124: 561–582.

Australian Bureau of Statistics (2006) *Survey of Motor Vehicle Use*, Canberra, Australia, *http: //www.abs.gov.au/ausstats/subscriber. nsf/log?openagent&92080_01%20nov%202004%20to%2031%20 oct%202005.pdf&9208.0&Publication&A2593C0478887F09C A2571E1001E8548&0&01%20Nov%202004%20to%2031%20 Oct%202005&07.09.2006&Latest*.

Australian Government (2008) *Sydney Opera House. http: //www. cultureandrecreation.gov.au/articles/sydneyoperahouse/*, accessed July 2, 2008.

Backlund, P., A. Janetos, and D. Schimel (2008) *The Effects of Climate Change on Agriculture, Land Resources, Water Resources, and Biodiversity in the United States*, Walsh, M., ed., U.S. Climate Change Science Program, Washington, D.C., *http://www.usda. gov/oce/global_change/files/CCSPFinalReport.pdf*.

Baer, F. (1992) Helmut E. Landsberg. *Memorial Tributes: National Academy of Engineering* 5: 153–158.

Bahn, Y. S. and F. A. Muhlschlegel (2006) CO_2 sensing in fungi and beyond. *Current Opinion in Microbiology* 9: 572–578.

Baker, C. S. and P. J. Clapham (2004) Modelling the past and future of whales and whaling. *Trends in Ecology & Evolution* 19: 365–371.

Baker, J. T. (2004) Yield responses of Southern US rice cultivars to CO_2 and temperature. *Agricultural and Forest Meteorology* 122: 129–137.

Baker, J. T., L. H. Allen, and K. J. Boote (1992) Response of rice to carbon dioxide and temperature. *Agricultural and Forest Meteorology* 60: 153–166.

Baker, T. L., S. R. Rae, J. E. Minor, and S. V. Connor (1973) *Water for the Southwest: Historical Survey and Guide to Historic Sites*, American Society of Civil Engineers, New York.

Balat, M. (2007) Production of biodiesel from vegetable oils: A survey. *Energy Sources Part a—Recovery Utilization and Environmental Effects* 29: 895–913.

Balmford, A., A. Manica, L. Airey, L. Birkin, A. Oliver, and J. Schleicher (2004) Hollywood, climate change, and the public. *Science* 305: 1713–1713.

Bannayan, M., K. Kobayashi, H. Y. Kim, M. Lieffering, M. Okada, and S. Miura (2005) Modeling the interactive effects of atmospheric CO_2 and N on rice growth and yield. *Field Crops Research* 93: 237–251.

Barbraud, C. and H. Weimerskirch (2006) Antarctic birds breed later in response to climate change. *Proceedings of the National Academy of Sciences of the United States of America* 103: 6248–6251.

Barrett, S. (2003) *Environment and Statecraft: the Strategy of Environmental Treaty-Making*, Oxford University Press, Oxford.

Barrett, S. (2008) Rethinking global climate change governance. *Economics Discussion Papers* 2: 2008–2031.

BassiriRad, H. (2000) Kinetics of nutrient uptake by roots: Responses to global change. *New Phytologist* 147: 155–169.

Basu, A. R., M. I. Petaev, R. J. Poreda, S. B. Jacobsen, and L. Becker (2003) Chondritic meteorite fragments associated with the Permian-Triassic boundary in Antarctica. *Science* 302: 1388–1392 DOI 10.1126/science.1090852.

Battelle Memorial Institute (2002) *Toward a Sustainable Cement Industry*, World Business Council for Sustainable Development, Geneva, *http://www.wbcsd.org/includes/getTarget. asp?type=d&id=ODU0NQ*.

BatteryUniversity.com (2007) *What's the Best Battery? http: //batteryuniversity.com/partone-3.htm*, accessed Aug. 18, 2007.

Battle, M., S. E. Mikaloff Fletcher, M. L. Bender, R. F. Keeling, A. C. Manning, N. Gruber, P. P. Tans, M. B. Hendricks, D. T. Ho, C. Simonds, R. Mika, and B. Paplawsky (2006) Atmospheric potential oxygen: New observations and their implications for some atmospheric and oceanic models. *Global Biogeochemical Cycles* 20: GB1010 1–15 DOI 10.1029/2005GB002534.

Baum, S. K. (2006) *Ocean/Atmosphere Circulation Modeling Projects. http: //stommel.tamu.edu/~baum/ocean_models.html*, accessed December 20, 2006.

Baumert, K. A., T. Herzog, and J. Pershing (2005) *Navigating the Numbers: Greenhouse Gas Data and International Climate Policy*, World Resources Institute, Washington, D.C., *http: //pdf.wri. org/navigating_numbers.pdf*.

BBC (2009) *Impartiality & Diversity of Opinion*. Editorial Guidelines in Full, *http: //www.bbc.co.uk/guidelines/editorialguidelines/ edguide/impariality/index.shtml*, accessed April 15, 2009.

BBC World Service Poll, GlobeScan, and Program on International Policy Attitudes (2007) *All Countries Need to Take Major Steps on Climate Change: Global Poll*, BBC World Service, London, *http: //news.bbc.co.uk/1/shared/bsp/hi/pdfs/25_09_07climatepoll. pdf*.

Bekker, A., H. D. Holland, P. L. Wang, D. Rumble, H. J. Stein, J. L. Hannah, L. L. Coetzee, and N. J. Beukes (2004) Dating the rise of atmospheric oxygen. *Nature* 427: 117–120 DOI 10.1038/nature02260.

Bellavance, F., G. Dionne, and M. Lebeau (2007) *The Value of a Statistical Life: A Meta-Analysis with a Mixed Effects Regression Model*, Canada Research Chair in Risk Management, Working paper 06-12. HEC Montreal, Montreal, *http: //neumann.hec.ca/gestiondesrisques/06-12.pdf*.

Below, F. and P. Brandau (2001) *How Much Nitrogen Does Corn Need?* Agronomy Day 2001, *http: //www.cropsci.uiuc.edu/agronomyday/2001/tours/nitrogen-need/*, accessed May 6, 2002.

Bemis, B. E., H. J. Spero, J. Bijma, and D. W. Lea (1998) Reevaluation of the oxygen isotopic composition of planktonic foraminifera: Experimental results and revised paleotemperature equations. *Paleoceanography* 13: 150–160.

Bender, F. A. M., H. Rodhe, R. J. Charlson, A. M. L. Ekman, and N. Loeb (2006) 22 views of the global albedo—comparison between 20 GCMs and two satellites. *Tellus Series a-Dynamic Meteorology and Oceanography* 58: 320–330.

Bender, M. L., D. T. Ho, M. B. Hendricks, R. Mika, M. O. Battle, P. P. Tans, T. J. Conway, B. Sturtevant, and N. Cassar (2005) Atmospheric O_2/N_2 changes, 1993–2002: Implications for the partitioning of fossil fuel CO_2 sequestration. *Global Biogeochemical Cycles* 19: Gb4017 1–16 DOI 10.1029/2004GB002410.

Benedick, R. E. (2007) Science, Diplomacy, and the Montreal Protocol. In: *Encyclopedia of Earth*, Cleveland, C. J., ed., Environmental Information Coalition, Washington, D.C., *http: //www.eoearth.org/article/Science,_diplomacy,_and_the_Montreal_Protocol*.

Berdichevsky, G., K. Kelty, J. Straubel, and E. Toomre (2007) *The Tesla Roadster Battery System*, Tesla Motors, San Carlos, CA, *http: //www.teslamotors.com/display_data/TeslaRoadsterBatterySystem.pdf*.

Berger, A. and M. F. Loutre (1991) Insolation values for the climate of the last 10000000 years. *Quaternary Science Reviews* 10: 297–317.

Berger, A. and M. F. Loutre (2002) An exceptionally long interglacial ahead? *Science* 297: 1287–1288.

Beringer, J., L. B. Hutley, N. J. Tapper, A. Coutts, A. Kerley, and A. P. O'Grady (2003) Fire impacts on surface heat, moisture and carbon fluxes from a tropical savanna in northern Australia. *International Journal of Wildland Fire* 12: 333–340.

Berner, R. A. (2003) The long-term carbon cycle, fossil fuels and atmospheric composition. *Nature* 426: 323–326.

Bernstein, L., J. Roy, K. C. Delhotal, J. Harnisch, R. Matsuhashi, L. Price, K. Tanaka, E. Worrell, F. Yamba, and Z. Fengqi (2007) Industry. In: *Climate Change 2007: Mitigation. Contribution of Working Group III to the Fourth Assessment Report of the Intergovernmental Panel on Climate Change*, Metz, B., O. R. Davidson, P. R. Bosch, R. Dave, and L. A. Meyer, eds. Cambridge University Press, Cambridge. pp. 447–496.

Bertani, R. (2005) World geothermal power generation in the period 2001–2005. *Geothermics* 34: 651–690.

Bertoldi, P. and B. Atanasiu (2007) *Electricity Consumption and Efficiency Trends in the Enlarged European Union*, European Commission, Ispra, Italy, *http: //sunbird.jrc.it/energyefficiency/pdf/EnEff%20Report%202006.pdf*.

Bindoff, N. L., J. Willebrand, V. Artale, A. Cazenave, J. Gregory, S. Gulev, K. Hanawa, C. Le Quéré, S. Levitus, Y. Nojiri, C. K. Shum, L. D. Talley, and A. Unnikrishnan (2007) Observations: Oceanic climate change and sea level. In: *Climate Change 2007: The Physical Science Basis. Contribution of Working Group I to the Fourth Assessment Report of the Intergovernmental Panel on Climate Change*, Solomon, S., D. Qin, M. Manning, Z. Chen, M. Marquis, K. B. Averyt, M. Tignor, and H. L. Miller, eds. Cambridge University Press, Cambridge. pp. 385–432.

Bjornstad, D. (2004) *Cost-Benefit Analysis*. National Center for Environmental Decision-Making Research, *http: //sunsite.utk.edu/ncedr/tools/othertools/costbenefit/lead.htm*, accessed August 4, 2008.

Bloom, A. J. (1997) Nitrogen as a limiting factor: crop acquisition of ammonium and nitrate. In: *Ecology in Agriculture*, Jackson, L. E., ed., Academic Press, San Diego. pp. 145–172.

Bloom, A. J. (2006) Rising carbon dioxide concentrations and the future of crop production. *Journal of the Science of Food and Agriculture* 86: 1289–1291.

Bloom, A. J., R. M. Caldwell, J. Finazzo, R. L. Warner, and J. Weissbart (1989) Oxygen and carbon dioxide fluxes from barley shoots depend on nitrate assimilation. *Plant Physiology* 91: 352–356.

Bloom, A. J., F. S. Chapin, III, and H. A. Mooney (1985) Resource limitation in plants: an economic analogy. *Annual Review of Ecology and Systematics* 16: 363–392.

Bloom, A. J., D. R. Smart, D. T. Nguyen, and P. S. Searles (2002) Nitrogen assimilation and growth of wheat under elevated carbon dioxide. *Proceedings of the National Academy of Sciences of the United States of America* 99: 1730–1735.

Bloom, A. J., S. S. Sukrapanna, and R. L. Warner (1992) Root respiration associated with ammonium and nitrate absorption and assimilation by barley. *Plant Physiology* 99: 1294–1301.

Bloom, A. J. and J. H. Troughton (1979) High productivity and photosynthetic flexibility in a CAM plant. *Oecologia* 38: 35–43.

Bloom, A. J., M. A. Zwieniecki, J. B. Passioura, L. B. Randall, N. M. Holbrook, and D. A. St. Clair (2004) Water relations under root chilling in a sensitive and tolerant tomato species. *Plant, Cell and Environment* 27: 971–979.

Bluestein, J. (2006) *Hearing on the Impact of Clean Air Regulations on Natural Gas Prices*, Committee on Environment and Public Works, U.S. Senate, Washington, D.C., *http: //epw.senate.gov/109th/Bluestein_Testimony.pdf*.

Bodiselitsch, B., C. Koeberl, S. Master, and W. U. Reimold (2005) Estimating duration and intensity of Neoproterozoic snowball glaciations from Ir anomalies. *Science* 308: 239–242.

Böning, C. W., M. Scheinert, J. Dengg, A. Biastoch, and A. Funk (2006) Decadal variability of subpolar gyre transport and its reverberation in the North Atlantic overturning. *Geophysical Research Letters* 33: L21S01 1–5 DOI 10.1029/2006GL026906.

Borenstein, S. (2008) An American life worth less today. *Associated Press*, July 10, 2008.

Bossel, U., B. Eliasson, and G. Taylor (2007) *The Future of the Hydrogen Economy: Bright or Bleak?*, European Sustainable Energy Forum, Lucerne, *http: //www.efcf.com/e/reports/E02_Hydrogen_Economy_Report.pdf*.

Bouvier-Soumagnac, Y. and J. C. Duplessy (1985) Carbon and oxygen isotopic composition of planktonic-foraminifera from laboratory culture, plankton tows and recent sediment: implications for the reconstruction of paleoclimatic conditions and of the global carbon-cycle. *Journal of Foraminiferal Research* 15: 302–320.

Bowles, C. (1971) *Promises to Keep: My Years in Public Life, 1941–1969*, 1st Edition. Harper & Row, New York.

Boyd, P. (2004) Ironing out algal issues in the southern ocean. *Science* 304: 396–397.

BP (2007) *BP Statistical Review of World Energy*, BP, London, *http: //www.bp.com/multipleimagesection.do?categoryId=9017892&contentId=7033503*.

BP (2008) *BP Statistical Review of World Energy, June 2008*, BP, London, *http: //www.bp.com/liveassets/bp_internet/globalbp/globalbp_uk_english/reports_and_publications/statistical_energy_review_2008/STAGING/local_assets/downloads/pdf/statistical_review_of_world_energy_full_review_2008.pdf*.

Bradshaw, W. E. and C. M. Holzapfel (2006) Evolutionary response to rapid climate change. *Science* 312: 1477–1478.

Branch, T. A., K. Matsuoka, and T. Miyashita (2004) Evidence for increases in Antarctic blue whales based on Bayesian modelling. *Marine Mammal Science* 20: 726–754.

Brandao-Burch, A., J. C. Utting, I. R. Orriss, and T. R. Arnett (2005) Acidosis inhibits bone formation by osteoblasts in vitro by preventing mineralization. *Calcified Tissue International* 77: 167–174.

Bray, D. and H. von Storch (2008) *A Response to RealClimate Concerning A New Survey of Climate Scientists.* Prometheus, *http://sciencepolicy.colorado.edu/prometheus/a-response-to-realclimate-concerning-a-new-survey-of-climate-scientists-4635*, accessed January 1, 2009.

Brewer, P. G., E. Peltzer, I. Aya, P. Haugan, R. Bellerby, K. Yamane, R. Kojima, P. Walz, and Y. Nakajima (2004) Small scale field study of an ocean CO_2 plume. *Journal of Oceanography* 60: 751–758.

Brinkman, N., M. Wang, T. Weber, and T. Darlington (2005) *Well-to-Wheels Analysis of Advanced Fuel/Vehicle Systems — A North American Study of Energy Use, Greenhouse Gas Emissions, and Criteria Pollutant Emissions*, General Motors, *http://www.transportation.anl.gov/pdfs/TA/339.pdf*.

Brohan, P., J. J. Kennedy, I. Harris, S. F. B. Tett, and P. D. Jones (2006) Uncertainty estimates in regional and global observed temperature changes: A new data set from 1850. *Journal of Geophysical Research-Atmospheres* 111:D12106 1–21 DOI 10.1029/2005JD006548.

Brook, E. J. (2005) Tiny bubbles tell all. *Science* 310: 1285–1287.

Brown, L. R. (2006) *Plan B 2.0: Rescuing a Planet under Stress and a Civilization in Trouble*, 1st Edition. W. W. Norton & Co., New York.

Bryden, H. L., H. R. Longworth, and S. A. Cunningham (2005) Slowing of the Atlantic meridional overturning circulation at 25 degrees N. *Nature* 438: 655–657.

Budd, G. E. and S. Jensen (2000) A critical reappraisal of the fossil record of the bilaterian phyla. *Biological Reviews* 75: 253–295.

Buesseler, K. O. and P. W. Boyd (2003) Will ocean fertilization work? *Science* 300: 67–68.

Buick, R. (2008) When did oxygenic photosynthesis evolve? *Philosophical Transactions of the Royal Society B-Biological Sciences* 363: 2731–2743 DOI 10.1098/rstb.2008.0041.

Bullock, M. A. and D. H. Grinspoon (2001) The recent evolution of climate on Venus. *Icarus* 150: 19–37 DOI 10.1006/icar.2000.6570.

Bureau of Economic Analysis (2008) *Gross Output by Industry.* U.S. Department of Commerce, Gross Domestic Product by Industry Accounts, *http://www.bea.gov/industry/gpotables/gpo_action.cfm?anon=70583&table_id=22079&format_type=0*, accessed May 20, 2008.

Bureau of Labor Statistics (2007) *Consumer Price Index.* U.S. Department of Labor, *http://data.bls.gov/cgi-bin/surveymost*, accessed Nov. 15, 2007.

Bureau of Transportation Statistics (2002) *Table 1-4: Public Road and Street Mileage in the United States by Type of Surface (Thousands of miles).* Department of Transportation, *http://www.bts.gov/publications/national_transportation_statistics/2002/html/table_01_04.html*, accessed July 5, 2007.

Bureau of Transportation Statistics (2007) *National Transportation Statistics*, U.S. Department of Transportation, Washington, D.C., *http://www.bts.gov/publications/national_transportation_statistics/pdf/entire.pdf*.

Bureau of Transportation Statistics (2009) *National Transportation Statistics*, U.S. Department of Transportation, Washington, D.C., *http://www.bts.gov/publications/national_transportation_statistics/pdf/entire.pdf*.

Burney, N. (2008) *International Law: a Brief Primer for Informational Purposes only.* The Burney Law Firm, *http://www.burneylawfirm.com/international_law_primer.htm#0_Top*, accessed Sept. 24, 2008.

Burnham, A., M. Wang, and Y. Wu (2007) *Development and Applications of GREET 2.7—The Transportation Vehicle-Cycle Model*, Argonne National Laboratories, Argonne, IL, *http://www.transportation.anl.gov/software/GREET/greet_2-7_beta.html*.

Bush, G. W. (2001) *Text of a Letter from the President to Senators Hagel, Helms, Craig, and Roberts.* Office of the Press Secretary for the President, *http://www.gcrio.org/OnLnDoc/pdf/bush_letter010313.pdf*, accessed April 10, 2009.

Butler, J. H., M. Battle, M. L. Bender, S. A. Montzka, A. D. Clarke, E. S. Saltzman, C. M. Sucher, J. P. Severinghaus, and J. W. Elkins (1999) A record of atmospheric halocarbons during the twentieth century from polar firn air. *Nature* 399: 749–755.

Butler, R. F. (1992) *Paleomagnetism: Magnetic Domains to Geologic Terranes*, Blackwell Scientific, Boston, *http://www.geo.arizona.edu/Paleomag/book/*.

Caldeira, K. and M. E. Wickett (2005) Ocean model predictions of chemistry changes from carbon dioxide emissions to the atmosphere and ocean. *Journal of Geophysical Research-Oceans* 110: C09s04 1–12 DOI 10.1029/2004JC002671.

California Department of Food and Agriculture (2007) *California Agricultural Resource Directory 2006*, CDFA, Sacramento, CA.

California Energy Commission (2006) *Inventory of California Greenhouse Gas Emissions and Sinks: 1990 to 2004*, State of California, Sacramento, CA, *http://www.energy.ca.gov/2005publications/CEC-600-2005-025/index.html*.

California Energy Commission (2007) *Levelized Costs of Electricity Generation by Resource Type. http://www.energy.ca.gov/electricity/levelized_costs.html*, accessed February 21, 2008.

California Energy Commission (2008) *California's Major Sources of Power. http://www.energy.ca.gov/html/energysources.html*, accessed June 12, 2008.

California Independent System Operator Corporation (2008) *Market Issues & Performance: 2007 Annual Report*, Folsom, CA, *http://www.caiso.com/1f9c/1f9c8b49e9f0.pdf*.

Callendar, G. S. (1938) The artificial production of carbon dioxide and its influence on temperature. *Quarterly Journal of the Royal Meteorological Society* 64: 223–240.

Camp, J. M. and C. B. Francis (1920) *The Making, Shaping and Treating of Steel*, 2nd Edition. Carnegie Steel Company, Pittsburgh, PA, *http://books.google.com/books?id=P9MxAAAAMAAJ*.

Campbell, G. (2007) *New Coal Power Plants—A Global View.* Commodities Now, *http://www.commodities-now.com/content/market-news/market-news-20070730141838.php?PHPSESSID=6c208de26b605cde1b43df3aa135d0ad*, accessed Oct. 1, 2007.

Canadian Centre for Climate Modelling and Analysis (2006) *Models. http://www.cccma.ec.gc.ca/models/models.shtml*, accessed December 30, 2006.

Canfield, D. E. (2005) The early history of atmospheric oxygen: Homage to Robert A. Garrels. *Annual Review of Earth and Planetary Sciences* 33: 1–36 DOI 10.1146/annurev.earth.33.092203.122711.

Capoor, K. and P. Ambrosi (2008) *State and Trends of the Carbon Market 2008*, The World Bank, Washington, D.C., *http://siteresources.worldbank.org/NEWS/Resources/State&Trendsformatted06May10pm.pdf*.

Carlin, A. (1968) Grand Canyon Controversy: Lessons for federal cost-benefit practices. *Land Economics* 44: 218–227.

Carr, L. and R. Mendelsohn (2003) Valuing coral reefs: A travel cost analysis of the Great Barrier Reef. *Ambio* 32: 353–357.

Cazenave, A. and R. S. Nerem (2004) Present-day sea level change: Observations and causes. *Reviews of Geophysics* 42: RG3001 DOI 10.1029/2003RG000139.

Cement Sustainability Initiative (2005) *Guidelines for the Selection and Use of Fuels and Raw Materials in the Cement Manufacturing Process*, World Business Council for Sustainable Development, Geneva, *http://www.wbcsd.org/DocRoot/Vjft3qGjo-1v6HREH7jM6/tf2-guidelines.pdf*.

Central Intelligence Agency (2007) *The World Factbook: U.S. https://www.cia.gov/library/publications/the-world-factbook/geos/us.html*, accessed July 5, 2007.

Cerling, T. E., J. M. Harris, B. J. MacFadden, M. G. Leakey, J. Quade, V. Eisenmann, and J. R. Ehleringer (1997) Global vegetation change through the Miocene/Pliocene boundary. *Nature* 389: 153–158.

Cerling, T. E., J. M. Harris, B. J. MacFadden, J. Quade, M. G. Leakey, V. Eisenmann, and J. R. Ehleringer (1998) Miocene/Pliocene shift: one step or several? Reply. *Nature* 393: 127.

Chalk, S. G. and J. E. Miller (2006) Key challenges and recent progress in batteries, fuel cells, and hydrogen storage for clean energy systems. *Journal of Power Sources* 159: 73–80.

Chaplin, M. (2004) *The Use of Enzymes in Starch Hydrolysis.* South Bank University, *http://www.lsbu.ac.uk/biology/enztech/starch.html*, accessed Sept. 4, 2007.

Chaudhuri, U. N., R. B. Burnett, M. B. Kirkham, and E. T. Kanemasu (1986) Effect of carbon dioxide on sorghum yield, root growth, and water use. *Agricultural and Forest Meteorology* 37: 109–122.

Chen, Y. H. and R. G. Prinn (2006) Estimation of atmospheric methane emissions between 1996 and 2001 using a three-dimensional global chemical transport model. *Journal of Geophysical Research-Atmospheres* 111: D10307 1–25 DOI 10.1029/2005JD006058.

Church, J. A. and N. J. White (2006) A 20th century acceleration in global sea-level rise. *Geophysical Research Letters* 33: L01602 1–4 DOI 10.1029/2005GL024826.

Circum-Arctic Resource Appraisal Assessment Team (2008) *Estimates of Undiscovered Oil and Gas North of the Arctic Circle*, Fact Sheet 2008-3049. U.S. Geological Survey, Menlo Park, CA, *http://pubs.usgs.gov/fs/2008/3049/fs2008-3049.pdf*.

Clawson, M. and J. L. Knetsch (1966) *Economics of Outdoor Recreation*, Published for Resources for the Future by Johns Hopkins Press, Baltimore.

Climate Action Network (2008) *Welcome to the Climate Action Network. http://www.climatenetwork.org/*, accessed November 19, 2008.

CNET (2008) *The Chart: HDTV Power Consumption Compared.* CNET Networks, *http://reviews.cnet.com/4520-6475_7-6400401-3.html?tag=arw*, accessed May 10, 2008.

CNW Marketing Research (2007) *"Dust to Dust" Automotive Energy Report*, Bandon, OR, *http://cnwmr.com/nss-folder/automotiveenergy/DUST%20PDF%20VERSION.pdf*.

Collins, W. D., C. M. Bitz, M. L. Blackmon, G. B. Bonan, C. S. Bretherton, J. A. Carton, P. Chang, S. C. Doney, J. J. Hack, T. B. Henderson, J. T. Kiehl, W. G. Large, D. S. McKenna, B. D. Santer, and R. D. Smith (2006) The Community Climate System Model version 3 (CCSM3). *Journal of Climate* 19: 2122–2143.

ColoradoENERGY (2008) *R-Value Table. http://www.coloradoenergy.org/procorner/stuff/r-values.htm*, accessed April 28, 2008.

Commission of the European Communities (2006) *Implementing the Community Strategy to Reduce CO$_2$ Emissions from Cars: Sixth annual Communication on the effectiveness of the strategy*, COM(2006) 463 final. Brussels, *http://ec.europa.eu/environment/co2/pdf/com_2006_463_en.pdf*.

Conant, R. T., R. A. Drijber, M. L. Haddix, W. J. Parton, E. A. Paul, A. F. Plante, J. Six, and J. M. Steinweg (2007) Sensitivity of decomposition to warming varies with organic matter quality. *Global Change Biology* 14: 868–877.

Congressional Budget Office (2003) *The Economics of Climate Change: A Primer*, U.S. Congressional Budget Office, Washington, D.C., *http://www.cbo.gov/ftpdocs/41xx/doc4171/04-25-ClimateChange.pdf*.

Consumer Energy Center (2006) *Compressed Natural Gas (CNG) as a Transportation Fuel.* California Energy Commission, *http://www.consumerenergycenter.org/transportation/afvs/cng.html*, accessed April 20, 2009.

Copper Development Association (2008) *Introduction to Premium Efficiency Motors. http://www.copper.org/applications/electrical/energy/motor_text.html*, accessed April 15, 2008.

Cornwall Alliance for the Stewardship of Creation (2009) *Evangelical Leaders Announce Landmark New Strategy for "Responsible & Balanced" Environmental Care http://www.cornwallalliance.org/press/read/evangelical-leaders-announce-landmark-new-strategy-for-responsible-and-balanced-environmental-care/*, accessed April 14, 2009.

Corrège, T. (2006) Sea surface temperature and salinity reconstruction from coral geochemical tracers. *Palaeogeography Palaeoclimatology Palaeoecology* 232: 408–428.

Corton, T. M., J. B. Bajita, F. S. Grospe, R. R. Pamplona, C. A. Assis, R. Wassmann, R. S. Lantin, and L. V. Buendia (2000) Methane emission from irrigated and intensively managed rice fields in Central Luzon (Philippines). *Nutrient Cycling in Agroecosystems* 58: 37–53.

Cotrufo, M. F., P. Ineson, and A. Scott (1998) Elevated CO$_2$ reduces the nitrogen concentration of plant tissues. *Global Change Biology* 4: 43–54.

Cousins, A. B. and A. J. Bloom (2003) Influence of elevated CO$_2$ and nitrogen nutrition on photosynthesis and nitrate photoassimilation in maize (*Zea mays* L.). *Plant, Cell and Environment* 26: 1525–1530.

Cousins, A. B. and A. J. Bloom (2004) Oxygen consumption during leaf nitrate assimilation in a C$_3$ and C$_4$ plant: the role of mitochondrial respiration. *Plant, Cell & Environment* 27: 1537–1545.

Cowan, I. R. and G. D. Farquhar (1977) Stomatal function in relation to leaf metabolism and environment. *Symposia of Society for Experimental Biology* 31: 471–505.

Cowen, T. (2007) Caring about the distant future: why it matters and what it means. *University of Chicago Law Review* 74: 5–40.

Cox, P. M., R. A. Betts, C. D. Jones, S. A. Spall, and I. J. Totterdell (2000) Acceleration of global warming due to carbon-cycle feedbacks in a coupled climate model. *Nature* 408: 184–187.

Craig, A. P., R. Jacob, B. Kauffman, T. Bettge, J. Larson, E. Ong, C. Dingo, and Y. He (2005) Cpl6: The new extensible, high performance parallel coupler for the Community Climate System Model. *International Journal of High Performance Computing Applications* 19: 309–327.

Crew, M. A., C. S. Fernando, and P. R. Kleindorfer (1995) The theory of peak-load pricing: a survey. *Journal of Regulatory Economics* 8: 215–248.

Criddle, E. J. (2004) The Vienna Convention on the Law of Treaties in U.S. treaty interpretation. *Virginia Journal of International Law* 44: 431–500.

Croxall, J. P., P. N. Trathan, and E. J. Murphy (2002) Environmental change and Antarctic seabird populations. *Science* 297: 1510–1514.

Crutzen, P. J. (2006) Albedo enhancement by stratospheric sulfur injections: A contribution to resolve a policy dilemma? *Climatic Change* 77: 211–219 DOI 10.1007/s10584-006-9101-y.

Curlin, J., G. Bankobeza, and S. Stone (2005) Summary of Workshop 3d: Analyzing the Compliance and Enforcement Mechanisms of the Montreal Protocol. In: *7th International Conference on Environmental Compliance and Enforcement*, International Network for Environmental Compliance and Enforcement,

Marrakech, Morroco. *http://inece.org/conference/7/vol2/41_Wkshop3D.pdf.*

Curry, J. A., R. J. Webster, and G. J. Holland (2006) Mixing politics and science in testing the hypothesis that greenhouse warming is causing a global increase in hurricane intensity. *Bulletin of the American Meteorological Society* 87: 1025–1037.

Curtis, P. S. and X. Z. Wang (1998) A meta-analysis of elevated CO_2 effects on woody plant mass, form, and physiology. *Oecologia* 113: 299–313.

Dai, A., K. E. Trenberth, and T. R. Karl (1998) Global variations in droughts and wet spells: 1900–1995. *Geophysical Research Letters* 25: 3367–3370.

Danish Wind Industry Association (2003) *Wind Speed Calculator.* *http://www.windpower.org/en/tour/wres/calculat.htm,* accessed February 17, 2008.

Darwin, C. (1875) *Insectivorous Plants,* John Murray, London.

Davidson, E. A. and I. A. Janssens (2006) Temperature sensitivity of soil carbon decomposition and feedbacks to climate change. *Nature* 440: 165–173.

Davis, C. (2007) *Sustainable Transport in the Developing World.* World Resources Institute, EarthTrends, *http://earthtrends.wri.org/updates/node/135,* accessed July 7, 2007.

Davis, S. C. and S. W. Diegel (2007) *Transportation Energy Data Book,* 26th Edition. Oak Ridge National Laboratory, Oak Ridge, TN, *http://cta.ornl.gov/data/tedb26/Edition26_Full_Doc.pdf.*

Davis, S. C., S. W. Diegel, and R. G. Boundy (2008) *Transportation Energy Data Book,* 27th Edition. U.S. Department of Energy, Oak Ridge, TN, *http://cta.ornl.gov/data/tedb27/Edition27_Full_Doc.pdf.*

de Almeida, A. T., P. Fonseca, H. Falkner, and P. Bertoldi (2003) Market transformation of energy-efficient motor technologies in the EU. *Energy Policy* 31: 563–575.

De Costa, W. A. J. M., W. M. W. Weerakoon, H. M. L. K. Herath, and R. M. I. Abeywardena (2003) Response of growth and yield of rice (*Oryza sativa*) to elevated atmospheric carbon dioxide in the subhumid zone of Sri Lanka. *Journal of Agronomy and Crop Science* 189: 83–95.

de Graaff, M. A., K. J. van Groenigen, J. Six, B. Hungate, and C. van Kessel (2006) Interactions between plant growth and soil nutrient cycling under elevated CO_2: a meta-analysis. *Global Change Biology* 12: 2077–2091.

de Tocqueville, A. (1835) *Democracy in America,* 2 vols, Saunders and Otley, London, *http://uclibs.org/PID/126171.*

Defa (2007) *Passenger Transport Emissions Factors,* Department for Environment, Food and Rural Affairs. Crown, London, *http://www.defra.gov.uk/environment/business/envrp/pdf/passenger-transport.pdf.*

DeLucia, E. H., D. J. Moore, and R. J. Norby (2005) Contrasting responses of forest ecosystems to rising atmospheric CO_2: Implications for the global C cycle. *Global Biogeochemical Cycles* 19: GB3006 1–9 DOI 10.1029/2004GB002346.

Demirbas, A. (2005) Estimating of structural composition of wood and non-wood biomass samples. *Energy Sources* 27: 761–767.

Denman, K. L., G. Brasseur, A. Chidthaisong, P. Ciais, P. M. Cox, R. E. Dickinson, D. Hauglustaine, C. Heinze, E. Holland, D. Jacob, U. Lohmann, S. Ramachandran, P. L. da Silva Dias, S. C. Wofsy, and X. Zhang (2007) Couplings between changes in the climate system and biogeochemistry. In: *Climate Change 2007: The Physical Science Basis. Contribution of Working Group I to the Fourth Assessment Report of the Intergovernmental Panel on Climate Change,* Solomon, S., D. Qin, M. Manning, Z. Chen, M. Marquis, K. B. Averyt, M. Tignor, and H. L. Miller, eds. Cambridge University Press, Cambridge. pp. 499–587.

Derocher, A. E., N. J. Lunn, and I. Stirling (2004) Polar bears in a warming climate. *Integrative and Comparative Biology* 44: 163–176.

Development Data Group. The World Bank (2007) *World Development Indicators Online,* World Resources Institute, EarthTrends. The World Bank, Washington, D.C., *http://earthtrends.wri.org/searchable_db/index.php?theme=5&variable_ID=640&action=select_countries.*

Diamond, J. M. (2006) *Collapse: How Societies Choose to Fail or Succeed,* Penguin, New York.

Division of Ocean Affairs and the Law of the Sea (2008) *Oceans and Law of the Sea.* United Nations, *http://www.un.org/depts/los/convention_agreements/convention_overview_convention.htm,* accessed Oct. 21, 2008.

Dixon, K. W., J. S. Pate, and W. J. Bailey (1980) Nitrogen nutrition of the tuberous sundew *Drosera erythrorhiza* Lindl. with special reference to catch of arthropod fauna by its glandular leaves. *Australian Journal of Botany* 28: 283–297.

Dockins, C., K. Maguire, N. Simon, and M. Sullivan (2004) *Value of Statistical Life Analysis and Environmental Policy: A White Paper,* U.S. Environmental Protection Agency, Washington, D.C., *http://yosemite.epa.gov/ee/epa/eerm.nsf/vwRepNumLookup/EE-0483?OpenDocument.*

Doorman, G. L. (2005) Capacity subscription: Solving the peak demand challenge in electricity markets. *IEEE Transactions on Power Systems* 20: 239–245 DOI 10.1109/Tpwrs.2004.841230.

Doran, P. T. and M. K. Zimmerman (2009) Examining the scientific consensus on climate change. *EOS Trans. AGU* 90: 21–22 DOI 10.1029/2009EO030002.

Dore, M. H. I. (2005) Climate change and changes in global precipitation patterns: What do we know? *Environment International* 31: 1167–1181.

Downs, H. W. and R. W. Hansen (2007) *Estimating Farm Fuel Requirements.* Colorado State U. Extension, *http://www.ext.colostate.edu/PUBS/FARMMGT/05006.html,* accessed Sept. 25, 2007.

Drake, J. B., P. W. Jones, and G. R. Carr (2005) Overview of the software design of the Community Climate System Model. *International Journal of High Performance Computing Applications* 19: 177–186.

Drennan, P. M. and P. S. Nobel (2000) Responses of CAM species to increasing atmospheric CO_2 concentrations. *Plant Cell and Environment* 23: 767–781.

Dukes, J. S., N. R. Chiariello, E. E. Cleland, L. A. Moore, M. R. Shaw, S. Thayer, T. Tobeck, H. A. Mooney, and C. B. Field (2005) Responses of grassland production to single and multiple global environmental changes. *Plos Biology* 3: 1829–1837.

Duvick, D. N. (2005) The contribution of breeding to yield advances in maize (*Zea mays* L.). *Advances in Agronomy* 86: 83–145.

Dykema, J. A. and J. G. Anderson (2006) A methodology for obtaining on-orbit SI-traceable spectral radiance measurements in the thermal infrared. *Metrologia* 43: 287–293.

Earth Observatory (2003) *NASA Satellite Measures Earth's Carbon Metabolism.* NASA, *http://earthobservatory.nasa.gov/Newsroom/NPP/npp.html,* accessed May 16, 2008.

Earth Observatory (2008) *Iron Enrichment Experiment.* NASA, *http://earthobservatory.nasa.gov/Newsroom/NewImages/images.php3?img_id=10793,* accessed May 20, 2008.

Ebert, J. R., N. A. Elliott, L. J. Hurteau, and A. E. Schulz (2005) *A Simple Dynamic Model for Paired Convection Cells and Rifting.* *http://www.bedford.k12.ny.us/flhs/science/stevek/DLESEworkshop/convectionmodel.doc,* accessed Dec. 10, 2006.

ecoAmerica (2008) *The America Climate Values Survey*, ecoAmerica, Washington, D.C., http: //www.ecoamerica.org/docs/ecoAmerica_ACVS_Summary.pdf.

Economic Research Service (2007) *Commodity Costs and Returns.* U.S. Department of Agriculture, http: //www.ers.usda.gov/Data/CostsAndReturns/testpick.htm, accessed Aug. 20, 2007.

Economic Research Service (2009) *Feed Grains Database.* U.S. Department of Agriculture, http: //www.ers.usda.gov/Data/FeedGrains/FeedGrainsQueriable.aspx, accessed February 15, 2009.

ECOtravel (2007) *Petrol vs. Diesel.* Merseytravel Environmental Information Office, http: //www.ecotravel.org.uk/fuels_5.html, accessed July 16, 2007.

Edwards, P. N. (2000) A brief history of atmospheric general circulation models. In: *General Circulation Development, Past, Present, and Future*, Randall, D. A., ed., Academic Press, New York, NY. pp. 67–90.

Eeckhoudt, L., C. Schieber, and T. Schneider (2000) Risk aversion and the external cost of a nuclear accident. *Journal of Environmental Management* 58: 109–117.

Ehleringer, J. R., T. E. Cerling, and M. D. Dearing (2002) Atmospheric CO_2 as a global change driver influencing plant-animal interactions. *Integrative and Comparative Biology* 42: 424–430.

Ehleringer, J. R., T. E. Cerling, and B. R. Helliker (1997) C_4 photosynthesis, atmospheric CO_2, and climate. *Oecologia* 112: 285–299.

Ehrenreich, D., G. Tinetti, A. L. des Etangs, A. Vidal-Madjar, and F. Selsis (2006) The transmission spectrum of Earth-size transiting planets. *Astronomy & Astrophysics* 448: 379–393.

Ellis, M. (2007) *Experience with Energy Efficiency Regulations for Electrical Equipment*, International Energy Agency, Paris, http: //www.iea.org/textbase/papers/2007/Appliances_Ellis.pdf.

Ellison, A. M. and N. J. Gotelli (2001) Evolutionary ecology of carnivorous plants. *Trends in Ecology & Evolution* 16: 623–629.

Ellsworth, D. S., P. B. Reich, E. S. Naumburg, G. W. Koch, M. E. Kubiske, and S. D. Smith (2004) Photosynthesis, carboxylation and leaf nitrogen responses of 16 species to elevated pCO_2 across four free-air CO_2 enrichment experiments in forest, grassland and desert. *Global Change Biology* 10: 2121–2138.

Emanuel, K. (2005) Increasing destructiveness of tropical cyclones over the past 30 years. *Nature* 436: 686–688 DOI 10.1038/Nature03906.

Emerson, R. W. (1889) Quoted in: *Borrowings*, Ladies of the First Unitarian Church of Oakland, California, C.A. Murdock, San Francisco.

Energy Efficiency and Renewable Energy (2009) *Alternative and Advanced Vehicles.* U.S. Department of Energy, http: //www.afdc.energy.gov/afdc/vehicles/flexible_fuel.html, accessed January 18, 2009.

Energy Information Administration (2000) *The Changing Structure of the Electric Power Industry 2000: An Update*, U.S. Department of Energy, Washington, D.C., http: //www.eia.doe.gov/cneaf/electricity/chg_stru_update/update2000.pdf.

Energy Information Administration (2006a) *Existing Generating Units in the United States by State, Company and Plant, 2005.* U.S. Department of Energy, http: //www.eia.doe.gov/cneaf/electricity/page/capacity/existingunits2005.xls, accessed Sept. 17, 2007.

Energy Information Administration (2006b) *International Carbon Dioxide Emissions and Carbon Intensity* Department of Energy, http: //www.eia.doe.gov/pub/international/iealf/tableh1.xls, accessed January 6, 2007.

Energy Information Administration (2006c) *U.S. Electric Power Industry Estimated Emissions by State.* U.S. Department of Energy, http: //www.eia.doe.gov/cneaf/electricity/epa/emission_state.xls, accessed Sept. 20, 2007.

Energy Information Administration (2006d) *U.S. Historical Data Series of Carbon Dioxide Emissions.* Department of Energy, http: //www.eia.doe.gov/oiaf/1605/ggrpt/excel/historical_co2.xls, accessed January 6, 2007.

Energy Information Administration (2006e) *World Consumption of Primary Energy by Energy Type and Selected Country* U.S. Department of Energy, http: //www.eia.doe.gov/pub/international/iealf/table18.xls, accessed June 22, 2007.

Energy Information Administration (2007a) *Annual Energy Outlook 2007*, U.S. Department of Energy, Washington, D.C., http: //www.eia.doe.gov/oiaf/aeo/pdf/0383(2007).pdf.

Energy Information Administration (2007b) *Annual Energy Review 2006*, U.S. Department of Energy, Washington, D.C., http: //www.eia.doe.gov/emeu/aer/pdf/aer.pdf.

Energy Information Administration (2007c) *World Per Capita Total Primary Energy Consumption,1980–2005.* U.S. Department of Energy, http: //www.eia.doe.gov/pub/international/iealf/tablee1c.xls, accessed August 5, 2008.

Energy Information Administration (2008a) *Annual Energy Review, 2007*, U.S. Department of Energy, Washington, D.C., http: //www.eia.doe.gov/emeu/aer/pdf/aer.pdf.

Energy Information Administration (2008b) *Coal Prices.* U.S. Department of Energy, http: //www.eia.doe.gov/emeu/aer/txt/stb0708.xls, accessed January 18, 2009.

Energy Information Administration (2008c) *Commercial Buildings Energy Consumption Survey.* U.S. Department of Energy, http: //www.eia.doe.gov/emeu/cbecs/, accessed May 12, 2008.

Energy Information Administration (2008d) *International Carbon Dioxide Emissions and Carbon Intensity.* U.S. Department of Energy, http: //www.eia.doe.gov/pub/international/iealf/tableh1co2.xls, accessed December 27, 2008.

Energy Information Administration (2008e) *International Energy Outlook 2008*, U.S. Department of Energy, Washington, D.C., http: //www.eia.doe.gov/oiaf/ieo/pdf/0484(2008).pdf.

Energy Information Administration (2009a) *Annual Energy Outlook 2009*, U.S. Department of Energy, Washington, D.C., http: //www.eia.doe.gov/oiaf/aeo/pdf/0383(2009).pdf.

Energy Information Administration (2009b) *International Energy Outlook, 2009*, U.S. Department of Energy, Washington, D.C., http: //www.eia.doe.gov/oiaf/ieo/pdf/0484(2009).pdf.

Energy Star (2008) *Life Cycle Cost Estimate for 100 Energy Star Qualified Compact Fluorescent Lamps.* U.S. Environmental Protection Agency and U.S. Department of Energy, http: //www.energystar.gov/ia/business/bulk_purchasing/bpsavings_calc/CalculatorCFLsBulk.xls, accessed May 8, 2008.

Environment Canada (2007) *A Climate Change Plan for the Purposes of the Kyoto Protocol Implementation Act–2007*, Gatineau, Quebec, http: //www.ec.gc.ca/doc/ed-es/p_123/CC_Plan_2007_e.pdf.

Environmental Studies Laboratory (2007) *Climate Change and Sea Level: Maps of Susceptible Areas.* Department of Geosciences, University of Arizona, http: //www.geo.arizona.edu/dgesl/research/other/climate_change_and_sea_level/sea_level_rise/sea_level_rise.htm, accessed February 2, 2007.

EoroQol (2008) *EQ-5D: An Instrument to Describe and Value Health.* http: //www.euroqol.org/, accessed July 22, 2008.

EPA Newsroom (2008) *EPA and DOE Spread a Bright Idea: Energy Star Light Bulbs Are Helping to Change the World.* U.S. Environmental Protection Agency, http: //yosemite.epa.gov/opa/admpress.nsf/dc57b08b5acd42bc852573c90044a9c4/970f05bf0bc5d9aa852573d10055b38d!OpenDocument, accessed May 1, 2008.

Epstein, E. and A. J. Bloom (2005) *Mineral Nutrition of Plants: Principles and Perspectives*, 2nd Edition. Sinauer Associates, Sunderland, MA.

Epstein, S., R. Buchsbaum, H. A. Lowenstam, and H. C. Urey (1953) Revised carbonate-water isotopic temperature scale. *Geological Society of America Bulletin* 64: 1315–1325.

ERDC/CERL Fuel Cell Team (2007) *Fuel Cell Information Guide.* U.S. Department of Defense, *http://dodfuelcell.cecer.army.mil/proton.html*, accessed Aug. 8, 2007.

Erez, J. and B. Luz (1983) Experimental paleotemperature equation for planktonic-foraminifera. *Geochimica Et Cosmochimica Acta* 47: 1025–1031.

Ericson, J. P., C. J. Vorosmarty, S. L. Dingman, L. G. Ward, and M. Meybeck (2006) Effective sea-level rise and deltas: Causes of change and human dimension implications. *Global and Planetary Change* 50: 63–82 DOI 10.1016/j.gloplacha.2005.07.004.

Esper, J., E. R. Cook, and F. H. Schweingruber (2002) Low-frequency signals in long tree-ring chronologies for reconstructing past temperature variability. *Science* 295: 2250–2253.

Esty, D. C., M. Levy, T. Srebotnjak, and A. de Sherbinin (2005) *2005 Environmental Sustainability Index: Benchmarking National Environmental Stewardship.*, Yale Center for Environmental Law & Policy, New Haven, CT, *http://www.yale.edu/esi/ESI2005.pdf*.

EurObserv'ER (2007) Biogas Barometer. *Systèmes solaires - le journal des énergies renouvelables* 179: 51–61.

European Automobile Manufacturers Association (2007) *Highly Efficient Cars are Available, but not much Loved. http://www.acea.be/examples_of_low_demand_for_fuel-efficiency*, accessed July 16, 2007.

European Automobile Manufacturers Association and European Commission Services (2006) *Monitoring of ACEA's Commitment on CO_2 Emission Reductions from Passenger Cars*, SEC(2006) 1078. Commission of the European Communities, Brussels, *http://ec.europa.eu/environment/co2/pdf/sec_2006_1078.pdf*.

European Commission (2002) *Study on Vehicle Taxation in the Member States of the European Union*, DG Taxation and Customs Union, *http://ec.europa.eu/taxation_customs/resources/documents/vehicle_tax_study_15-02-2002.pdf*.

Eurostat (2003) *Panorama of Transport: Part 2*, European Communities, Luxembourg, *http://epp.eurostat.ec.europa.eu/cache/ITY_OFFPUB/KS-DA-04-001-2/EN/KS-DA-04-001-2-EN.PDF*.

Eurostat (2006) *Energy and Transport in Figures*, European Commission, Directorate of General Energy and Transport, Brussels, *http://ec.europa.eu/dgs/energy_transport/figures/pocketbook/2006_en.htm*.

Evangelical Climate Initiative, The (2009) *Christians and Climate. http://christiansandclimate.org/*, accessed April 12, 2009.

Evans, D. A. D. (2000) Stratigraphic, geochronological, and paleomagnetic constraints upon the neoproterozoic climatic paradox. *American Journal of Science* 300: 347-433.

ExxonMobil (2005) *Corporate Citizenship Report*, ExxonMobil, Irving, Texas.

Fairhurst, T. H., C. Witt, R. J. Buresh, and A. Dobermannand, eds. (2007) *Rice: A Practical Guide to Nutrient Management*, 2nd Edition. International Plant Nutrition Institute & International Potash Institute, Singapore, *http://www.ipni.net/ppiweb/seasia.nsf/$webindex/C76C1813872632A9482570760008A05F*.

Fangmeier, A., L. De Temmerman, C. Black, K. Persson, and V. Vorne (2002) Effects of elevated CO_2 and/or ozone on nutrient concentrations and nutrient uptake of potatoes. *European Journal of Agronomy* 17: 353–368.

Fangmeier, A., L. De Temmerman, L. Mortensen, K. Kemp, J. Burke, R. Mitchell, M. van Oijen, and H. J. Weigel (1999) Effects on nutrients and on grain quality in spring wheat crops grown under elevated CO_2 concentrations and stress conditions in the European, multiple-site experiment 'ESPACE-wheat'. *European Journal of Agronomy* 10: 215–229.

FAOSTAT (2009) *Agricultural Data.* Food and Agricultural Organization of the United Nations, *http://faostat.fao.org/*, accessed June 8, 2009.

Farrell, A. E., R. J. Plevin, B. T. Turner, A. D. Jones, M. O'Hare, and D. M. Kammen (2006a) *Energy and Resources Group Biofuels Analysis Meta-Model. Release 1.1.* Regents of the University of California, *http://rael.berkeley.edu/EBAMM/*, accessed Aug. 26, 2007.

Farrell, A. E., R. J. Plevin, B. T. Turner, A. D. Jones, M. O'Hare, and D. M. Kammen (2006b) Ethanol can contribute to energy and environmental goals. *Science* 311: 506–508.

Fehr, E. and S. Gachter (2000a) Cooperation and punishment in public goods experiments. *American Economic Review* 90: 980–994.

Fehr, E. and S. Gachter (2000b) Fairness and retaliation: The economics of reciprocity. *Journal of Economic Perspectives* 14: 159–181.

FERC Staff Report (2000) *Staff Report to the Federal Energy Regulatory Commission on Western Markets and the Causes of the Summer 2000 Price Abnormalities*, Federal Energy Regulatory Commission, Washington, D.C., *http://www.utilityregulation.com/content/reports/section3.pdf*.

Fioletov, V. E., G. E. Bodeker, A. J. Miller, R. D. McPeters, and R. Stolarski (2002) Global and zonal total ozone variations estimated from ground-based and satellite measurements: 1964–2000. *Journal of Geophysical Research-Atmospheres* 107: 4647 1–14 DOI 10.1029/2001jd001350.

Flory, A. (2005) *Legal Aspects of Storing CO_2*, International Energy Agency, Paris, *http://www.iea.org/textbase/nppdf/free/2005/co2_legal.pdf*.

Flyvbjerg, B. (2007) Policy and planning for large-infrastructure projects: problems, causes, cures. *Environment and Planning B-Planning & Design* 34: 578–597.

Flyvbjerg, B. and COWI (2004) *Procedures for Dealing with Optimism Bias in Transport Planning: Guidance Document*, The British Department for Transport, London, *http://www.hm-treasury.gov.uk/media/B/9/greenbook_optimism_bias_guidance_june04.pdf*.

Flyvbjerg, B., M. K. S. Holm, and S. L. Buhl (2005) How (in)accurate are demand forecasts in public works projects? The case of transportation. *Journal of the American Planning Association* 71: 131–146.

Flyvbjerg, B., M. S. Holm, and S. Buhl (2002) Underestimating costs in public works projects: Error or lie? *Journal of the American Planning Association* 68: 279–295.

Fontaine, K. R., D. T. Redden, C. Wang, A. O. Westfall, and D. B. Allison (2003) Years of life lost due to obesity. *Journal of the American Medical Association* 289: 187–193 10.1001/jama.289.2.187.

Food and Agriculture Organization of the United Nations (2006) *Global Forest Resources Assessment 2005: Progress Towards Sustainable Forest Management*, FAO Forestry Paper 147. Rome, *ftp://ftp.fao.org/docrep/fao/008/A0400E/A0400E00.pdf*.

Forster, P., V. Ramaswamy, P. Artaxo, T. Berntsen, R. Betts, D. W. Fahey, J. Haywood, J. Lean, D. C. Lowe, G. Myhre, J. Nganga, R. Prinn, G. Raga, M. Schulz, and R. Van Dorland (2007) Changes in atmospheric constituents and in radiative forcing. In: *Climate Change 2007: The Physical Science Basis. Contribution of Working Group I to the Fourth Assessment Report of the Intergovernmental Panel on Climate Change*, Solomon, S., D. Qin, M. Manning, Z. Chen, M. Marquis, K. B. Averyt, M. Tignor, and H. L. Miller, eds. Cambridge University Press, Cambridge. pp. 129–234.

Fouet, A. and M. Mock (2006) Regulatory networks for virulence and persistence of *Bacillus anthracis. Current Opinion in Microbiology* 9: 160–166.

Foukal, P., C. Frohlich, H. Spruit, and T. M. L. Wigley (2006) Variations in solar luminosity and their effect on the Earth's climate. *Nature* 443: 161–166.

Foyer, C. H., A. J. Bloom, G. Queval, and G. Noctor (2009) Photorespiratory metabolism: Genes, mutants, energetics, and redox signaling. *Annual Review of Plant Biology* 60: 455–484.

Freeman, A. M. (2006) Valuing environmental health effects—An economic perspective. *Environmental & Resource Economics* 34: 347–363 DOI 10.1007/s10640-006-9002-6.

Friedlingstein, P., P. Cox, R. Betts, L. Bopp, W. Von Bloh, V. Brovkin, P. Cadule, S. Doney, M. Eby, I. Fung, G. Bala, J. John, C. Jones, F. Joos, T. Kato, M. Kawamiya, W. Knorr, K. Lindsay, H. D. Matthews, T. Raddatz, P. Rayner, C. Reick, E. Roeckner, K. G. Schnitzler, R. Schnur, K. Strassmann, A. J. Weaver, C. Yoshikawa, and N. Zeng (2006) Climate-carbon cycle feedback analysis: Results from the (CMIP)-M-4 model intercomparison. *Journal of Climate* 19: 3337–3353.

Fuel Cells 2000 (2009) *Worldwide Hydrogen Fueling Stations*. *http://www.fuelcells.org/info/charts/ h2fuelingstations.pdf*, accessed April 20, 2009.

Galloway, J. N., F. J. Dentener, D. G. Capone, E. W. Boyer, R. W. Howarth, S. P. Seitzinger, G. P. Asner, C. C. Cleveland, P. A. Green, E. A. Holland, D. M. Karl, A. F. Michaels, J. H. Porter, A. R. Townsend, and C. J. Vorosmarty (2004) Nitrogen cycles: past, present, and future. *Biogeochemistry* 70: 153–226.

Galmes, J., J. Flexas, A. J. Keys, J. Cifre, R. A. C. Mitchell, P. J. Madgwick, R. P. Haslam, H. Medrano, and M. A. J. Parry (2005) Rubisco specificity factor tends to be larger in plant species from drier habitats and in species with persistent leaves. *Plant Cell and Environment* 28: 571–579.

Gamble, J. L., ed. (2008) *Analyses of the Effects of Global Change on Human Health and Welfare and Human Systems, Final Report, Synthesis and Assessment Product 4.6*. U.S. Climate Change Science Program, U.S. Environmental Protection Agency, Washington, D.C., *http://downloads.climatescience.gov/sap/ sap4-6/sap4-6-final-all.pdf*.

GE (2008) *GE Ecomagination Report 2007*, General Electric, *http://ge.ecomagination.com/site/#media/2007ecoreport*.

Geisenberg, C. and K. Stewart (1986) Field crop management. In: *The Tomato Crop: a Scientific Basis for Improvement*, Atherton, J. G. and J. Rudich, eds. Chapman & Hall, London. pp. 511–557.

Gerrard, M. (2007) *Global Climate Change and U.S. Law*, American Bar Association Section of Environment Energy and Resources, Chicago, Ill., *http://www.abanet.org/abapubs/ globalclimate*.

Getzner, M., C. L. Spash, and S. Stagl (2005) *Alternatives for Environmental Valuation*, Routledge explorations in environmental economics ; 4. Routledge, London, *http://www.netLibrary.com/ urlapi.asp?action=summary&v=1&bookid=115117*.

Gibbon, E. (1776) *The History of the Decline and Fall of the Roman Empire*, 6 vols, Printed for W. Strahan and T. Cadell in the Strand, London.

Global Hydrology & Climate Center (2007) *Globally-Averaged Atmospheric Temperatures*. *http://www.ghcc.msfc.nasa.gov/MSU/ msusci.html*, accessed January 25, 2007.

Global Monitoring Division (2007) *Halocarbons and other Atmospheric Trace Species*. National Oceanic & Atmospheric Administration, *ftp://ftp.cmdl.noaa.gov/hats/*, accessed January 18, 2007.

Global Wind Energy Council (2008) *Global Installed Wind Power Capacity*. *http://www.gwec.net/uploads/media/gwec-table-2008. pdf*, accessed February 15, 2008.

Goddard Institute for Space Studies (2009) *Global Temperature Trends: 2008 Annual Summation*. *http://data.giss.nasa.gov/ gistemp/2008/*, accessed May 26, 2009.

Gouveia, N., S. Hajat, and B. Armstrong (2003) Socioeconomic differentials in the temperature-mortality relationship in Sao Paulo, Brazil. *International Journal of Epidemiology* 32: 390–397.

Gowdy, J. and J. D. Erickson (2005) The approach of ecological economics. *Cambridge Journal of Economics* 29: 207–222 DOI 10.1093/cje/bei033.

Graham, J. D. (2007) Valuing the future: OMB's refined position. *University of Chicago Law Review* 74: 51–57.

Gray, W. M. (1979) Hurricanes: Their formation, structure and likely role in the tropical circulation. In: *Meteorology Over Tropical Oceans*, Shaw, D. B., ed., Royal Meteorological Society, Bracknell, UK. pp. 155–218.

Green, M. A., K. Emery, Y. Hishikawa, and W. Warta (2008) Solar cell efficiency tables (Version 31). *Progress in Photovoltaics: Research and Applications* 16: 61–67 DOI 10.1002/pip.808.

Greenberg, A. and J. Berktold (2004) *Evangelicals in America*. Religion & Ethics Newsweekly, U.S. News and World Report, *http://www.greenbergresearch.com/articles/1230/758_Evangelicals%20in%20America%20(Survey).pdf*, accessed March 30, 2009.

Greene, D. L. (2005) *Improving the Nation's Energy Security: Can Cars and Trucks Be Made More Fuel Efficient?*, Committee on Science, U.S. House of Representatives, Washington, D.C., *http://www-cta.ornl.gov/cta/Publications/Reports/US%20Science%20Committee_DavidGreene.pdf*.

Griffin, J. M. and S. L. Puller (2005) *Electricity Deregulation: Choices and Challenges*, Bush School series in the economics of public policy. Vol. 4, University of Chicago Press, Chicago.

Groisman, P. Y., R. W. Knight, T. R. Karl, D. R. Easterling, B. M. Sun, and J. H. Lawrimore (2004) Contemporary changes of the hydrological cycle over the contiguous United States: Trends derived from in situ observations. *Journal of Hydrometeorology* 5: 64–85.

Grossman, C. L. (2008) Evangelicals less worried about global climate change. *USA Today*, 9/18/08.

Haines, A., R. S. Kovats, D. Campbell-Lendrum, and C. Corvalan (2006) Harben Lecture—Climate change and human health: impacts, vulnerability, and mitigation. *Lancet* 367: 2101–2109.

Häkkinen, S. and P. B. Rhines (2004) Decline of subpolar North Atlantic circulation during the 1990s. *Science* 304: 555–559.

Hales, S., N. de Wet, J. Maindonald, and A. Woodward (2002) Potential effect of population and climate changes on global distribution of dengue fever: an empirical model. *Lancet* 360: 830–834.

Hall, C. T. (1998) Staying alive: High in California's White Mountains grows the oldest living creature ever found. *San Francisco Chronicle*, San Francisco, August 23, 1998.

Hammitt, J. K. (2002) QALYs versus WTP. *Risk Analysis* 22: 985–1001.

Han, G. Q. and C. L. Tang (2001) Interannual variations of volume transport in the western Labrador Sea based on TOPEX/ Poseidon and WOCE data. *Journal of Physical Oceanography* 31: 199–211.

Hansen, J., M. Sato, R. Ruedy, L. Nazarenko, A. Lacis, G. A. Schmidt, G. Russell, I. Aleinov, M. Bauer, S. Bauer, N. Bell, B. Cairns, V. Canuto, M. Chandler, Y. Cheng, A. Del Genio, G. Faluvegi, E. Fleming, A. Friend, T. Hall, C. Jackman, M. Kelley, N. Kiang, D. Koch, J. Lean, J. Lerner, K. Lo, S. Menon, R. Miller, P. Minnis, T. Novakov, V. Oinas, Ja. Perlwitz, Ju. Perlwitz, D. Rind, A. Romanou, D. Shindell, P. Stone, S. Sun, N. Tausnev, D. Thresher, B. Wielicki, T. Wong, M. Yao, and S. Zhang (2005) Efficacy of climate forcings. *Journal of Geophysical Research-Atmospheres* 110: D18104 1–45 DOI 10.1029/2005JD005776.

Hansen, J. E. and M. Sato (2001) Trends of measured climate forcing agents. *Proceedings of the National Academy of Sciences of the United States of America* 98: 14778–14783.

Harden, B. (2007) *Fluorescent Bulbs Are Known to Zap Domestic Tranquillity*. Washington Post, *http://www.washingtonpost. com/wp-dyn/content/article/2007/04/29/AR2007042901500.html*, accessed May 1, 2008.

Hardin, G. (1968) Tragedy of Commons. *Science* 162: 1243–1248.

Harley, C. D. G., A. R. Hughes, K. M. Hultgren, B. G. Miner, C. J. B. Sorte, C. S. Thornber, L. F. Rodriguez, L. Tomanek, and S. L. Williams (2006) The impacts of climate change in coastal marine systems. *Ecology Letters* 9: 228–241.

Harris Interactive (2007) *U.S. Adults Less Likely Than Europeans to Think Humans Are Contributing to the Increase in Global Temperatures*. The Harris Poll *http://www.harrisinteractive.com/ harris_poll/index.asp?PID=750*, accessed March 22, 2009.

Harris, J. M. and B. Roach (2007) *The Economics of Global Climate Change*, Global Development and Environment Institute, Tufts U., Medford, MA, *http: //www.ase.tufts.edu/gdae/education_materials/modules/The_Economics_of_Global_Climate_Change.pdf*.

Harvell, C. D., C. E. Mitchell, J. R. Ward, S. Altizer, A. P. Dobson, R. S. Ostfeld, and M. D. Samuel (2002) Ecology—Climate warming and disease risks for terrestrial and marine biota. *Science* 296: 2158–2162.

Harvey, L. D. D. (2007) Net climatic impact of solid foam insulation produced with halocarbon and non-halocarbon blowing agents. *Building and Environment* 42: 2860-2879 DOI 10.1016/j.buildenv.2006.10.028.

Hayhoe, K., D. Cayan, C. B. Field, P. C. Frumhoff, E. P. Maurer, N. L. Miller, S. C. Moser, S. H. Schneider, K. N. Cahill, E. E. Cleland, L. Dale, R. Drapek, R. M. Hanemann, L. S. Kalkstein, J. Lenihan, C. K. Lunch, R. P. Neilson, S. C. Sheridan, and J. H. Verville (2004) Emissions pathways, climate change, and impacts on California. *Proceedings of the National Academy of Sciences of the United States of America* 101: 12422–12427.

Haynes, R. J. (1986a) Origin, distribution, and cycling of nitrogen in terrestrial ecosystems. In: *Mineral Nitrogen in the Plant-Soil System*, Haynes, R. J., ed., Academic Press, Orlando. pp. 1–51.

Haynes, R. J. (1986b) Uptake and assimilation of mineral nitrogen by plants. In: *Mineral Nitrogen in the Plant Soil System*, Haynes, R. J., ed., Academic Press, Orlando. pp. 303–378.

Head, J. W., D. R. Marchant, and M. A. Kreslavsky (2008) Formation of gullies on Mars: Link to recent climate history and insolation microenvironments implicate surface water flow origin. *Proceedings of the National Academy of Sciences of the United States of America* 105: 13258–13263 DOI 10.1073/pnas.0803760105.

Heal, G. (2007) Discounting: A review of the basic economics. *University of Chicago Law Review* 74: 59–77.

Healey, J. R. (2007) Highlander packs a lot in a tidy package. *USA Today*, August 31, 2007.

HEATCO (2007) *Developing Harmonised European Approaches for Transport Costing and Project Assessment*, Institute of Energy Economics and the Rational Use of Energy, Stuttgart, *http: //heatco.ier.uni-stuttgart.de/HEATCO_D5.pdf*.

Heavenrich, R. M. (2006) *Light-Duty Automotive Technology and Fuel Economy Trends: 1975 Through 2006*, U.S. Environmental Protection Agency, Office of Transportation and Air Quality, Washington, D.C., *http: //www.epa.gov/otaq/cert/mpg/fetrends/420r06011.pdf*.

Hemming, N. G. and G. N. Hanson (1992) Boron isotopic composition and concentration in modern marine carbonates. *Geochimica Et Cosmochimica Acta* 56: 537–543.

Henderson, B. (2006) *The Gospel of the Flying Spaghetti Monster*, Villard Books, New York.

Hendy, E. J., M. K. Gagan, C. A. Alibert, M. T. McCulloch, J. M. Lough, and P. J. Isdale (2002) Abrupt decrease in tropical Pacific Sea surface salinity at end of Little Ice Age. *Science* 295: 1511–1514.

Herter, K., P. McAuliffe, and A. Rosenfeld (2007) An exploratory analysis of California residential customer response to critical peak pricing of electricity. *Energy* 32: 25–34.

Highum, E. G. (2006) *Success vs. Failure in International Environmental Treaty Negotiations: A Comparative Analysis of the Montreal Protocol and Global Climate Change Negotiations*, International Studies Association, San Diego, *http: //www.allacademic.com/one/isa/isa06/index.php?cmd=Download+Document&key=unpublished_manuscript&file_index=2&pop_up=true&no_click_key=true&attachment_style=attachment&PHPSESSID=c123141638d43d6ceec7b220c27e33d7*.

Himmelstein, D. U., E. Warren, D. Thorne, and S. Woolhandler (2005) Market watch: Illness and injury as contributors to bankruptcy. *Health Affairs* 24: W563–W573 DOI 10.1377/hlthaff.W5.63.

Hitt, K. J. and B. T. Nolan (2005) *Nitrate in Ground Water: Using a Model to Simulate the Probability of Nitrate Contamination of Shallow Ground Water in the Conterminous United States*. U.S. Geological Survey, *http: //pubs.usgs.gov/sim/2005/2881/*, accessed Sept. 20, 2007.

HM Treasury (2003) *The Green Book: Appraisal and Evaluation in Central Government*, The Stationery Office, London, *http: //www.hm-treasury.gov.uk/media/3/F/green_book_260907.pdf*.

Hodum, R. (2007) *Kunming Heats Up as China's "Solar City"*. Worldwatch Institute, *http: //www.worldwatch.org/node/5105*, accessed March 28, 2008.

Hofmann, D. J., J. H. Butler, E. J. Dlugokencky, J. W. Elkins, K. Masarie, S. A. Montzka, and P. Tans (2006) The role of carbon dioxide in climate forcing from 1979 to 2004: introduction of the Annual Greenhouse Gas Index. *Tellus Series B-Chemical and Physical Meteorology* 58: 614–619.

Holland, H. D. (2006) The oxygenation of the atmosphere and oceans. *Philosophical Transactions of the Royal Society B-Biological Sciences* 361: 903–915.

Hopkins, M., T. M. Harrison, and C. E. Manning (2008) Low heat flow inferred from >4 Gyr zircons suggests Hadean plate boundary interactions. *Nature* 456: 493–496.

Hopp, M. J. and J. A. Foley (2003) Worldwide fluctuations in dengue fever cases related to climate variability. *Climate Research* 25: 85–94.

Horowitz, L. W. (2006) Past, present, and future concentrations of tropospheric ozone and aerosols: Methodology, ozone evaluation, and sensitivity to aerosol wet removal. *Journal of Geophysical Research-Atmospheres* 111: 16 D22211 DOI 10.1029/2005jd006937.

Hoyos, C. D., P. A. Agudelo, P. J. Webster, and J. A. Curry (2006) Deconvolution of the factors contributing to the increase in global hurricane intensity. *Science* 312: 94–97.

Hoyt, D. V. and K. H. Schatten (1998) Group Sunspot Numbers: A new solar activity reconstruction. *Solar Physics* 181: 491–512.

Hubbell, B. J. (2006) Implementing QALYs in the analysis of air pollution regulations. *Environmental & Resource Economics* 34: 365–384 DOI 10.1007/s10640-004-7437-1.

Huebler, F. (2008) *National Wealth and Years of Education*. International Education Statistics, *http: //huebler.blogspot.com/2008/08/wealth.html*, accessed April 18, 2009.

Hungate, B. A., P. D. Stiling, P. Dijkstra, D. W. Johnson, M. E. Ketterer, G. J. Hymus, C. R. Hinkle, and B. G. Drake (2004) CO_2 elicits long-term decline in nitrogen fixation. *Science* 304: 1291.

Huntington, T. G. (2006) Evidence for intensification of the global water cycle: Review and synthesis. *Journal of Hydrology* 319: 83–95.

Huybers, P. (2006) Early Pleistocene glacial cycles and the integrated summer insolation forcing. *Science* 313: 508–511.

HVACWebsite.com (2008) *HVAC Efficiency Finder*. US Airconditioning Distributors, *http: //www.hvacopcost.com/EfficiencyFinder.html*, accessed May 1, 2008.

Hyder, A. A. and M. Peden (2003) Inequality and road-traffic injuries: call for action. *Lancet* 362: 2034–2035.

ICLEI (2009) *About ICLEI*. ICLEI Local Governments for Sustainability, *http: //www.iclei.org/index.php?id=global-about-iclei*, accessed April 10, 2009.

Idaho Transportation Department (2006) *State-by-State Comparison of Annual Motor Vehicle Registration Fees and Fuel Taxes*, Economic and Research Section, *http: //itd.idaho.gov/econ/MiscReports/ComparisonofAnnualMotorVehicleOperatingCosts.pdf*.

Industrial Technologies Program (2004) *Improving Steam System Performance: A Sourcebook for Industry*, U.S. Department of

Energy, Washington, D.C., *http: //www1.eere.energy.gov/indus-try/bestpractices/pdfs/steamsourcebook.pdf*.

Institute of Medicine (2004) *Insuring America's Health: Principles and Recommendations*, The National Academies Press, Washington, D.C.

Insurance Institute for Highway Safety (2006) Minicars: first crash test results. *Status Report (Dec. 19, 2006)* 41: 1–4.

Insurance Institute for Highway Safety (2007) Driver deaths by make and model. *Status Report (Apr. 19, 2007)* 42: 1–7.

Interfaith Stewardship Alliance (2006) *A Call to Truth, Prudence, and Protection of the Poor: An Evangelical Response to Global Warming*. The Ethics & Religious Liberty Commission of the Southern Baptist Convention, *http: //www.sbc.net/redirect. asp?url=http%3A%2F%2Ferlc.com%2Farticle%2Fa-call-to-truth-prudence-and-protection&key=climate&title=A+Call+to+Truth%2 C+Prudence%2C+and+Protection+of+the+Poor%3A+An+Evang elical+Response+to+Global+Warming&ndx=SBC%2C+IMB%2C +NAMB%2C+ANNUITY%2C+LIFEWAY%2C+WMU%2C+E RLC%2C+SEMINARIES*, accessed April 12, 2009.

International Association of Natural Gas Vehicles (2007) *International Statistics. http: //www.iangv.org/content/view/17/35/*, accessed Aug. 5, 2007.

International Atomic Energy Agency (2008) *Energy, Electricity and Nuclear Power Estimates for the Period up to 2030*, 2008 Edition. Vienna, *http://www-pub.iaea.org/MTCD/publications/PDF/ RDS1-28_web.pdf*.

International Court of Justice (1945) *Statute of the Court. http:// www.icj-cij.org/documents/index.php?p1=4&p2=2&p3=0*, accessed September 20, 2008.

International Energy Agency (2004a) *Prospects for CO$_2$ Capture and Storage*, Organization for Economic Cooperation and Development, Paris, *http://www.iea.org/textbase/nppdf/free/2004/ prospects.pdf*.

International Energy Agency (2004b) *World Energy Outlook 2004*, Organisation for Economic Cooperation and Development, Paris, *http: //www.iea.org//textbase/nppdf/free/2004/weo2004.pdf*.

International Energy Agency (2005) *Prospects for Hydrogen and Fuel Cells*, Organization for Economic Cooperation and Development, Paris, *http: //www.iea.org/textbase/nppdf/free/2005/hydro-gen2005.pdf*.

International Energy Agency (2006a) *Energy Technology Perspective 2006*, Organisation for Economic Co-operation and Development, Paris.

International Energy Agency (2006b) *Light's Labour's Lost: Policies for Energy-efficient Lighting*, Organization for Economic Co-operation and Development, Paris, *http: //www.iea.org/textbase/ nppdf/free/2006/light2006.pdf*.

International Energy Agency (2006c) *World Energy Outlook 2006*, Organisation for Economic Co-operation and Development, Paris.

International Energy Agency (2006d) *World Energy Outlook 2007*, Organisation for Economic Cooperation and Development, Paris, *http://www.worldenergyoutlook.org/2006.asp*.

International Energy Agency (2007a) *Biofuel Production*, Organization for Economic Cooperation and Development, Paris, *http: //www.iea.org/Textbase/techno/essentials2.pdf*.

International Energy Agency (2007b) *Hydrogen Production and Distribution*, Organization for Economic Cooperation and Development, Paris, *http: //www.iea.org/Textbase/techno/essen-tials5.pdf*.

International Energy Agency (2007c) *Renewables in Global Energy Supply*, Organisation for Economic Cooperation and Development, Paris, *http: //www.iea.org/textbase/papers/2006/renew-able_factsheet.pdf*.

International Energy Agency (2008) *Key World Energy Statistics 2008*, Organisation for Economic Cooperation and Development, Paris, *http: //www.iea.org/textbase/nppdf/free/2008/ key_stats_2008.pdf*.

International Nuclear Safety Center (2009) *Maps of Nuclear Power Reactors*, Argonne National Laboratory *http: //www.insc.anl. gov/pwrmaps/map/world_map.php*.

IPCC (2001) The Scientific Basis. Contribution of Working Group I to the Third Assessment Report. In: *Climate Change 2001*, Houghton, J. T., Y. Ding, D. J. Griggs, M. Noguer, P. J. van der Linden, X. Da, K. Maskell, and C. A. Johnson, eds. Cambridge University Press, Cambridge. p. 881.

IPCC (2004) *16 Years of Scientific Assessment in Support of the Climate Convention*, Geneva, *http: //www.ipcc.ch/pdf/10th-anniversary/ anniversary-brochure.pdf*.

IPCC (2005) *IPCC Special Report on Carbon Dioxide Capture and Storage prepared by Working Group III of the Intergovernmental Panel on Climate Change*, Metz, B., O. H. Davidson, C. de Coninck, M. Loos, and L. A. Meyer, eds. Cambridge University Press, New York.

IPCC (2007a) *Climate Change 2007: Impacts, Adaptation and Vulnerability. Summary for policymakers*. World Meteoroligical Organization, Working Group II, *http: //www.ipcc.ch/SPM13apr07. pdf*, accessed May 1, 2007.

IPCC (2007b) *Climate Change 2007: Mitigation of Climate Change*, World Meteriological Organization, Working Group III, *http: //www.mnp.nl/ipcc/pages_media/AR4-chapters.html*.

IPCC (2007c) Summary for Policymakers. In: *Climate Change 2007: The Physical Science Basis. Contribution of Working Group I to the Fourth Assessment Report of the Intergovernmental Panel on Climate Change*, Solomon, S., D. Qin, M. Manning, Z. Chen, M. Marquis, K. B. Averyt, M. Tignor, and H. L. Miller, eds. Cambridge University Press, Cambridge. pp. 1–18, *http:// www.ipcc.ch/pdf/assessment-report/ar4/wg1/ar4-wg1-spm.pdf*.

IPCC (2009) *About IPCC. http: //www.ipcc.ch/about/index.htm*, accessed April 10, 2009.

IPCC Working Group III (2000) *Emissions Scenarios*. Intergovernmental Panel on Climate Change, *http: //www.grida.no/climate/ ipcc/spmpdf/sres-e.pdf*, accessed January 1, 2007.

Israeli Central Bureau of Statistics (2008) *Summary of Israel's Energy Balances. http: //www1.cbs.gov.il/www/yarhon/n1_e.xls*, accessed March 26, 2008.

Jacobson, M. Z. (2005) Studying ocean acidification with conservative, stable numerical schemes for nonequilibrium air-ocean exchange and ocean equilibrium chemistry. *Journal of Geophysical Research-Atmospheres* 110: D07302.

Jaffe, A. B., R. G. Newell, and R. N. Stavins (2005) A tale of two market failures: Technology and environmental policy. *Ecological Economics* 54: 164–174 10.1016/j.ecolecon.2004.12.027.

Japanese Automobile Manufacturers Association (2007) *Active Matrix Database System. http: //jamaserv.jama.or.jp/newdb/eng/ index.html*, accessed July 11, 2007.

Jeffries, T. W. and Y. S. Jin (2004) Metabolic engineering for improved fermentation of pentoses by yeasts. *Applied Microbiology and Biotechnology* 63: 495–509.

Joint Transport Research Centre (2007) *Managing Urban Traffic Congestion: Summary Document*, European Conference of Transport Ministers, *http: //www.cemt.org/JTRC/Working-Groups/TrafficCongestion/CongestionSummary.pdf*.

Jones, P. D. and A. Moberg (2003) Hemispheric and large-scale surface air temperature variations: An extensive revision and an update to 2001. *Journal of Climate* 16: 206–223.

Jones, W. D., P. Cayirlioglu, I. G. Kadow, and L. B. Vosshall (2007) Two chemosensory receptors together mediate carbon dioxide detection in Drosophila. *Nature* 445: 86–90.

Kanipe, J. (2006) A cosmic connection. *Nature* 443: 141–143.

Kapp, C. (2003) WHO acts on road safety to reverse accident trends. *Lancet* 362: 1125–1125.

Kasting, J. F. and S. Ono (2006) Palaeoclimates: the first two billion years. *Philosophical Transactions of the Royal Society B-Biological Sciences* 361: 917–929.

Katz, M. E., D. K. Pak, G. R. Dickens, and K. G. Miller (1999) The source and fate of massive carbon input during the latest Paleocene thermal maximum. *Science* 286: 1531–1533.

Kaufman, D., A. Krayy, and M. Mastruzzi (2008) *Governance Matters VII: Aggregate and Individual Governance Indicators, 1996–2007*, World Bank Policy Research Working Paper No. 4654. The World Bank, Washington, D.C., *http: //papers. ssrn.com/sol3/Delivery.cfm/SSRN_ID1151867_code332182. pdf?abstractid=1148386&mirid=5.*

Kazmerski, L., D. Gwinner, and A. Hicks (2008) National Renewable Energy Laboratory, *http://en.wikipedia.org/wiki/ Image:PVeff%28rev110707%29d.jpg*, accessed August 28, 2008.

Keeling, C. D. (1978) The influence of Mauna Loa observatory on the development of atmospheric CO_2 research. In: *Mauna Loa Observatory: A 20th Anniversary Report*, Miller, J., ed., NOAA Special Report, Silver Springs, MD. *http: //www.mlo.noaa.gov/ HISTORY/Fhistory.htm.*

Keeling, R. F. (2005) Comment on "The ocean sink for anthropogenic CO_2". *Science* 308: 1743c DOI 10.1126/science.1109620.

Keeling, R. F. and H. E. Garcia (2002) The change in oceanic O_2 inventory associated with recent global warming. *Proceedings of the National Academy of Sciences of the United States of America* 99: 7848–7853.

Keitel, H. G., H. Berman, H. Jones, and E. MacLachlan (1955) The chemical composition of normal human red blood cells, including variability among centrifuged cells. *Blood* 10: 370–376.

Keller, G., T. Adatte, G. Baum, and Z. Berner (2008) Reply to 'Chicxulub impact predates K-T boundary: New evidence from Brazos, Texas' Comment by Schulte et al.—Discussion. *Earth and Planetary Science Letters* 269: 620–628 DOI 10.1016/j. epsl.2007.12.025.

Keller, G., T. Adatte, Z. Berner, M. Harting, G. Baum, M. Prauss, A. Tantawy, and D. Stueben (2007) Chicxulub impact predates K-T boundary: New evidence from Brazos, Texas. *Earth and Planetary Science Letters* 255: 339–356 10.1016/j. epsl.2006.12.026.

Kenkel, D. (2006) WTP- and QALY-based approaches to valuing health for policy: Common ground and disputed territory. *Environmental & Resource Economics* 34: 419–437 DOI 10.1007/ s10640-006-9003-5.

Kesselaar, H., I. Moreno, and R. Sturges (1996) Montreal Protocol: Enforcement of Chlorofluorocarbons (CFC) and Related Requirements. In: *4th International Conference on Environmental Compliance and Enforcement*, International Network for Environmental Compliance and Enforcement, Chiang Mai, Thailand. pp. 1–3, *http://www.inece.org/4thvol2/mpws.pdf.*

Keynes, J. M. (1936) *The General Theory of Employment, Interest, and Money*, Harcourt Brace, New York, *http://www.marxists.org/ reference/subject/economics/keynes/general-theory/index.htm.*

Kheshgi, H. S. and D. E. Archer (2004) A nonlinear convolution model for the evasion of CO_2 injected into the deep ocean. *Journal of Geophysical Research-Oceans* 109: C02007 DOI 10.1029/2002JC001489.

Kiehl, J. T. and K. E. Trenberth (1997) Earth's annual global mean energy budget. *Bulletin of the American Meteorological Society* 78: 197–208.

Kim, S. T. and J. R. O'Neil (1997) Equilibrium and nonequilibrium oxygen isotope effects in synthetic carbonates. *Geochimica Et Cosmochimica Acta* 61: 3461–3475.

Kimball, B. A., S. B. Idso, S. Johnson, and M. C. Rillig (2007) Seventeen years of carbon dioxide enrichment of sour orange trees: final results. *Global Change Biology* 13: 2171–2183.

Kimball, B. A., C. F. Morris, P. J. Pinter, G. W. Wall, D. J. Hunsaker, F. J. Adamsen, R. L. LaMorte, S. W. Leavitt, T. L. Thompson, A. D. Matthias, and T. J. Brooks (2001) Elevated CO_2, drought and soil nitrogen effects on wheat grain quality. *New Phytologist* 150: 295–303.

Kirschbaum, M. U. F. (2006) The temperature dependence of organic-matter decomposition—still a topic of debate. *Soil Biology & Biochemistry* 38: 2510–2518.

Knepp, R. G., J. G. Hamilton, J. E. Mohan, A. R. Zangerl, M. R. Berenbaum, and E. H. DeLucia (2005) Elevated CO2 reduces leaf damage by insect herbivores in a forest community. *New Phytologist* 167: 207–218.

Knowlton, N. (2001) The future of coral reefs. *Proceedings of the National Academy of Sciences of the United States of America* 98: 5419–5425.

Kochi, I., B. Hubbell, and R. Kramer (2006) An empirical Bayes approach to combining and comparing estimates of the value of a statistical life for environmental policy analysis. *Environmental & Resource Economics* 34: 385–406 DOI 10.1007/s10640-006-9000-8.

Kopp, R. E., J. L. Kirschvink, I. A. Hilburn, and C. Z. Nash (2005) The paleoproterozoic snowball Earth: A climate disaster triggered by the evolution of oxygenic photosynthesis. *Proceedings of the National Academy of Sciences of the United States of America* 102: 11131–11136.

Korner, C. (2006) Plant CO_2 responses: an issue of definition, time and resource supply. *New Phytologist* 172: 393–411.

Kosatsky, T. (2005) The 2003 European heat wave. *Eurosurveillance* 10: 148–149.

Krauss, C. (2008) Move over, oil, there's money in Texas wind. *The New York Times*, New York, February 23, 2008.

Kristjánsson, J. E., C. W. Stjern, F. Stordal, A. M. Fjæraa, G. Myhre, and K. Jónasson (2008) Cosmic rays, cloud condensation nuclei and clouds—a reassessment using MODIS data. *Atmospheric Chemistry and Physics* 8: 7373–7387.

Kroeze, C., J. Vlasblom, J. Gupta, C. Boudri, and K. Blok (2004) The power sector in China and India: greenhouse gas emissions reduction potential and scenarios for 1990–2020. *Energy Policy* 32: 55–76.

Krosnick, J. A., A. L. Holbrook, L. Lowe, and P. S. Visser (2006) The origins and consequences of democratic citizens' policy agendas: A study of popular concern about global warming. *Climatic Change* 77: 7–43.

Krull, E. S. and G. J. Retallack (2000) $\delta^{13}C$ depth profiles from paleosols across the Permian-Triassic boundary: Evidence for methane release. *Geological Society of America Bulletin* 112: 1459–1472.

Kruse, B., S. Grinna, and C. Buch (2002) *Hydrogen: Status of Production*, Bellona Foundation, Olso, *http://www.bellona.org/ filearchive/fil_Hydrogen_6-2002.pdf.*

Ku, M. S. B., J. R. Wu, Z. Y. Dai, R. A. Scott, C. Chu, and G. E. Edwards (1991) Photosynthetic and photorespiratory characteristics of *Flaveria* species. *Plant Physiology* 96: 518–528.

Kues, U. and Y. Liu (2000) Fruiting body production in basidiomycetes. *Applied Microbiology and Biotechnology* 54: 141–152.

Kurbatov, A. V., G. A. Zielinski, N. W. Dunbar, P. A. Mayewski, E. A. Meyerson, S. B. Sneed, and K. C. Taylor (2006) A 12,000 year record of explosive volcanism in the Siple Dome Ice Core, West Antarctica. *Journal of Geophysical Research-Atmospheres* 111: D12307 1–18 DOI 10.1029/2005JD006072.

Land and Plant Nutrition Management Service (2004) *Fertilizer Use by Crop in Brazil*, Food and Agriculture Organization of the United Nations, Rome, *http://www.fao.org/docrep/007/y5376e/ y5376e00.htm.*

Land Protection (2006) *The Prickly Pear Story*, Vol. PP62, Natural Resources and Water, Queensland Government, Australia, *http://www.nrw.qld.gov.au/factsheets/pdf/pest/pp62.pdf.*

Landsberg, H. (1946) Climate as a natural resource. *The Scientific Monthly* 63: 293–298.

Landsberg, H. E. (1958) Trends in Climatology. *Science* 128: 749–758.

Larsen, M. J. and S. J. Jensen (1989) The hydroxyapatite solubility product of human dental enamel as a function of pH in the range 4.6–7.6 at 20°C. *Archives of Oral Biology* 34: 957–961.

Laskar, J., P. Robutel, F. Joutel, M. Gastineau, A. C. M. Correia, and B. Levrard (2004) A long-term numerical solution for the insolation quantities of the Earth. *Astronomy & Astrophysics* 428: 261–285.

Latif, M., C. Boning, J. Willebrand, A. Biastoch, J. Dengg, N. Keenlyside, U. Schweckendiek, and G. Madec (2006) Is the thermohaline circulation changing? *Journal of Climate* 19: 4631–4637.

Leakey, A. D. B., C. J. Bernacchi, F. G. Dohleman, D. R. Ort, and S. P. Long (2004) Will photosynthesis of maize (*Zea mays*) in the US Corn Belt increase in future [CO_2] rich atmospheres? An analysis of diurnal courses of CO_2 uptake under free-air concentration enrichment (FACE). *Global Change Biology* 10: 951–962.

Leakey, A. D. B., M. Uribelarrea, E. A. Ainsworth, S. L. Naidu, A. Rogers, D. R. Ort, and S. P. Long (2006) Photosynthesis, productivity, and yield of maize are not affected by open-air elevation of CO_2 concentration in the absence of drought. *Plant Physiology* 140: 779–790.

Lean, J. L. and D. H. Rind (2008) How natural and anthropogenic influences alter global and regional surface temperatures: 1889 to 2006. *Geophysical Research Letters* 35: L18701 DOI 10.1029/2008gl034864.

Leaver, R. H. (2007) *Wood Pellet Fuel and the Residential Market* Northeast Regional Biomass Program, Washington, D.C., *http://www.nrbp.org/papers/032.pdf*.

Leemans, R. and A. M. Solomon (1993) Modeling the potential change in yield and distribution of the earth's crops under a warmed climate. *Climate Research* 3: 79–96.

Lefèvre, N., P. de T'Serclaes, and P. Waide (2006) *Barriers to Technology Diffusion: The Case of Compact Fluorescent Lamps*, Organization for Economic Co-operation and Development, Paris, *http://regserver.unfccc.int/seors/file_storage/ytnsn6qk5npfqie.pdf*.

Leiserowitz, A. (2006) Climate change risk perception and policy preferences: The role of affect, imagery, and values. *Climatic Change* 77: 45–72 DOI 10.1007/s10584-006-9059-9.

Lemoine, N., H. C. Schaefer, and K. Bohning-Gaese (2007) Species richness of migratory birds is influenced by global climate change. *Global Ecology and Biogeography* 16: 55–64.

Levine, M., D. Ürge-Vorsatz, K. Blok, L. Geng, D. Harvey, S. Lang, G. Levermore, A. Mongameli Mehlwana, S. Mirasgedis, A. Novikova, J. Rilling, and H. Yoshino (2007) Residential and commercial buildings. In: *Climate Change 2007: Mitigation. Contribution of Working Group III to the Fourth Assessment Report of the Intergovernmental Panel on Climate Change*, Metz, B., O. R. Davidson, P. R. Bosch, R. Dave, and L. A. Meyer, eds. Cambridge U. Press, Cambridge. pp. 387–446, *http://www.ipcc.ch/pdf/assessment-report/ar4/wg3/ar4-wg3-chapter6.pdf*.

Levinson, D. H. and J. H. Lawrimore (2008) State of the Climate in 2007. *Bulletin of the American Meteorological Society* 89: 1–181.

Lewis, O. T. (2006) Climate change, species-area curves and the extinction crisis. *Philosophical Transactions of the Royal Society B-Biological Sciences* 361: 163–171.

Lichter, S. R. (2008) *Climate Scientists Agree on Warming, Disagree on Dangers, and Don't Trust the Media's Coverage of Climate Change.* Statistical Assessment Service, *http://stats.org/stories/2008/global_warming_survey_apr23_08.html*, accessed January 1, 2009.

Lind, R. C., ed. (1982) *Discounting for Time and Risk in Energy Policy*, Resources for the Future & Johns Hopkins U. Press, Washington, D.C. & Baltimore.

Lips, K. R., F. Brem, R. Brenes, J. D. Reeve, R. A. Alford, J. Voyles, C. Carey, L. Livo, A. P. Pessier, and J. P. Collins (2006) Emerging infectious disease and the loss of biodiversity in a Neotropical amphibian community. *Proceedings of the National Academy of Sciences of the United States of America* 103: 3165–3170.

Lockwood, M. (2008) Recent changes in solar outputs and the global mean surface temperature. III. Analysis of contributions to global mean air surface temperature rise. *Proceedings of the Royal Society A-Mathematical Physical and Engineering Sciences* 464: 1387–1404 DOI 10.1098/rspa.2007.0348.

Loisos, G. and D. Springer (2000) *Alternatives to Compressor Cooling*, California Energy Commission, Sacramento, CA, *http://www.energy.ca.gov/reports/2002-01-10_600-00-003.PDF*.

Long, S. P., E. A. Ainsworth, A. D. B. Leakey, J. Nosberger, and D. R. Ort (2006) Food for thought: Lower-than-expected crop yield stimulation with rising CO_2 concentrations. *Science* 312: 1918–1921.

Lorenzoni, I., N. F. Pidgeon, and R. E. O'Connor (2005) Dangerous climate change: The role for risk research. *Risk Analysis* 25: 1387–1398 DOI 10.1111/j.1539-6925.2005.00686.x.

Lovallo, D. and D. Kahneman (2003) Delusions of success: How optimism undermines executives' decisions. *Harvard Business Review* 81: 56–63.

Lovelock, J. E. (1979) *Gaia, a New Look at Life on Earth*, Oxford University Press, Oxford.

Lovelock, J. E. (1988) *The Ages of Gaia: A Biography of our Living Earth*, Norton, New York.

Lucas, R. E. and J. F. Davis (1961) Relationships between pH values of organic soils and availabilities of 12 plant nutrients. *Soil Science* 92: 177–182.

Lynch, J. and A. McGoldrick (2005) *Peace Journalism*, Hawthorn Press, Stroud, UK, *http://www.uq.edu.au/journ-comm/docs/shortcourse/Revised%20Annabel%20+%20Jake%20chptr.pdf*.

Ma, F. R. and M. A. Hanna (1999) Biodiesel production: a review. *Bioresource Technology* 70: 1–15.

Macedo, I. C., M. R. L. V. Leal, and J. E. A. R. da Silva (2004) *Assessment of Greenhouse Gas Emissions in the Production and Use of Fuel Ethanol in Brazil*, Government of the State of São Paulo, *http://www.unica.com.br/i_pages/files/pdf_ingles.pdf*.

Mackie, P. J., M. Wadman, A. S. Fowkes, G. Whelan, J. Nellthorp, and J. Bates (2003) *Values of Travel Time Savings in the UK*, Vol. Working Paper 567, Institute of Transport Studies, University of Leeds, Leeds, *http://eprints.whiterose.ac.uk/2079/2/Value_of_travel_time_savings_in_the_UK_protected.pdf*.

MacLeod, K. G., B. T. Huber, and C. Isaza-Londono (2005) North Atlantic warming during global cooling at the end of the Cretaceous. *Geology* 33: 437–440.

Makino, A. and T. Mae (1999) Photosynthesis and plant growth at elevated levels of CO_2. *Plant and Cell Physiology* 40: 999–1006.

Malte, P. (2001) *Improving IC Engine Efficiency.* University of Washington, *http://courses.washington.edu/me341/oct22v2.htm*, accessed Aug. 7, 2007.

Malthus, T. R. (1798) *An Essay on the Principle of Population, as it affects the future improvement of society with remarks on the speculations of Mr. Godwin, M. Condorcet and other writers*, Printed for J. Johnson, London, *http://galenet.galegroup.com/servlet/EC CO?c=1&stp=Author&ste=11&af=BN&ae=T095674&tiPG=1&d d=0&dc=flc&docNum=CW104783583&vrsn=1.0&srchtp=a&d4= 0.33&n=10&SU=0LRL+OR+0LRI; http://uclibs.org/PID/126171*.

Manabe, S., K. Bryan, and M. J. Spelman (1975) Global Ocean-Atmosphere Climate Model.1. Atmospheric Circulation. *Journal of Physical Oceanography* 5: 3–29.

Mankiw, G. (2007) Unemployment Benefits around the World. *The New York Times*, New York, June 15, 2007.

Mankinen, E. A. and C. M. Wentworth (2003) *Preliminary Paleomagnetic Results from the Coyote Creek Outdoor Classroom Drill Hole, Santa Clara Valley, California*, U.S. Geological Survey Open-

File Report 03-187. U.S. Department of the Interior, Menlo Park, CA, *http://geopubs.wr.usgs.gov/open-file/of03-187/*.

Mann, M. E., R. S. Bradley, and M. K. Hughes (1998) Global-scale temperature patterns and climate forcing over the past six centuries. *Nature* 392: 779–787.

Mann, M. E., R. S. Bradley, and M. K. Hughes (1999) Northern hemisphere temperatures during the past millennium: Inferences, uncertainties, and limitations. *Geophysical Research Letters* 26: 759–762.

Mansfield, E. and G. W. Yohe (2004) *Microeconomics: Theory & Applications*, 11th Edition. Norton, New York.

Markandya, A. and D. W. Pearce (1991) Development, the environment, and the social rate of discount. *The World Bank Research Observer* 6: 137–152.

Marshall, C. R. (2006) Explaining the Cambrian "explosion" of animals. *Annual Review of Earth and Planetary Sciences* 34: 355–384.

Marshall, P. and H. Schuttenberg (2006) *A Reef Manager's Guide to Coral Bleaching*, Great Barrier Reef Marine Park Authority, Townsville, Australia.

Matson, P. A., R. Naylor, and I. Ortiz-Monasterio (1998) Integration of environmental, agronomic, and economic aspects of fertilizer management. *Science* 280: 112–115.

Mazzetti, M. (2007) Spy Chief Backs Study of Impact of Warming. *The New York Times*, New York, May 12, 2007.

McKibben, B., C. Mooney, and R. Gelbspan (2005) Put a tiger in your think tank. *Mother Jones, http://www.motherjones.com/news/featurex/2005/05/exxon_chart.html*, accessed July 1, 2006.

McMichael, A. J., R. E. Woodruff, and S. Hales (2006) Climate change and human health: present and future risks. *Lancet* 367: 859–869.

McNamara, W. (2002) *The California Energy Crisis: Lessons for a Deregulating Industry*, PennWell, Tulsa, OK.

McNeil, M. A. and V. E. Letschert (2005) *Forecasting Electricity Demand in Developing Countries: A Study of Household Income and Appliance Ownership*. Lawrence Berkeley National Laboratories, *http://efficiency.lbl.gov/drupal.files/ees/6259%20paper%20McNeil%20final.pdf*, accessed May 10, 2008.

Measure DHS (2009) *STATcompiler*. Macro International, *http://www.statcompiler.com/index.cfm*, accessed April 18, 2009.

Mechoso, C. R., J. D. Farrara, and J. A. Spahr (1994) Achieving superlinear speedup on a heterogeneous, distributed system. *IEEE Parallel & Distributed Technology* 2: 57–61.

Mehra, R. (2006) The equity premium puzzle: a review. *Foundations and Trends in Finance* 2: 1–81 DOI: 10.1561/0500000006.

Melis, A. and T. Happe (2001) Hydrogen production. Green algae as a source of energy. *Plant Physiology* 127: 740–748.

Mendelsohn, R., W. Morrison, M. E. Schlesinger, and N. G. Andronova (2000) Country-specific market impacts of climate change. *Climatic Change* 45: 553–569.

Metschies, G. P. (2005) *International Fuel Prices*, 4th Edition. Deutsche Gesellschaft für Technische Zusammenarbeit, Eschborn, Germany, *http://www.international-fuel-prices.com/downloads/FuelPrices2005.pdf*.

Metschies, G. P. (2007) *International Fuel Prices*, 5th Edition. Deutsche Gesellschaft für Technische Zusammenarbeit, Eschborn, Germany, *http://www.gtz.de/de/dokumente/en-international-fuelprices-final2007.pdf*.

Meure, C. M., D. Etheridge, C. Trudinger, P. Steele, R. Langenfelds, T. van Ommen, A. Smith, and J. Elkins (2006) Law Dome CO_2, CH_4 and N_2O ice core records extended to 2000 years BP. *Geophysical Research Letters* 33: L14810.

Migliore, H.E. Msgr. Celestino (2008) Intervention by the Holy See. In: *UN General Assembly debating on the theme "Addressing Climate Change"*, The Holy See, New York. February 11, 2008.

Miller, K. G., M. A. Kominz, J. V. Browning, J. D. Wright, G. S. Mountain, M. E. Katz, P. J. Sugarman, B. S. Cramer, N. Christie-Blick, and S. F. Pekar (2005) The phanerozoic record of global sea-level change. *Science* 310: 1293–1298.

Miller, L. and B. C. Douglas (2004) Mass and volume contributions to twentieth-century global sea level rise. *Nature* 428: 406–409.

Miller, L. and B. C. Douglas (2006) On the rate and causes of twentieth century sea-level rise. *Philosophical Transactions of the Royal Society A-Mathematical Physical and Engineering Sciences* 364: 805–820.

Milne, A. (1990) Gaian Test. *Nature* 345: 27.

Ministry of the Environment (2007) *CO_2 Emissions in Japan*. Government of Japan, *http://www.env.go.jp/en/statistics/global/img/g_8.gif*, accessed January 6, 2007.

MIT Study Group (2006) *The Future of Geothermal Energy*, Massachusetts Institute of Technology, Cambridge, MA, *http://www1.eere.energy.gov/geothermal/pdfs/future_geo_energy.pdf*.

MIT Study Group (2007) *The Future of Coal*, Massachusetts Institute of Technology, Cambridge, MA, *http://web.mit.edu/coal/The_Future_of_Coal.pdf*.

Moberg, A., P. D. Jones, D. Lister, A. Walther, M. Brunet, J. Jacobeit, L. V. Alexander, P. M. Della-Marta, J. Luterbacher, P. Yiou, D. L. Chen, A. Tank, O. Saladie, J. Sigro, E. Aguilar, H. Alexandersson, C. Almarza, I. Auer, M. Barriendos, M. Begert, H. Bergstrom, R. Bohm, C. J. Butler, J. Caesar, A. Drebs, D. Founda, F. W. Gerstengarbe, G. Micela, M. Maugeri, H. Osterle, K. Pandzic, M. Petrakis, L. Srnec, R. Tolasz, H. Tuomenvirta, P. C. Werner, H. Linderholm, A. Philipp, H. Wanner, and E. Xoplaki (2006) Indices for daily temperature and precipitation extremes in Europe analyzed for the period 1901–2000. *Journal of Geophysical Research-Atmospheres* 111: D22106 1–25 DOI 10.1029/2006JD007103.

Moneychimp (2009) *Compound Annual Growth Rate (Annualized Return)*. *http://www.moneychimp.com/features/market_cagr.htm*, accessed May 8, 2009.

Monterey County Water Resources Agency (2006) *Salt Water Intrusion 180 ft. Aquifer*. *http://www.mcwra.co.monterey.ca.us/SVWP/01swi180.pdf*, accessed August 15, 2009.

Moore, B. D., S. H. Cheng, J. Rice, and J. R. Seemann (1998) Sucrose cycling, Rubisco expression, and prediction of photosynthetic acclimation to elevated atmospheric CO_2. *Plant, Cell and Environment* 21: 905–915.

Moore, T. (2005) Coal-based generation at the crossroads. *EPRI Journal*, Summer, 2005: 6–15.

Morgan, J. (2006) It's reno time for the city's grandest house. *The Sydney Morning Herald*, Sydney, November 10, 2006.

Morgan, P. B., G. A. Bollero, R. L. Nelson, F. G. Dohleman, and S. P. Long (2005) Smaller than predicted increase in aboveground net primary production and yield of field-grown soybean under fully open-air [CO_2] elevation. *Global Change Biology* 11: 1856–1865.

Morgenstern, O., P. Braesicke, M. M. Hurwitz, F. M. O'Connor, A. C. Bushell, C. E. Johnson, and J. A. Pyle (2008) The world avoided by the Montreal Protocol. *Geophysical Research Letters* 35: L16811 DOI 10.1029/2008GL034590.

Morse, J. W., R. S. Arvidson, and A. Luttge (2007) Calcium carbonate formation and dissolution. *Chemical Reviews* 107: 342–381.

Mortimer, D. and L. Segal (2008) Comparing the incomparable? A systematic review of competing techniques for converting descriptive measures of health status into QALY-weights. *Medical Decision Making* 28: 66–89.

Morton, O. (2007) Is this what it takes to save the world? *Nature* 447: 132–136 DOI 10.1038/447132a.

Multilateral Fund (2008) *Multilateral Fund for the Implementation of the Montreal Protocol*. *http://www.multilateralfund.org/homepage.htm*, accessed November 16, 2008.

Mundil, R., K. R. Ludwig, I. Metcalfe, and P. R. Renne (2004) Age and timing of the Permian mass extinctions: U/Pb dating of closed-system zircons. *Science* 305: 1760–1763 DOI 10.1126/science.1101012.

Murphy, E. J., J. L. Watkins, P. N. Trathan, K. Reid, M. P. Meredith, S. E. Thorpe, N. M. Johnston, A. Clarke, G. A. Tarling, M. A. Collins, J. Forcada, R. S. Shreeve, A. Atkinson, R. Korb, M. J. Whitehouse, P. Ward, P. G. Rodhouse, P. Enderlein, A. G. Hirst, A. R. Martin, S. L. Hill, I. J. Staniland, D. W. Pond, D. R. Briggs, N. J. Cunningham, and A. H. Fleming (2007) Spatial and temporal operation of the Scotia Sea ecosystem: a review of large-scale links in a krill centred food web. *Philosophical Transactions of the Royal Society B-Biological Sciences* 362: 113–148.

Nabuurs, G. J., O. Masera, K. Andrasko, P. Benitez-Ponce, R. Boer, M. Dutschke, E. Elsiddig, J. Ford-Robertson, P. Frumhoff, T. Karjalainen, O. Krankina, W. A. Kurz, M. Matsumoto, W. Oyhantcabal, N. H. Ravindranath, M. J. Sanz Sanchez, and X. Zhang (2007) Forestry. In: *Climate Change 2007: Mitigation. Contribution of Working Group III to the Fourth Assessment Report of the Intergovernmental Panel on Climate Change*, Metz, B., O. R. Davidson, P. R. Bosch, R. Dave, and L. A. Meyer, eds. Cambridge University Press, New York. pp. 541–584.

Nadel, S. (2002) Appliance and equipment efficiency standards. *Annual Review of Energy and the Environment* 27: 159–192.

Nadel, S., R. N. Elliott, M. Shepard, S. Greenberg, G. Katz, and A. T. de Almeida (2002) *Energy-Efficient Motor Systems: a Handbook on Technology, Program, and Policy Opportunities*, 2nd Edition. American Council for an Energy-Efficient Economy, Washington, D.C.

National Climatic Data Center (2007a) *Global Climate at a Glance.* NOAA Satellite and Information Service, *http://www.ncdc. noaa.gov/gcag/gcag.html*, accessed January 25, 2007.

National Climatic Data Center (2007b) *Global Surface Temperature Anomalies.* NOAA Satellite and Information Service, *http:// www.ncdc.noaa.gov/oa/climate/research/anomalies/anomalies.html*, accessed January 25, 2007.

National Energy Technology Laboratory (2007a) *Carbon Sequestration Program Environmental Reference Document*, U.S. Department of Energy, Albany, OR, *http://www.netl.doe.gov/ technologies/carbon_seq/refshelf/nepa/AA%20-%20Assembled%20 Document.pdf.*

National Energy Technology Laboratory (2007b) *Fossil Energy Power Plant Desk Reference*, U.S. Department of Energy, *http: //204.154.137.14/energy-analyses/pubs/Cost%20and%20Perfor- mance%20Baseline%20for%20Fossil%20Plants%20Desk%20 Referen.pdf.*

National Park Service (2008) *NPS Stats.* U.S. Department of the Interior, *http://www.nature.nps.gov/stats/park.cfm*, accessed July 8, 2008.

National Public Relations (2008) *Spanning the Great Divide: A Report on the Relationship Between Canadian NGOs and Corporations*, Montreal, *http://www.national.ca/en/news/2008-nov24.PDF.*

National Research Council (1982) *Causes and Effects of Stratospheric Ozone Reduction: an Update*, National Academy Press, Washington, D.C., *http://www.nap.edu/catalog.php?record_id=319.*

National Research Council (2001) *Energy Research at DOE: Was It Worth It? Energy Efficiency and Fossil Energy Research 1978 to 2000*, National Academy Press, Washington, D.C., *http://www. nap.edu/catalog/10165.html.*

National Research Council (2002) *Effectiveness and Impact of Corporate Average Fuel Economy (CAFE) Standards*, National Academy of Sciences, Washington, D.C., *http://books.nap.edu/ execsumm_pdf/10172.pdf.*

National Research Council (2004) *The Hydrogen Economy: Opportunities, Costs, Barriers, and R&D Needs*, The National Academies Press, Washington, D.C., *http://books.nap.edu/execsumm_ pdf/10922.pdf.*

National Research Council (2005) *Earth Science and Applications from Space: Urgent Needs and Opportunities to Serve the Nation*, National Academies Press, Washington, D.C., *http://www.nap. edu/catalog/11281.html.*

National Research Council (2006) *Surface Temperature Reconstructions for the Last 2,000 Years*, The National Academies Press, Washington, D.C., *http://www.nap.edu/catalog/11676.html.*

National Research Council (2007a) *Emergency and Continuous Exposure Guidance Levels for Selected Submarine Contaminants*, The National Academies Press, Washington, D.C.

National Research Council (2007b) *Environmental Impacts of Wind-Energy Projects*, National Academies of Sciences, Washington, D.C.

National Rural Electric Cooperative Association (2008) *Co-ops by the Numbers. http://www.nreca.org/AboutUs/Co-op101/Coopera- tiveFacts.htm*, accessed June 1, 2008.

National Statistics (2007) *Family Spending. http://www.statistics.gov. uk/cci/nugget.asp?id=284*, accessed Aug. 3, 2007.

NEA and IAEA (2006) *Uranium 2005: Resources, Production and Demand*, Organization for Economic Co-operation and Development, Paris, *http://www.oecdbookshop.org/oecd/get-it. asp?REF=6606031E.PDF&TYPE=browse.*

Nealson, K. (2006) Lakes of liquid CO_2 in the deep sea. *Proceedings of the National Academy of Sciences of the United States of America* 103: 13903–13904.

Neelis, M., Y. Hofman, and E. Worrell (2007) Energy efficient motors. In: *UNDO seminar on Energy Efficiency Project in CDM*, Vienna. March 19, 2007, 16 pp. *http://www.unido.org/fileadmin/ import/64788_Maartin_Neels_20March2007.pdf.*

Neue, H. U. (1997) Fluxes of methane from rice fields and potential for mitigation. *Soil Use and Management* 13: 258–267.

Newell, P. (2008) Civil society, corporate accountability and the politics of climate change. *Global Environmental Politics* 8: 122–153.

Newell, R. G., A. B. Jaffe, and R. N. Stavins (2006) The effects of economic and policy incentives on carbon mitigation technologies. *Energy Economics* 28: 563–578 10.1016/j. eneco.2006.07.004.

Nilson, S. E. and S. M. Assmann (2007) The control of transpiration. Insights from Arabidopsis. *Plant Physiology* 143: 19–27.

Nobelprize.org (2009) *The Nobel Peace Prize 2007.* Nobel Foundation, *http://nobelprize.org/nobel_prizes/peace/laureates/2007/*, accessed April 15, 2009.

Nolan, B. T. and K. J. Hitt (2006) Vulnerability of shallow groundwater and drinking-water wells to nitrate in the United States. *Environmental Science & Technology* 40: 7834–7840.

Nordhaus, W. D. (2008a) *DICE Model Discussion as of March 19, 2008. http://nordhaus.econ.yale.edu/DICE2007.htm*, accessed Aug. 28, 2008.

Nordhaus, W. D. (2008b) *A Question of Balance: Weighing the Options on Global Warming Policies*, Yale University Press, New Haven, CT.

Nordhaus, W. D. and J. G. Boyer (1999) Requiem for Kyoto: an assessment of the economics of the Kyoto Protocol. *The Energy Journal* Special Issue: 93–130. *http://www.econ.yale. edu/~nordhaus/homepage/Kyoto.pdf.*

Norris, J. R. (2005) Multidecadal changes in near-global cloud cover and estimated cloud cover radiative forcing. *Journal of Geophysical Research-Atmospheres* 110: D08206 1–17 DOI 10.1029/2004jd005600.

Nowak, R. S., D. S. Ellsworth, and S. D. Smith (2004) Functional responses of plants to elevated atmospheric CO_2: do photosynthetic and productivity data from FACE experiments support early predictions? *New Phytologist* 162: 253–280.

NOVA (2007) *Secrets of the Samurai Sword.* PBS, *http://www.pbs.org/ wgbh/nova/samurai/*, accessed April 26, 2009.

Nuclear Energy Agency and International Energy Agency (2005) *Projected Costs of Generating Electricity, 2005 Update*, Organization for Economic Cooperation and Development, Paris, *http: //www.iea.org/Textbase/npsum/ElecCostSUM.pdf.*

Nye, P. H. and P. B. Tinker (1977) *Solute Movement in the Soil-Root System*, University of California Press, Berkeley, CA.

O'Dell, J. (2005) Prius Keeps Car Dealers—but Not Repairmen—Busy. *L.A. Times*, Los Angeles, June 1, 2005.

O'Neil, J. R., R. N. Clayton, and T. K. Mayeda (1969) Oxygen isotope fractionation in divalent metal carbonates. *Journal of Chemical Physics* 51: 5547–5558.

Oak Ridge National Laboratory (2004) *ORNL Online Simple Whole Wall R-Value Calculator. http: //www.ornl.gov/sci/roofs+walls/AWT/InteractiveCalculators/NS/SimCalc.htm*, accessed April 30, 2008.

OECD (2007) *Health Care at a Glance*, Organisation for Economic Co-operation and Development, Paris, *http: //titania.sourceoecd.org/vl=112823/cl=15/nw=1/rpsv/health2007/index.htm*.

OECD (2008a) *OECD Factbook 2008: Economic, Environmental and Social Statistics; Macroeconomic Trends*, Organisation for Economic Co-operation and Development, Paris, *http: //dx.doi.org/10.1787/272560812251*.

OECD (2008b) *OECD Factbook 2008: Economic, Environmental and Social Statistics; Quality of Life*, Organisation for Economic Co-operation and Development, Paris, *http: //dx.doi.org/10.1787/275676256538*.

Oerlemans, J. (2005) Extracting a climate signal from 169 glacier records. *Science* 308: 675–677.

Office of Enforcement (1992) *Principles of Environmental Enforcement*, U.S. Environmental Protection Agency, Washington, D.C., *http: //nepis.epa.gov/Exe/ZyPURL.cgi?Dockey=500003C8.txt*.

Old Farmer's Almanac, The (2006) *How The Old Farmer's Almanac Predicts The Weather*. Yankee Publishing Inc., *http: //www.almanac.com/weathercenter/howwepredict.php*, accessed December 9, 2006.

Olivier, J. G. J., J. A. Van Aardenne, F. J. Dentener, V. Pagliari, L. N. Ganzeveld, and J. A. H. W. Peters (2005) Recent trends in global greenhouse gas emissions: regional trends 1970–2000 and spatial distribution of key sources in 2000. *Environmental Sciences* 2: 81–99.

Olshansky, S. J., D. J. Passaro, R. C. Hershow, J. Layden, B. A. Carnes, J. Brody, L. Hayflick, R. N. Butler, D. B. Allison, and D. S. Ludwig (2005) A potential decline in life expectancy in the United States in the 21st century. *New England Journal of Medicine* 352: 1138-1145 DOI 10.1056/NEJMsr043743.

Organisation for Economic Co-operation and Development (2007) *OECD/EEA Database on Instruments used for Environmental Policy and Natural Resources Management. http: //www2.oecd.org/ecoinst/queries/index.htm*, accessed July 14, 2007.

Orr, J. C., V. J. Fabry, O. Aumont, L. Bopp, S. C. Doney, R. A. Feely, A. Gnanadesikan, N. Gruber, A. Ishida, F. Joos, R. M. Key, K. Lindsay, E. Maier-Reimer, R. Matear, P. Monfray, A. Mouchet, R. G. Najjar, G. K. Plattner, K. B. Rodgers, C. L. Sabine, J. L. Sarmiento, R. Schlitzer, R. D. Slater, I. J. Totterdell, M. F. Weirig, Y. Yamanaka, and A. Yool (2005) Anthropogenic ocean acidification over the twenty-first century and its impact on calcifying organisms. *Nature* 437: 681–686.

Ortiz-Monasterio-R, J. I., K. D. Sayre, S. Rajaram, and M. McMahon (1997) Genetic progress in wheat yield and nitrogen use efficiency under four nitrogen rates. *Crop Science* 37: 898–904.

Ottman, M. J., B. A. Kimball, P. J. Pinter, G. W. Wall, R. L. Vanderlip, S. W. Leavitt, R. L. LaMorte, A. D. Matthias, and T. J. Brooks (2001) Elevated CO_2 increases sorghum biomass under drought conditions. *New Phytologist* 150: 261–273.

Ozone Secretariat (2006) *Handbook for the Montreal Protocol on Substances that Deplete the Ozone Layer*, 7th Edition. United Nations Environment Programme, Nairobi, *http: //ozone.unep.org/Publications/Handbooks/MP_Handbook_2006.pdf*.

Pagani, M., D. Lemarchand, A. Spivack, and J. Gaillardet (2005) A critical evaluation of the boron isotope-pH proxy: The accuracy of ancient ocean pH estimates. *Geochimica Et Cosmochimica Acta* 69: 953–961.

Panchuk, K., A. Ridgwell, and L. R. Kump (2008) Sedimentary response to Paleocene-Eocene Thermal Maximum carbon release: A model-data comparison. *Geology* 36: 315–318 DOI 10.1130/g24474a.1.

Parmesan, C. (2006) Ecological and evolutionary responses to recent climate change. *Annual Review of Ecology Evolution and Systematics* 37: 637–669.

Parmesan, C. and G. Yohe (2003) A globally coherent fingerprint of climate change impacts across natural systems. *Nature* 421: 37–42.

Pathak, H. and R. Wassmann (2007) Introducing greenhouse gas mitigation as a development objective in rice-based agriculture: I. Generation of technical coefficients. *Agricultural Systems* 94: 807–825.

Patz, J. A., D. Campbell-Lendrum, T. Holloway, and J. A. Foley (2005) Impact of regional climate change on human health. *Nature* 438: 310–317.

Patzek, T. W. (2004) Thermodynamics of the corn-ethanol biofuel cycle. *Critical Reviews in Plant Sciences* 23: 519–567; updated July 22, 2006.

Patzek, T. W. (2006) *The Real Biofuel Cycles*, Online supporting material for Science 312: 1747 (2006), Berkeley, CA, *http: //petroleum.berkeley.edu/patzek/BiofuelQA/Materials/RealFuelCycles-Web.pdf*.

Patzek, T. W. and D. Pimentel (2005) Thermodynamics of energy production from biomass. *Critical Reviews in Plant Sciences* 24: 327–364.

Pearce, D. and E. Özdemiroglu (2002) *Economic Valuation with Stated Preference Techniques: Summary Guide*, Department for Transport, Local Government and the Regions, London, *http: //www.communities.gov.uk/documents/corporate/pdf/146871.pdf*.

Pearson, P. N. and M. R. Palmer (2000) Atmospheric carbon dioxide concentrations over the past 60 million years. *Nature* 406: 695–699.

Pell, Cardinal George (2008) *Global Warming*. Archdiocese of Sydney, Our People, *http: //www.sydney.catholic.org.au/people/archbishop/stc/2008/2008420_382.shtml*, accessed April 14, 2009.

Pessarakli, M. and I. Szabolcs (1999) Soil salinity and sodicity as particular plant/crop stress factors. In: *Handbook of Plant and Crop Stress*, 2nd Edition. Pessarakli, M., ed., Dekker, New York. pp. 1–15.

Petty, S. and S. Porro (2007) *Updated U.S. Geothermal Supply Characterization*, Conference Paper NREL/CP-640-41073. National Renewable Energy Laboratory, Golden, CO, *http: //www.nrel.gov/docs/fy07osti/41073.pdf*.

Pew Research Center (2006a) *America's Image Slips, But Allies Share U.S. Concerns Over Iran, Hamas; No Global Warming Alarm in the U.S., China*. Pew Global Attitudes Project, *http: //pewglobal.org/reports/display.php?ReportID=252*, accessed November 21, 2008.

Pew Research Center (2006b) *Little Consensus on Global Warming*. The Pew Research Center for the People & the Press, *http: //people-press.org/report/280/little-consensus-on-global-warming*, accessed November 21, 2008.

Pew Research Center (2007) *Global Warming: A Divide on Causes and Solutions*. The Pew Research Center for the People & the Press, *http: //people-press.org/report/303/global-warming-a-divide-on-causes-and-solutions*, accessed November 21, 2008.

Pew Research Center (2008) *A Deeper Partisan Divide over Global Warming*. The Pew Research Center for the People & the Press, *http: //people-press.org/report/417/a-deeper-partisan-divide-over-global-warming*, accessed November 21, 2008.

Pew Research Center (2009a) *The Global Middle Class: Views on Democracy, Religion, Values, and Life Satisfaction in Emerging*

Nations Pew Global Attitudes Project, Washington, D.C., *http://pewglobal.org/commentary/pdf/1051.pdf*.

Pew Research Center (2009b) *On Obama's Desk: Economy, Jobs Trump All Other Policy Priorities*. The Pew Research Center for the People & the Press, *http://pewresearch.org/pubs/1087/economy-jobs-top-public-priorities-2009*, accessed March 15, 2009.

Philibert, C. (2007) *Technology Penetration and Capital Stock Turnover: Lessons from IEA Scenario Analysis*, International Energy Agency, Paris, *http://www.iea.org/textbase/papers/2007/capital_stock.pdf*.

Phillips, C. and G. Thompson (2001) *What Is a Qualy?*, Vol. 1, No. 6, Hayward Medical Communications, *http://www.medicine.ox.ac.uk/bandolier/painres/download/whatis/QALY.pdf*.

Pilson, M. E. Q. (1998) *An Introduction to the Chemistry of the Sea*, Prentice Hall, Upper Saddle River, NJ.

Pimentel, D. and T. W. Patzek (2005) Ethanol production using corn, switchgrass, and wood; biodiesel production using soybean and sunflower. *Natural Resources Research* 14: 65–76.

Pollack, H. N. and J. E. Smerdon (2004) Borehole climate reconstructions: Spatial structure and hemispheric averages. *Journal of Geophysical Research-Atmospheres* 109: D11106 1–9 DOI 10.1029/2003jd004163.

Pollard, D. and J. F. Kasting (2005) Snowball Earth: A thin-ice solution with flowing sea glaciers. *Journal of Geophysical Research-Oceans* 110: C07010 1–16 DOI 10.1029/2004JC002525.

Pollitt, M. G. (2005) *Electricity Reform in Chile: Lessons for Developing Countries*, Cambridge Working Papers in Economics, CWPE 0448. Dept. of Applied Economics, University of Cambridge, Cambridge, *http://www.econ.cam.ac.uk/electricity/publications/wp/ep51.pdf*.

Poorter, H. and M. L. Navas (2003) Plant growth and competition at elevated CO_2: on winners, losers and functional groups. *New Phytologist* 157: 175–198.

Poorter, H. and M. Perez-Soba (2001) The growth response of plants to elevated CO_2 under non-optimal environmental conditions. *Oecologia* 129: 1–20.

Poorter, H., Y. VanBerkel, R. Baxter, J. DenHertog, P. Dijkstra, R. M. Gifford, K. L. Griffin, C. Roumet, J. Roy, and S. C. Wong (1997) The effect of elevated CO_2 on the chemical composition and construction costs of leaves of 27 C_3 species. *Plant Cell and Environment* 20: 472–482.

Population Division (2009) *World Population Prospects: The 2008 Revision*. Department of Economic and Social Affairs of the United Nations Secretariat, *http://esa.un.org/unpp*, accessed April 18, 2009.

Portney, P. R. and J. P. Weyant, eds. (1999) *Discounting and Intergenerational Equity*, Resources for the Future, Washington, D.C.

Pounds, J. A., M. R. Bustamante, L. A. Coloma, J. A. Consuegra, M. P. L. Fogden, P. N. Foster, E. La Marca, K. L. Masters, A. Merino-Viteri, R. Puschendorf, S. R. Ron, G. A. Sanchez-Azofeifa, C. J. Still, and B. E. Young (2006) Widespread amphibian extinctions from epidemic disease driven by global warming. *Nature* 439: 161–167.

Powe, N. A. (2007) *Redesigning Environmental Valuation: Mixing Methods within Stated Preference Techniques*, Edward Elgar, Cheltenham, UK, *http://www.loc.gov/catdir/toc/ecip077/2006102433.html*.

PowerStream (2007) *Battery Technology Comparison: Rechargeable. http://www.powerstream.com/Compare.htm*, accessed Aug. 18, 2007.

Price, L., S. de la Rue du Can, J. Sinton, E. Worrell, Z. Nan, J. Sathaye, and M. Levine (2006) *Sectoral Trends in Global Energy Use and Greenhouse Gas Emissions*, Lawrence Berkeley National Laboratory, Berkeley, CA, *http://ies.lbl.gov/iespubs/56144.pdf*.

Price, P. B. and T. Sowers (2004) Temperature dependence of metabolic rates for microbial growth, maintenance, and survival. *Proceedings of the National Academy of Sciences of the United States of America* 101: 4631–4636.

Pring, G. W. (2001) The United States Perspective. In: *Kyoto: from Principles to Practice*, Cameron, P. D. and D. N. Zillman, eds. Vol. 60, International Environmental Law and Policy. Kluwer Law International, The Hague. pp. 185–224.

Prins, G. and S. Rayner (2007) Time to ditch Kyoto. *Nature* 449: 973–975 10.1038/449973a.

Public Service of New Hampshire (2008) *Motor Efficiency. http://www.psnh.com/Business/SmallBusiness/Motor.asp*, accessed April 15, 2008.

Rachmilevitch, S., A. B. Cousins, and A. J. Bloom (2004) Nitrate assimilation in plant shoots depends on photorespiration. *Proceedings of the National Academy of Sciences of the United States of America* 101: 11506–11510.

Ragland, K. W., D. J. Aerts, and A. J. Baker (1991) Properties of wood for combustion analysis. *Bioresource Technology* 37: 161–168.

Rahmstorf, S. (2007) A semi-empirical approach to projecting future sea-level rise. *Science* 315: 368–370.

Ramanathan, V., R. D. Cess, E. F. Harrison, P. Minnis, B. R. Barkstrom, E. Ahmad, and D. Hartmann (1989) Cloud-radiative forcing and climate: Results from the earth radiation budget experiment. *Science* 243: 57–63.

Ramankutty, N., J. A. Foley, J. Norman, and K. McSweeney (2002) The global distribution of cultivable lands: current patterns and sensitivity to possible climate change. *Global Ecology and Biogeography* 11: 377–392.

Randall, D. A., T. D. Ringler, R. P. Heikes, P. Jones, and J. Baumgardner (2002) Climate modeling with spherical geodesic grids. *Computing in Science & Engineering* 4: 32–41.

Rapp, D. (2008) *Assessing Climate Change: Temperatures, Solar Radiation, and Heat Balance*, Springer, Berlin.

Rasmussen, B., I. R. Fletcher, J. J. Brocks, and M. R. Kilburn (2008) Reassessing the first appearance of eukaryotes and cyanobacteria. *Nature* 455: 1101-U9 DOI 10.1038/nature07381.

Rasse, D. P., G. Peresta, and B. G. Drake (2005) Seventeen years of elevated CO_2 exposure in a Chesapeake Bay Wetland: sustained but contrasting responses of plant growth and CO_2 uptake. *Global Change Biology* 11: 369–377.

Read, A. J., P. N. Halpin, L. B. Crowder, K. D. Hyrenbach, B. D. Best, E. Fujioka, and M. S. Coyne, Editors (2007) *OBIS-SEAMAP: Mapping Marine Mammals, Birds and Turtles. http://seamap.env.duke.edu*, accessed May 12, 2007.

Reed, P. and M. Hudson (2005) *We Test the Tips: What Really Saves Gas? And How Much?* edmunds.com, *http://www.edmunds.com/advice/fueleconomy/articles/106842/article.html*, accessed July 8, 2007.

Regional Greenhouse Gas Initiative (2008) *http://www.rggi.org/home*, accessed November 18, 2008.

Reich, P. B., S. E. Hobbie, T. Lee, D. S. Ellsworth, J. B. West, D. Tilman, J. M. H. Knops, S. Naeem, and J. Trost (2006a) Nitrogen limitation constrains sustainability of ecosystem response to CO_2. *Nature* 440: 922–925.

Reich, P. B., B. A. Hungate, and Y. Q. Luo (2006b) Carbon-nitrogen interactions in terrestrial ecosystems in response to rising atmospheric carbon dioxide. *Annual Review of Ecology Evolution and Systematics* 37: 611–636.

Reiter, P., C. J. Thomas, P. M. Atkinson, S. I. Hay, S. E. Randolph, D. J. Rogers, G. D. Shanks, R. W. Snow, and A. Spielman (2004) Global warming and malaria: a call for accuracy. *Lancet Infectious Diseases* 4: 323–324.

REN21 (2008) *Renewables 2007 Global Status Report*, Worldwatch Institute and Deutsche Gesellschaft für Technische Zusammenarbeit, Paris and Washington, D.C., *http://www.ren21.net/pdf/RE2007_Global_Status_Report.pdf*.

Renewable Fuels Association (2007) *U.S. Ethanol Biorefinery Locations http://www.ethanolrfa.org/objects/documents/plant-map_040307.pdf*, accessed Aug. 31, 2007.

Report of the Secretary-General (2008) *Sustainable fisheries, including through the 1995 Agreement for the Implementation of the Provisions of the United Nations Convention on the Law of the Sea of 10 December 1982 relating to the Conservation and Management of Straddling Fish Stocks and Highly Migratory Fish Stocks, and related instruments*, United Nations, New York, *http://daccess-dds.un.org/doc/UNDOC/GEN/N08/422/64/PDF/N0842264.pdf*.

Revelle, R. and H. E. Suess (1957) Carbon dioxide exchange between atmosphere and ocean and the question of an increase of atmospheric CO_2 during the past decades. *Tellus* 9: 18–27.

Revenga, C., S. Murray, J. Abramovitz, and A. Hammond (1998) *Watersheds of the World: Ecological Value and Vulnerability*, World Resources Institute, Washington, D.C., *http://earthtrends.wri.org/pdf_library/maps/P1_27.pdf*.

Revenoor Company (2007) *Revenoor Alcohol Stilles*, Yamhill, OR, *http://www.revenoor.com/pdf/Stills_Catalog_2006.pdf*.

Revkin, A. C. (2007) A team of 2, following the scent of polar bears. *The New York Times*, New York, June 5, 2007.

Revkin, A. C. (2008) News Coverage of Climate Entering 'Trance'? citing Maxwell Boykoff and Maria Mansfield. *The New York Times*, New York, December 5, 2008.

Rind, D. (2002) Climatology—The sun's role in climate variations. *Science* 296: 673–677.

Ringler, T. D., R. P. Heikes, and D. A. Randall (2000) Modeling the atmospheric general circulation using a spherical geodesic grid: A new class of dynamical cores. *Monthly Weather Review* 128: 2471–2490.

Rogers, H. H. and R. C. Dahlman (1993) Crop responses to CO_2 enrichment. *Vegetatio* 104: 117–131.

Rogner, H.-H., D. Zhou, B. R., P. Crabbé, O. Edenhofer, B. Hare, L. Kuijpers, and M. Yamaguchi (2007) Introduction. In: *Climate Change 2007: Mitigation. Contribution of Working Group III to the Fourth Assessment Report of the Intergovernmental Panel on Climate Change*, Metz, B., O. R. Davidson, P. R. Bosch, R. Dave, and L. A. Meyer, eds. Cambridge University Press, New York. pp. 95–116.

Root, T. L., J. T. Price, K. R. Hall, S. H. Schneider, C. Rosenzweig, and J. A. Pounds (2003) Fingerprints of global warming on wild animals and plants. *Nature* 421: 57–60.

Rothwell, G. S. (2002) *Does the US Subsidize Nuclear Power Insurance?*, Stanford Institute for Economic Policy Research, Stanford, CA, *http://siepr.stanford.edu/papers/briefs/policybrief_jan02.pdf*.

Royal Society, The (2005) *Ocean Acidification Due to Increasing Atmospheric Carbon Dioxide*, The Royal Society, London, *http://www.royalsoc.ac.uk/displaypagedoc.asp?id=13539*.

Royal Society, The (2006) *Letter to Exxon Mobil*, London, *http://royalsociety.org/displaypagedoc.asp?id=23780*.

Royer, D. L., R. A. Berner, and D. J. Beerling (2001) Phanerozoic atmospheric CO_2 change: evaluating geochemical and paleobiological approaches. *Earth-Science Reviews* 54: 349–392.

Royer, D. L., R. A. Berner, I. P. Montañez, N. J. Tabor, and D. J. Beerling (2004) CO_2 as a primary driver of Phanerozoic climate. *GSA Today* 14: 4–10 DOI 10.1130/1052-5173(2004)014<4: CAAPDO>2.0.CO;2.

Rubin, E. S., C. Chen, and A. B. Rao (2007) Cost and performance of fossil fuel power plants with CO_2 capture and storage. *Energy Policy* 35: 4444–4454 DOI 10.1016/j.enpol.2007.03.009.

Rupprecht, J., B. Hankamer, J. H. Mussgnug, G. Ananyev, C. Dismukes, and O. Kruse (2006) Perspectives and advances of biological H_2 production in microorganisms. *Applied Microbiology and Biotechnology* 72: 442–449.

Saad, L. (2008) *Americans' Weight Issues Not Going Away*. Gallup Poll, *http://www.gallup.com/poll/112426/Americans-Weight-Issues-Going-Away.aspx*, accessed June17, 2009.

Saad, L. (2009) *Increased Number Think Global Warming is "Exaggerated"*. Gallup Poll, *http://www.gallup.com/poll/116590/Increased-Number-Think-Global-Warming-Exaggerated.aspx*, accessed March 15, 2009.

Sage, R. F. (1999) Why C_4 photosynthesis? In: C_4 *Plant Biology*, Sage, R. F. and R. K. Monson, eds. Academic Press, San Diego. pp. 3–16.

Sage, R. F. (2002) How terrestrial organisms sense, signal, and respond to carbon dioxide. *Integrative and Comparative Biology* 42: 469–480.

Salter, S., G. Sortino, and J. Latham (2008) Sea-going hardware for the cloud albedo method of reversing global warming. *Philosophical Transactions of the Royal Society A Mathematical Physical and Engineering Sciences* 366: 3989–4006.

Sands, P. (1988) *Chernobyl, Law and Communication: Transboundary Nuclear Air Pollution, the Legal Materials*, Grotius Publications, Cambridge.

Sands, P. and R. Y. Jennings (2003) *Principles of International Environmental Law*, 2nd Edition. Cambridge University Press, Cambridge.

Sarma, K. M. (2005) Compliance with the Montreal Protocol. In: *7th International Conference on Environmental Compliance and Enforcement*, International Network for Environmental Compliance and Enforcement, Marrakech, Morroco, pp. 1–12, *http://inece.org/conference/7/vol1/Sarma.pdf*.

SASI Group and M. Newman (2006) *Hydroelectric Power*. University of Sheffield, *http://www.worldmapper.org/posters/worldmapper_map110_ver5.pdf*, accessed February 8, 2008.

Satoh, M., H. Tomita, T. Nasuno, S.-I. Iga, K. Goto, Y. Tanaka, M. Tsugawa, M. Sakasita, and M. Kogi (2004) Development of super high-resolution atmospheric and oceanic general circulation models on quasi-uniform grids. In *Annual Report of the Japanese Earth Simulator Center*.

Scharf, P. and B. Wiebold (2001) Nitrogen prices—how do they affect optimum N management? *Integrated Pest & Crop Management Newletter, University of Missouri-Columbia* 11: Article 1 of 9.

Scherer, K., H. Fichtner, T. Borrmann, J. Beer, L. Desorgher, E. Flukiger, H. J. Fahr, S. E. S. Ferreira, U. W. Langner, M. S. Potgieter, B. Heber, J. Masarik, N. J. Shaviv, and J. Veizer (2006) Interstellar-terrestrial relations: Variable cosmic environments, the dynamic heliosphere, and their imprints on terrestrial archives and climate. *Space Science Reviews* 127: 327–465 DOI10.1007/s11214-006-9126-6.

Schiermeier, Q. (2006) A sea change. *Nature* 439: 256–260.

Schievano, E., T. Calisti, I. Menegazzo, R. Battistutta, E. Peggion, S. Mammi, G. Palu, and A. Loregian (2004) pH-dependent conformational changes and topology of a herpesvirus translocating peptide in a membrane-mimetic environment. *Biochemistry* 43: 9343–9351.

Schipper, L. (2007) *Vehicle Efficiency and CO_2 Emissions: Troubling Trends*, World Resources Institute, Center for Sustainable Transport, Washington, D.C., *http://embarq.wri.org/documents/Schipper-VehicEfficiency.pdf*.

Schipper, L., C. Marie-Lilliu, and L. Fulton (2002) Diesels in Europe - Analysis of characteristics, usage patterns, energy savings and CO_2 emission implications. *Journal of Transport Economics and Policy* 36: 305–340.

Schipper, M. (2006) *Energy-related Carbon Dioxide Emissions in U.S. Manufacturing*, Energy Information Administration, U.S.

Department of Energy, Washington, D.C., *http://www.eia.doe. gov/oiaf/1605/ggrpt/pdf/industry_mecs.pdf*.

Schneider, M. K., A. Luscher, M. Richter, U. Aeschlimann, U. A. Hartwig, H. Blum, E. Frossard, and J. Nosberger (2004) Ten years of free-air CO_2 enrichment altered the mobilization of N from soil in *Lolium perenne* L. swards. *Global Change Biology* 10: 1377–1388.

Schoeberl, M. R. (1999) Ozone and stratospheric chemistry. In: *Earth Observing System (EOS) Science Plan*, King, M. D., ed., NASA/Goddard Space Flight Center, Greenbelt, MD. pp. 309–337, *http://eospso.gsfc.nasa.gov/science_plan/Ch7.pdf*.

Schrag, D. P., J. F. Adkins, K. McIntyre, J. L. Alexander, D. A. Hodell, C. D. Charles, and J. F. McManus (2002a) The oxygen isotopic composition of seawater during the Last Glacial Maximum. *Quaternary Science Reviews* 21: 331–342.

Schrag, D. P., R. A. Berner, P. F. Hoffman, and G. P. Halverson (2002b) On the initiation of a snowball Earth. *Geochemistry Geophysics Geosystems* 3: 1–21 DOI 10.1029/2001GC000219.

Schrank, D. and T. Lomax (2007) *The 2007 Urban Mobility Report*, Texas Transportation Institute, The Texas A&M University System, College Station, TX, *http://tti.tamu.edu/documents/ mobility_report_2007_wappx.pdf*.

Schulte, P., R. P. Speijer, H. Brinkhuis, A. Kontny, P. Claeys, S. Galeotti, and J. Smit (2008) Comment on the paper "Chicxulub impact predates K-T boundary: New evidence from Brazos, Texas" by Keller et al. (2007)—Discussion. *Earth and Planetary Science Letters* 269: 613-619 DOI 10.1016/j.epsl.2007.11.066.

Schulze, E. D., G. Gebauer, W. Schulze, and J. S. Pate (1991) The utilization of nitrogen from insect capture by different growth forms of *Drosera* from Southwest Australia. *Oecologia* 87: 240–246.

Schulze, W. and E. D. Schulze (1990) Insect capture and growth of the insectivorous *Drosera rotundifolia* L. *Oecologia* 82: 427–429.

Schwartz, M. W., L. R. Iverson, A. M. Prasad, S. N. Matthews, and R. J. O'Connor (2006) Predicting extinctions as a result of climate change. *Ecology* 87: 1611–1615.

Science Applications International Corporation (2002) *Greenhouse Gas Emission Reductions and Natural Gas Vehicles*, National Energy Technology Laboratory, Pittsburgh, PA, *http://www. osti.gov/bridge/servlets/purl/816573-V0Mlif/native/816573.PDF*.

Searles, P. S. and A. J. Bloom (2003) Nitrate photoassimilation in tomato leaves under short-term exposure to elevated carbon dioxide and low oxygen. *Plant, Cell and Environment* 26: 1247–1255.

Sharkey, T. D. (1988) Estimating the rate of photorespiration in leaves. *Physiologia Plantarum* 73: 147–152.

Shaviv, N. J. and J. Veizer (2003) Celestial driver of Phanerozoic climate? *GSA Today* 13: 4–10.

Shaw, M. R., E. S. Zavaleta, N. R. Chiariello, E. E. Cleland, H. A. Mooney, and C. B. Field (2002) Grassland responses to global environmental changes suppressed by elevated CO_2. *Science* 298: 1987–1990.

Sheehan, P. M. (2001) The Late Ordovician mass extinction. *Annual Review of Earth and Planetary Sciences* 29: 331–364.

Shein, K. A. and 116 other authors (2006) State of the climate in 2005. *Bulletin of the American Meteorological Society* 87: S6–S102.

Sherry, R. A., X. H. Zhou, S. L. Gu, J. A. Arnone, D. S. Schimel, P. S. Verburg, L. L. Wallace, and Y. Q. Luo (2007) Divergence of reproductive phenology under climate warming. *Proceedings of the National Academy of Sciences of the United States of America* 104: 198–202.

Siebert, S., P. Doll, J. Hoogeveen, J. M. Faures, K. Frenken, and S. Feick (2005) Development and validation of the global map of irrigation areas. *Hydrology and Earth System Sciences* 9: 535–547.

Sims, R. E. H., R. N. Schock, A. Adegbululgbe, J. Fenhann, I. Konstantinaviciute, W. Moomaw, H. B. Nimir, B. Schlamadinger,

J. Torres-Martínez, C. Turner, Y. Uchiyama, S. J. V. Vuori, N. Wamukonya, and X. Zhang (2007) Energy supply. In: *Climate Change 2007: Mitigation. Contribution of Working Group III to the Fourth Assessment Report of the Intergovernmental Panel on Climate Change*, Metz, B., O. R. Davidson, P. R. Bosch, R. Dave, and L. A. Meyer, eds. Cambridge University Press, New York. pp. 251–322.

Six, J. and J. D. Jastrow (2002) Organic matter turnover. In: *Encyclopedia of Soil Science*, Lal, R., ed., Marcel Dekker, New York. pp. 936–942.

Sloan, T. and A. W. Wolfendale (2008) Testing the proposed causal link between cosmic rays and cloud cover. *Environmental Research Letters* 3: 024001 1–6 DOI 10.1088/1748-9326/3/2/024001.

Small, J., S. J. Goetz, and S. I. Hay (2003) Climatic suitability for malaria transmission in Africa, 1911–1995. *Proceedings of the National Academy of Sciences of the United States of America* 100: 15341–15345.

Smart, D. R. and A. J. Bloom (2001) Wheat leaves emit nitrous oxide during nitrate assimilation. *Proceedings of the National Academy of Sciences of the United States of America* 98: 7875–7878.

Smetacek, V. and S. Nicol (2005) Polar ocean ecosystems in a changing world. *Nature* 437: 362–368.

Smith, P., D. Martino, Z. Cai, D. Gwary, H. Janzen, P. Kumar, B. McCarl, S. Ogle, F. O'Mara, C. Rice, B. Scholes, and O. Sirotenko (2007) Agriculture. In: *Climate Change 2007: Mitigation. Contribution of Working Group III to the Fourth Assessment Report of the Intergovernmental Panel on Climate Change*, Metz, B., O. R. Davidson, P. R. Bosch, R. Dave, and L. A. Meyer, eds. Cambridge University Press, New York. pp. 497–540.

Smokers, R., R. Vermeulen, R. van Mieghem, and R. Gense (2006) *Review and Analysis of the Reduction Potential and Costs of Technological and other Measures to Reduce CO_2-Emissions from Passenger Cars*, European Commission Report. TNO Science and Industry, Delft, Netherlands.

Snow, C. P. (1959) *The Two Cultures and the Scientific Revolution*, Rede Lecture 1959. Cambridge University Press, Cambridge.

Solanki, S. K., I. G. Usoskin, B. Kromer, M. Schussler, and J. Beer (2004) Unusual activity of the Sun during recent decades compared to the previous 11,000 years. *Nature* 431: 1084–1087.

Somerton, C. W. (2005) *Wind Energy Calculations*, Mechanical Engineering 417. Michigan State U., East Lansing, MI, *http://www. egr.msu.edu/classes/me491/benard/WindCalc.pdf*.

Spahni, R., J. Chappellaz, T. F. Stocker, L. Loulergue, G. Hausammann, K. Kawamura, J. Fluckiger, J. Schwander, D. Raynaud, V. Masson-Delmotte, and J. Jouzel (2005) Atmospheric methane and nitrous oxide of the late Pleistocene from Antarctic ice cores. *Science* 310: 1317–1321.

Spotnitz, R. (2005) Advanced EV and HEV Batteries. *IEEE Xplore*, Sept. 7–9, 334–337, *http://ieeexplore.ieee.org/ iel5/10416/33078/01554579.pdf?arnumber=1554579*.

Stainless Steel Information Center (2008) *Chemical and Physical Properties of Stainless Steel*. Specialty Steel Industry of North America, *http://www.ssina.com/composition/chemical.html*, accessed April 21, 2008.

Stange, G. (1996) Sensory and behavioural responses of terrestrial invertebrates to biogenic carbon dioxide gradients. *Advances in Bioclimatology* 4: 223–253.

Stange, G. and S. Stowe (1999) Carbon-dioxide sensing structures in terrestrial arthropods. *Microscopy Research and Technique* 47: 416–427.

Steele, R. V. (2007) The story of a new light source. *Nature Photonics* 1: 25–26.

Stehfest, E. and L. Bouwman (2006) N_2O and NO emission from agricultural fields and soils under natural vegetation: summarizing available measurement data and modeling of global annual emissions. *Nutrient Cycling in Agroecosystems* 74: 207–228.

Steinfeld, H., P. Gerber, T. Wassenaar, V. Castel, M. Rosales, and C. de Haan (2006) *Livestock's Long Shadow: Environmental Issues and Options*, Food and Agriculture Organization of the United Nations, Rome, *ftp: //ftp.fao.org/docrep/fao/010/a0701e/a0701e. pdf*.

Stellwagen Bank National Marine Sanctuary (2001) *New Whales of 2000 - Part 2*. *http: //stellwagen.nos.noaa.gov/gallery/ newwhales/2000tails/2000tails2.html*, accessed May 12, 2007.

Stern, N. (2008) The economics of climate change. *American Economic Review* 98: 1–37 DOI 10.1257/aer.98.2.1.

Stern, N. H. (2007) *The Economics of Climate Change: the Stern Review*, Great Britain Treasury, Cambridge University Press, Cambridge, *http: //www.hm-treasury.gov.uk/independent_ reviews/stern_review_economics_climate_change/stern_review_ report.cfm*.

Stevick, P. T., J. Allen, P. J. Clapham, N. Friday, S. K. Katona, F. Larsen, J. Lien, D. K. Mattila, P. J. Palsboll, J. Sigurjonsson, T. D. Smith, N. Oien, and P. S. Hammond (2003) North Atlantic humpback whale abundance and rate of increase four decades after protection from whaling. *Marine Ecology-Progress Series* 258: 263–273.

Stirling, I. and C. L. Parkinson (2006) Possible effects of climate warming on selected populations of polar bears (*Ursus maritimus*) in the Canadian Arctic. *Arctic* 59: 261–275.

Stirton, R. A. (1947) Observations on evolutionary rates in hypsodonty. *Evolution* 1: 32–41.

Stoeve, J., M. M. Holland, W. Meir, T. Scambos, and M. Serreze (2007) Arctic sea ice decline: Faster than forecast. *Geophysical Research Letters* 34, L09501: DOI 10.1029/2007GL029703.

Stoney, W. E. (2008) *ASPRS Guide to Land Imaging Satellites*, American Society for Photogrammetry & Remote Sensing, Bethesda, MD, *http: //www.asprs.org/news/satellites/ASPRS_ DATABASE_021208.pdf*.

Stott, P. A., G. S. Jones, J. A. Lowe, P. Thorne, C. Durman, T. C. Johns, and J. C. Thelen (2006) Transient climate simulations with the HadGEM1 climate model: Causes of past warming and future climate change. *Journal of Climate* 19: 2763–2782.

Stott, P. A., D. A. Stone, and M. R. Allen (2004) Human contribution to the European heatwave of 2003. *Nature* 432: 610–614.

Stuart, S. N., J. S. Chanson, N. A. Cox, B. E. Young, A. S. L. Rodrigues, D. L. Fischman, and R. W. Waller (2004) Status and trends of amphibian declines and extinctions worldwide. *Science* 306: 1783–1786.

Suarez, F., J. Furne, J. Springfield, and M. Levitt (1997) Insights into human colonic physiology obtained from the study of flatus composition. *American Journal of Physiology-Gastrointestinal and Liver Physiology* 35: G1028–G1033.

Sunstein, C. R. (2007) Of Montreal and Kyoto: A Tale of Two Protocols. *Harvard Environmental Law Review* 31: 1–65.

Supreme Court of the U.S. (2007) *Massachusetts et al. v. Environmental Protection Agency et al., No. 05-1120*, Washington, D.C., *http: //www.supremecourtus.gov/opinions/06pdf/05-1120.pdf*.

Sustainability@MIT (2007) *Revolving Door*. Massachusetts Institute of Technology, *http: //sustainability.mit.edu/Revolving_Door*, accessed April 30, 2008.

Taguchi, T. (2008) Present status of energy saving technologies and future prospect in white LED lighting. *IEEJ Transactions on Electrical and Electronic Engineering* 3: 21–26 DOI 10.1002/ Tee.20228.

Taiz, L. and E. Zeiger (2006) *Plant Physiology*, 4th Edition. Sinauer Associates, Sunderland, MA.

Tanser, F. C., B. Sharp, and D. le Sueur (2003) Potential effect of climate change on malaria transmission in Africa. *Lancet* 362: 1792–1798.

Taylor, T. N., P. M. Schwarz, and J. E. Cochell (2005) 24/7 hourly response to electricity real-time pricing with up to eight summers of experience. *Journal of Regulatory Economics* 27: 235–262.

Techconsult (2007) *Analysis of Wind Power in the Danish Electricity Supply in 2005 and 2006*. Reel Energi Oplysning, *http: //www. wind-watch.org/documents/wp-content/uploads/dk-analysis-wind. pdf*, accessed February 20, 2008.

Terao, T., S. Miura, T. Yanagihara, T. Hirose, K. Nagata, H. Tabuchi, H. Y. Kim, M. Lieffering, M. Okada, and K. Kobayashi (2005) Influence of free-air CO_2 enrichment (FACE) on the eating quality of rice. *Journal of the Science of Food and Agriculture* 85: 1861–1868.

Terrachoice Group Inc. (2009) *The Seven Sins of Greenwashing: Environmental Claim in Consumer Markets*, TerraChoice Environmental Marketing, Philadelphia, *http: //sinsofgreenwashing. org/?dl_id=4*.

Thomas, C. D., A. Cameron, R. E. Green, M. Bakkenes, L. J. Beaumont, Y. C. Collingham, B. F. N. Erasmus, M. F. de Siqueira, A. Grainger, L. Hannah, L. Hughes, B. Huntley, A. S. van Jaarsveld, G. F. Midgley, L. Miles, M. A. Ortega-Huerta, A. T. Peterson, O. L. Phillips, and S. E. Williams (2004) Extinction risk from climate change. *Nature* 427: 145–148.

Thompson, L. G., E. Mosley-Thompson, H. Brecher, M. Davis, B. Leon, D. Les, P. N. Lin, T. Mashiotta, and K. Mountain (2006) Abrupt tropical climate change: Past and present. *Proceedings of the National Academy of Sciences of the United States of America* 103: 10536–10543.

Thompson, L. G., E. Mosley-Thompson, M. E. Davis, K. A. Henderson, H. H. Brecher, V. S. Zagorodnov, T. A. Mashiotta, P. N. Lin, V. N. Mikhalenko, D. R. Hardy, and J. Beer (2002) Kilimanjaro ice core records: Evidence of Holocene climate change in tropical Africa. *Science* 298: 589–593.

Thompson, M. (2007) *Generic Battery Technology Comparison*. *http: //www.madkatz.com/ev/batteryTechnologyComparison.html*, accessed Aug. 18, 2007.

Tiax LLC (2007) *Full Fuel Cycle Assessment: Tank to Wheels Emissions and Energy Consumption*, California Energy Commission, Sacramento, *http: //www.energy.ca.gov/2007publications/CEC-600-2007-003/CEC-600-2007-003-D.PDF*.

Tobin, J. (1974) What is permanent endowment income? *American Economic Review* 64: 427–432.

Tolbert, N. E. (1994) Role of photosynthesis and photorespiration in regulating CO_2 and O_2. In: *Regulation of CO_2 and O_2 by Photosynthetic Carbon Metabolism*, Tolbert, N. E. and J. Preiss, eds. Oxford University Press, New York. pp. 8–33.

Torsvik, T. H. (2003) The Rodinia jigsaw puzzle. *Science* 300: 1379–1381.

Torsvik, T. H. and L. R. M. Cocks (2004) Earth geography from 400 to 250 Ma: a palaeomagnetic, faunal and facies review. *Journal of the Geological Society* 161: 555–572.

Travis, D. J., A. M. Carleton, and R. G. Lauritsen (2002) Climatology: Contrails reduce daily temperature range—A brief interval when the skies were clear of jets unmasked an effect on climate. *Nature* 418: 601.

Travis, D. J., A. M. Carleton, and R. G. Lauritsen (2004) Regional variations in US diurnal temperature range for the 11–14 September 2001 aircraft groundings: Evidence of jet contrail influence on climate. *Journal of Climate* 17: 1123–1134.

Trenberth, K. E., J. T. Fasullo, and J. Kiehl (2008) Earth's global energy budget. *Bulletin of the American Meteorological Society* 90: 311–324 DOI 10.1175/2008BAMS2634.1.

Trenberth, K. E., P. D. Jones, P. Ambenje, R. Bojariu, D. Easterling, A. Klein Tank, D. Parker, F. Rahimzadeh, J. A. Renwick, M. Rusticucci, B. Soden, and P. Zhai (2007) Observations: Surface and atmospheric climate change. In: *Climate Change 2007: The Physical Science Basis. Contribution of Working Group I to the Fourth Assessment Report of the Intergovernmental Panel on*

Climate Change, Solomon, S., D. Qin, M. Manning, Z. Chen, M. Marquis, K. B. Averyt, M. Tignor, and H. L. Miller, eds. Cambridge University Press, Cambridge, pp. 237–336.

Trindade, J. A. and C. Fiolhais (1998) *Virtual Water. http: //nautilus.fis.uc.pt/softc/Read_c/RV/virtual_water/virtual_water.htm*, accessed April 12, 2009.

Tufte, E. R. (2001) *The Visual Display of Quantitative Information*, 2nd Edition. Graphics Press, Cheshire, Conn.

Turner, F. J. (1921) *The Frontier in American History*, Henry Holt, New York.

U.K. Department for Transport (2007) *2005 Valuation of the Benefits of Prevention of Road Accidents and Casualties*, Highways Economics Note No. 1. London, *http: //www.dft.gov.uk/pgr/ roadsafety/ea/pdfeconnote105.pdf*.

UNESCO Institute of Statistics (2009) *Regional average of enrollment ratios for pre-primary to tertiary education (ISCED 0-6)*. UNESCO, *http: //stats.uis.unesco.org/unesco/TableViewer/ tableView.aspx?ReportId=194*, accessed April 16, 2009.

UNFCCC (2008) *An Introduction to the Kyoto Protocol Compliance Mechanism. http: //unfccc.int/kyoto_protocol/compliance/introduction/items/3024.php*, accessed November 7, 2008.

United Nations (1945) *Charter of the United Nations. http: //www. un.org/aboutun/charter/*, accessed Sept. 22, 2008.

United Nations (1969) *Vienna Convention on the Law of Treaties*, United Nations, New York, *http: //untreaty.un.org/ilc/texts/ instruments/english/conventions/1_1_1969.pdf*.

United Nations (1982) *United Nations Convention on the Law of the Sea*, New York, *http: //www.un.org/depts/los/convention_agreements/texts/unclos/unclos_e.pdf*.

United Nations (1986) *Vienna Convention on Early Notification of a Nuclear Accident*. International Atomic Energy Agency, *http: // www.iaea.org/Publications/Documents/Infcircs/Others/infcirc335. shtml*, accessed Sept. 24, 2008.

United Nations (1992) *United Nations Framework Convention on Climate Change*, New York, *http: //unfccc.int/resource/docs/convkp/conveng.pdf*.

United Nations (1998) *Kyoto Protocol to the United Nations Framework Convention on Climate Change*, New York, *http: //unfccc. int/resource/docs/convkp/kpeng.pdf*.

United Nations (2007) *National Greenhouse Gas Inventory Data for the Period 1990–2005*, New York, *http: //unfccc.int/resource/ docs/2007/sbi/eng/30.pdf*.

United Nations Environmental Programme (2007) *Key Achievements of Montreal Protocol to Date*. United Nations, *http: //ozone. unep.org/Publications/MP_Key_Achievements-E.pdf*, accessed October 24, 2008.

United Nations Forum on Forests (2004) *An Overview of International Law. www.un.org/esa/forests/pdf/aheg/param/background-3. pdf* accessed Sept. 25, 2008.

United Nations Population Division (2008) *World Population Prospects: The 2006 Revision Population Database*. UN Department of Economic and Social Affairs, *http: //esa.un.org/unpp/*, accessed April 14, 2008.

United Nations Statistics Division (2007) *Motor Vehicles in Use*. United Nations, *http: //millenniumindicators.un.org/unsd/cdb/ cdb_series_xrxx.asp?series_code=29500*, accessed Aug. 5, 2007.

United Nations Treaty Collection (2008) *Treaty Reference Guide*. United Nations, *http: //untreaty.un.org/English/guide.asp*, accessed Sept. 24, 2008.

U.S. Bureau of Labor Statistics (2007) *Consumer Expenditures in 2005*, U.S. Department of Labor, Washington, D.C., *http: // www.bls.gov/cex/csxann05.pdf*.

U.S. Census Bureau (2007) *Health Insurance Coverage: 2007. http: // www.census.gov/hhes/www/hlthins/hlthin07/fig06.pdf*, accessed December 12, 2008.

U.S. Census Bureau (2008) *Characteristics of New Housing*. U.S. Census Bureau, *http: //www.census.gov/const/www/charindex. html#singlecomplete*, accessed May 10, 2008.

U.S. Census Bureau (2009) *International Data Base. http: //www.census.gov/ipc/www/idb/summaries.html*, accessed April 16, 2009.

U.S. Department of Energy (2007a) *2007 Buildings Energy Data Book*. U.S. Department of Energy, *http: //buildingsdatabook.eere. energy.gov/?id=view_book*, accessed May 8, 2008.

U.S. Department of Energy (2007b) *Fuel Economy Guide. http: // www.fueleconomy.gov/feg/findacar.htm*, accessed Aug. 8, 2007.

U.S. Department of Energy (2008a) *Assessments Reveal Steam Energy Savings Opportunities for Chemical, Automotive, and Paper Manufacturing Companies*. Industrial Technology Program, Energy Matters, *http: //www.eere.energy.gov/industry/ bestpractices/energymatters/articles.cfm/article_id=273*, accessed May 25, 2008.

U.S. Department of Energy (2008b) *Commercial Buildings Total Energy Consumption*. U.S. Department of Energy, *http: //www1. eere.energy.gov/ba/pba/intensityindicators/total_commercial.html*, accessed May 12, 2008.

U.S. Department of Energy (2008c) *Residential Buildings Total Energy Consumption*. U.S. Department of Energy, *http: //www1. eere.energy.gov/ba/pba/intensityindicators/total_residential.html*, accessed May 11, 2008.

U.S. Environmental Protection Agency (1999) *Modeling Emission Factors for Compressed Natural Gas Vehicles*, Washington, D.C., *http: //www.epa.gov/otaq/models/mobile6/m6ful004.pdf*.

U.S. Environmental Protection Agency (2005a) *Emission Facts: Average Carbon Dioxide Emissions Resulting from Gasoline and Diesel Fuel. http: //www.epa.gov/otaq/climate/420f05001. htm#calculating*, accessed July 17, 2007.

U.S. Environmental Protection Agency (2005b) *Regulatory Impact Analysis for the Final Clean Air Interstate Rule*, Washington, D.C., *http: //www.epa.gov/cair/pdfs/finaltech08.pdf*.

U.S. Environmental Protection Agency (2006a) *Global Anthropogenic Non-CO_2 Greenhouse Gas Emissions: 1990–2020*, Office of Atmospheric Programs, Washington, D.C., *http: //www.epa. gov/nonco2/econ-inv/downloads/GlobalAnthroEmissionsReport. pdf*.

U.S. Environmental Protection Agency (2006b) *Global Mitigation of Non-CO_2 Greenhouse Gases*, Office of Atmospheric Programs, Washington, D.C., *http: //www.epa.gov/nonco2/econ-inv/downloads/GlobalMitigationFullReport.pdf*.

U.S. Environmental Protection Agency (2007a) *Testing and Measuring Emissions: Dynamometer Driver's Aid. http: //www.epa.gov/ nvfel/testing/dynamometer.htm*, accessed July 7, 2007.

U.S. Environmental Protection Agency (2007b) *TRI Explorer. http: // www.epa.gov/triexplorer/industry.htm*, accessed Sept. 11, 2007.

U.S. Environmental Protection Agency (2008) *Light-Duty Automotive Technology and Fuel Economy Trends: 1975 through 2008*. U.S. Environmental Protection Agency, *http: //www.epa.gov/ otaq/fetrends.htm*, accessed February 12, 2009.

U.S. Geological Survey (2008a) *Cement: Statistics and Information. http: //minerals.usgs.gov/minerals/pubs/commodity/cement/*, accessed April 25, 2008.

U.S. Geological Survey (2008b) *Iron and Steel Scrap Statistics and Information. http: //minerals.usgs.gov/minerals/pubs/commodity/ iron_&_steel_scrap/*, accessed April 22, 2008.

U.S. Government Accountability Office (2008) *Climate Change: Expert Opinion on the Economics of Policy Options to Address Climate Change*, Vol. GAO-08-605, Washington, D.C., *http: // www.gao.gov/new.items/d08605.pdf*.

U.S. Nuclear Regulatory Commission (2008) *Fact Sheet on Nuclear Insurance and Disaster Relief Funds. http: //www.nrc.gov/reading-rm/doc-collections/fact-sheets/funds-fs.html*, accessed July 28, 2008.

U.S. Office of Management and Budget (2003) *Circular A-4, Regulatory Analysis*, Washington, D.C., *http: //www.whitehouse.gov/ omb/circulars/a004/a-4.pdf*.

U.S. Office of Management and Budget (2008) *Discount Rates for Cost-Effectiveness, Lease Purchase, and Related Analyses. http:*

//www.whitehouse.gov/omb/circulars/a094/a94_appx-c.html, accessed August 2, 2008.

U.S. Office of Management and Budget (2009) *Table of Past Years Discount Rates from Appendix C of OMB Circular No. A-94 (12/12/2008)*. *http: //www.whitehouse.gov/omb/assets/omb/circulars/a094/dischist.pdf*, accessed May 5, 2009.

U.S. Supreme Court Center (2008) *International Agreements without Senate Approval*. Justia.com, *http: //supreme.justia.com/constitution/article-2/22-international-agreements-by-authorization.html*, accessed Sept. 24, 2008.

Universal Traffic Management Society of Japan (2004) *Traffic Conditions in Japan*. *http: //www.utms.or.jp/english/condi/index.html*, accessed July 6, 2007.

Uranium Info (2007) *NUEXCO Exchange Value (Monthly Uranium Spot)*. TradeTech, *http: //www.uranium.info/prices/monthly.html*, accessed Nov. 15, 2007.

Usoskin, I. G., M. Schussler, S. K. Solanki, and K. Mursula (2005) Solar activity, cosmic rays, and Earth's temperature: A millennium-scale comparison. *Journal of Geophysical Research-Space Physics* 110: A10102 1–10 DOI 10.1029/2004JA010946.

van Groenigen, K. J., J. Six, B. A. Hungate, M. A. de Graaff, N. van Breemen, and C. van Kessel (2006) Element interactions limit soil carbon storage. *Proceedings of the National Academy of Sciences of the United States of America* 103: 6571–6574.

van Vuuren, D., F. Q. Zhou, B. de Vries, K. J. Jiang, C. Graveland, and L. Yun (2003) Energy and emission scenarios for China in the 21st century—exploration of baseline development and mitigation options. *Energy Policy* 31: 369–387.

Veenstra, J. J., W. R. Horwath, and J. P. Mitchell (2007) Tillage and cover cropping effects on aggregate-protected carbon in cotton and tomato. *Soil Science Society of America Journal* 71: 362–371.

Vehicle Certification Agency (2007) *New Car Fuel Consumption and Exhaust Emissions*. *http: //www.vcacarfueldata.org.uk/search/search.asp*, accessed Aug. 7, 2007.

Victor, D. G. (1998) The operation and effectiveness of the Montreal Protocol's non-compliance procedure. In: *The Implementation and Effectiveness of International Environmental Commitments: Theory and Practice*, Victor, D. G., K. Raustiala, and E. B. Skolnikoff, eds. MIT Press, Cambridge, MA. pp. 137–176, *http: //books.google.com/books?id=nHZAyKrn6QQC&pg=PA137 &source=gbs_toc_r&cad=0_0#PPA138,M1*.

Victor, D. G. and L. A. Coben (2005) A herd mentality in the design of international environmental agreements. *Global Environmental Politics* 5: 24–57.

Victor, D. G., J. C. House, and S. Joy (2005) A Madisonian approach to climate policy. *Science* 309: 1820–1821 DOI 10.1126/science.1113180.

Viscusi, W. K. and J. E. Aldy (2003) The value of a statistical life: A critical review of market estimates throughout the world. *Journal of Risk and Uncertainty* 27: 5–76.

von Helmolt, R. and U. Eberle (2007) Fuel cell vehicles: Status 2007. *Journal of Power Sources* 165: 833–843.

Wachs, M. (1989) When planners lie with numbers. *Journal of the American Planning Association* 55: 476–479.

Wand, S. J. E., G. F. Midgley, M. H. Jones, and P. S. Curtis (1999) Responses of wild C_4 and C_3 grass (Poaceae) species to elevated atmospheric CO_2 concentration: a meta-analytic test of current theories and perceptions. *Global Change Biology* 5: 723–741.

Wang, J. C. F. (1992) *Handbook on Ocean Politics and Law*, Greenwood Publishing Group, Westport, CT.

Wang, M. (2005) *Updated Energy and Greenhouse Gas Emission Results of Fuel Ethanol*, The 15th International Symposium on Alcohol Fuels, San Diego, CA, *http: //www.transportation.anl. gov/pdfs/TA/375.pdf*.

Wang, S. S., A. P. Trishchenko, K. V. Khlopenkov, and A. Davidson (2006) Comparison of International Panel on Climate Change Fourth Assessment Report climate model simulations of surface albedo with satellite products over northern latitudes. *Journal of Geophysical Research-Atmospheres* 111: D21108 1–17 DOI 10.1029/2005JD006728.

Wang, T. and J. Watson (2007) *Who Owns China's Carbon Emissions?*, Tyndall Briefing Note No. 23. Tyndall Centre for Climate Change Research, Norwich, UK, *http: //tyndall.webapp1. uea.ac.uk/publications/briefing_notes/bn23.pdf*.

Wang, X. J. and J. R. Key (2003) Recent trends in arctic surface, cloud, and radiation properties from space. *Science* 299: 1725–1728.

Wang, X. J. and J. R. Key (2005) Arctic surface, cloud, and radiation properties based on the AVHRR Polar Pathfinder dataset. Part II: Recent trends. *Journal of Climate* 18: 2575–2593.

Wang, Z. Y., Y. C. Xu, Z. Li, Y. X. Guo, R. Wassmann, H. U. Neue, R. S. Lantin, L. V. Buendia, Y. P. Ding, and Z. Z. Wang (2000) A four-year record of methane emissions from irrigated rice fields in the Beijing region of China. *Nutrient Cycling in Agroecosystems* 58: 55–63.

Warner, J. L. (1995) Selecting windows for energy efficiency. *Home Energy Magazine Online* July/August 1995, *http: //www. homeenergy.org/archive/hem.dis.anl.gov/eehem/95/950708.html*, accessed April 30, 2008.

Washington, W. M. and C. L. Parkinson (2005) *An Introduction to Three-Dimensional Climate Modeling*, 2nd Edition. University Science Books, Sausalito, CA.

Wassmann, R., M. S. Aulakh, R. S. Lantin, H. Rennenberg, and J. B. Aduna (2002) Methane emission patterns from rice fields planted to several rice cultivars for nine seasons. *Nutrient Cycling in Agroecosystems* 64: 111–124.

Wassmann, R. and H. Pathak (2007) Introducing greenhouse gas mitigation as a development o ective in rice-based agriculture: II. Cost-benefit assessment for different technologies, regions and scales. *Agricultural Systems* 94: 826–840.

Weart, S. R. (2003) *The Discovery of Global Warming*, Harvard University Press, Cambridge, MA., *http: //www.aip.org/history/climate/*.

Weber, E. U. (2006) Experience-based and description-based perceptions of long-term risk: Why global warming does not scare us (yet). *Climatic Change* 77: 103–120 DOI 10.1007/s10584-006-9060-3.

Webster, P. J., G. J. Holland, J. A. Curry, and H. R. Chang (2005) Changes in tropical cyclone number, duration, and intensity in a warming environment. *Science* 309: 1844–1846.

Weerakoon, W. M. W., K. T. Ingram, and D. N. Moss (2005) Atmospheric CO_2 concentration effects on N partitioning and fertilizer N recovery in field grown rice (*Oryza sativa* L.). *Agriculture Ecosystems & Environment* 108: 342–349.

Weisbach, D. A. and C. R. Sunstein (2007) Symposium on Intergenerational Equity and Discounting. *University of Chicago Law Review* 74: 1–3.

Weisser, D. (2007) A guide to life-cycle greenhouse gas (GHG) emissions from electric supply technologies. *Energy* 32: 1543–1559.

Weitzman, M. L. (2001) Gamma discounting. *American Economic Review* 91: 260–271.

Welch, I. (2000) Views of financial economists on the equity premium and on professional controversies. *Journal of Business* 73: 501–537.

Westerling, A. L., H. G. Hidalgo, D. R. Cayan, and T. W. Swetnam (2006) Warming and earlier spring increase western US forest wildfire activity. *Science* 313: 940–943.

Western Climate Initiative (2008) *Design Recommendations for the WCI Regional Cap-and-Trade Program*, *http: //www.westernclimateinitiative.org/ewebeditpro/items/O104F20432.PDF*.

Western GeoPower Corporation (2009) *Western GeoPower Unit 1—The Geysers.* http://www.geopower.ca/geysersdesc.htm, accessed April 20, 2009.

Wheelock College (2007) *WhaleNet Data Search.* http://whale.wheelock.edu/whalenet-stuff/humpcat.html, accessed May 12, 2007.

White, M. A., N. S. Diffenbaugh, G. V. Jones, J. S. Pal, and F. Giorgi (2006) Extreme heat reduces and shifts United States premium wine production in the 21st century. *Proceedings of the National Academy of Sciences of the United States of America* 103: 11217–11222.

White, W. M. (2005) *Geochemistry.* http://www.geo.cornell.edu/geology/classes/geo455/Chapters.HTML, accessed August 5, 2006.

Whittaker, R. H. (1975) *Communities and Ecosystems*, Macmillan, New York.

Wielicki, B. A., T. M. Wong, N. Loeb, P. Minnis, K. Priestley, and R. Kandel (2005) Changes in Earth's albedo measured by satellite. *Science* 308: 825–825.

Wigley, T. M. L. (2006) A combined mitigation/geoengineering approach to climate stabilization. *Science* 314: 452–454 DOI 10.1126/science.1131728.

Wignall, P. (2005) The link between large igneous province eruptions and mass extinctions. *Elements* 1: 293–297.

Wikipedia (2007) *Fuel Efficiency in Transportation.* http://en.wikipedia.org/wiki/Fuel_efficiency_in_transportation, accessed July 26, 2007.

Wikipedia (2008a) *Building Insulation Materials.* Wikimedia Foundation, Inc., http://en.wikipedia.org/wiki/Building_insulation_materials, accessed April 27, 2008.

Wikipedia (2008b) *R-value (Insulation).* Wikimedia Foundation, Inc, http://en.wikipedia.org/wiki/R-value_%28insulation%29, accessed April 27, 2008.

Willett, K. M., N. P. Gillett, P. D. Jones, and P. W. Thorne (2007) Attribution of observed surface humidity changes to human influence. *Nature* 449: 710–712.

Wingler, A., P. J. Lea, W. P. Quick, and R. C. Leegood (2000) Photorespiration: metabolic pathways and their role in stress protection. *Philosophical Transactions of the Royal Society of London Series B-Biological Sciences* 355: 1517–1529.

Wipke, K., S. Sprik, H. Thomas, C. Welch, and J. Kurtz (2007) *Controlled Hydrogen Fleet and Infrastructure Validation and Demonstration Project: Learning Demonstration Interim Progress Report, Summer 2007*, National Renewable Energy Laboratory, Golden, CO, http://www.nrel.gov/hydrogen/pdfs/41848.pdf.

Wiser, R., M. Bolinger, P. Cappers, and R. Margolis (2006) *Letting the Sun Shine on Solar Costs: An Empirical Investigation of Photovoltaic Cost Trends in California*, Lawrence Berkeley National Laboratory, Berkeley, CA, http://eetd.lbl.gov/EA/EMP.

Witze, A. (2006) Tempers flare at hurricane meeting. *Nature* 441: 11–11.

Woo, C. K., E. Kollman, R. Orans, S. Price, and B. Horii (2008) Now that California has AMI, what can the state do with it? *Energy Policy* 36: 1366–1374.

Woodworth, P. L. (2006) Some important issues to do with long-term sea level change. *Philosophical Transactions of the Royal Society A-Mathematical Physical and Engineering Sciences* 364: 787–803.

Woolsey, M. (2008) World's Best Places for Unemployment Pay. *Forbes*, June 27, 2008.

World Business Council for Sustainable Development (2004) *Mobility 2030: Meeting the Challenges to Sustainability*, WBCSD, Hertfordshire, England, http://www.wbcsd.org/includes/getTarget.asp?type=d&id=OTQ4Mw.

World Coal Institute (2004) *Clean Coal - Building a Future through Technology*, Richmond, UK, http://www.worldcoal.org/assets_cm/files/PDF/clean_coal_building_a_future_thro_tech.pdf.

World Energy Council (2007) *2007 Survey of Energy Resources*, London, http://www.worldenergy.org/documents/ser2007_final_online_version_1.pdf.

World Health Organization (2002) *Fact Sheet: Dengue and Dengue Haemorrhagic Fever.* http://www.who.int/mediacentre/factsheets/fs117/en/index.html, accessed May 22, 2007.

World Meteorological Organization (2007) *Scientific Assessment of Ozone Depletion: 2006*, Global Ozone Research and Monitoring Project—Report No. 50. Geneva, http://www.esrl.noaa.gov/csd/assessments/2006/report.html.

World Nuclear Association (2007) *Advanced Nuclear Power Reactors.* http://www.world-nuclear.org/info/inf08.html, accessed Nov. 19, 2007.

World Resources Institute (2007) *Climate Analysis Indicators Tool, version 4.0.* http://cait.wri.org/cait.php, accessed Oct. 4, 2007.

WSU CE Energy Program (2003) *Adjustable Speed Motor Drives*, Energy Efficiency Factsheet. Washington State University Cooperative Extension www.energy.wsu.edu/documents/engineering/motors/MotorDrvs.pdf.

Wuebbles, D. J. and K. Hayhoe (2002) Atmospheric methane and global change. *Earth-Science Reviews* 57: 177–210.

Yablokov, A. V. (1994) Validity of whaling data. *Nature* 367: 108.

Yelle, S., R. C. Beeson, M. J. Trudel, and A. Gosselin (1989) Acclimation of 2 Tomato Species to High Atmospheric CO_2 .1. Sugar and Starch Concentrations. *Plant Physiology* 90: 1465–1472.

Yihui, D., R. Guoyu, S. Guangyu, G. Peng, Z. Xunhua, Z. Panmao, Z. De'er, Z. Zongci, W. Shaowu, W. Huijun, L. Yong, C. Deliang, G. Xuejie, and D. Xiaosu (2007) China's National Assessment Report on Climate Change (I): Climate change in China and the future trend. *Advances in Climate Change Research* 3 (supplement): 1–5.

Yusuf, S., K. Nabeshima, and D. Perkins (2007) China and India reshape global industrial geography. In: *Dancing with Giants: China, India, and the Global Economy*, Winters, L. A. and S. Yusuf, eds. World Bank Publications, Singapore. pp. 27–55, http://siteresources.worldbank.org/INTCHIINDGLOECO/Resources/CE_Ch02pp.027-56_FINAL.pdf.

Zachos, J., M. Pagani, L. Sloan, E. Thomas, and K. Billups (2001) Trends, rhythms, and aberrations in global climate 65 Ma to present. *Science* 292: 686–693.

Zaelke, D., D. Kaniaru, and E. Kruzíková, eds. (2005) *Making Law Work: Environmental Compliance & Sustainable Development*, 2 vols, International Network for Environmental Compliance and Enforcement, Institute for Governance & Sustainable Development, Cameron May, London, http://www.inece.org/makinglawwork.html.

Zamparini, L. and A. Reggiani (2007) Meta-analysis and the value of travel time savings: A transatlantic perspective in passenger transport. *Networks & Spatial Economics* 7: 377–396 DOI 10.1007/s11067-007-9028-5.

Zielinski, G. A. (2000) Use of paleo-records in determining variability within the volcanism-climate system. *Quaternary Science Reviews* 19: 417–438.

Ziska, L. H., O. Namuco, T. Moya, and J. Quilang (1997) Growth and yield response of field-grown tropical rice to increasing carbon dioxide and air temperature. *Agronomy Journal* 89: 45–53.

Zoettl, G. (2008) Investment decisions in liberalized electricity markets: A framework of peak load pricing with strategic firms. *University of Cologne: Working Papers in Economics* 38: 1–49.

Zvereva, E. L. and M. V. Kozlov (2006) Consequences of simultaneous elevation of carbon dioxide and temperature for plant-herbivore interactions: a metaanalysis. *Global Change Biology* 12: 27–41.

INDEX

Page numbers followed by the letter *f* denote figures; those followed by the letter *t* denote tables.

ILLUSTRATION CREDITS

CHAPTER 1

1.1: Courtesy of NOAA; imagery from the GOES-12 weather satellite. 1.5B: Courtesy of the NOAA Coastal Services Center. 1.5C: Photograph by Commander John Bortniak, NOAA Corps (ret.). 1.10A: Images courtesy of Jim Williams, NASA GSFC Scientific Visualization Studio, and the Landsat 7 Science Team. 1.11: Courtesy of NASA. The data used to create these images were collected by the Defense Meteorological Satellite Program (DMSP) Special Sensor Microwave Imager (SSMI).

CHAPTER 2

2.1: © Ian Francis/Alamy. 2.3: Courtesy of J. Meeks, UC Davis. 2.4: Courtesy of André Karwath. 2.12: Photograph by David McIntyre; drill cores courtesy of Dr. Laurie Brown, UMass, Amherst. 2.15: Courtesy of Kai Liu, UC Davis. 2.18B: Painting by Mary Parrish; courtesy of Smithsonian Institution. 2.24: Painting by Don Davis, courtesy of NASA. Inset: NASA image. 2.27A: David McIntyre. 2.27B: Courtesy of Emily Stone/NSF. 2.28A: Photograph courtesy of E. Aristidi, LUAN/IPEV. 2.28B: From NASA satellite images. 2.30A: ©Jim Barber/Shutterstock. 2.30B: Courtesy of Richard Keim. 2.31A: Courtesy of Jonathan Pilcher, Palaeoecology Centre, Queen's University, Belfast. 2.31B: Courtesy of Peter Brown/Rocky Mountain Tree-Ring Research and NOAA. 2.32: David McIntyre. 2.34: Image courtesy of B. Williams, NOAA, HBOI. 2.36: Courtesy of Dr. Mauri S. Pelto, Nichols College.

CHAPTER 3

3.1: Photograph by NASA's Spitzer Space Telescope. 3.15A: Courtesy of NASA; image from the Michelson Doppler Imager on the Solar and Heliospheric Observatory satellite launched in 1995. 3.15B: Courtesy of Göran Scharmer/the Royal Swedish Academy of Sciences; image from the Swedish Solar Telescope on the Canary Island of La Palma. 3.20: Photograph by Dave Harlow, USGS/Cascades Volcano Observatory. 3.21: Courtesy of NASA. 3.24: Courtesy of Crystal Schaaf, Boston University, based upon data processed by the MODIS Land Science Team. 3.28: © Edward Kinsman/Photo Researchers, Inc. 3.30: Courtesy of the U.S. Fish and Wildlife Service. 3.40: Courtesy of NASA. 3.46: Courtesy of Twentieth-Century Fox. 3.52: Created using User:jdorje/ Tracks by Nilfanion on August 5, 2006. 3.53: The source of this image is the COMET® Web site at http:// meted.ucar.edu/ of the University Corporation for Atmospheric Research (UCAR) pursuant to a Cooperative Agreements with the National Oceanic and Atmospheric Administration, U.S. Department of Commerce. ©1997–2007 University Corporation for Atmospheric Research. All Rights Reserved.

CHAPTER 4

4.1A: Courtesy of Douglas Millard. 4.1B: Courtesy of the UC Davis Center for Geotechnical Modeling. 4.1C: © Gary Crabbe/Alamy. 4.3: Courtesy of NOAA. 4.4: U.S. Army photo. 4.6: © JAMSTEC. 4.9: From Randall et al. 2002 © IEEE. 4.19: Chen, Y. H. and R. G. Prinn 2006. Estimation of atmospheric methane emissions between 1996 and 2001 using a three-dimensional global chemical transport model. *Journal of Geophysical Research-Atmospheres* 111: DOI:D10307. © 2006 American Geophysical Union. Reproduced by permission of American Geophysical Union. 4.24: Stehfest, E. and L. Bouwman. 2006. N_2O and NO emission from agricultural fields and soils under natural vegetation. *Nutrient Cycling in Agroecosystems* 74: 207–228. 4.28: Image based on a U.S. Fish and

Wildlife Service photo. 4.31: Courtesy of NOAA. 4.34, 4.38: Climate Change 2007: The Physical Science Basis. Working Group I Contribution to the Fourth Assessment Report of the Intergovernmental Panel on Climate Change, Figure SPM.6 and SPM.7 Cambridge University Press. 4.36: Courtesy of Weiss and Overpeck, The University of Arizona. 4.37: Courtesy of the World Food Program. 4.41: Courtesy of NOAA.

CHAPTER 5

5.1: Courtesy of NASA. 5.11: © Ray Simons/Photo Researchers, Inc. 5.14A: Courtesy of Controlled Environments Limited of Winnipeg, Canada. 5.14B: Courtesy of Fitzgerald Booker, USDA, Agricultural Research Service. 5.14C: Courtesy of Michelle N. Tremblay and Edward B. Mondor, Georgia Southern University. 5.19: Yelle, S., R. C. Beeson, M. J. Trudel, and A. Gosselin. 1989. Acclimation of 2 Tomato Species to High Atmospheric CO_2. 1. Sugar and Starch Concentrations. *Plant Physiology* 90: 1465–1472. 5.22: © Gary Lee/Zuma Press. Box 5.1 Figure B: From Stirton, R. A. (1947) Observations on evolutionary rates in hypsodonty. Evolution 1:3 2–41. 5.30: Courtesy of Laura Rose/CDC. 5.31: Courtesy of Stephen Davis/USDA APHIS PPQ, Bugwood.org. 5.32: Courtesy of James Gathany/CDC. Box 5.2: Courtesy of the Department of Natural Resources, Queensland, Australia.

CHAPTER 6

6.3: Courtesy of the National Arbor Day Foundation. 6.5A: Photograph by Scott Schielbe, U.S. Fish and Wildlife Service. 6.5B: NASA Satellite image. 6.7A: Courtesy of Patrick Rowe/NSF. 6.7B: Courtesy of Beth Minneci/NSF. 6.8A: NASA Satellite image. 6.9A: Courtesy of Florida Fish and Wildlife Conservation Commission/NOAA. 6.9B: Courtesy of NOAA. 6.12: © Øystein Paulsen. 6.13: Courtesy of SeaWiFS Project, NASA Goddard Space Flight Center and ORBIMAGE. 6.14: Courtesy of Peter Stevick. 6.17: Courtesy of Stellewagen Bank National Marine Sanctuary/NOAA. 6.19 United States Department of Labor/Mine Safety & Health Administration. 6.20: Lemoine, N., H. C. Schaefer, and K. Bohning-Gaese. 2007. Species richness of migratory birds is influenced by global climate change. *Global Ecology and Biogeography* 16: 55–64. 6.21, 6.46: http://

earthguide.ucsd.edu/earthguide/maps/reliefmapca.html. 6.22: White, M. A., N. S. Diffenbaugh, G. V. Jones, J. S. Pal, and F. Giorgi. 2006. Extreme heat reduces and shifts United States premium wine production in the 21st century. *Proceedings of the National Academy of Sciences of the United States of America* 103: 11217–11222. © 2006 National Academy of Sciences, U.S.A. 6.23: Courtesy of Florida Keys Marine Sanctuary/NOAA. 6.24: Courtesy of the NASA/GSFC/LaRC/JPL, MISR Team. 6.25: Courtesy of NOAA Center for Coastal Monitoring and Assessment. 6.28: Courtesy of Charles H. Smith, U.S. Fish and Wildlife Service. 6.38: Ramankutty, N., J. A. Foley, J. Norman, and K. McSweeney. 2002. The global distribution of cultivable lands: current patterns and sensitivity to possible climate change. *Global Ecology and Biogeography* 11: 377–392. 6.42: NASA image by Robert Simmon and Reto Stöckli. 6.43: Courtesy of the United Nations Environment Programme. 6.46: Courtesy of Monterey County Water Resource Agency. 6.47: Courtesy of Jorge A. Trindade.

CHAPTER 7

7.9: Courtesy of Warren K. Leffler. 7.33: Courtesy of National Park Service. http://www.nps.gov/archive/grca/photos/image s/T053.jpg. 7.40: Courtesy of HyFLEET:CUTE. 7.48: © The Revenoor Company, Inc. P.O. Box 602, Yamhill, OR 97148 www.revenoor.com

CHAPTER 8

8.9, 8.17: © CO2CRC online image library *www.cp2crc.com.au*. 8.14: IPCC 2005: IPCC Special Report on Carbon Dioxide Capture and Storage. Prepared by Working Group III of the Intergovernmental Panel on Climate Change. Figure 6.1. Cambridge University Press. 8.22: Courtesy of National Nuclear Security Administration/Nevada Site Office. 8.23: Based on a U.S. Department of Energy drawing. 8.24: © 2006 SASI Group (University of Sheffield) and Mark Newman (University of Michigan). 8.25: The data for this special collection is provided in full by *Watersheds of the World Online*, a collaborative product of four organizations: IUCN-The World Conservation Union, the International Water Management Institute (IWMI), the Ramsar Convention Bureau, and

the World Resources Institute (WRI). 8.29B: Courtesy of Julie Stone, U.S. Army Corps of Engineers. 8.29D: Courtesy of Hector Campbell. 8.31: U.S. Department of Energy, National Renewable Energy Laboratory. 8.34: Courtesy of Mark Rodgers/Cape Wind. 8.40B: NASA photo ED01-020905 courtesy of Nick Galante. 8.41B: Ausra Solar, http://en.wikipedia.org/wiki/Image:Fresnel_reflectors_ausra.jpg 8.42: Courtesy of Schott Solar, Inc. 8.51: Courtesy of NASA/Goddard Space Flight Center. 8.52: Courtesy of Idaho National Laboratory. 8.56: http://en.wikipedia.org/wiki/File:Rance_tidal_power_plant.JPG. 8.57: © Ocean Power Technologies, Inc.

CHAPTER 9

9.6: Jialiang Gao/http://en.wikipedia.org/wiki/File:Terrace_field_yunnan_china.jpg. 9.7: Courtesy of Fachagentur Nachwachsende Rohstoffe e.V. (FNR). 9.8: Food and Agriculture Organization of the United Nations 2006. 9.12A: Courtesy of Andrew Greenberg, Portland State Aerospace Society. 9.12B: Courtesy of Brian Murray. Box 9.1 Figure B: Courtesy of Skip Williams and Lee Sauder, The Rockridge Bloomery, Lexington, VA. 9.17: Courtesy of Rodney J. Scott, Invisions, Inc. 9.20B: Courtesy of Jeff Endow, Microscopebulbsplus.com. 9.21: Courtesy of Texas Energy Partnership. 9.22: http://en.wikipedia.org/wiki/File:CFBulbs.jpg. 9.24: © David J. Green/Alamy. 9.25: Courtesy of LEDtronics, Inc. 9.28B: © DK Images. 9.29: Courtesy of James Terry. 9.36B: Courtesy of Roger Angel, University of Arizona Steward University. 9.37, 9.38: Courtesy of NASA. 9.39: © John MacNeill Illustration. 9.40: Earth Observatory 2003. 9.41: Earth Observatory 2008.

CHAPTER 10

10.1A: Courtesy of Chris Lattner. 10.1B: © Patrick Ward/Alamy. 10.12: © Ken Avidor. Box 10.2: Courtesy of Christian Mehlführer. 10.18: Courtesy of Sotheby's. 10.19A: http://en.wikipedia.org/wiki/File:Canyon_midday.jpg 10.19B: http://en.wikipedia.org/wiki/File:Yosemite.JPG 10.21A: Courtesy of Seaview Diving Contractors, Inc. 10.21B: Courtesy of Scott Wilson/CancunAssist.com.

CHAPTER 11

11.4: Courtesy of Philippe Rekacewicz, UNEP/GRID-Arendal. 11.6: Martin Richards, http://www.battleships-cruisers.co.uk/s+t_class.htm. 11.7: Prince Collins, VLIZ Maritime Boundaries Database 2009, Google Earth. 11.9: Circum-Arctic Resource Appraisal Assessment Team 2008.

CHAPTER 12

12.13: Courtesy of T. Lindsay Baker. 12.14: Courtesy of John Doebely. 12.15: Courtesy of the NASA/Goddard Space Flight Center Scientific Visualization Studio.